69.00 70E

Methods of
Experimental Physics

VOLUME 21

SOLID STATE: NUCLEAR METHODS

METHODS OF EXPERIMENTAL PHYSICS

Robert Celotta and Judah Levine, *Editors-in-Chief*

Founding Editors

L. MARTON
C. MARTON

Volume 21

Solid State: Nuclear Methods

Edited by

J. N. Mundy, S. J. Rothman,
M. J. Fluss, and L. C. Smedskjaer

Materials Science and Technology Division
Argonne National Laboratory
Argonne, Illinois

1983

ACADEMIC PRESS, INC.
(Harcourt Brace Jovanovich, Publishers)

Orlando San Diego San Francisco New York London
Toronto Montreal Sydney Tokyo São Paulo

COPYRIGHT © 1983, BY ACADEMIC PRESS, INC.
ALL RIGHTS RESERVED.
NO PART OF THIS PUBLICATION MAY BE REPRODUCED OR
TRANSMITTED IN ANY FORM OR BY ANY MEANS, ELECTRONIC
OR MECHANICAL, INCLUDING PHOTOCOPY, RECORDING, OR ANY
INFORMATION STORAGE AND RETRIEVAL SYSTEM, WITHOUT
PERMISSION IN WRITING FROM THE PUBLISHER.

ACADEMIC PRESS, INC.
Orlando, Florida 32887

United Kingdom Edition published by
ACADEMIC PRESS, INC. (LONDON) LTD.
24/28 Oval Road, London NW1 7DX

Library of Congress Cataloging in Publication Data
Main entry under title:

Solid state: Nuclear methods

   (Methods of experimental physics ; v. 21)
   Bibliography: p.
   1. Lattice dynamics--Measurement. 2. Crystals--
Defects--Measurement. 3. Nuclear physics--Measurement.
I. Mundy, J. N. II. Rothman, S. J. III. Fluss, M. J.
IV. Title. V. Series.
QC176.8.L3S65 1983   530.4'1   83-22460
ISBN 0-12-475963-7 (alk. paper)

PRINTED IN THE UNITED STATES OF AMERICA

83 84 85   9 8 7 6 5 4 3 2 1

CONTENTS

CONTRIBUTORS xi
PREFACE xiii
LIST OF VOLUMES IN TREATISE xvii

1. Radiotracer Techniques

   1.1. Introduction 1
       by J. N. MUNDY AND S. J. ROTHMAN

       1.1.1. The Use of Radiotracers 1
       1.1.2. Introduction to Radioactivity 1

   1.2. Production of Radioisotopes 11
       by J. N. MUNDY AND S. J. ROTHMAN

       1.2.1. Principles of Isotope Production 11
       1.2.2. Determination of Yields and Specific Activities . . . 17
       1.2.3. Isotope Production Reactions 18
       1.2.4. Chemical Processing in Radioisotope Production . . 22
       1.2.5. Parameter Selection in Isotope Production 25
       1.2.6. Purchase of Radioisotopes 26

   1.3. Handling of Radioisotopes 27

       1.3.1. Safety 27
          by M. ROBINET
       1.3.2. Source Preparation 40
          by S. J. ROTHMAN AND J. N. MUNDY

   1.4. Detection and Assay 44

       1.4.1. Autoradiography 44
          by S. J. ROTHMAN AND J. N. MUNDY
       1.4.2. Radiation Detectors for Pulse Counting 50
          by J. N. MUNDY AND S. J. ROTHMAN
       1.4.3. Pulse-Counting Systems 55
          by M. J. FLUSS, J. N. MUNDY, AND S. J. ROTHMAN
       1.4.4. Radioactive Particle Counting 60
          by M. J. FLUSS, J. N. MUNDY, AND S. J. ROTHMAN

1.4.5. Total Relative Counting. 64
by J. N. MUNDY AND S. J. ROTHMAN
1.4.6. Isotope Effects 70
by S. J. ROTHMAN AND J. N. MUNDY

Bibliography 74

2. Experimental Methods of Positron Annihilation for the Study of Defects in Metals
by L. C. SMEDSKJAER AND M. J. FLUSS

2.1. Introduction. 77
2.2. Basic Theory 78

    2.2.1. The Positron. 78
    2.2.2. Positronium 78
    2.2.3. Electron Momenta 79
    2.2.4. Positrons in Defects. 80
    2.2.5. Positron Sources 81
    2.2.6. Thermalization of the Positron 82
    2.2.7. Trapping of Positrons 82
    2.2.8. The Trapping Model 85

2.3. Positron Studies and Metal Physics 87

    2.3.1. Surface Studies. 89
    2.3.2. Vacancy Formation Enthalpy 89
    2.3.3. Vacancy–Impurity Binding 92
    2.3.4. Annealing Studies of Defects. 93
    2.3.5. Electron Structure 94

2.4. Detection of Annihilation Radiation 95

    2.4.1. Resolution Considerations. 95
    2.4.2. Counting Rate and Resolution 97
    2.4.3. Detector Systems 97
    2.4.4. Doppler Broadening 98
    2.4.5. Angular Correlation 106
    2.4.6. Lifetime. 122

2.5. Instabilities (Detectors and Electronics) 139
2.6. Some Final Remarks 143

3. Neutron Scattering Studies of Lattice Defects

3.1. Static Properties of Defects 147
  by W. SCHMATZ

  3.1.1. Introduction. 147
  3.1.2. Theoretical Background. 149
  3.1.3. Experimental Techniques 158
  3.1.4. Typical Results. 167

3.2. Dynamic Properties of Defects 172
  by R. M. NICKLOW

  3.2.1. Introduction. 172
  3.2.2. Perfect Crystals. 173
  3.2.3. Defect Dynamics: Theory and Experiment. 182

3.3. Diffusion Studies. 194
  by N. WAKABAYASHI

  3.3.1. Introduction. 194
  3.3.2. Neutron Scattering Cross Sections and Correlation Functions. 195
  3.3.3. Incoherent Scattering and the Jump Model of Diffusion 198
  3.3.4. Information Obtained from Coherent Scattering Processes 206
  3.3.5. Samples and Instruments 211
  3.3.6. Data Reduction 217
  3.3.7. Future Directions. 219

4. Ion Beam Interactions with Solids

4.0. Introduction. 221
  by L. M. HOWE, M. L. SWANSON, AND J. A. DAVIES

4.1. Compositional Studies 222
  by N. CUE

  4.1.1. Ion Backscattering Spectrometry 223
  4.1.2. Nuclear Reaction Analysis. 244
  4.1.3. Particle-Induced X-Ray Emission. 250
  4.1.4. Secondary Particle Emission 256

CONTENTS

4.2. Channeling Studies of Lattice Defects 275
by L. M. Howe, M. L. Swanson, and J. A. Davies

    4.2.1. Introduction 275
    4.2.2. Basic Channeling Theory 283
    4.2.3. Experimental Techniques 297
    4.2.4. Investigation of Lattice Disorder 303
    4.2.5. Lattice Site Location of Solute Atoms 322
    4.2.6. Surface Studies 345
    4.2.7. Conclusions 358

5. Magnetic Resonance Methods for Studying Defect Structure in Solids
by David C. Ailion and William D. Ohlsen

    5.1. Introduction 361

        5.1.1. Basic Phenomenon of Magnetic Resonance 361
        5.1.2. Relaxation Times T_1 and T_2 364
        5.1.3. Bloch Equations 367
        5.1.4. The Nature of Magnetic Resonance Information . . 371

    5.2. Nuclear Magnetic Resonance Spectrometer Design 373

        5.2.1. Pulse versus Continuous-Wave Spectrometers . . . 374
        5.2.2. Pulse NMR Spectrometer Design 375

    5.3. Applications of Nuclear Magnetic Resonance 391

        5.3.1. Line-Shape Studies 391
        5.3.2. Quadrupole Resonance 394
        5.3.3. Electron–Nuclear Magnetic Interactions 398
        5.3.4. NMR Imaging 403

    5.4. Electron Paramagnetic Resonance 405

        5.4.1. Electron Resonance Spectra 405
        5.4.2. Electron–Crystal Interactions 406
        5.4.3. Spectrometer Design Considerations 407
        5.4.4. Representative Studies 423

    5.5. Additional Perturbations 428

        5.5.1. Light 428
        5.5.2. Stress 429
        5.5.3. Electric Fields 430

5.6. Double-Resonance Techniques. 430

    5.6.1. Nuclear–Nuclear Double Resonance 430
    5.6.2. Electron–Nuclear Double Resonance 432
    5.6.3. Dynamic Nuclear Polarization 435
    5.6.4. Other Double-Resonance Techniques 436

6. Nuclear Magnetic Resonance Relaxation Time Methods for Studying Atomic and Molecular Motions in Solids
by DAVID C. AILION

6.1. Introduction. 439

6.2. Conventional Relaxation Time Methods 441

    6.2.1. Motional Narrowing of the Linewidth or the Spin–Spin Relaxation Time (T_2) 441
    6.2.2. Spin–Lattice Relaxation Time (T_1) 447

6.3. Ultraslow Motion Techniques 453

    6.3.1. Spin–Lattice Relaxation in the Rotating Frame ($T_{1\rho}$) 454
    6.3.2. Dipolar Relaxation (T_{1D}) 460
    6.3.3. Dipolar Relaxation in the Rotating Frame ($T_{1D\rho}$) . . 465

6.4. Methods for Calibration and Other Auxiliary Experiments. 469

    6.4.1. Determination of Exact Resonance 469
    6.4.2. Calibrated H_1 Monitor 470
    6.4.3. Determination of the Local Field in the Rotating Frame 471

6.5. Applications of Nuclear Magnetic Resonance Techniques to Motional Studies in Solids 472

    6.5.1. Diffusion 472
    6.5.2. Molecular Reorientations 479

AUTHOR INDEX. 483

SUBJECT INDEX 494

CONTRIBUTORS

Numbers in parentheses indicate the pages on which the authors' contributions begin.

DAVID C. AILION (361, 439), *Department of Physics, The University of Utah, Salt Lake City, Utah 84112*

N. CUE (222), *Department of Physics, State University of New York at Albany, Albany, New York 12222*

J. A. DAVIES (221, 275), *Atomic Energy of Canada Limited Research Company, Chalk River Nuclear Laboratories, Chalk River, Ontario, Canada K0J 1J0*

M. J. FLUSS (55, 60, 77), *Materials Science and Technology Division, Argonne National Laboratory, Argonne, Illinois 60439*

L. M. HOWE (221, 275), *Atomic Energy of Canada Limited Research Company, Chalk River Nuclear Laboratories, Chalk River, Ontario, Canada K0J 1J0*

J. N. MUNDY (1, 11, 40, 44, 50, 55, 60, 64, 70), *Materials Science and Technology Division, Argonne National Laboratory, Argonne, Illinois 60439*

R. M. NICKLOW (172), *Solid State Division, Oak Ridge National Laboratory, Oak Ridge, Tennessee 37830*

WILLIAM D. OHLSEN (361), *Department of Physics, The University of Utah, Salt Lake City, Utah 84112*

M. ROBINET (27), *Occupational Health and Safety Division, Argonne National Laboratory, Argonne, Illinois 60439*

S. J. ROTHMAN (1, 11, 40, 44, 50, 55, 60, 64, 70), *Materials Science and Technology Division, Argonne National Laboratory, Argonne, Illinois 60439*

W. SCHMATZ (147), *Kernforschungszentrum Karlsruhe, Karlsruhe, Federal Republic of Germany*

L. C. SMEDSKJAER (77), *Materials Science and Technology Division, Argonne National Laboratory, Argonne, Illinois 60439*

M. L. SWANSON (221, 275), *Atomic Energy of Canada Limited Research Company, Chalk River Nuclear Laboratories, Chalk River, Ontario, Canada K0J 1J0*

N. WAKABAYASHI* (194), *Solid State Division, Oak Ridge National Laboratory, Oak Ridge, Tennessee 37830*

---

*Present address: Department of Physics, Keio University, Yokohama, Japan.

PREFACE

The application of nuclear methods in solid state physics is far too large a topic for one volume. We have therefore limited this volume to some aspects of the defect solid state. One of the main topics is the formation and motion of point defects, which in turn define the magnitude of solid state diffusion. The earliest measurements of solid state diffusion were made in 1896 by Sir William Roberts-Austen, using analytical chemical techniques to follow stable isotopes. In the same year, Becquerel discovered radioactivity, and the use of radioactive isotopes has made analysis orders of magnitude more sensitive. Since that time the study of diffusion and lattice defects has continually benefited from the development and improvement of nuclear methods and technology. This volume provides a detailed description of some of, but not all, the experimental nuclear methods that have been applied to the study of lattice defects in solids.

The application of radiotracer techniques to classical tracer diffusion studies is discussed in Part 1. This part includes a section on measurements of the isotope effect of the tracer diffusion coefficient D^T. Such measurements can yield values of the tracer correlation factor f, which in turn yields information on the details of the atomic jump process.

This volume also discusses the more recent methods for measuring diffusion: quasi-elastic neutron scattering (Chapter 3.3) and NMR (Part 6). These techniques measure the macroscopic diffusion coefficient D^{SD}, and comparisons between D^T and D^{SD} can, through a determination of f, lead to the identification of diffusion mechanisms. Another technique used to measure D^{SD} is the Mössbauer effect for which the experimental techniques were discussed in Volume 11 of this treatise.

A variety of nuclear techniques are applied to study other aspects of point defects. Positron annihilation (Part 2) is used to determine the formation enthalpy of vacancies and vacancy–impurity binding enthalpies. Recent developments in two-dimensional angular correlation positron spectroscopy may allow the symmetry of defect structures to be examined. Another technique that has the possibility of examining the symmetry of defect structures is channeling (Chapter 4.2). Neutron diffraction (Chapter 3.1) and NMR (Part 5) can determine certain static properties. Changes of vibration frequencies can also be determined by neutron diffraction techniques (Chapter 3.2). In addition, perturbed angular correlation and positive muon spin resonance may also be used as complementary techniques to study point defects. These two techniques are in a stage of rapid development and are therefore not covered in this volume.

The majority of the studies of point defects and diffusion have involved measurements of the entropies and enthalpies of defect formation and migration or, in the case of diffusion, their sums. In the past 15 years the increased accuracy of these studies has shown that in many systems more than one defect mechanism may be operating in a particular temperature range. In such cases, precise definition of the entropies and enthalpies of the two different processes becomes difficult, and the data analysis may depend on which defects are assumed to be responsible for each process. A further complication arises because in most metals the interstitial impurity content may be higher than the equilibrium vacancy concentration. These problems emphasize the need to apply several techniques to the same material. One technique may then be employed to define the enthalpy and entropy of the defect and another the symmetry or vibration frequency. One technique might be more sensitive to vacancylike defects (positron annihilation) while another measures properties of the interstitial population (e.g., channeling studies).

Radiation damage, as well as point defects, is increasingly often studied by ion beam techniques (Part 4). Depth profiling to show impurity concentration in a near-surface region or to reveal the symmetry of impurity defect clusters are two examples of these techniques. Again, impurities as defects are also studied by NMR and by neutron diffraction. The nonequilibrium defect ensembles resulting from radiation damage have been effectively studied using positron annihilation, and the motion of vacancylike defects has been identified by use of the specificity of positron annihilation toward vacancylike defects.

An important aspect of several parts is the limitations of the various techniques. These limitations, which can be either technical or theoretical, emphasize the importance of complementary techniques. They also provide a challenge for improvements in the various specialties.

The notation used in the various parts is common (as much as possible) to the particular specialty and thus inconsistencies between chapters will be noted.

The editors take pleasure in acknowledging the support and encouragement of Dr. L. Marton, the late treatise editor, and the staff of Academic Press. We thank our wives and children and the contributors for their patience and cooperation, Dr. S. Sinha, Dr. J. Faber, Jr., Dr. P. Pronko, and Dr. C. Wiley for useful conversations in planning this volume, and Ms. Helen Mirenic and Mrs. Bonnie Russell for their skillful secretarial services.

We gratefully acknowledge the partial support of the Division of Basic Energy Sciences, U.S. Department of Energy, through the Argonne National Laboratory.

J. N. MUNDY
S. J. ROTHMAN
M. J. FLUSS
L. C. SMEDSKJAER

METHODS OF EXPERIMENTAL PHYSICS

Editors-in-Chief
Robert Celotta and Judah Levine

Volume 1. Classical Methods
Edited by Immanuel Estermann

Volume 2. Electronic Methods, Second Edition (in two parts)
Edited by E. Bleuler and R. O. Haxby

Volume 3. Molecular Physics, Second Edition (in two parts)
Edited by Dudley Williams

Volume 4. Atomic and Electron Physics—Part A: Atomic Sources and Detectors, Part B: Free Atoms
Edited by Vernon W. Hughes and Howard L. Schultz

Volume 5. Nuclear Physics (in two parts)
Edited by Luke C. L. Yuan and Chien-Shiung Wu

Volume 6. Solid State Physics—Part A: Preparation, Structure, Mechanical and Thermal Properties, Part B: Electrical, Magnetic, and Optical Properties
Edited by K. Lark-Horovitz and Vivian A. Johnson

Volume 7. Atomic and Electron Physics—Atomic Interactions (in two parts)
Edited by Benjamin Bederson and Wade L. Fite

Volume 8. Problems and Solutions for Students
Edited by L. Marton and W. F. Hornyak

Volume 9. Plasma Physics (in two parts)
Edited by Hans R. Griem and Ralph H. Lovberg

Volume 10. Physical Principles of Far-Infrared Radiation
By L. C. Robinson

Volume 11. Solid State Physics
Edited by R. V. Coleman

Volume 12. Astrophysics—Part A: Optical and Infrared Astronomy
Edited by N. Carleton
Part B: Radio Telescopes, Part C: Radio Observations
Edited by M. L. Meeks

Volume 13. Spectroscopy (in two parts)
Edited by Dudley Williams

Volume 14. Vacuum Physics and Technology
Edited by G. L. Weissler and R. W. Carlson

Volume 15. Quantum Electronics (in two parts)
Edited by C. L. Tang

Volume 16. Polymers — Part A: Molecular Structure and Dynamics; Part B: Crystal Structure and Morphology; Part C: Physical Properties
Edited by R. A. Fava

Volume 17. Accelerators in Atomic Physics
Edited by P. Richard

Volume 18. Fluid Dynamics (in two parts)
Edited by R. J. Emrich

Volume 19. Ultrasonics
Edited by Peter D. Edmonds

Volume 20. Biophysics
Edited by Gerald Ehrenstein and Harold Lecar

Volume 21. Solid State: Nuclear Methods
Edited by J. N. Mundy, S. J. Rothman, M. J. Fluss, and L. C. Smedskjaer

1. RADIOTRACER TECHNIQUES

1.1. Introduction

By J. N. Mundy and S. J. Rothman

Materials Science and Technology Division
Argonne National Laboratory
Argonne, Illinois

1.1.1. The Use of Radiotracers

Radioisotopes are the basis for two different approaches to the study of the defect solid state. The defect may perturb the emission of radiation from the radioisotope, e.g., in the Mössbauer effect, positron annihilation, and $\gamma-\gamma$ angular correlation. Alternatively, the radioactive isotope may be used as a tracer, chemically indistinguishable from the stable isotopes of the same element, to study atomic motion, which usually takes place by means of defects. Research has been greatly facilitated by the commercial availability of many radioisotopes, as well as of health physics and waste-removal services and of off-the-shelf equipment for "counting" nuclear radiation (i.e., determining the number of particles emitted per unit time); this equipment is adequate for all but the most specialized experiments. The purpose of this chapter is to summarize the physics underlying the techniques used in radiotracer studies of atomic motion and to provide guidance for useful laboratory procedures.

1.1.2. Introduction to Radioactivity

1.1.2.1. Unstable Nuclei. The nuclei of radioisotopes are unstable because (a) their mass is too great, (b) they have excess neutrons, or (c) they are neutron deficient. Nuclei of type (b) and type (c) lie, respectively, below and above the line of β stability on a plot of proton number versus neutron number.[1] Nuclei with excess mass decay by emission of an alpha particle (α). Nuclei with excess neutrons decay by emission of a beta particle (β^-), neutron-deficient nuclei either by positron (β^+) emission or electron capture (EC). These processes change the atomic number of the atom, converting it into a different element. The radioactive decay process frequently leaves the

[1] R. D. Evans, "The Atomic Nucleus." McGraw-Hill, New York, 1955.

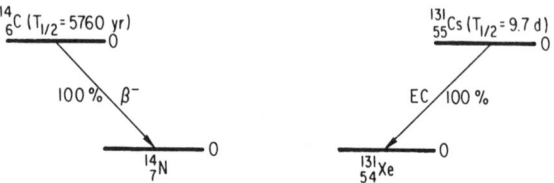

Fig. 1. Decay schemes for ^{14}C and ^{131}Cs.

resulting nuclei in an excited state, in which case the excess energy is almost immediately given up either by the emission of a γ ray or by internal conversion.

1.1.2.2. Decay Modes, Characteristics and Ranges. Decay modes are frequently represented by energy-level diagrams (Figs. 1 and 2). Ground states and excited states are indicated by thick and thin horizontal lines, respectively. Nuclei with higher atomic numbers are to the right, and diagonal arrows between the two states represent either the emission of β^- or β^+ particles or EC. In the simplest case (e.g., Fig. 1), the radioisotope emits an α, β^-, or β^+ particle, and the resulting nucleus is both stable and in its lowest energy state. However, radioisotopes do not always decay in such a simple manner; rather, the daughter nuclei are typically in an excited state, the energy of which is carried away by the emission of one or more γ rays (Fig. 2). The vertical arrows in Fig. 2 indicate γ transitions; the energy of a γ transition is the energy difference between the two levels. Many radioisotopes have much more complex decay schemes.*,[2]

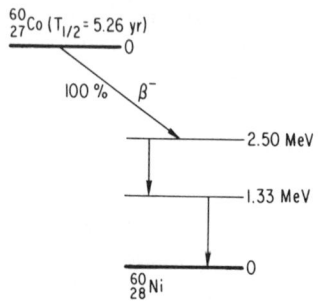

Fig. 2. Decay scheme for ^{60}Co.

---

[2] C. M. Lederer, J. M. Hollander, and I. Perlman, "Table of Isotopes," 7th ed. Wiley, New York, 1978.

* See Bibliography at the end of Part 1 (Nuclear Data Sources section).

1.1. INTRODUCTION

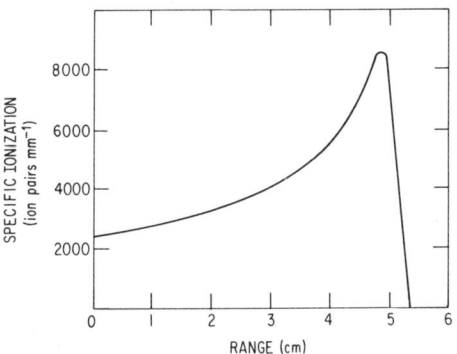

FIG. 3. Specific ionization of an α particle in air as a function of distance from the source. (From Carswell.[3])

1.1.2.2.1. ALPHA DECAY. Alpha particles (^{4}He nuclei) are emitted from nuclei with masses greater than 200, such as the heavy elements of the natural radioactive series and transuranic elements. The particles are monoenergetic, with energies E ranging from 4 MeV for the longest-lived emitters, which have a half-life $T_{1/2}$ (the time required for the total activity of the isotope to fall to half its initial value) of $\sim 10^{10}$ yr, to 9 MeV for the shortest-lived emitters ($T_{1/2} \sim 10^{-7}$ s). When the daughter nuclei are produced in an excited state, the α-particle emission may be accompanied by γ emission.

When α particles pass through matter, they lose energy mainly by excitation and ionization of atoms and molecules. The density of ionization increases as the α particle slows down, reaching a maximum near the end of its range R (Fig. 3). An α particle with a few MeV of energy will undergo $\sim 10^5$ collisions before coming to rest. In air, $R \sim 0.3E^{3/2}$ cm, where E is in MeV. R is small because of the large mass and charge of the α particle. The corresponding range in solids is at least a hundred times smaller [$(2-8) \times 10^{-3}$ cm in aluminum]. The short range of α particles means that for accurate radioactive counting, and especially for spectrometry, the α sources must be exceedingly thin ("weightless").

1.1.2.2.2. EMISSION OF BETA PARTICLES OR POSITRONS. Nuclei with excess neutrons (n) relative to protons (p), such as nuclei formed by neutron capture, decay by emission of a β^- and an antineutrino ($\bar{\nu}$):

$$n \rightarrow p + \beta^- + \bar{\nu}. \quad (1.1.1)$$

The β^- particles are emitted with a continuous range of energies from zero to a maximum energy $E_{\max}$, equal to the difference between the nuclear

[3] D. J. Carswell, "Introduction to Nuclear Chemistry." Elsevier, Amsterdam, 1967.

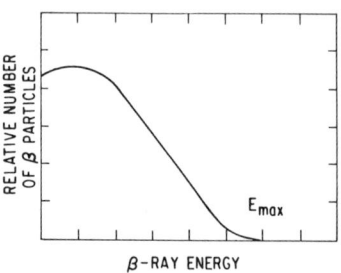

FIG. 4. β^--particle spectrum. (From Carswell.[3])

energy levels (Fig. 1); this is because the β^- particle shares the transition energy with the antineutrino, which is necessary to balance the nuclear spin. An example of a β^- spectrum is shown in Fig. 4. The form is the same for all β^--emitting isotopes, but the value of E_{max} ranges from 50 keV to 5 MeV, with most values between 0.5 and 2 MeV.

Nuclei that have excess protons relative to neutrons decay by either β^+ emission or EC. The reaction of β^+ emission is

$$p \to n + \beta^+ + \nu, \quad (1.1.2)$$

with the positron and neutrino (ν) sharing the energy of the nuclear transition. A β^+ energy spectrum is similar to Fig. 4, with smaller intensities at low energies owing to the Coulomb repulsion of the nucleus. Positron emission occurs only when the energy difference of the nuclear transition is greater than 1.022 MeV (twice the rest mass of the electron); decay by EC is also possible under these conditions.

When the transition energy is less than 1.022 MeV, decay is possible only by EC. The unstable nucleus captures an orbital electron by the reaction

$$p + e^- \to n + \nu, \quad (1.1.3)$$

where e^- denotes an electron originating from outside the nucleus. The rearrangement of the orbital electrons following the EC gives rise to characteristic x rays and Auger electrons with discrete energies corresponding to the atomic structure of the daughter isotope; these energies are much lower than those of most decay β^- particles or γ rays. As neutrinos are not readily detected, pure EC is observable only by detection of the x rays or Auger electrons. In applications where only low-energy emissions are required, pure EC isotopes are particularly useful.

When passing through matter, β^- particles cause electronic excitation and ionization in the same manner as α particles, but the small mass of the β^- particle results in both higher velocities and greater scattering angles than those found for α particles. The energy loss by inelastic scattering is accompanied by a loss through x radiation (bremsstrahlung). The radiation is

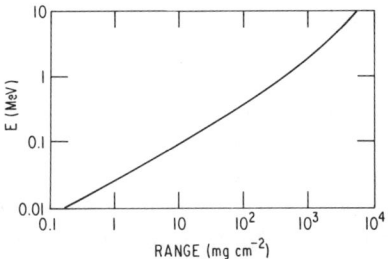

FIG. 5. Range-energy relation for β^- particles in aluminum. (From Clark.[4])

produced by the interaction of the fast-moving β^- particle with the positive field of either the β^--emitting source ("inner" bremsstrahlung) or nuclei along the path of the β ray. The less intense "inner" bremsstrahlung is also emitted in the EC process. The bremsstrahlung has a continuous spectrum of energies, with zero intensity at the energy of the incident β^- particle and with intensity increasing approximately exponentially as the energy decreases. At energies below 100 keV, β^- particles do not produce bremsstrahlung.

The stopping of β^- particles by matter is thus the result of several mechanisms. Because these mechanisms depend on the energy of the β^- particles and β^- particles are emitted with a spectrum of energies (Fig. 4), the exact value of the absorption is best determined experimentally. The transmission for the first part of the β^--particle range follows an exponential relation,

$$A = A_0 \exp(-\mu_m d), \tag{1.1.4}$$

where A is the activity transmitted from a point source of radioactivity A_0 after passing through an absorber of thickness d (measured in mg cm^{-2}). The mass absorption coefficient μ_m is a function of $E_{\max}$, but is relatively insensitive to the atomic number Z of the absorber and can be estimated from the relation

$$\mu_m = 0.017 E_{\max}^{-1.43}, \tag{1.1.5}$$

where μ_m is in cm^2 mg^{-1} and E is in MeV.

The range of β^- particles is two orders of magnitude greater than that of α particles of similar energy, and the specific ionization is correspondingly less (10 ion pairs mm^{-1} in air). The range R of β^- particles in aluminum is shown in Fig. 5; as the absorption is not very sensitive to atomic number, the plot

---

[4] H. M. Clark, in "Techniques of Chemistry" (B. Rossiter, ed.), Vol. I, Part III D, Chap. IX. Wiley (Interscience), New York, 1971.

also applies to other absorbers. The greater range makes source preparation easier than for α sources, but self-absorption is still a problem with low-energy β^- emitters. Equation (1.1.4) predicts that for solid sources of thickness 10 mg cm^{-2} (equivalent to 4.4×10^{-3} cm for carbon) the absorption losses will be $\sim 61\%$ for 0.1-MeV and $\sim 8\%$ for 1.0-MeV β^- particles.

The scattering of β^- particles is much greater than that of α particles and causes two types of problems with the measurement of β activity. The first problem is self-scattering: When the same quantity of radioisotope is contained in sources of different thickness, the count rate (number of β^- particles observed per unit time) will increase with source thickness up to the thickness at which self-absorption becomes the stronger effect because of the increased scattering in the forward direction. The second problem is backscattering of the β^- particles from the material on which the source is deposited: β^- particles are reflected back through the source to the detector and effectively enhance the count rate. The enhancement by backscattering is a function of both the thickness of the scattering material (at small thickness $\leq \frac{1}{5}R$) and its atomic number. Thus a count rate for a thin source may exceed the true value by as much as 75% if the source is deposited on a high-Z material.

The range of β^+ particles is the same order of magnitude as that of β^- particles, but is $\sim 80\%$ greater because of the positive charge. However, β^+ particles have the additional property that at the end of their range they annihilate with an electron. For two-γ annihilation, the energy of 1.022 MeV is divided between two 0.511-MeV γ rays that are emitted in opposite directions, consistent with the conservation of linear momentum. In most tracer experiments one does not detect the β^+ particle itself but rather the annihilation γ rays.

1.1.2.2.3. GAMMA DECAY. In both α and β decay, the daughter nuclei are often left in an excited state from which they decay, usually instantaneously, by one of three processes. The most common process is the emission of one or more γ rays. Generally, a γ-ray spectrum consists of one or more discrete lines formed from transitions between different excited levels (Fig. 2); the common energy range is 0.1–3 MeV. De-excitation of metastable states (noted by the addition of an "m" to the mass number), referred to as an isomeric transition, is usually by γ emission, but can be by particle emission or EC.

The processes of internal conversion or electron–positron pair production de-excite the nucleus without γ-ray emission. The latter process requires a transition energy of greater than 1.022 MeV and is not commonly found. The internal-conversion process transfers the de-excitation energy directly to an orbital electron, leading to the expulsion of this electron from the atom. The expelled electron has a kinetic energy equal to the de-excita-

1.1. INTRODUCTION

tion energy less E_B, the binding energy of the K or L shell of the daughter nucleus. Internal-conversion electrons are more common when the transition energy is of the same order as the shell binding energy. The ratio of the number of conversion electrons to the number of γ rays emitted is called the conversion coefficient.[2] Isotopes undergoing internal conversion complicate measurements of radioactivity because of the different efficiencies of counting systems for electrons and γ rays. After the emission of the conversion electron, the orbital electrons are rearranged, and characteristic x rays and Auger electrons are emitted.

Gamma rays and x rays penetrate matter deeply and have a specific ionization two orders of magnitude lower than that of β rays of the same energy. Their range is a function of the three competing processes that cause energy loss. These processes are the photoelectric effect, the Compton effect, and pair production.

Low-energy γ rays and x rays can lose all of their energy (E_γ) by transferring it totally to an electron. Most of these photoelectric interactions are with the K-shell electrons. Interaction with "free electrons" is impossible, as the recoil of the atom must exist for the conservation of energy and momentum. The photoelectron is expelled with an energy $E_\gamma - E_B$, which is dissipated by ionization and electronic excitation. The orbital electrons of the original positive ion are rearranged, and characteristic x rays and Auger electrons are emitted. These x rays undergo photoelectric absorption, and the process is repeated until the photons have only thermal energy. The probability of photoelectric absorption is approximately proportional to Z^4 of the absorber and E_γ^{-3} of the incident photon. The photoelectric effect is thus most effective for low photon energies interacting in high-Z absorbers.

The Compton effect is the process by which the γ rays interact with loosely bound (outer orbital) electrons and transfer only part of their original energy E_γ in the ejection of the electron. The γ ray is scattered through an angle θ relative to its original direction and has an energy after the collision given by

$$E'_\gamma = \frac{E_\gamma}{1 + E_\gamma(1 - \cos\theta)/E_e}, \qquad (1.1.6)$$

where $E_e = 0.511$ MeV is the rest mass of the electron. From Eq. (1.1.6) it is clear that E'_γ varies from a minimum at $\theta = 180°$ to a maximum for $\theta = 0°$; E'_γ can be lost by successive Compton interactions and finally by photoelectric absorption. The Compton electron has an energy spectrum from zero to a maximum determined by Eq. (1.1.6) ($E_\gamma - E'_\gamma$, for $\theta = 0°$); it loses its energy by ionization and electronic excitation. The probability of a photon undergoing a Compton interaction is proportional to Z for the absorber and E_γ^{-1} of the incident photon energy.

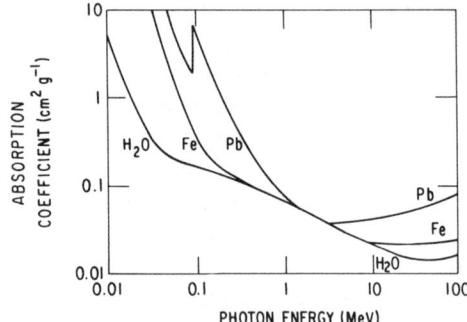

FIG. 6. Mass absorption coefficients for three materials as a function of energy. (From Clark.[4])

Gamma rays that have energies greater than 1.022 MeV can interact with the field of the absorbing nuclei and create an electron–positron pair. The nucleus participates in the conservation of momentum and energy. The energy of the incident photon in excess of 1.022 MeV is shared by the pair and is in turn lost by ionization and electronic excitation. When the positron's energy has been degraded to thermal energy, the positron annihilates, and the two 0.511-MeV γ rays lose their energy by Compton interactions and photoelectric absorption. The probability of pair production by γ rays is proportional to Z^2 for the absorber and increases with increasing γ-ray energy.

The attenuation of γ rays follows an exponential law as given by Eq. (1.1.4), where μ_m is the sum of terms from the photoelectric interactions, the Compton-effect interaction, and pair production. A similar expression determines the energy absorbed; however, the mass absorption coefficient is smaller because the attenuation by scattering is not included. The absorption coefficient is a measure of the number of primary photons that interact with the absorber and is a function of both the absorber and the energy of the γ ray, as shown for three materials in Fig. 6. The initial, steep part of each curve is where the γ rays interact mainly by photoelectric absorption. The discontinuity observable for lead is a result of γ-ray energies exceeding the binding energy of the K-shell electrons. This is followed by a region where the Compton-effect interaction dominates; at still higher energies, pair production is the dominant mechanism for energy loss.

Self-absorption is not usually a problem with the preparation of γ sources unless the photon energy is very low. For accurate comparative measurements, the effects of self-absorption can easily be eliminated by preparing each source with the same geometry. Sample preparation with similar

1.1. INTRODUCTION

geometry can be achieved by dissolving each source in equal volumes of solution. The attenuation of 5 mm of water is less than 10% for a γ-ray energy of 0.1 MeV. The greater attenuation for α and β^- particles makes similar sample preparation considerably more difficult.

1.1.2.3. Half-Lives and Decay Chains. The probability of the decay of a radioactive nucleus is independent of the environment of the nucleus except in the case of EC. This concept is expressed by first-order kinetics such that the activity or number of disintegrations per unit time, A, is proportional to the number of radioactive nuclei present, N:

$$A = \frac{dN}{dt} = -\lambda N, \qquad (1.1.7)$$

where A is commonly expressed in disintegrations per second (dps) [the SI unit of radioactivity, the becquerel (Bq), equals 1 dps; the older unit, the curie (Ci), equals 3.7×10^{10} Bq] and λ is the decay constant. When the decay constant is known, the maximum specific activity (activity per unit mass) of that radioisotope is

$$S_{\max} = (\lambda/W) \times 6.023 \times 10^{23} \quad \text{dps/g}, \qquad (1.1.8)$$

where W is the atomic weight of the element. Alternatively, when the overall efficiency ϵ of the radioactive-source–detector system is known, Eq. (1.1.7) allows one to determine the number of radioactive atoms present at a given time from the measured activity $C = \epsilon A$. Equation (1.1.7) can be integrated to give

$$A = A_0 \exp(-\lambda t), \qquad (1.1.9)$$

where A_0 is the activity at $t = 0$. From Eq. (1.1.9) it follows that the half-life $T_{1/2}$ is given by

$$T_{1/2} = (\ln 2)/\lambda \qquad (1.1.10)$$

It follows from Eq. (1.1.9) that a single parent radioisotope decaying to a stable daughter isotope will yield a straight-line plot of $\log(A/A_0)$ versus t with a slope that gives the half-life of the isotope. Half-lives that are convenient for defect solid state applications range from 10^4 to 10^{12} s.

When n independent radioisotopes are present, observed activity varies with time as

$$C(t) = \sum_{j=1}^{j=n} A_j^0 \, \epsilon_j \exp(-\lambda_j t), \qquad (1.1.11)$$

where $A_j^0 \epsilon_j \exp(-\lambda_j t)$ is the contribution of the jth component at time t. In this case a least-squares fitting of the decay data to Eq. (1.1.11) allows one to determine the specific activities A_j^0 at a particular time. The decay constants need to be well known; if absolute values of the A_j^0 are required, the efficiencies ϵ_j must also be well established.

When the daughter isotope is also unstable, the measured activity is given by

$$C(t) = \lambda_1 \epsilon_1 A_1^0 \exp(-\lambda_1 t) + \frac{\lambda_1 \lambda_2}{\lambda_2 - \lambda_1} \epsilon_2 A_1^0 [\exp(-\lambda_1 t) + \exp(-\lambda_2 t)]$$
$$+ \lambda_2 \epsilon_2 A_2^0 \exp(-\lambda_2 t), \quad (1.1.12)$$

where the subscripts 1 and 2 refer to the parent and daughter radioisotopes, respectively. The last term is the contribution from the daughter atoms present at $t = 0$, which usually results from incomplete chemical separation of the parent isotope. The forms of the decay curves will vary depending on the relative magnitudes of the decay constants λ_1 and λ_2. If $\lambda_1 \ll \lambda_2$, i.e., if the parent is much longer lived than the daughter, both activities decay with the half-life of the parent. If $\lambda_1 \gg \lambda_2$, i.e., if the daughter is much longer lived, then after several half-lives of the parent, the activity will decay with the half-life of the daughter. This is called secular equilibrium.

1.2. Production of Radioisotopes

By J. N. Mundy and S. J. Rothman

Materials Science and Technology Division
Argonne National Laboratory
Argonne, Illinois

Prior to the discovery of artificial radioactivity in 1934, only the radioisotopes of the naturally occurring radioactive chains, uranium, thorium, and actinium, were known. At present, radioisotopes of almost every element can be produced and approximately 200 radioisotopes are commercially available (Table I). For most uses, purchase from a commercial supplier is the easiest and most economical way of obtaining radioisotopes; special production should be considered only when the desired characteristics (e.g., short half-life, nonstandard chemical form, exceptional radiochemical purity, or very high specific activity) demand it. Some examples of specially produced isotopes are listed in Table II.

Nevertheless, an understanding of the limitations of radioisotope production is helpful even when purchasing commercial radioisotopes. Once the experimenter has decided which radioisotope to use, the most important characteristics to consider are (a) specific activity, (b) radiochemical purity, (c) chemical and physical form, and (d) chemical purity.

For many experiments, such as straightforward self-diffusion measurements, radioactive impurities of the same chemical species can be tolerated (e.g., ^{55}Fe with ^{59}Fe, ^{108}Ag with ^{110m}Ag). In other experiments a pure radioisotope may be desired. In almost all experiments a radioactive impurity of another chemical species (e.g., ^{54}Mn with ^{59}Fe) is highly undesirable.

1.2.1. Principles of Isotope Production

1.2.1.1. Nuclear Reactions.
Artificial radioisotopes are produced by nuclear reactions. One of the reactants is commonly a light particle such as a neutron (n), proton (p) deuteron (d), triton (t), helium ion (α particle or ^{3}He), electron (e), or photon (γ). The products of the reaction are unstable nuclei and one or more light particles. The total number of hadrons is always conserved at these energies. Such reactions are frequently written analogously to chemical reactions, with mass numbers indicated by super-

TABLE I. Commercially Available Radioisotopes of Interest for Diffusion Studies

Isotope	$T_{1/2}$	Type of decay	Specific activity[a]
^{3}H	12.3 yr	β	90% pure
^{7}Be	53 d	EC,γ	CF
^{14}C	5640 yr	β	—
^{22}Na	2.6 yr	β^+,EC,γ	CF
^{24}Na	15.0 h	β, γ	5 Ci g^{-1}
^{26}Al	1×10^6 yr	β^+,EC,γ	low
^{32}P	14.2 d	β	CF
^{35}S	89 d	β	CF
^{36}Cl	3×10^5 yr	β,γ	10 mCi g^{-1}
^{42}K	12.5 h	β,γ	10 Ci g^{-1}
^{45}Ca	165 d	β	40 Ci g^{-1}
^{46}Sc	84 d	β,γ	500 Ci g^{-1}
^{48}V	16.2 d	β^+,EC,γ	CF
^{51}Cr	27.7 d	EC,γ	500 Ci g^{-1}
^{54}Mn	313 d	EC,γ	CF
^{55}Fe	2.7 yr	EC	CF
^{59}Fe	45 d	β,γ	40 Ci g^{-1}
^{56}Co	77 d	β^+,EC,γ	CF
^{57}Co	270 d	EC,γ	CF
^{58}Co	71 d	β^+,EC,γ	CF
^{60}Co	5.3 yr	β,γ	300 Ci g^{-1}
^{63}Ni	92 yr	β	20 Ci g^{-1}
^{64}Cu	12.8 h	β,β^+,EC,γ	10 Ci g^{-1}
^{65}Zn	245 d	β^+,EC,γ	10 Ci g^{-1}
^{67}Ga	14.1 h	EC,γ	CF
^{68}Ge	280 d	EC	CF
^{74}As	18 d	β,β^+,EC,γ	1 Ci g^{-1}
^{75}Se	121 d	EC,γ	500 Ci g^{-1}
^{82}Br	35.6 h	β,γ	3 Ci g^{-1}
^{85}Kr	10.6 yr	β,γ	20 Ci g^{-1}
^{86}Rb	18.7 d	β,γ	10 Ci g^{-1}
^{85}Sr	65 d	EC,γ	30 Ci g^{-1}
^{89}Sr	51 d	β,γ	100 Ci g^{-1}
^{90}Sr	28 yr	β	75 Ci g^{-1}
^{88}Y	108 d	β^+,γ	CF
^{90}Y	64 h	β	10 Ci g^{-1}
^{95}Zr	65 d	β,γ	CF
^{95}Nb	35 d	β,γ	CF
^{99}Mo	66.7 d	β,γ	0.1 Ci g^{-1}
^{99}Tc	6 d	γ	CF
^{103}Ru	40 d	β,γ	3 Ci g^{-1}
^{106}Ru	371 d	β	CF
^{103}Pd	17 d	EC,γ	CF

TABLE I. (Continued)

Isotope	$T_{1/2}$	Type of decay	Specific activity[a]
^{110}Ag	253 d	β,γ	20 Ci g^{-1}
^{111}Ag	7.5 d	β,γ	2 Ci g^{-1}
^{109}Cd	1.3 yr	EC,γ	3 Ci g^{-1}
$^{115m-115}$Cd	43 d	β,γ	1 Ci g^{-1}
$^{114m-114}$In	50 d	β,γ	60 Ci g^{-1}
^{113}Sn	119 d	EC,γ	20 Ci g^{-1}
^{124}Sb	60 d	β,γ	10 Ci g^{-1}
^{125}Sb	2.7 yr	β,γ	100 Ci g^{-1}
^{125m}Te	58 d	β,γ	100 Ci g^{-1}
^{125}I	60 d	EC,γ	CF
^{131}I	8 d	β,γ	CF
^{137}Cs	30 yr	β,γ	CF
^{133}Ba	10.7 yr	EC,γ	10 Ci g^{-1}
^{140}La	40.2 h	β,γ	1 Ci g^{-1}
^{141}Ce	32.5 d	β,γ	5 Ci g^{-1}
^{147}Nd	11.1 d	β,γ	1 Ci g^{-1}
^{147}Pm	2.5 yr	β	CF
$^{152m-152}$Eu	9.3 h	β,EC,γ	10 Ci g^{-1}
^{154}Eu	16 y	β,γ	500 mCi g^{-1}
^{153}Gd	236 d	EC,γ	25 Ci g^{-1}
^{160}Tb	73 d	β,γ	1.5 Ci g^{-1}
^{166m}Ho	1.2×10^3 yr	β,γ	0.75 mCi g^{-1}
^{169}Er	9.4 d	β	0.1 Ci g^{-1}
^{170}Tm	127 d	β,EC,γ	1 Ci g^{-1}
^{169}Yb	31 d	EC,γ	10 Ci g^{-1}
^{177}Lu	6.8 d	β,γ	10 Ci g^{-1}
^{175}Hf	70 d	EC,γ	200 Ci g^{-1}
^{181}Hf	42.5 d	β,γ	50 Ci g^{-1}
^{182}Ta	115 d	β,γ	10 Ci g^{-1}
^{185}W	73 d	β	10 Ci g^{-1}
^{186}Re	3.7 d	β,EC,γ	5 Ci g^{-1}
^{192}Ir	74.2 d	β,EC,γ	10 Ci g^{-1}
^{195m}Pt	4.1 d	β,γ	0.1 Ci g^{-1}
^{195}Au	185 d	EC,γ	CF
^{198}Au	64.8 h	β,γ	40 Ci g^{-1}
$^{197m-197}$Hg	24 h	EC,γ	10 Ci g^{-1}
^{203}Hg	47 d	β,γ	20 Ci g^{-1}
^{204}Tl	3.6 yr	β,EC	15 Ci g^{-1}
^{210}Pb	21 yr	β,γ	60 Ci g^{-1}
^{206}Bi	6.3 d	EC,γ	CF
^{207}Bi	30.2 yr	β,γ	CF
^{210}Po	138 d	α	CF
^{241}Am	470 yr	α,γ	3 Ci g^{-1}

[a] CF = carrier-free.

TABLE II. Some Specially Produced Radioisotopes

Isotope	$T_{1/2}$	Type of decay	Production reaction	Activity or cross section (σ)
^{28}Mg	20.93 h[a]	β,γ	^{26}Mg(T,p)^{28}Mg	—
^{31}Si	2.6 h	β,γ	^{30}Si(n,γ)^{31}Si	$\sigma = 0.11$ b
^{52}Mn	5.7 d	β^+,γ	^{51}V(α,3n)^{52}Mn	CF
^{52}Fe	8.275 h[a]	β^+,γ	^{52}Cr(^{3}He,3n)^{52}Fe	CF
^{67}Cu	62.0 h[a]	β,γ	^{67}Zn(n,p)^{67}Cu	CF
$^{69m-69}$Zn	13.76 h[a]	γ	^{68}Zn(n,γ)^{69}Zn	$\sigma = 0.1$ b
^{77}Ge	11.3 h	β,γ	^{76}Ge(n,γ)^{77}Ge	$\sigma = 0.2$ b
^{105}Ag	41.29 d[a]	EC,γ	^{103}Rh(α,2n)^{105}Ag	CF
^{187}W	24 h	β,γ	^{186}W(n,γ)^{187}W	$\sigma = 40$ b

[a] S. J. Rothman et. al., *Phys. Rev.* C **9**, 2272 (1974).

scripts and atomic numbers by subscripts. For example, the bombardment of natural sodium with neutrons produces the reaction

$$^{23}_{11}\text{Na} + ^{1}_{0}\text{n} \rightarrow ^{24}_{11}\text{Na} + Q, \quad (1.2.1)$$

where Q is the energy given off as a γ ray, and is equal to the mass difference between the reactants and products, by means of the Einstein formula $E = mc^2$. Equation (1.2.1) may also be written in the abbreviated form

$$^{23}\text{Na}(n,\gamma)^{24}\text{Na}. \quad (1.2.2)$$

Most nuclear data compilations will give the principal means of isotope production in the format of Eq. (1.2.2).

A number of different reactions can give rise to the same isotope. A useful map for these is the chart of the nuclides,[1] which plots all known isotopes with Z as the ordinate and the number of neutrons in the nucleus as the abscissa. An illustration of part of such a chart is shown in Fig. 1a, with stable isotopes marked by heavy borders. Nuclear reactions that correspond to movement from one nuclide to another in Fig. 1a are shown in Fig. 1b. As an example, consider the manufacture of the radionuclide ^{51}Cr. The possible reactions to produce ^{51}Cr are ^{48}Ti(α,n)^{51}Cr, ^{49}Ti(α,2n)^{51}Cr, ^{50}Ti(α,3n)^{51}Cr, ^{51}V(d,2n)^{51}Cr, ^{51}V(p,n)^{51}Cr, and ^{50}Cr(n,γ)^{51}Cr. Which of these reactions gives the highest yield is determined by the cross section for each reaction, as discussed below. However, the other considerations listed above, as well as the cost, also influence the choice of reaction.

It is important to note that the desired reaction is not necessarily the only

[1] N. E. Holden and F. W. Walker, "Chart of the Nuclides," 11th ed. General Electric Co., Schenectady, New York, 1972.

1.2. PRODUCTION OF RADIOISOTOPES

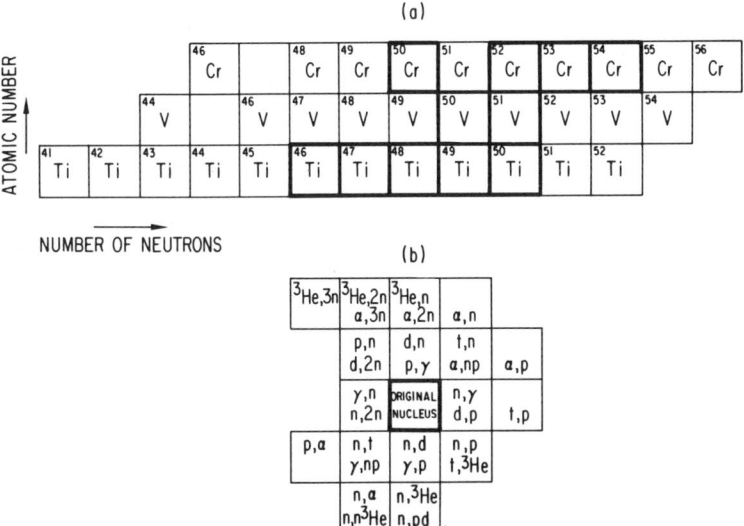

FIG. 1. (a) Partial chart of the nuclides. (b) Displacements caused by nuclear bombardment reactions.

one taking place. If, for instance, the first reaction is chosen and natural Ti is irradiated with α particles, other stable Ti isotopes will also undergo (α,n) reactions, yielding other Cr isotopes [^{46}Ti(α,n)^{49}Cr]. Also, (α,p) and (α,2n) reactions, e.g., ^{46}Ti(α,p)^{49}V and ^{47}Ti(α,2n)^{49}Cr can occur, and these can yield impurity radionuclides such as ^{49}V. Further, if impurities are present in the starting material, these will also undergo nuclear reactions and yield radioactive impurities. Again, the relative rate at which each radionuclide is produced is proportional to the fraction of the reactant nuclide present in the target and to the cross section of the reaction in question.

1.2.1.2. Nuclear Cross Sections. The probability that a nuclear reaction will occur is usually expressed in terms of a cross section σ, which has the dimension of an area. The unit of area originates from the simple idea that the probability of reaction between a bombarding particle and a nucleus should be proportional to the effective cross-sectional area that the nucleus presents as a target. For heavy nuclei, the fast-neutron cross section is, in fact, of the same order as the geometrical cross-sectional area, $\sim 10^{-24}$ cm^2; therefore, this area has been chosen as the unit of nuclear cross section [10^{-24} cm^2 = 1 barn (b)]. However, this picture does not hold for charged particles, which have to overcome the repulsive Coulomb field of the nucleus, or for slow neutrons, which penetrate a nucleus easily. In these cases the cross sections can range from small fractions of a barn to 10^5 b.

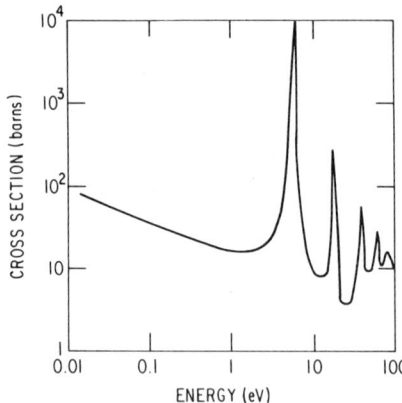

FIG. 2. Schematic of energy dependence of neutron capture cross section, showing resonances.

For a thin target where there is no significant attenuation of the flux of bombarding particles, the number of radioactive nuclei formed in each square centimeter per second is given by

$$\frac{dN}{dt} = \phi \sigma n_d x, \quad (1.2.3)$$

where ϕ is the particle flux (number of particles in each square centimeter per second), n_d is the number of target nuclei per cubic centimeter, and x is the thickness of the target in centimeters. When the target is thick, Eq. (1.2.3) can be written

$$\frac{dN}{dt} = \phi_0 (1 - e^{-n_d \sigma x}), \quad (1.2.4)$$

where ϕ_0 is the flux of the incident beam. The cross section here is the total cross section for all possible reactions. Partial cross sections need to be specified for each particular reaction.

The nuclear cross section as a function of the incident energy is called the excitation function. The excitation functions are seldom smooth over an extended range of energy and usually contain resonances, where for a small range of energies the nuclear cross section is considerably higher (Fig. 2). Therefore, the energy of the incident particle has to be adjusted carefully to give the maximum yield of the desired reaction and minimum yield of the undesired reactions. Nuclear reactions that require energy (i.e., endoenergetic reactions, in which the mass of the products is heavier than the reactants) have a threshold energy below which the reaction will not take place.

1.2.2. Determination of Yields and Specific Activities

The increase of the concentration of a particular radioisotope under irradiation depends on (a) flux ϕ, (b) number of target nuclei n_d present at the start of the irradiation, (c) cross section σ_1 of the target nuclei for the desired reaction, (d) reaction cross section σ_2 of the product nuclei, which may be destroyed by the capture of further bombarding particles, (e) decay constant λ_1 of the product nuclei, and (f) irradiation time t.

The activity A built up during an irradiation of duration t is given (in becquerels) by

$$A = \frac{n_d \sigma_1 \phi \, \lambda_1}{\lambda_1 + \sigma_2 \phi - \sigma_1 \phi} \{e^{-\sigma_1 \phi t} - e^{-(\lambda_1 + \sigma_2 \phi)t}\}. \quad (1.2.5)$$

Allowance for the reaction cross section of the product nucleus is only important in cases where $\sigma_2 \gg \sigma_1$. Such second-order reactions are a problem when the second product has a similar decay mode and a decay constant equal to or greater than λ_1, e.g., ^{181}Ta(n,γ)^{182}Ta(n,γ)^{183}Ta, but in most cases such reactions are not significant. Allowance for the small decrease in the number of target nuclei during irradiation (burnup) is usually neglected, as $\sigma_1 \phi \ll \lambda_1$. For most cases, Eq. (1.2.5) may be rewritten

$$S = \frac{0.6 \phi \sigma_1 (1 - e^{-\lambda_1 t})}{W} f_i, \quad (1.2.6)$$

where S is the specific activity (activity per unit weight) in becquerels per gram of target material, W is the atomic weight of the target element, and f_i is the atomic fraction of the nucleus in question in the target.

Examination of Eq. (1.2.6) shows that for $t \ll 1/\lambda_1$, S increases linearly with t; for $t \gg 1/\lambda_1$, S reaches a saturation value $(0.6 \phi \sigma / W) f_i$ (Fig. 3). The

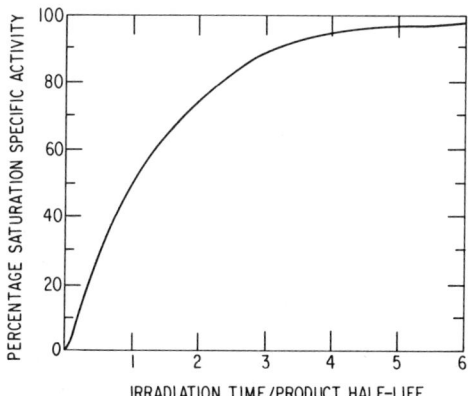

FIG. 3. Buildup of radioactivity as a function of irradiation time.

asymptotic form of the curve shows clearly that when the desired nuclear reaction is predominant, little is gained by irradiating for times greater than a few half-lives of the product radioisotope. On the other hand, because of impurities in the target material, other nuclear reactions are occurring simultaneously, so there are advantages to varying the irradiation time. For example, when the product of the competing nuclear reaction has a longer half-life, relatively more of the desired isotope will be produced by irradiation at higher fluxes and for shorter times. Thus appropriate choice of flux and irradiation time can optimize the relative yield of the desired radioisotope.

Equation (1.2.6) can be used to estimate the parameters involved in an irradiation. Assuming a nucleus with $W = 100$ and $\sigma = 1$ b, irradiation to saturation in a flux of 10^{12} n cm^{-2} will yield an activity of 0.6×10^{10} Bq g^{-1}, or about 0.2 Ci g^{-1}. To obtain activities in the millicurie or microcurie range, samples weighing a few milligrams must usually be used.

As shown in Eq. (1.2.6) the specific activity of a radioisotope depends on the fraction of the reacting nucleus present in the target. It is obviously easier to irradiate an element in its natural isotopic constitution; however, the specific activity of an isotope for given irradiation conditions can be increased by using a target enriched in the desired nucleus. On the other hand, isotopically enriched materials are often less pure chemically than commercially obtainable high-purity material and therefore may yield products with undesirable radioactive impurities.

1.2.3. Isotope Production Reactions

1.2.3.1. *Natural Radioactivity.* A few radioactive nuclides are found in nature. These include the uranium, thorium, and actinium chains, i.e., the elements Po through U, all the isotopes of which are radioactive. Among the useful isotopes in these series are ^{210}Pb (RaD), the best tracer for Pb, and ^{210}Po, which is an α source with a useful half-life. Some of these isotopes are available commercially. The naturally occurring nuclides ^{40}K and ^{87}Rb have half-lives greater than 10^9 yr; more useful artificially produced isotopes of these elements are available. Some nuclides, such as ^{14}C, are formed from reactions with cosmic radiation.

1.2.3.2. *Thermal-Neutron Capture.* The great majority of useful radioisotopes are produced by thermal-neutron capture in a nuclear reactor because the cross sections for these reactions and the availability and cost of thermal neutrons are favorable. The entry of a neutron into a nucleus is not opposed by a Coulomb barrier, and so neutrons with very low energy react readily with all nuclei to form a radioactive isotope of the same element. The excitation energy contributed by the incident thermal neutron is usually not sufficient for the emission of a charged nuclear particle, and the resulting compound nucleus usually gives up its energy as photons or neutrons. The

1.2. PRODUCTION OF RADIOISOTOPES

few exceptions to this type of (n,γ) reaction (referred to as radioactive capture) are found among the lighter nuclei, where the binding energy and Coulomb barrier of a proton or α particle are considerably lower than those of a neutron. Examples of such reactions are ^{6}Li(n,α)^{3}H, ^{14}N(n,p)^{14}C, and ^{35}Cl(n,p)^{35}S.

Thermal neutrons are produced by the slowing down of fission neutrons in a reactor. The average fission-neutron energy is approximately 1.5 MeV. The neutrons are slowed down by collisions in a moderator, a material of low mass number, such as graphite, water, or heavy water, until they reach thermal energies (0.025 eV). The neutron flux ϕ and neutron spectrum vary with position in the reactor. In the core of the reactor, $10^{12} < \phi < 10^{15}$ n cm^{-2} s^{-1}, up to 20% of which can be fast neutrons ($E > 10$ keV). In the moderator, $10^{11} < \phi < 5 \times 10^{12}$ n cm^{-2} s^{-1}, and the fraction of fast neutrons is small. Each reactor has its own neutron spectrum, and the only way to select the neutron energy is by varying the location of the sample. A pure fast flux can be obtained only in special facilities[2]; pure thermal fluxes can be obtained in most research reactors.

Most research reactors have two types of facilities available for the irradiation of samples: those that can be loaded and unloaded while the reactor is in operation and those that are accessible only when the reactor is closed down. In the first type, the sample is either placed in a transfer container ("rabbit") and driven into or out of the reactor by a pneumatic transfer system or placed in a can and pushed into the moderator region of the reactor. Facilities of the second type are usually located in channels between the reactor shielding and core or between the fuel elements; in some cases the samples may actually be placed in hollow fuel elements. Samples are suitably encapsulated and hand-loaded when the reactor is down. These facilities are useful when long irradiations, very high fluxes, or a higher proportion of fast neutrons is required.

The requirements for sample encapsulation vary from one reactor to another, and the potential user must refer to the user's guide for the particular reactor. Samples are frequently sealed in fused silica tubes under vacuum or a small pressure of argon. Encapsulation in quartz minimizes the risk of radioactive contamination and serves to localize the frequently very small samples. Pyrex is not used because it contains boron, which has a large thermal-neutron absorption cross section (762 b). Polyethylene capsules can be used if the reactor ambient temperature is low enough. Samples are, as a rule, irradiated in the solid state. The chemical and physical form of the sample should be chosen so as to make subsequent handling as easy as possible. Welding into the secondary container, which is usually aluminum,

---

[2] A. C. Klank, T. H. Blewitt, J. Minarik, and T. L. Scott, *Annexe Bull. Inst. Int. Froid* **5**, 373 (1966).

may be advisable if the sample has a high vapor pressure at the reactor temperature, which may reach 100–200°C.

The transfer tube and in turn the sample are cooled by the reactor coolant. The sample is heated by γ rays produced by both the fission process and its own irradiation. The heat developed is dependent on the γ-ray energies and the mass absorption coefficient of the sample. The amount of gamma heating produced in a 1-g sample at the center of a 10-MW nuclear reactor is 2–3 W. The space between the sample and the walls of the secondary container must be filled with material of high thermal conductivity, e.g., aluminum foil, in order to conduct away the heat from the gamma heating. Because the outer can usually weighs more than the sample, it may be very active. (A gram of material with $\sigma = 10$ b and $\phi = 10^{12}$ n cm^{-2} s^{-1} produces 5 Ci of activity.) The outer can should be removed behind lead shields (≥ 5 cm thick), preferably in a glove box. The quartz capsule itself may be broken open in a hood. Samples resulting from long irradiations are best handled in hot cells (Chapter 1.3).

At least 75% of the elements have at least one isotope that undergoes an n–γ reaction with thermal neutrons. Examples of such reactions useful for isotope production are ^{23}Na(n,γ)^{24}Na, $\sigma = 0.53$ b, and ^{59}Co(n,γ)^{60}Co, $\sigma = 37$ b. As the isotope produced has an extra neutron, it is frequently a β-emitting isotope of the target element. The isotopic composition of the target element is not significantly changed during irradiation. For example, for the nuclear reaction ^{63}Cu(n,γ)^{64}Cu, an irradiation of the target to saturation with a flux of 10^{13} n cm^{-2} s^{-1} would result in conversion of only 3×10^{-6} of the initial ^{63}Cu nuclei. The product of the irradiation is thus a small concentration of radioisotope in a matrix of an isotope of the same element. A chemical separation cannot be performed, and so the specific activities of isotopes generated by neutron capture are limited. However, the reaction sometimes forms a short-lived isotope that undergoes beta decay. An example is the reaction ^{198}Pt(n,γ)^{199}Pt $\xrightarrow{\beta^+}$ ^{199}Au; the half-life of ^{199}Pt is 30 min, so in half a day the majority of the ^{199}Pt has become the radioisotope ^{199}Au, which can be separated "carrier-free" (i.e., free of nonradioactive nuclei of the same element; see Section 1.2.4) by solvent extraction.

1.2.3.3. Fast-Neutron Reactions. Radioisotopes can also be made through transmutation, i.e., absorption of a neutron by the nucleus with the subsequent release of a charged particle. The most common transmutation reaction is the (n,p) reaction, e.g., $^{35}_{17}$Cl(n,p)$^{35}_{16}$S, although (n,d), (n,α), (n,t), and (n,^{3}He) are possible, e.g., $^{27}_{13}$Al(n,α)$^{24}_{11}$Na. Some neutron-rich isotopes, e.g., ^{67}Cu, can best be produced by these reactions. Transmutation reactions result in product isotopes that are chemically different from the target nucleus and can thus be separated from the target material to provide carrier-free isotopes (see Section 1.2.4.).

In general, fast neutrons (>0.5 MeV) are required for transmutation

1.2. PRODUCTION OF RADIOISOTOPES

because the excitation energy must exceed the Coulomb barrier for charged-particle emission. The yield of fast neutron reactions is low, as the cross sections are usually small and the neutrons lose energy rapidly as they penetrate the target.

Fast-neutron reactions can also produce undesirable contaminants. For example, irradiation of natural iron in the core of a research reactor produces as much ^{54}Mn by the reaction ^{54}Fe(n,p)^{54}Mn (natural abundance of ^{54}Fe: 5.84%) as ^{59}Fe by the reaction ^{58}Fe (n,γ)^{59}Fe (natural abundance of ^{58}Fe: 0.31%). In most cases, however, the fast-neutron flux and cross section are low enough so that no unwanted impurities are formed.

1.2.3.4. Fission. Fission products ranging in mass number from 72 to 161 (zinc to dysprosium) are separated during the reprocessing of spent reactor fuel elements; a dozen of the longer-lived isotopes, such as ^{85}Kr, ^{95}Zr, and ^{137}Cs, are available commercially. Some short-lived isotopes, such as ^{133}Xe and ^{133}I, are also prepared from fission products; a specially prepared uranium target is used to generate such isotopes.

1.2.3.5. Charged-Particle Reactions. Neutron-deficient radionuclides can be produced by bombarding suitable nuclei with protons, deuterons, tritons, α particles, or ^{3}He ions, usually in a cyclotron. A common maximum energy for protons accelerated in a small isotope-production cyclotron is 25–30 MeV; the beam intensity can vary from microamperes to milliamperes. For small targets the beam intensities can be comparable to the neutron fluxes available in the core of nuclear reactors, but the cross sections for isotope production are commonly smaller than those for neutron-capture reactions. Many different reactions are possible during charged-particle bombardment, such as (α,n), (α,2n) and (α,d). As a rule, neutrons are more easily ejected than protons, which must overcome a Coulomb barrier. However, it takes more energy to eject two neutrons than one. For example, Fig. 4 shows the cross sections as a function of energy for the reactions of α particles on ^{60}Ni.[3] Evidently, the energy of the bombarding particle must be chosen to maximize the cross section for the desired reaction and to minimize the others, although traces of unwanted radioisotopes may nevertheless be present. As charged-particle bombardment produces a species that is chemically different from the target, a chemical separation is necessary, and carrier-free radioisotopes can often be produced.

Bombardment with charged particles can also result in spallation or fission reactions. Examples are, respectively, the production of ^{28}Mg by 350-MeV H$^+$ bombardment of NaCl[4] and the fission of ^{238}U by 40-MeV α particles to produce ^{112}Pd.[5]

---

[3] S. N. Ghoshal, *Phys. Rev.* **80**, 939 (1950).
[4] P. G. Shewmon and F. N. Rhines, *Trans. Am. Inst. Min. Met. Eng.* **200**, 1021 (1954).
[5] N. L. Peterson, *Phys. Rev.* **136**, A568 (1964).

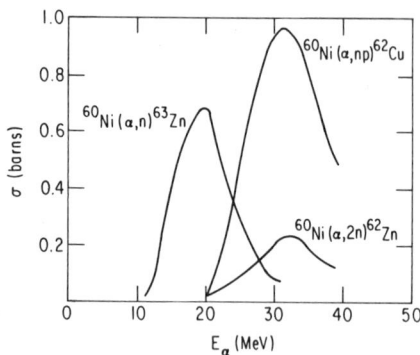

FIG. 4. Reaction cross sections for α irradiation of nickel. (From Ghoshal.[3])

1.2.4. Chemical Processing in Radioisotope Production

In the case of transmutation reactions, induced by either charged-particle or fast-neutron bombardment, the desired radioisotope is a chemically different species from the target. It must, in most cases, be separated chemically from the relatively large amount of inert target material and the radioisotopes of other elements that have been formed (see Fig. 4). In the case of radioisotopes formed by thermal-neutron capture, the isotope being produced is chemically the same as the target material. However, radioisotopes of other elements (radioimpurities) can be formed, either by fast-neutron-induced transmutation reactions on the main constituent of the target or by thermal-neutron capture by an impurity. If these radioimpurities cannot be discriminated out in the counting scheme (see Chapter 1.4), they have to be separated chemically. Chemical processing not involving separations is sometimes necessary to change the chemical form of the radioisotope; examples of this are given in Section 1.3.2.2.

The chemical processes used in the separation of radioisotopes cover the whole range of techniques developed for classical chemical analysis; only the most common methods will be outlined here. First, it is important to note the special limitations imposed by the nature of radioisotopes. These include the requirement for speed in the chemical separation of short-lived isotopes and the necessity for remote handling of high activity levels. The small scale of radioisotope separation is another important factor, as both the desired radionuclide in the case of a transmutation reaction and the undesirable impurity in the case of a neutron-capture reaction are present in very low concentrations.

A nonradioactive carrier that is chemically the same as the radioisotope to be separated may be added. This facilitates the use of several purification techniques, as discussed below. This procedure changes only the radiochem-

1.2. PRODUCTION OF RADIOISOTOPES

ical purity, not the chemical purity, of the final product. (In evaluating the stated purity of radioisotopes, it is important to note whether chemical or radionuclidic purity is quoted.) Care must be taken to add a carrier in the same valence state as the radioisotope. In those cases where the radioisotope can exist in several valence states, the system (radioisotope plus carrier) is usually taken through a suitable oxidation–reduction cycle before the separation process is started. Carriers are not always necessary for all of the interfering radioisotopes in some chemical processes; several elements may behave sufficiently alike that traces of one will be carried by macroscopic quantities of another.

Carrier-free isotopes are useful for experiments that require high dilution factors because of the low solubility of the radioisotope in a particular matrix and for applications in which the layer thickness must be minimized. The term "carrier-free" strictly means that no carriers, i.e., inactive isotopes of the same element, are present. However, in real life, small amounts of the inactive element may be present in the target or reagents; in practice, radioisotopes are classified as "carrier-free" when they contain no weighable amounts (i.e., < 1 μg) of stable isotopes of the same element.

If carriers cannot be added, as in the case of a transmutation-reaction product that is to be kept carrier-free, the range of available purification techniques is more limited. Another problem with carrier-free isotopes is that because of their very low concentrations, undesirable reactions not encountered at higher concentrations may become important. An example is absorption of carrier-free ^{105}Ag from dilute HCl solution by exchange with Na$^+$ ions in the walls of the glass container holding the solution. The following summary of chemical separation techniques for radioisotopes distinguishes those which are and are not applicable to the purification of carrier-free radioisotopes.

Filtration. Simple filtration through filter paper or sintered glass can remove carrier-free radioisotopes from a neutral solution. The method has been used in the purification of ^{7}Be, ^{47}Ca, and ^{206}Bi.

Precipitation. Solid radioisotopes can be deposited from solution either by the addition of a chemical reagent or by exposure to an external energy source, such as heat. Since precipitation usually requires addition of a carrier, this method is used mainly for removal of radioimpurities. Normal analytical techniques use slow precipitation so that large, well-shaped crystals are formed and separation is easily accomplished by filtration and washing. Small precipitates (colloids) are not so easy to separate, but can absorb undesirable radioisotopes that are present in extremely low concentrations (coprecipitation). Ferric hydroxide is used as a "scavenger" of radioisotopic impurities because of its ability to cause coprecipitation. The precipitate can be removed by filtration, ion exchange, solvent extraction, or centrifugation.

Solvent Extraction. The carrier element is converted from ionic form in aqueous solution into a nonpolar form that can be extracted into an organic solvent. The method is simple, specific, and fast. A large variety of extraction systems have been developed for radiochemical separations. (The NAS series on the radiochemistry of the elements includes solvent extraction procedures for almost all the elements; see "Source Preparation" in the Bibliography after Chapter 1.4.)

Ion Exchange. The target is dissolved, and the resulting solution is passed through a column of either cation- or anion-exchange resin; the various ions are eluted from the column with suitable solvents. A high separation efficiency can be obtained with slow flow rates and a high resin-to-ion ratio. In practice, a compromise is made between the efficiency and the speed of the operation. The method can be used for the preparation of carrier-free radioisotopes. The ion-exchange technique is suitable for a wide variety of separations and can be readily automated, giving improved reproducibility and reliability.

Electrodeposition. By passage of current through an electrolytic cell, particular radioisotopes (either the desired species or the contaminants) are plated out. The concentration of ions in the electrolytic cell must be sufficient for electrolysis to occur. For this reason, a carrier for the ions to be plated must usually be added, and carrier-free radioisotopes can only be purified if they remain in the electrolyte. To minimize the volume of added carriers, small-volume electrolytic cells have been used (see Chapter 1.3). The separation of two cations can be achieved by controlling the applied potential so that the deposition potential of only one of the cations is exceeded.

Distillation. This technique only applies to those targets in which the desired element or compound is much more volatile than the matrix in which it is embedded. It can be used to prepare carrier-free radioisotopes if they meet this criterion.

Chemical processing may also be needed in the initial preparation of the target. The target element may be highly chemically reactive, may have impurities with undesirable nuclear cross sections, or may be in an undesirable chemical compound. Processing may be necessary to produce an irradiation target of the appropriate shape and size or to prepare it for encapsulation. The procedures required for the production of ^{24}Na provide an example of simple processing. Metallic Na (monoisotopic ^{23}Na) is too reactive for easy handling throughout the entire irradiation and postirradiation process. However, high-purity sodium chloride, free of significant quantities of contaminants with large neutron cross sections, is commercially available. Irradiation of 2 mg of NaCl in a thermal-neutron flux of 10^{13} n cm^{-2} s^{-1} for two days will produce ~ 2 mCi of ^{24}Na and ~ 0.5 mCi of ^{38}Cl. The ^{38}Cl has a short half-life ($T_{1/2} = 37$ min) and will be undetectable after a few hours. The 2 mg of NaCl have dimensions $1 \times 1 \times 1$ mm and so can be cleaved from a single crystal. However, when microcurie quantities of ^{24}Na are required, it is easier to dissolve the NaCl in high-purity water and take an aliquot. The solution can be evaporated to dryness in a quartz tube, and the tube can be sealed in an argon atmosphere.

In general, it is advantageous to carry out as much of the processing as possible on inactive material. On the other hand, great care has to be taken not to introduce impurities during preirradiation processing that would form undesirable radioisotopes during the irradiation.

1.2.5. Parameter Selection in Isotope Production

It is clear that the production of a pure radioisotope involves the careful choice of many parameters. In order to focus attention on these choices, they are listed below.

(a) The nuclear reaction to be produced in the available target material must be chosen so that a minimum quantity of interfering isotopes is formed, whether from other elements in the target compound, other isotopes of the required target element, or trace impurities. Nuclear reactions that result in a chemical change of the nucleus are advantageous in the production of "carrier-free" radioisotopes.

(b) The irradiation facility must be chosen relative to the physical properties of the desired isotope. When short-lived isotopes are required, the proximity of a particular irradiation facility can compensate for the lower production efficiency of a given nuclear reaction. Some irradiation facilities do not permit the irradiation of liquids, powders, or chemically reactive materials.

(c) The target material must be chosen to minimize handling problems and maximize the production of the desired radioisotope. Ideal target materials are highly pure, and any additional elements they contain have low activation cross sections. The yield of the desired radioisotope both in an absolute sense and relative to interfering nuclear reactions may be improved by the use of isotopically enriched targets. Target materials are also chosen so that the total bulk (including that added by encapsulation, if applicable) is minimized to avoid the handling of excess activity.

(d) The irradiation time, together with the particle flux, must be chosen to produce the desired level of radioactivity as well as the maximum yield of the desired radioisotope relative to other competing nuclear reactions. When the desired radioisotope has a long half-life relative to the other radioisotopes, irradiation to saturation followed by a long cooling period in a shielded cell can enhance the radioactive purity.

(e) The means of separation of the desired radioisotope from the target materials must be chosen. The quality of the chemical separation will depend on the requirements of subsequent experimentation and the ability of the detection system to observe only the desired radioisotope. The choice of chemical separation is also influenced by the time needed for the separation and the physical and chemical form required for the final product.

As an example, we return to the production of ^{51}Cr mentioned earlier. If carrier-free ^{51}Cr is not needed, the obvious choice is the (n,γ) reaction on ^{50}Cr. The natural abundance of ^{50}Cr is 4.3% and $\sigma = 16$ b, so high specific activities can be obtained. Also, high-purity Cr is available. Alternatively, ^{51}Cr is commercially available as chromate solution from a number of suppliers, and this is the material of choice in many cases. If carrier-free ^{51}Cr is needed, the ^{51}V(p,n) reaction is best, as it gives a considerably higher yield than the α-particle bombardment of Ti (600 μCi μA^{-2} h^{-1} for 21-MeV protons versus 22 μCi μA^{-1} h^{-1} for 40-MeV α particles).[6] Also, vanadium is 99.75% ^{51}V, so no reactions on other vanadium isotopes are expected, and the (p,2n) reaction would produce stable ^{50}Cr. The ^{51}V(d,2n) reaction, which also gives a favorable yield, is a possible alternative.

1.2.6. Purchase of Radioisotopes

About 200 radioisotopes are commercially available, and many of these may be obtained in different chemical forms. Tabulations of these isotopes together with the names and addresses of the producing companies are available. Clearly, the ability to irradiate and purify relatively large quantities usually reduces the unit cost considerably relative to production in local facilities. Radioisotopes that are short-lived, or for which there is little demand, may not be supplied commercially. In such cases the information given in Sections 1.2.1 – 1.2.5 will help either in defining the specifications of the desired radioisotope for special production by a commercial company or in producing the isotope using local facilities.

---

[6] P. P. Dmitriev, I. O. Konstantinov, and N. N. Krasnov, *At. Energ.* **29**, 206 (1970).

1.3. Handling of Radioisotopes

1.3.1. Safety

By M. Robinet*

The purpose of this section is to put the hazards and risks associated with handling radioisotopes in perspective, to discuss radiation protection philosophy and standards, and to provide practical suggestions for safe handling of radioisotopes.

1.3.1.1. Hazards of Radiation. Radiation interacts with all matter in essentially the same manner and is hazardous to man because it can damage the cells of the body by ionization and excitation of molecules (including bond breaking). Cells may be damaged directly by ionization in critical molecules, such as DNA, and indirectly by chemical interactions involving highly reactive free radicals produced by the radiolytic breakdown of water. Over 60% of the molecules in the body are water, so the latter effect is probably the major mechanism of damage. The transfer of about 33 eV of energy is required to ionize a water molecule, so absorption of the energy from a single 0.5-MeV particle or photon could produce $(0.5 \times 10^6)/33 = 1.5 \times 10^4$ ionizations. In addition to producing damage by emitting radiation, radioisotopes may produce harmful effects as a result of chemical changes caused by transmutation, but this is believed to present a far smaller risk than the radiation effects.

Two general types of radiation effects are seen in humans: *Somatic effects* are injuries that occur only in individuals who are themselves exposed, and *genetic effects* are injuries that occur in descendants of exposed persons.

Nonacute somatic effects are usually delayed for 10 to 20 years. The most common manifestation is the induction of runaway growth of cells in some part of the body, commonly referred to by the broad term "cancer." The radiation-induced death of a few thousand cells in an adult body may not be critical, but mutations that survive the radiation exposure are believed to cause delayed effects such as cancer.

The fetus is especially vulnerable to somatic radiation injury. Such injuries are sometimes seen at birth, and sometimes show the usual delay. Female scientists should therefore make an extra effort to minimize their

---

* Occupational Health and Safety Division, Argonne National Laboratory, Argonne, Illinois.

exposure. It is now generally agreed that there is no safe period for minimizing the risk to a fetus and all unnecessary exposure during pregnancy should be avoided. A detailed review of the implications of fetal irradiation has been published by the National Council on Radiation Protection.[1]

Genetic effects cause injury in future generations. The cells of concern in the body of an exposed person are the reproductive cells in the testes and ovaries. Radiation-induced genetic damage may occur in cells throughout the body, but from the standpoint of heredity it is only the reproductive cells that matter. The most tangible product of total genetic damage is probably "ill health" in future generations because of chromosomal defects, recessive diseases, and disabilities. It is thought that a significant percentage of ill health is proportional to the mutation rate.[2]

The current thinking with regard to radiation protection is that since "any" amount of radiation can produce biological damage, any level of radiation may carry some risk, and all unnecessary radiation exposure should be avoided. A corollary to this is the linear nonthreshold dose-response hypothesis, which states that there is no threshold for radiation damage and that the degree of damage is a linear function of the dose. Most of the long-term radiation effects in man are from low doses; direct experimental data are very difficult to obtain for such doses, and estimates of expected low-dose response must be extrapolated from high doses for which more data are available. The assumption of a linear dose response is usually considered conservative, but in fact, a linear extrapolation from high doses could actually underestimate low-dose effects.[3,4]

1.3.1.2. Radiation Protection Units. The quantities used in radiation protection are activity (defined in Section 1.1.2), *exposure, dose,* and *dose equivalent.* The definitions of these quantities in old and new (SI) units are as follows:

Exposure. The roentgen (R) is a unit of exposure defined as the quantity of γ or x rays that produces ions totaling a charge of 2.58×10^{-7} coulomb per gram of air. The SI unit for exposure is coulombs per kilogram, $1 \text{ C kg}^{-1} = 3.88 \times 10^3$ R.

Dose. The old dose unit is the rad, which is defined as 100 ergs of

---

[1] National Council on Radiation Protection and Measurements (NCRP), "Review of NCRP Radiation Dose Limit for Embryo and Fetus in Occupationally Exposed Women," Rep. No. 53. NCRP, Washington, D.C., 1977.

[2] National Research Council Advisory Committee on the Biological Effects of Ionizing Radiation (BEIR), "The Effects on Populations of Exposure to Low Levels of Ionizing Radiation," p. 3. Nat. Acad. Sci., Washington, D.C., 1972.

[3] J. M. Brown, *Health Phys.* **31,** 231 (1976).

[4] A. M. Brues, *Science (Washington, D.C.)* **128,** 693 (1958).

absorbed energy per gram of irradiated material. The SI dose unit is the gray (Gy), defined as one joule absorbed per kilogram (J kg^{-1}). A dose of 1 Gy is equal to 100 rads. Note that identical exposures of, say 1 R may produce quite different doses in different materials if their energy absorption coefficients are different. A roentgen produces 87.7 ergs per gram of air or 96 ergs per gram of tissue. This is equivalent to 0.877 rads (8.77 × 10^{-3} Gy) in air or 0.96 rads (9.6 × 10^{-3} Gy) in tissue.

Dose Equivalent. A rad of 250-keV x rays may be 10 times less effective in producing a given biological effect than a rad of fast neutrons. The concept of dose equivalent allows one to compare different types of radiation on a common basis. Dose equivalent is defined as the product of the dose and an assigned dimensionless constant called a quality factor (Q), which is related to the biological effectiveness of the radiation. (Additional modifying factors may be necessary.) Quality factors for different types of radiation at various energies are listed in a report by the NCRP.[5] The Q factor is ~ 1 for x and γ radiation and ~ 20 for α radiation. When the dose is in rads, the unit of dose equivalent is the rem (roentgen equivalent man). When the dose is in grays, the corresponding SI unit of dose equivalent is the sievert (Sv). One sievert is equal to 100 rems.

1.3.1.3. Quantifying the Risk. True quantification of the risk associated with the use of radioisotopes is somewhat elusive, but it is possible to gain a "feel" for what might be considered low or high exposures by comparing them with well-known reference points, remembering of course that "low" does not mean insignificant. A lower-limit reference point is background radiation, to which we are exposed our entire lives. An upper-limit reference point is the acute somatic radiation injury dose that causes death within days or a few weeks. The radiation protection standards serve as an intermediate reference point between these two extremes. These three reference points are summarized in Table I. Note that current radiation protection standards are about two orders of magnitude above background and two orders of magnitude below the lethal doses.

Radiation protection standards used in the United States are primarily those developed by the International Commission on Radiological Protection (ICRP) and the National Council on Radiation Protection and Measurements (NCRP).[6] Both the ICRP and the NCRP are private groups whose recommendations are, at least in theory, purely advisory. Promulgation of "official" radiation protection guidelines in the United States is

---

[5] National Council on Radiation Protection and Measurements (NCRP), "Basic Radiation Protection Criteria," Rep. No. 39, p. 80. NCRP, Washington, D.C., 1971.

[6] L. S. Taylor, "Radiation Protection Standards." Chem. Rubber Publ. Co., Cleveland, Ohio, 1971.

TABLE I. Reference Points for Evaluation of Radiation Exposure Levels in Humans

Reference point	Dose equivalent	Reference point	Dose equivalent
Natural background in United States in millirems per year[a]		*Radiation protection standards in rems per year*[b]	
Cosmic radiation		Whole body	5
Sea level	26	Skin (ICRP limit is 30 rem/yr)	15
1.6-km altitude	50	Hands	75
External terrestrial radiation		Forearms (ICRP limit is 75 rem/yr)	30
Atlantic and Gulf coastal plains	15	Other organs, tissues, and organ systems	15
Majority of United States	30	Pregnant women during entire gestation	0.5
Rocky Mountains (some areas)	55	*Acute lethal radiation in rads*	
Natural radioisotopes deposited in the body (predominantly from ^{40}K)	20	LD-50 (whole-body lethal dose for 50% of exposed persons)[c]	300
Total from all sources		LD-100 (whole-body dose that causes certain death within a week for all exposed persons, regardless of medical treatment)[d]	1000
Atlantic and Gulf coastal plains	65		
Majority of United States	80		
Rocky Mountains (some areas)	125		

[a] Whole-body averages; from National Council on Radiation Protection and Measurements (NCRP), "Natural Background Radiation in the United States," Rep. No. 5, p. 108. NCRP, Washington, D.C., 1975.

[b] NCRP dose limits for occupational exposure; from National Council on Radiation Protection and Mesaurements (NCRP), "Review of the Current State of Radiation Protection Philosophy," Rep. No. 43, p. 34. Washington, D.C., 1975. Note that the standards in Table I apply to persons who work with radioactivity and not to the general population.

[c] C. Lushbaugh, F. Comas, C. L. Edwards, and G. A. Andrews, *Proc. Symp. Dose Rate Mamm. Radiat. Biol., Oak Ridge, Tenn.* **CONF-680410**, p. 17.1 (1968).

[d] C. Lushbaugh, *Proc. Int. Symp. Biol. Interpret. Dose Accel.-Produced Radiat., Berkeley, Calif.* **CONF-670305**, Sect. 111–3, p. 100 (1967).

currently the responsibility of the Environmental Protection Agency, but its guidelines are essentially identical to those of the ICRP and NCRP. The recommendations of the ICRP are generally accepted throughout the world. However, most major countries have private or governmental agencies that tailor radiation protection standards to their specific conditions.

Radiation protection standards are intended to represent risks that are not unacceptable to the individual or to the population at large and serve

1.3. HANDLING OF RADIOISOTOPES

primarily as guides for administrative action within the framework of the philosophy of minimizing exposure. The so-called permissible levels were never intended to be magic numbers below which radiation exposures are harmless and above which exposures are dangerous. Current standards were derived primarily from genetic considerations that take into account the amount of exposure per generation associated with a doubling of the natural mutation rate. The associated somatic risk can be evaluated only by statistical means.

Radiation risks are usually expressed as the number of specific disabilities expected per unit dose of radiation in the lifetime of a million members of the population. The risk estimate for leukemia, for example, is approximately 20 cases per rad for each million exposed.[7] This can also be expressed as $20/10^6$ or 2×10^{-5} per rad and is thus referred to as a fifth-order risk, implying that the probability of an injury to an individual is in the range of 10^{-5} per rad.[8] Note that the risk per rad is extrapolated from higher levels by assuming a linear dose response; expressing risk in terms of incidence per rad does not necessarily imply that there is direct evidence of an effect at 1 rad.

From Table I, the maximum permissible whole-body dose is 5 rems per year, so we would expect an exposure of no more than 250 rads over a 50-year working life. Assuming a risk of 10^{-5} per rad, this exposure could approximate a risk of the third order per working life. Experience shows, however, that with good radiation protection measures, the majority of those directly engaged in radiation work have whole-body exposures that are only a small percentage of the maximum permissible dose.[9]

1.3.1.3.1. DOSE-RATE EFFECT. It is now known that both somatic and genetic effects are dependent on dose rate as well as total dose. That is, the effect produced by a given dose delivered in a short time is usually reduced if the same dose is distributed over a longer time. Radioisotope users should therefore distribute their exposures over as long a time period as possible in addition to keeping the total dose as low as possible.

In addition to exposure from radiation sources outside the body, there is the possibility of exposure from radioisotopes that inadvertently enter the body by inhalation, absorption through skin or a wound, or ingestion. The basic radiation protection standards listed in Table I apply to both types of

---

[7] International Commission on Radiological Protection (ICRP), "The Evaluation of Risk from Radiation," Publ. No. 8, p. 5. Pergamon, Oxford, 1965.

[8] International Commission on Radiological Protection (ICRP), "The Evaluation of Risk from Radiation," Publ. No. 8, p. 3. Pergamon, Oxford, 1965.

[9] National Research Council Advisory Committee on the Biological Effects of Ionizing Radiation (BEIR), "The Effects on Populations of Exposure to Low Levels of Ionizing Radiation," p. 18. Nat. Acad. Sci., Washington, D.C., 1972.

exposure. Standards for internal exposure can also be expressed in alternative, but equivalent, ways, such as maximum body burden and maximum permissible concentration (MPC) in air and water. Unlike the exposure from an external source, which can be terminated at will by shielding the source or simply walking away from it, the exposure from an internal source is fixed by the combined biological and physical half-lives of the given radioisotope. The retention period in the body may be a few days or many years, and the body is committed to accumulate a fixed dose over that period equal to the time integral of the dose rate from the radioisotope retained in the body. This dose is called the dose commitment.

Maximum body burden is the quantity of a specific radioisotope that delivers the annual maximum permissible dose to the organ whose damage would result in greatest damage to the body (critical organ). Similarly, the MPC of a radioisotope in air or water is the concentration that if breathed or drunk "continuously" for 50 years, assuming best data for uptake and retention, would deliver the maximum permissible dose rate at the end of 50 years. That is, we assume 50 working years of continuous intake such that the dose-equivalent rate to the critical organ increases until it reaches the maximum permissible at the end of the 50th year. Tables of maximum body burden and MPCs as well as more details on their implications are listed in ICRP Publication 2 and other references.[10-12]

1.3.1.3.2. RELATIVE HAZARDS OF RADIOISOTOPES. The same activity of different radioisotopes may represent quite different degrees of hazard, even when they emit the same types of radiation. Radioisotopes are usually grouped into hazard categories according to the potential dose commitment that would result from accidental inhalation of radioactive material. Assuming the same physical and chemical form, the major factors that determine relative hazard are the energy of emitted particles and photons, physical and biological half-lives, specific activity, and the fraction of the radioisotope inhaled that becomes deposited in the critical organ. There are schemes that take into account some or all of these factors to group radioisotopes into hazard categories. The simplest direct-hazard indices are specific activity and MPC. A high-specific-activity source is potentially more hazardous than one of lower specific activity.[13] Similarly, radioisotopes with low MPC

---

[10] International Commission on Radiological Protection (ICRP), "Report of Committee II on Permissible Dose for Internal Radiation," Publ. No. 2. Pergamon, Oxford, 1959.

[11] National Council on Radiation Protection and Measurements (NCRP), "Maximum Permissible Body Burdens and Maximum Permissible Concentrations of Radionuclides in Air and Water for Occupational Exposure," Rep. No. 22. NCRP, Washington, D.C., 1963.

[12] A. Brodsky, in "Handbook of Radioactive Nuclides" (Y. Wang, ed.), p. 616. Chem. Rubber Publ. Co., Cleveland, Ohio, 1969.

[13] P. S. Baker, in "Handbook of Radioactive Nuclides" (Y. Wang, ed.), p. 34. Chem. Rubber Co., Cleveland, Ohio, 1969.

1.3. HANDLING OF RADIOISOTOPES

values are more hazardous than those with high MPC values. Both the IAEA[14] and Morgan et al.[15] have classified radioisotopes according to relative hazard per unit activity.

1.3.1.4. Practical Methods of Radiation Protection. Radiation exposure and contamination should be minimized by making use of the techniques described in the following sections.

1.3.1.4.1. FUME HOODS AND GLOVE BOXES. Small quantities of radioisotopes that have low relative radiotoxicity can be handled safely on an open bench. However, most operations with unsealed radioisotopes should be done in a properly designed fume hood with air velocity at the hood face of at least 5 mm s^{-1}. High-efficiency particulate air (HEPA) filters should be used on hoods to prevent release of airborne activity to the outside environment. Volatile compounds require activated charcoal or zeolite beds for filtration. Maintenance and replacement of filtration systems can be complicated and expensive, so requirements should be considered realistically. If the amount of activity used and the flow rate through the hood is such that the concentration outside the building would be well below the applicable MPC value, one might reason that high-efficiency filtration is unnecessary. If the probability of generating airborne activity in the hood is extremely small, such as in simple operations with liquids, then this reasoning is acceptable. However, if the operation is likely to produce airborne activity, such as in work with powders, it is unacceptable because it implies potential exposure of others without their permission.

One means to reducing filtration cost is to confine the operation to as small a volume as possible so that less air flow is required and a smaller filter can be used. Hoods in the same laboratory must be interlocked so that if flow rate in one hood is reduced, the other hood will not cause a flow reversal into the laboratory.

If it is virtually impossible to perform an operation without serious airborne contamination, or if an inert atmosphere is required, or if filtering a high-flow-rate hood is too costly, a glove box or gloved bags must be used. The pressure in both should be negative with respect to the laboratory. A negative pressure equal to a few millimeters of water is usually adequate. Commercial lightweight plastic glove boxes may be placed on a lab bench top or even mounted on wheels. Glove boxes and hoods with built-in exhaust fans should be used with caution because a leak in the positive-pressure exhaust duct would allow the effluent to reenter the laboratory area.

1.3.1.4.2. SURVEY INSTRUMENTATION. No work with radioisotopes should be attempted without adequate instrumentation to check for con-

---

[14] International Atomic Energy Agency (IAEA), "Safe Handling of Radioisotopes," Safety Series, No. 1, p. 34. Vienna, 1958.
[15] K. Z. Morgan, W. S. Snyder, and M. R. Ford, *Health Phys.* **10**, 151 (1964).

tamination and to measure exposure rates. One of the simplest yet most efficient survey instruments is a portable end-window Geiger-Müller (GM) meter. It is useful for checking contamination as well as measuring exposure rates. A thin (<3 mg cm^{-2}) end-window permits detection of β particles with $E_{\max}$ down to about 0.05 MeV;[16] audio output eliminates the need to watch the meter, allowing the user to concentrate on where he is putting the detector. Miniature pocket-size GM meters provide an audible signal proportional to exposure rate. Some warn the wearer when a preset exposure is exceeded.

Ion-chamber-type survey meters are preferred for accurate exposure rate measurements, but their response, especially at low levels, is slow compared to that of GM meters, which makes them unacceptable for finding contaminated spots.

Large-area gas-flow proportional counters with 0.8 mg cm^{-2} windows are excellent for surveying α contamination even in high γ backgrounds. Dual-voltage models are available that permit operation on the α or β plateau.

Low levels of removable surface contamination (less than a few disintegrations per minute for each square centimeter) may be statistically very difficult to detect with conventional rate meter survey instruments and are best measured by taking a smear with a thin, absorbent material and counting the smear with a low-background counting system. Smear paper in the form of small disks is available in convenient pocket-size dispensers. Smears may be counted in special-purpose detector systems designed to accept standard smear disks or in a liquid scintillation counter.

Diligent use of hand and shoe monitors provides an early indication of a flaw in the handling procedures or the containment system. GM survey instruments could of course be used, but they are less efficient and certainly not as convenient. Most hand and shoe monitors capable of detecting α, low-energy β, and γ rays use gas-flow proportional detectors. Sophisticated instruments with background compensation and lighted messages, as well as simpler units consisting of a single rate meter with two gas-flow proportional detectors, are available. Unfortunately, gas-flow proportional detectors require a continuous flow of counting gas, which appears to run out when it is needed most. A possible solution to this problem is a new hand and shoe monitor that uses several fast-response air ion chambers.

Air monitors are used in many radioisotope laboratories, especially those that handle large quantities of α activity. However, surface contamination is usually the initial source of airborne contamination, so careful checks for surface contamination may be more effective.

---

[16] "Radiological Health Handbook," p. 123. U.S. Dep. Health, Educ. Welfare, Washington, D.C., 1970.

1.3. HANDLING OF RADIOISOTOPES

1.3.1.4.3. PERSONNEL MONITORING. Some means of measuring individual personnel exposure is necessary to ensure that protection measures are effective in keeping exposures as low as practical, as well as to satisfy licensing requirements. Dosimeters are personnel monitoring devices worn on the body to record the time-integrated exposure (to the device). The four most common types are described below.

The *pocket integrating ion chamber dosimeter* is essentially a capacitor connected to a quartz fiber electroscope mounted in a pen-size air-filled chamber. The chamber has a transparent scale and an eyepiece for viewing the deflection of the fiber. A special charger is used to "zero" the dosimeter by applying about 100 V to the capacitor. The total charge produced in the chamber and the corresponding deflection of the fiber are proportional to the exposure. Pocket ion chamber dosimeters are convenient and inexpensive and provide a direct continuous readout, but they cannot be expected to retain exposure data for long periods of time.

Film dosimeters consist of a photographic emulsion partially covered by different types of materials called "filters." The optical density of the developed emulsion is proportional to the dose. The filters are used to estimate the average photon energy so that film dose can be converted to tissue dose. Dose retention by film dosimeters is sensitive to environmental changes.

Thermoluminescent dosimeters (TLDs) use crystals such as LiF in the form of small wafers or chips that contain trace amounts of impurities. Radiation gives the electrons energy to move from the valence band to the conduction band; some are trapped in intermediate energy levels created by the impurities in the crystal. The number of trapped electrons is proportional to the radiation dose to the crystal. The dose is read out by heating the crystal chips so that the trapped electrons are freed and emit light when they return to the valence band. The light output, measured with a photomultiplier tube, is proportional to the radiation dose. TLDs can reliably hold dose data longer than film dosimeters and are less sensitive to environmental changes. Becker has compared the characteristics of film, thermoluminescent, and radiophotoluminescent dosimeters.[17]

Integrating digital GM dosimeters use a miniature GM detector with a pocket-size scaler that stores the total counts. The counts are electronically converted to approximate equivalent exposure for some assumed energy range and read out on a digital solid state display.

It is usually assumed that the exposure to a dosimeter is representative of the exposure to the whole body of the wearer, but it is not surprising that

---

[17] K. Becker, *Health Phys.* **12**, 955 (1966).

finger exposures are usually many times greater than shown by dosimeters on the torso. In such cases finger-ring TLDs provide a better clue to potential exposure problems than do body dosimeters. Film dosimeters and TLDs are currently the only meters "accepted" for official records.

Measurement of internal dose (dose commitment) requires bioassay of urine and feces and perhaps whole-body counting. The frequency of bioassay is determined by the physical, chemical, and radiological characteristics of the radioisotopes being used. Tritium users, for example, may require bioassay after every procedure. Personnel monitoring and bioassay services can be obtained commercially.[18]

1.3.1.4.4. SHIELDING. To keep external exposures as low as practical, most radioisotope operations require some shielding. Ordinary lead bricks, which are easily adapted to specific needs, are the most common shield material. Interlocking lead bricks make it feasible to erect relatively large enclosures without support structures. They also eliminate radiation leakage between flat surfaces. High-density shielding materials such as lead should be avoided when using high-energy β emitters because bremsstrahlung production is proportional to atomic number.[19] For the same reason, containers such as beakers and bottles should not be made of high-Z materials. Aluminum and most plastics are good low-Z β-shield materials. A composite shield of plastic or aluminum backed by lead is used when high-energy β and γ rays must be blocked. Ordinary prescription glasses can reduce by several orders of magnitude the β dose to the eyes, which are particularly vulnerable. In addition to shielding, tongs and other remote handling devices should be used as much as possible to take advantage of the inverse-square law.

1.3.1.4.5. DRY RUNS. For even the simplest radioisotope procedure, dry runs can save much time and prevent unnecessary contamination and exposure. Procedures should be written to facilitate careful review. Every step from receipt of the radioisotope shipment to packaging of the waste should be considered. Dry runs should simulate the actual operation as closely as possible, including actual pipetting, cutting, weighing, and heating; also, gloves and other protective clothing should be worn, as in a real experiment. It is very helpful to have a colleague watch the dry-run performance; he may see potential contamination or exposure problems that were overlooked by the participants.

1.3.1.4.6. RADIOACTIVE WASTE AND DECONTAMINATION. Disposal procedures for radioactive waste, including methods of packaging and shipping to approved disposal sites, are regulated by several government

---

[18] *Nucl. News* **24**, No. 4 (1981).
[19] R. D. Evans, "The Atomic Nucleus," p. 615. McGraw-Hill, New York, 1955.

agencies. The procedures to be used to dispose of radioactive waste should be determined early in the planning stage so that special requirements can be dealt with. For example, containers for shipping radioactive waste must be approved by the Department of Transportation. Licensed firms specializing in disposal of radioactive waste are listed in an issue of *Nuclear News*.[18]

Waste producers are usually required to document the identity and quantity of radioisotopes in waste and to adsorb liquids on materials such as vermiculite. The cost of waste disposal is usually based on volume, so compaction machines may provide long-term savings as well as convenience.

Decontamination of items is often a potential source of additional contamination and may generate waste that will cost more to dispose of than the item itself is worth. Therefore, disposable equipment should be used whenever possible. If contaminated equipment is too valuable to dispose of, it should be cleaned with special commercial decontamination solutions that combine a number of different chemical and physical properties, such as surface wetting, and chelating. Ultrasonic cleaners used with these special solutions can significantly reduce decontamination time and cost.

All operations with radioactive liquids should be done in stainless steel or plastic trays to contain possible spills. Use of standard plastic-backed absorbent paper in trays and other potential spill areas can practically eliminate the need for extensive decontamination. Equipment may be protected from contamination with absorbent strippable skin coatings that may be peeled away at any time.

1.3.1.4.7. PERSONNEL DECONTAMINATION. Proper use of protective clothing such as gloves, lab coats, and shoe covers should preclude the need for decontamination of bare skin, but if the need does arise, special care must be taken not to cause injury. Acids, organic solvents, or harsh detergents should never be used. Plain water may be sufficient, but if necessary, the special multiproperty decontamination solutions mentioned above are usually safe for use on skin.

Decontamination cannot be done in an ordinary sink unless provisions have been made to collect all waste. One technique of decontaminating personnel without using large quantities of liquid is to apply liquid with a hand spray bottle or a plastic squeeze bottle and absorb the runoff with paper towels. The skin is flushed with clean water after all contamination is removed. Stubborn contamination under finger nails may require scrubbing with a brush in a small plastic tub or heavy plastic bag. When contamination covers a large portion of the body, the larger volume of liquid needed may be collected by placing the person in an inflatable swimming pool or large plastic garbage can and gently sprinkling the affected area.

If extensive showering is required, arrangements must be made to use a special decontamination facility. A physician should be consulted on decontamination of open or abraded skin. Contamination in a cut or puncture may require surgical removal.

1.3.1.4.8. RECEIVING THE SOURCE. Radioisotopes received in approved ready-to-dispense containers are not likely to be contaminated, but the outer packing and the primary container should nevertheless be smear tested for contamination. Also, the exposure rate should be measured to check that it is consistent with the activity marked on the label. Exposure rate at a given distance per unit activity can be calculated or obtained from tables.[20,21]

Radioisotopes that are received in their original irradiation cans are often a serious potential hazard of contamination and exposure if the can itself is highly radioactive. It is not uncommon, for example, for aluminum irradiation cans used in some reactors to have exposure rates that are thousands of times greater than those of the contents. Loose contamination is frequently found on the outside of the can. All containers of radioactive material should be clearly labeled using standard labels that bear the international radiation symbol (three-bladed magenta propeller on yellow background). So-called warning tape in roll form is convenient for labeling small items. Labels should list the radioisotope, the activity, the date, and the exposure rate at a specified distance. Labels should also be placed on shielding pots and storage areas. Radiation warning signs reading "CAUTION RADIOACTIVE MATERIALS" should be posted outside each entrance to the laboratory.

For safety reasons, and to meet waste disposal requirements, a dated log should be kept for each radioisotope, listing activity in use, approximate activity in waste, and total activity remaining in storage.

Radioactive material that must be moved from one laboratory to another should be in a sealed container surrounded by absorbent packing and appropriate shielding. Shipment or transfer of radioactive materials by any means must comply with state and federal packaging and transportation regulations.[22]

1.3.1.4.9. SPILLS. A spill is the most probable mishap associated with handling of radioisotopes. However, if dealt with properly, it need not be a catastrophe with serious health consequences.

---

[20] "Radiological Health Handbook," p. 131. U.S. Dep. Health, Educ. Welfare, Washington, D.C., 1970.

[21] J. R. Cameron, in "Handbook of Radioactive Nuclides" (Y. Wang, ed.), p. 70. Chem. Rubber Publ. Co., Cleveland, Ohio, 1969.

[22] F. B. Conlon and G. L. Pettigrew, "Summary of Federal Regulations for Packaging and Transportation of Radioactive Materials." Bur. Radiol. Health, Rockville, Maryland, 1971.

1.3. HANDLING OF RADIOISOTOPES

The most important action for anyone who finds himself and his surroundings grossly contaminated is to call for help from health physicists immediately. The possibility of a spill should be considered before use of radioisotopes is begun, and a plan of action developed and posted in the area. The plan should list what to do, whom to call, etc.

The user of radioisotopes would usually be the first to discover a spill, detecting contamination by means of hand and shoe monitoring, a smear test of lab benches or floor, or an area survey showing an abnormal increase in radiation background.

Handling a spill usually involves the following steps: (a) identifying the origin of contamination, (b) sealing off the primary source of contamination, (c) tracing all possible routes of contamination and preventing further spread, (d) decontamination, and (e) assessment of dose commitment for all persons involved.

1.3.1.5. Licensing Requirements. In order to purchase or use radioisotopes in the United States, a license must be obtained from the Nuclear Regulatory Commission (NRC) or from an Agreement State, i.e. a state that exercises regulating and licensing authority by agreement with the NRC, unless the kind and quantity of material is covered by an exemption. Similar agencies exist in other countries. Licenses may be granted to an individual, firm, corporation, association, or public or private institution.

1.3.2. Source Preparation

By S. J. Rothman and J. N. Mundy*

The preparation of Mössbauer sources is discussed by Cohen and Wertheim.[23] Sources for positron annihilation are described by Smedskjaer and Fluss in Part 2 of this volume. This section discusses sources for diffusion samples.

1.3.2.1. Introduction.
The most common geometry used in studying the diffusion of radioactive tracers is the infinitesimal source (source thickness less than $0.1\sqrt{Dt}$) on a semi-infinite sample (sample thickness greater than $10\sqrt{Dt}$), where D is the diffusion coefficient and t the diffusion time. The thickness of the radioisotope layer ranges from a fraction of a monolayer to a few thousand angstroms; the sample is several millimeters thick. (Naturally, the thinner the layer, the higher the radioisotope specific activity must be.) This layer is mathematically described as a Dirac δ function, and the solution of the diffusion equation for this initial condition is

$$C = (M/\sqrt{\pi Dt})\exp(-x^2/4Dt), \qquad (1.3.1)$$

where C is the concentration of the diffusing species at a distance x from the surface and M is the strength of the source. If a radioactive tracer is used, C is simply the specific activity. When the layer thickness is less than $\sqrt{Dt}/10$, the D obtained using Eq. (1.3.1) is accurate to better than 1%.

The characteristic of the deposited layer that is most important for the success of the experiment is the nature of its bonding to the sample. Not only is adhesion necessary, but even more importantly, there must be no reaction at the sample–layer interface that slows down the dissolution of the tracer in the sample: If the entry of the tracer into the sample rather than its diffusion therein is the rate-limiting reaction, the simple Gaussian equation (1.3.1) is no longer valid, and it becomes difficult to obtain the value of the diffusion coefficient from the data. Examples of situations to be avoided include oxide layers or an oxidized radioisotope on a metallic sample; it is often essential to deposit the radiotracer on metallic samples as a metal rather than as an oxide. An oxide is, of course, a perfectly acceptable chemical form of radiotracer on an oxide sample.

[23] R. L. Cohen and G. K. Wertheim, *in* "Solid State Physics" (R. V. Coleman, ed.), Methods of Experimental Physics, Vol. 11, p. 307. Academic Press, New York, 1974.

* Materials Science and Technology Division, Argonne National Laboratory, Argonne, Illinois.

To deposit a radioisotope as an oxide, one uses a solution of the sulfate, nitrate, or oxalate of the element in question that is soluble in water. The solution is dried on the sample surface with a heat lamp, and the sulfate, nitrate, or oxalate is decomposed to the oxide during heating to the diffusion temperature. Other nonmetallic compounds of radioisotopes (e.g., ^{22}NaCl) can often be deposited by vacuum evaporation. The technique used in an actual experiment depends of course on the chemical characteristics and form of the radiotracer and of the sample.

1.3.2.2. *Deposition of Metallic Layers on Metallic Samples.* If the radioisotope is in metallic form (e.g., a chunk of Cr or a gold wire that has been irradiated in a reactor), it must be deposited by evaporation in vacuum or by sputtering. A high-vacuum system is often adequate, but for refractory metal samples an ultrahigh-vacuum (UHV) system may have to be used. The sample is placed in a boat or basket of high-melting-temperature metal, and this is Joule-heated to the desired temperature. A hairpin-type filament can also be used; this is convenient for wire-shaped sources, which can be wound around the tip. Boats and baskets can be purchased commercially, but they are easily made in the laboratory. Boats can be cut from Pt, Ta, or W foil, and a spiral or basket can be wound on a machine or wood screw. The empty boat or basket is placed between the electrodes in the vacuum system and heated briefly to a white heat under vacuum to clean it; the isotope is subsequently loaded. We have found Ta to be the best material, as it has a high melting temperature and is not embrittled by high-temperature heating as W is. Tantalum foil 125 μm thick is convenient for boats, and 250–125-μm wire for baskets. Low voltages and large currents are needed, so the evaporator should have high ($\sim$ 100 A) current feed throughs, and a step-down transformer should be placed after the variable transformer. Manual control of the temperature is usually adequate. The filament is brought to the desired temperature reasonably rapidly and is held there until the material is evaporated or the filament breaks owing to a chemical reaction with the source. Rapid evaporation (high temperature) is recommended to avoid heating the sample. The samples can be placed under or over the source (obviously, they must be above a boat). The yield is determined by the geometry, so it is advantageous to have the sample as close to the source as possible; on the other hand, heating of the sample must be avoided. Yields run from a few percent to 20% at best. A glass chimney or a large beaker should be placed around the source–sample configuration to prevent, as far as possible, contamination of the evaporator.

Evaporation sources for metals are sometimes produced on the filament itself. Sources can be made by electroplating on Pt; Ta or W are difficult to electroplate on. Care must be taken to ensure that the deposit is perfectly dry, as the evolution of water vapor during heating of the filament will spatter the radioisotope everywhere in the evaporator except on the sample. An exchange reaction can be used if the metal to be evaporated has a much higher vapor pressure than the compound in which it is irradiated; ^{22}Na has been deposited as metal by heating irradiated NaCl with unirradiated Na metal in a Ta boat.[24] A chemical reduction by heating of the filament is also possible, e.g., the reduction of ^{49}VCl$_3$ by Ca and the subsequent evaporation of the ^{49}V.[25]

A metallic source can also be evaporated by an electron gun, and this may be necessary in the case of very-high-melting-temperature metals, such as tungsten. Sputtering can also be used in special cases,[26] but the smallness of the radioactive sources makes it a less useful technique than evaporation. The yield, at best, is the same. It may be useful to clean the sample by sputtering in the evaporator before depositing the radioisotope by evaporation,[27] especially in the case of metals that form oxides during the surface preparation.

An elegant but expensive technique for isotope deposition is ion implantation. The yield is $\leq 2\%$, and the rest of the isotope is left in the separator. This method offers two advantages: The isotope is deposited carrier-free, as the carrier atoms have a different mass, and surface barriers are penetrated. The depths and widths of the deposits can be calculated.[28]

As commercial radioisotopes often come in an aqueous solution, they must be reduced to metallic form. For most elements not too far above hydrogen on the electromotive series, say, zinc or below, the easiest way to do this is by electroplating. If the sample is not attacked by the electrolyte, the isotope can be plated directly onto the sample; this is the cheapest and easiest method and the most efficient in terms of yield. Electroplating does not work on certain metals, e.g., aluminum,[29] and in such cases the isotope must be plated onto a filament and evaporated.

Because of the small amount of tracer involved in an experiment, the volume of plating solution must be kept small so that enough ions are present to conduct the current. One way to do this is to mask about 1 mm on the outside of the sample surface with Tygon paint and place the drop of plating solution in the middle. Surface tension will keep up to 200 μL of solution in a drop, and the positive electrode, a Pt wire loop, is gently lowered into the top of the solution. Plating for 1 min at the appropriate

[24] J. N. Mundy, *Phys. Rev. B* **3**, 2431 (1970).
[25] J. Stanley and C. Wert, *J. Appl. Phys.* **32**, 267 (1961).
[26] R. Weil, S. J. Rothman, and L. T. Lloyd, *Rev. Sci. Instrum.* **30**, 541 (1959).
[27] S. J. Rothman, *J. Nucl. Mater.* **3**, 77 (1961).
[28] U. Litmark and J. F. Ziegler, "Handbook of Range Distributions for Energetic Ions in All Elements." Pergamon, New York, 1980.
[29] N. L. Peterson and S. J. Rothman, *Phys. Rev. B* **1**, 3264 (1970).

1.3. HANDLING OF RADIOISOTOPES

voltage (10 V is usually enough) will often give 50% yield. Alternatively, an O-ring ball joint can be put on the sample to hold the solution, or the sample can be masked, except on the desired surface, and immersed in a beaker of solution. An element with a very high vapor pressure, such as Zn, should be plated on all surfaces of the sample, as it redistributes to all surfaces through the vapor phase during the anneal.

Use of anodes of the material to be plated has the same effect as adding carrier, so Pt anodes are usually used. Recipes used in commercial electroplating often do not work because of the small concentration of radiotracer. We have found that Ag, Cu, Ni, and Zn plate well from a neutral ammonia solution (pH $\approx$ 7). Gold is usually plated from an alkaline cyanide solution, and Fe seems to plate best from $FeCl_3$ in an ammonium oxalate-oxalic acid solution at pH 5.5–6.0.[30] Palladium has been plated from an acidic chloride solution containing NH_4Cl,[31] and Pb from a nitrate solution with $HClO_4$ added to pH 2.[32]

Elements of groups 1A, 2A, and 3B–6B (including the rare earths and the actinides), aluminum, and the nonmetals of groups 3A–7A of the periodic table cannot be plated from an aqueous solution. Some of these elements can be deposited by drying the solution onto the sample surface and letting the sample reduce the radioisotope (e.g., Nb on U, CrCl on Cr).[33,34] This is generally a risky procedure as there is often isotope holdup on the surface.

---

[30] J. G. Mullen, *Phys. Rev.* **121**, 1949 (1961).
[31] N. L. Peterson, *Phys. Rev.* **136**, A568 (1964).
[32] S. J. Rothman, Ph.D. Thesis, Stanford Univ., Stanford, California, 1954.
[33] N. L. Peterson and S. J. Rothman, *Phys. Rev.* **136**, A842 (1964).
[34] J. N. Mundy, C. W. Tse, and W. D. McFall, *Phys. Rev. B* **13**, 2349 (1976).

1.4. Detection and Assay

1.4.1. Autoradiography

By S. J. Rothman and J. N. Mundy*

1.4.1.1. Introduction. In some assays of radioactivity, it is desired to show the distribution of the radioactive isotope in a sample so as to correlate it with features of the sample, for example, to show the presence of grain-boundary diffusion, or to show that the radioelement is concentrated in certain places in a sample, e.g., at inclusions, or to show the homogeneity of a radioactive deposit. Autoradiography, which lends itself well to this type of application, consists in principle of bringing a film into contact with the sample, letting the radiation from the isotope blacken the film, and developing and viewing the film. The same sort of images can also be obtained in a two-dimensional scan with an electron microprobe or with a scanning electron microscope equipped with an x-ray analyzer. At present, autoradiography is applied more frequently in biological research than in solid state science, and most of the available texts are for biologists,[1] with only a few reports on metallurgical studies.[2,3]

1.4.1.2. Interaction of Radiation with Photographic Emulsions.[4] A photographic emulsion used for autoradiography consists of crystals of AgBr (0.02–0.5 μm in diameter for x-ray film) suspended in gelatin. If radiation sensitizes a AgBr crystal, subsequent development reduces that entire crystal to silver metal, making a dark spot on the developed film.[5] A series of such grains in a film marks out the track of the emitted particle.

Different types of radiation leave different types of tracks. An α particle suffers a relatively high specific energy loss, so that it leaves a short (a few tens of micrometers), dense, straight track. The specific energy loss of β particles is lower, and they are more easily scattered through high angles.

---

[1] A. W. Rogers, "Techniques of Autoradiography." Elsevier, Amsterdam, 1979.

[2] C. Leymonie, "Radioactive Tracers in Physical Metallurgy." Chapman & Hall, London, 1963.

[3] H. A. Fischer and G. Werner, "Autoradiography." de Gruyter, Berlin, 1971.

[4] M. Blau, *in* "Nuclear Physics" (L. Yuan and C.-S. Wu, eds.), Methods of Experimental Physics, Vol. 5, Part A, p. 208. Academic Press, New York, 1961.

[5] J. F. Hamilton and F. Urbach, *in* "The Theory of the Photographic Process" (T. H. James, ed.), Chapters 5, 9, and 10. Macmillan, New York, 1966.

---

* Materials Science and Technology Division, Argonne National Laboratory, Argonne, Illinois.

Their tracks are thus less dense, of variable length depending on the energy ($\sim 3\,\mu$m for 20 keV, 10 mm for 6 MeV), and not necessarily straight. Branching can be encountered if a photoelectron is knocked out. The interaction of x and γ rays with the AgBr grains occurs only at large intervals, so they do not leave a track, but only background fog; in general, autoradiography is not a good technique for γ emitters.

In the application of autoradiography to a bulk sample, radiation from below the surface will also strike the emulsion, and, if the radiation is penetrating, resolution will be lost.

1.4.1.3. Methods. The exact technique used to make an autoradiograph depends on the scale of the phenomenon to be studied. We may divide the scale into three parts, encompassing inspection of the autoradiograph with (a) the naked eye (including enlargement up to, say, 20×), (b) an optical microscope, or (c) an electron microscope.

For (a) and (b), and sometimes (c), the sample is polished by standard metallographic techniques.[6] For (a), it is often sufficient simply to place the sample on a strip of x-ray film with fiducial markers to locate the film with respect to the sample, and place the assembly in a light-tight box. Exposures are usually several hours to several days. Figure 1 illustrates such an autoradiograph of the grain-boundary penetration of ^{59}Fe in a 65% Fe–15% Cr–20% Ni alloy.[7]

For higher resolution, stripping films or liquid emulsions are used. In the stripping-film technique, the emulsion is usually carried on a layer of plain gelatin mounted on a glass support. An appropriate area of emulsion is cut out, stripped off the glass, and floated on water, emulsion side down. The emulsion expands as it soaks up water. After a few minutes, the emulsion stops expanding and is then picked up on the sample by slipping the sample beneath it. The emulsion makes close contact with the specimen as it dries, and the two are left together through exposure and development. In the liquid-emulsion technique, the sample is actually dipped into liquid emulsion, which is allowed to dry on the sample. The sample and emulsion remain in contact during development.

The advantage of the stripping-film technique is reproducibility. It is adequate for most autoradiographic applications. The liquid-emulsion technique offers the opportunity to use a much wider variety of emulsions, but involves some difficulty in producing an emulsion of uniform thickness. Details of both techniques and the properties of different emulsions are discussed by Rogers[1] and Fischer and Werner.[3]

---

[6] G. Kehl, "Principles of Metallographic Practice," 3rd ed. McGraw-Hill, New York, 1949.
[7] L. J. Nowicki and S. J. Rothman, unpublished work, 1980.

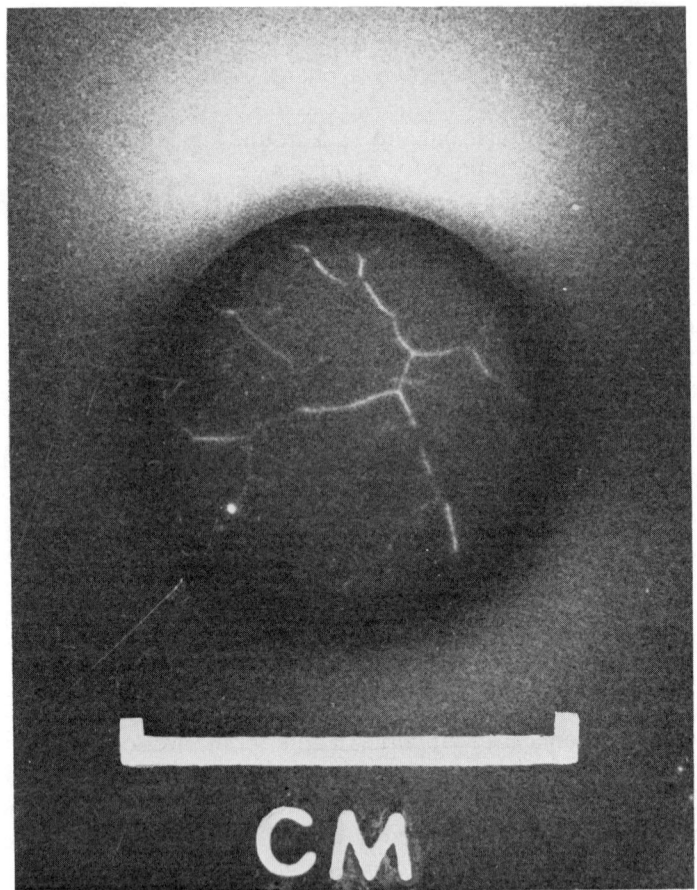

FIG. 1. Autoradiograph showing grain-boundary diffusion of ^{59}Fe in Fe–Cr–Ni alloy.

If the sample is likely to react chemically with either the film or the processing solutions, it may first be covered with a layer of plastic, though with some loss in resolution. Dipping in a solution of 2% Vynilite VYNS (90% vinyl chloride, 10% vinyl acetate) in methyl ethyl ketone and drying under a heat lamp for 30 min produces an ~ 1-μm-thick plastic film.[8]

1.4.1.4. Microscopic Observation. The autoradiograph can be observed and photographed on the sample in a standard metallurgical microscope. At

[8] M. D. Adams and R. K. Steunenberg, "Some Metallurgical Applications of Autoradiography," Rep. No. 6412. Argonne Natl. Lab., Argonne, Illinois, 1961.

low magnification both the silver grains and the sample surface can be in focus, but at high magnification they must be observed separately. Darkfield illumination helps in the observation of the developed silver grains.[1]

The use of autoradiography with electron-microscope examination is a difficult and time-consuming technique.[1] Resolution is $\geqslant 700$ Å at best, and exposures must last for months in order to produce a detectable amount of blackening. The emulsion must be removed from the sample for observation, so correlation of the autoradiograph and the sample is impossible. To our knowledge, the technique has not been applied in solid state studies.

1.4.1.5. Quantitative Evaluation. In some cases, it is possible to get all the information one needs from a visual inspection of an autoradiograph. In other cases, one wants to compare more quantitatively the amounts of radioisotope present in the different parts of the sample or in different samples. The last is more difficult, as all the handling of the emulsion and its processing must be carefully standardized. Three techniques are available for quantitative autoradiography: measurement of photographic density, counting of developed silver grains, and counting of individual tracks.

Measurements of the density are useful when fairly large, homogeneous areas of emulsion are to be compared, and in such a case, this is a rapid and relatively simple method. Fine resolution is not possible, as the magnification must be low enough so as not to resolve individual silver grains. The emulsion must be separated from the sample for the measurement. The density is measured in a microdensitometer. If I_0 is the light illuminating the film and I the light transmitted, then the density D is $\log(I_0/I)$. In a film exposed to light, the density is proportional to log (exposure) over a large range. However, for β particles, the density is proportional to the exposure at low densities. For high accuracy, calibrated density strips, exposed to the same type of radiation, should be prepared. A comparison of densitometers is given by Lutze-Birk et al.[9]

Grain counting is a simple but tedious process: One simply counts the grains of exposed silver in the areas under question. Problems may be encountered in defining what is a silver grain, especially near the end of a track, where the particle loses energy more rapidly than earlier and the silver grains lie closer together and tend to be larger. The counting is done in a microscope at high enough magnification to resolve the grains; the resolution of this technique is innately higher than that of density measurements.

At somewhat lower magnifications, one can count entire tracks rather than grains. This is an especially suitable technique for α particles or heavier

---

[9] A. Lutze-Birk, L. Wallis, T. Chajechi, K. Freyer, H. C. Trentler, W. Birkholz, and B. Sturak, *Isotopenraxis* **14**, 177 (1978).

fragments that leave short, thick tracks, although the angle of entry will obviously affect the length of the track. Track counting for β particles is more difficult, as the tracks are even more variable in length.

1.4.1.6. Recording Tracks in Nonphotographic Materials.[10] In certain plastics (e.g., Lexan, cellulose nitrate, polypropylene) and the mineral mica, the radiation damage caused by a heavy particle (α particle, proton, fission fragment, or recoil nucleus) can be etched to reveal the particle's track. These materials can thus be used to map the distribution of α-active radioisotopes. Also, as the detectors are not sensitive to β or γ radiation, they can be used to detect materials by induced nuclear reactions, e.g., to map boron in steel by the reaction $^{10}B(n,\alpha)^{7}Li$, or to determine uranium distribution from fission in a nuclear reactor. The material used for a detector is chosen according to its sensitivity; e.g., 0.55-MeV protons leave tracks in cellulose nitrate, but full-energy (≥ 30 MeV) fission fragments are needed to leave tracks in amber.

Techniques for the use of track detectors are described completely in the book by Fleischer et al.,[10] and somewhat resemble the stripping-film technique. The plastic film is pressed onto the sample and formed on it by heating. After exposure, it is etched in situ, and the sample and detector are examined together in a microscope.

1.4.1.7. The Use of Autoradiography in Diffusion Studies. Autoradiography has been used for quantitative measurements of both volume and grain-boundary diffusion coefficients. One usually takes the autoradiograph from a flat surface cut at a small angle to the plane on which the radioisotope was originally deposited (taper section), so as to spread out the diffusion zone over a larger distance, and then takes a densitometer scan in the diffusion direction. This is not the most precise technique available, even though some workers have been successful.[11-14] The problems are the following: (a) The relation between distance in the film and distance on the sample is inexact. Films tend to change dimensions during processing and an accurate marking technique must therefore be used. The measurement of the taper angle can introduce a further inaccuracy. (b) The relation between amount of exposure and density is complicated and probably should be determined for each radioisotope. (c) The sensitivity of the film does not extend over nearly as large a range as that of a counter, and only about a

[10] R. L. Fleischer, P. B. Price, and R. M. Walker, "Nuclear Tracks in Solids." Univ. of California Press, Berkeley, 1975.
[11] A. J. Mortlock and D. H. Tomlin, *Philos. Mag.* **4**, 628 (1959).
[12] D. Rebout, J. F. Stohr, and M. Aucouturier, *J. Mater. Sci.* **13**, 2333 (1978).
[13] S. Yukawa and M. J. Sinnott, *Trans. Am. Inst. Min. Metall. Eng.* **203**, 996 (1955).
[14] T. J. Renouf, *Philos. Mag.* **22**, 359 (1970).

factor of ~ 30 change in specific activity can be measured. (d) Radiation from deep inside the sample further complicates the measurement.

As mentioned above, autoradiography is ideal for showing the presence of grain-boundary diffusion qualitatively, although the absence of darkened boundaries does not necessarily mean that grain-boundary diffusion is not present. Quantitative measurements of grain-boundary diffusion have also been made, but the sectioning technique seems preferable for this application. In this method, one removes thin sections perpendicular to the diffusion direction and counts the radiation from the sections (Section 1.4.5).

1.4.2. Radiation Detectors for Pulse Counting

By J. N. Mundy and S. J. Rothman*

Most radiation detectors produce pulse heights proportional to the energy deposited in the detector by the impinging radiation. The three most commonly used "proportional" counters are the gas-flow proportional counter, which is suitable for γ rays, x rays, and low-energy (<0.15 MeV) electrons; the NaI(Tl) scintillator, suitable for γ rays; and a variety of solid state detectors produced from silicon or germanium single crystals and suitable for both γ rays and x rays, or for charged particles. Of these types, NaI(Tl) scintillators and the solid state detectors produced from Ge are the most widely used detectors for determining the abundance of radionuclides used in radiotracer experiments.

1.4.2.1. Gas-Flow Proportional Counters. The counter is a vessel containing gas and two electrodes that separate and collect the ion–electron pairs (about one pair for every 30 eV of energy) produced by the passage of ionizing radiation through the gas. In many designs the electrons, under the influence of the applied field, migrate to a central-wire electrode. The small diameter of the central wire creates a large field gradient, and the accelerating electrons produce secondary ionization. The amplification produced in this way is usually between 10^2 and 10^4. The most common gas mixture is 90% argon and 10% methane. The methane as well as other polyatomic gases are present so as to remove the effect of after-pulsing from long-lived excited states. The purity of the gas is the limiting factor for pulse resolution, and so most counters are operated at atmospheric pressure and continuously flushed. Because of experimental problems associated with gas-flow proportional counters, they are not the method of choice except for those cases where low-energy radiation or high space angles are needed. The expected energy resolution for these detectors will vary approximately as the inverse square root of the deposited energy. At a few keV, the expected resolution will be of the order of 20% of the mean energy of the peak.

1.4.2.2. Scintillation Detectors. The detector system consists of a scintillator (phosphor) that is optically coupled to a photomultiplier. Radiation impinging on the scintillator produces light pulses that are converted by the photomultiplier to electrical pulses with amplitude proportional to the amount of absorbed energy. Anthracene, stilbene, and "plastic" scintillators

---

* Materials Science and Technology Division, Argonne National Laboratory, Argonne, Illinois.

1.4. DETECTION AND ASSAY

are used particularly for γ radiation but also for β and x radiation. The decay time of the pulse is short (5 ns) for "plastic" scintillators. Zinc sulfide activated with silver, copper, or manganese is the common scintillator for α particles. Sodium iodide activated with $\sim 1.3 \times 10^{-3}$ mole fraction of thallium is a highly efficient detector of γ rays. NaI(Tl) is almost 100% efficient for 200-keV γ rays and about 20% efficient for 1-MeV γ rays. The pulse decay time is ~ 250 ns. Although NaI is hygroscopic, it is easily machined into a variety of shapes; the frequently used well counter is particularly valuable because of an almost full 4π geometry.

The observed resolution of the NaI(Tl)–photomultiplier system is the result of the cascading of five separate processes: (a) light production in the scintillator, (b) arrival of photons at the photocathode, (c) production of photoelectrons at the photocathode, (d) collection of photoelectrons on the first dynode, and (e) the multiplication of successive dynode stages. The linewidth expected for the 662-keV γ ray from ^{137}Cs (used as a common standard) is of the order of 5–7%, and the best detector systems yield just about this value. A complete discussion of the statistical properties of phototubes can be found elsewhere.[15-19] Although the multiplication factor of the photomultiplier can be as high as 10^8, the total yield of the scintillation process is low. NaI(Tl) produces ~ 30 photons keV,$^{-1}$ and the efficiency of photoelectric production is of the order of 0.3%. The statistical variation in the number of photons explains the low resolving power of scintillation detectors.

A well-designed NaI(Tl) scintillator and phototube will use a phosphor that is thick with respect to its diameter and will have highly reflective surfaces. To overcome edge effects of the photocathode, the phosphor is somewhat smaller in area than the photocathode. A stable high-voltage supply is provided for the dynode chain, and the voltage divider is designed with regard to amplification stability over a wide range of counting rates. The count rate is limited primarily by the rate at which the tube anode can be discharged. This RC time constant is constrained, however, by the desire to integrate the current pulse for a period of the order of 1 μs, as the pulse decay time is ~ 0.25 μs. To avoid pulse pileup at the anode, an RC time constant of 10 μs is used; this will limit the count rate to 10 kHz, at which point gain shifts will start to be observed. Shorter integration times will allow higher count rates but at the expense of linearity of the response to the energy deposited in the crystal. This may not be a significant problem in

[15] W. Shockley and J. R. Pierce, *Proc. IRE* **26**, 321 (1933).
[16] T. Jorgensen, *Am. J. Phys.* **16**, 285 (1948).
[17] E. Bretenberger, *Prog. Nucl. Phys.* **4**, 56 (1955).
[18] J. R. Prescot, *Nucl. Instrum. Methods* **39**, 173 (1966).
[19] G. A. Morton, H. M. Smith, and H. R. Krall, *Appl. Phys. Lett.* **13**, 356 (1968).

many counting experiments, particularly for integral counting or where timing is more important than energy resolution.

1.4.2.3. Liquid Scintillation Detectors. Liquid organic scintillators are useful for counting radioisotopes like ^3H, ^{14}C, or ^{63}Ni, which emit β particles of such low energy (18, 159, and 67 keV, respectively) that the particles are absorbed by the window of a conventional gas-flow or scintillation counter. The samples are dissolved either directly in the organic scintillator or in a miscible toluene solution. When the emission band of the scintillator does not coincide with the absorption band of the photomultiplier, a second scintillator or wave shifter is added. If the samples are not soluble, they can be finely ground and suspended in a gel. Careful sample preparation is required to prevent "quenching," that is, any process that interferes with the conversion of the β-particle energy to scintillations or with the light transmission to the photomultiplier. Quenching decreases the number of scintillations per disintegration and thus shifts the pulse spectrum to lower energies. The energy of the β particle can be absorbed by an interfering substance and dissipated as heat (chemical quenching), or the energy may not reach a scintillator molecule because of the dilution effect of the other solvents (water, alkalis, and alcohols are strong quenching agents). Thus the chemical contents of all sample vials must be kept the same if the comparison of different counting rates is to be meaningful.

Liquid scintillation counting is generally done in a large commercial unit, which usually consists of two photomultiplier tubes placed opposite each other around a well that holds the vial containing the scintillating mixture. The units are commonly refrigerated to reduce the noise level of the photomultipliers. The outputs of the photomultipliers are fed into separate coincidence and summing circuits. The coincidence circuit discriminates against exterior background and noise; it gates the summing circuit, the purpose of which is to increase the size of the pulse and eliminate the effect of the position of the radionuclide in the vial. The output of the summing circuit is fed to a pulse-height analyzer. Most commercial counters have at least two channels, making possible the simultaneous determination of the activity from two different isotopes (e.g., ^3H and ^{14}C). They also commonly have automatic sample changers.

Liquid scintillation counting can be reproducible if sufficient care is taken to overcome such problems as quenching, chemiluminescence produced from solubilizers, temperature variations, and dehomogenization of the sample in the scintillating mixture. Reproducibility can be checked by the counting of calibration standards. Liquid scintillation can also be used for α emitters, but is not useful for γ emitters. Because of the problems of sample preparation, liquid scintillation should be used only when it is especially suitable.

1.4.2.4. Semiconductor Detectors.

Semiconductor (Si and Ge) detectors, in common with gas counters, obtain their signal from direct ionization. In a gas counter, the electron–ion pairs produced by radiation are separated and counted so that the electrical signal obtained is proportional to the energy received by the detector. In a semiconductor detector, electron–hole pairs are produced in much greater number than the electron–ion pairs in a gas. This effect, together with the higher density of the solid, results in higher efficiency and improved energy resolution.

Semiconductors used as detectors must be good insulators, relatively free of impurities or defects, and have high crystal uniformity throughout the detection volume. The insulating properties are required so that the electrical field for charge collection is not overwhelmed by excessive electronic noise from leakage currents. The "purity" of the detector is required so that the charges produced by irradiation are not trapped at impurities or other defects before their collection by the electric field. High crystal uniformity is needed so that the detector behaves ideally throughout the detection volume.

The three types of semiconductor detectors presently in use are the junction detector, the surface-barrier detector, and the lithium-drifted detector. All types make a sandwich of a p-type semiconductor, which contains an excess of free holes and very few free electrons, and an n-type semiconductor, which has an excess of free electrons and almost no free holes. The junction detector is simply a junction between n- and p-type semiconductors made by diffusing one into the other to form a depleted region that is the sensitive detection volume. Larger depleted regions can be formed by creating a p-type surface layer on an n-type silicon or germanium wafer by oxidation of the surface. The largest-volume detectors are produced by diffusing lithium into p-type material; a reverse bias is then used to make the lithium ions drift to form a thick layer (~ 1 cm) in which the concentration of free electrons and free holes is the same. Germanium has a higher absorption coefficient for γ rays than silicon and so is frequently used for such detectors. However, Ge(Li) detectors have to be maintained at low temperature (commonly liquid-nitrogen temperature) to prevent outward diffusion of the lithium. Although Si(Li) detectors suitable for x-ray detection can be operated at room temperature, the leakage currents of all semiconductor detectors are reduced by operating at lower temperatures. Silicon surface-barrier detectors are well suited to the routine detection and spectroscopy of α particles.

The average energy to produce an electron–hole pair in a semiconductor (2.95 eV for Ge, 3.65 eV for Si) is much lower than for an electron–ion pair in a gas (~ 30 eV) or a photoelectron in a photomultiplier (300–500 eV). The available signal is thus 10 to 1000 times larger, and the statistical fluctuations that affect the energy resolution could be expected to be 3 to 30 times smaller. However, because the production of electron–ion pairs (n_p) in a solid is highly correlated, the statistical variance (σ_n) is reduced from n_p to Fn_p, where F is the Fano factor. Measured values of the Fano factor for germanium gives $F = 0.12$. The resolution of a Ge semiconductor detector can be as good as 1.5 keV in 1 MeV.

1.4.3. Pulse-Counting Systems

By M. J. Fluss, J. N. Mundy, and S. J. Rothman*

The output of the detectors discussed in Section 1.4.2 is current pulses. To measure a counting rate directly, these pulses are integrated to provide a dc current; such systems are frequently used in monitoring devices. The more common method of operation is to couple the detector with a pulse-counting system. This section outlines the nature of the components of a "simple" counting system. It is important to recognize that although the primary goal of a nuclear-counting laboratory is the determination of the activity of a particular radioisotope, the equipment necessary to develop the final counting strategy may be considerably more complex than that used in the final experiments. This section reviews some counting techniques that are of particular importance to radiotracer work.

1.4.3.1. A Simple System. The components of a simple system are shown in Fig. 2. The high-voltage supply and the preamplifier are matched to the detector being used. The amplifier both amplifies and shapes the incoming pulses so that the pulses are readily resolved on the basis of energy or time. These criteria can be in conflict, and different amplifiers may be necessary. A variety of different amplifier designs are commercially available.

The pulse-height analysis carried out by the single-channel analyzer (SCA) is usually one of two types, integral or differential. In the integral mode all pulses greater than a minimum lower level of discrimination (LLD) result in the generation of a standard logic signal. When the SCA is operated in the differential mode, an upper level of discrimination (ULD) is also chosen. All pulses with maximum amplitude between the LLD and ULD result in the production of a logic pulse. These pulses are counted by a scaler whose digital output can be fed directly to a printer or to a computer for data processing. The scaler is gated by a timing system and gathers pulses until either a preset time or a preset number of pulses is reached.

1.4.3.2. Analog-to-Digital Converters. It is possible to obtain a histogram of a pulse-height spectrum from a proportional detector by using only an SCA. To accomplish this, the ULD and LLD are moved simultaneously such that their difference is kept constant (corresponding to a constant ΔE, for example). The histogram is constructed by measuring the count rate in a

---

* Materials Science and Technology Division, Argonne National Laboratory, Argonne, Illinois.

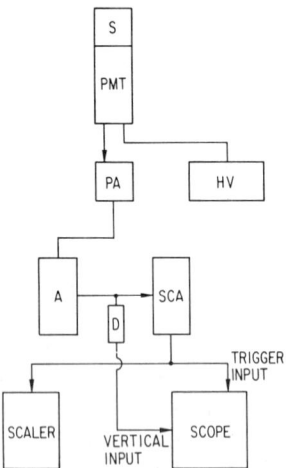

FIG. 2. Block diagram for single-channel spectrometer. S = scintillator, PMT = photomultiplier tube, PA = preamplifier, HV = high voltage, A = amplifier, SCA = single-channel analyzer, D = delay.

scaler as a function of the LLD setting. However, even if one had many SCAs and scalers and could accumulate several channels simultaneously, this would be a tedious and inefficient approach.

In contrast, the analog-to-digital converter (ADC) functions to record the entire spectrum. Unlike the SCA, which provides only one bit of information, the ADC will typically be able to digitize signals from 0 to 10 V into 10 (1024) to 13 (8196) bits or channels of information. The most commonly used ADCs are the Wilkinson type. These devices work on the principle of charging a capacitor to a level proportional to the sampled peak height of the pulses presented to them. The capacitor is then discharged to a reference zero level, during which time a high-frequency clock (10–200 MHz) is run into an address register. The contents of this register are the converted digital information proportional to the observed peak height of the pulse. An SCA circuit and linear gate are often constructed as an integral part of the ADC.

1.4.3.3. *Multichannel Analyzers.* The ultimate construction of the pulse-height histogram is accomplished with some type of "intelligent" component, such as a hardwired sorter or with a microprocessor or computer system. In most cases, the multichannel analyzer (MCA) will contain a memory that is sufficiently large to allow the scaling of all the channels that can possibly be addressed by the ADC. The arrival of converted data in the output buffer of the ADC signals a storage algorithm that results in the incrementing by one of the address in the storage memory, which has a one-to-one correspondence with the digital value in the ADC buffer. The

1.4. DETECTION AND ASSAY

cost of memory today is relatively low; thus one should expect most MCAs to allow for counts in excess of 10^6.

The MCA provides a quick and quantitative way to set the windows of an SCA. An example is the distribution of pulse heights, shown as oscilloscope traces in Fig. 3. To "replace" the scope with an MCA requires that the linear gate in front of the MCA be triggered by the logic signal from the SCA. This gating then reveals the gated spectrum, which may be compared with the total spectrum. With such a tool, as shown in Fig. 3, the efficiency or sharpness of the LLD and ULD levels can be studied. As we shall discuss below, other properties such as random coincidences between the counters are easily studied and quantitatively counted with the aid of an MCA.

The memory storage unit of the MCA can also be directly addressed by the SCA logic pulses. In this mode each channel of the memory serves as a true scaler for a period of time (dwell time). The dwell time is usually derived from a high-precision clock, which, when it times out, directs the SCA pulses to the next channel in sequence in the memory. Such multichannel scaling (MCS) is well suited to multiple counting, automatic sample changing, and on-line analysis of decay curves.

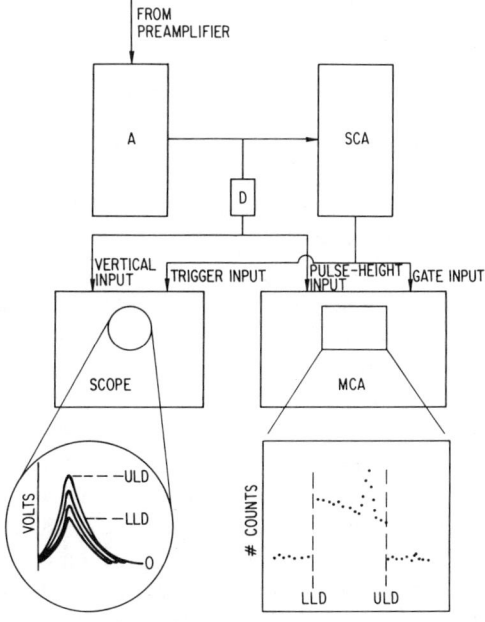

FIG. 3. Block diagram for setting SCA window levels with an MCA. LLD and ULD are lower- and upper-level discriminators; other abbreviations as in Fig. 2.

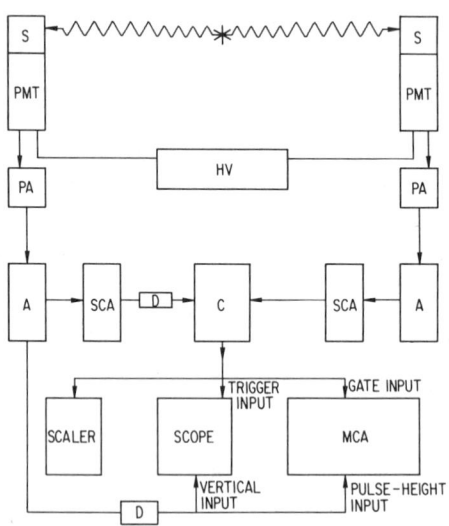

FIG. 4. Block diagram for coincidence counting. C = coincidence unit; other abbreviations as in Fig. 2.

1.4.3.4. Coincidence-Counting Techniques. A β particle and a γ particle or two γ particles can be emitted from the same nuclide simultaneously. Another case of radiation coincident in time is the two 511-keV γ rays emitted in the annihilation of positrons. Coincident detection of such radiation is called coincidence counting. At times the coincidence method is the only technique that can be used to adequately separate a nuclide from other radionuclide emissions. Figure 4 shows the block diagram typical for coincidence experiments: SCAs are shown in each "arm" of the coincidence system. The SCAs are of the type suitable for broad-band timing resolution; that is, the timing is independent of amplitude. Since the 511–511 keV coincidence is spatially as well as temporally correlated, the detectors must be appropriately positioned.

The energy windows of the SCAs in Fig. 4 are set to bracket the 511-keV photopeak of the two detectors; the logic signals are then passed on to a coincidence counter, where they are compared with a timing resolution of ~ 1–10 ns. The associated coincidence pulse-height histogram can be viewed by using the coincidence logic signal to gate open the MCA.

Quantitative coincidence counting is complicated by the always-present random coincidences, that is, the coincident detection of independent radiation events. Since the phenomenon is always count-rate dependent, it is advisable to measure in situ the chance-coincidence rate for each sample. This is simply done by introducing a delay that displaces the true coinci-

1.4. DETECTION AND ASSAY

dence but preserves the count rates in each channel; hence the chance-coincidence rate is the same as without the delay.

Coincidence counting is usually appropriate when the radioisotope of interest has a poor signal-to-noise ratio because of the presence of other radioisotopes. Hence single-count rates will be orders of magnitude greater than the coincidence rate and dead-time problems may be serious. Other pitfalls in coincidence counting are considered by Remsberg.[20] Maintenance of high radiopurity or suitable decay-curve analysis may prove to be a more accurate method for studying relative concentrations of radioisotopes than the coincidence method, albeit the latter is a very selective method for temporally or spatially correlated radiations.

---

[20] L. P. Remsberg, *Annu. Rev. Nucl. Sci.* **17**, 347 (1967).

1.4.4. Radioactive Particle Counting

By M. J. Fluss, J. N. Mundy, and S. J. Rothman*

1.4.4.1. Counting Alpha Particles. Gas-filled counters or liquid scintillators can be used for α-particle detection; however, semiconductor detectors have the best energy resolution and efficiency. The high absorption rate of α particles requires that gas-filled counters have a thin (<1 mg cm^{-2}) window; alternatively, the source is placed inside the counter, which is then sealed and flushed with the counting gas. The self-absorption in the solid sources commonly used for α counting necessitates both the preparation of uniformly thin sources (a few μg cm^{-2}) and a knowledge of the absorption coefficient. Alternatively, standards and samples can be prepared in the same way as the sample, and the detector thus calibrated. Sample preparation methods include electroplating, electrospraying, vacuum sublimation, and simply evaporating a solution to dryness. Uniformity of deposition can be a problem with simple drying unless a wetting agent is used.

1.4.4.2. Counting Beta Particles. The small penetration of β particles creates problems similar to those for α particles, and similar solutions have been found through the use of thin-window gas-flow counters and the placing of the source within the detector either as a solid or, in the case of liquid scintillators, as a solution. The latter method has proved popular, and a large number of techniques and procedures have been described in the literature. Solid sources have self-absorption problems, and the sample support creates backscattering problems. Even for relative counting, both samples and standards need to be counted in the same chemical form and on the same support. Liquid scintillators avoid many of these self-absorption and scattering problems and can have high efficiencies (70–90% for ^{14}C). Beta emitters also give rise to a bremsstrahlung spectrum starting at the maximum energy, $E_{\max}$, of the β ray and increasing in intensity towards lower energies. The energy loss due to bremsstrahlung is relatively small ($Z_{\max}/300$, where Z is the atomic number of the material through which the β ray is passing). Nevertheless, corrections for bremsstrahlung are required when making an absolute β assay. The correction can be reduced by lining the β-counting apparatus with low-Z material.

1.4.4.3. Counting Gamma Rays. Gamma-ray-emitting sources in radioactive work areas are monitored by photographic film, ion chambers, and

---

* Materials Science and Technology Division, Argonne National Laboratory, Argonne, Illinois.

1.4. DETECTION AND ASSAY

Geiger–Müller counters. In precise counting, the detection of γ rays is usually by means of a NaI(Tl) scintillator or a lithium-drifted Ge(Li) or intrinsic Ge detector. The first has proved to be the workhorse for the past 25 years and is still much in use where relatively low cost and high efficiency are required. NaI(Tl) scintillators are available in a wide range of sizes (commonly in cylindrical shapes); the 3 × 3-in. cylinder has proved most popular and is used as a reference for gamma counting efficiencies. Well-type crystals increase the efficiency, but their energy resolution is a few percent lower. The Ge(Li) detectors have much better energy resolution, but they must be kept at liquid-nitrogen temperatures and for comparable cost have much lower efficiencies. High-efficiency well-type Ge(Li) detectors are now available at a cost many times more than NaI(Tl) detectors of similar efficiency.

Gamma rays are not easily absorbed and so sources are usually counted either in solid form or in a solution. The sources are placed in standard glass or plastic vials and positioned in a fixed geometry relative to the detector. Such simple procedures are usually sufficient for reliable relative counting because absorption within the source or vial is small enough that reproducibility is not a problem.

The low absorption of γ rays leads to a complex energy spectrum as seen by the detector. As discussed above, the γ rays will interact with the scintillator by the photoelectric effect, the Compton effect, and pair production. These three interactions allow the incident γ ray to be either totally or partially absorbed.

At low energies the γ ray may be totally absorbed by ejecting a photoelectron, and one would expect to find a single-line spectrum (counts versus photon energy) with a Gaussian shape from statistical variations in the detector process. The atom from which the photoelectron is ejected will emit x rays or Auger electrons, and if these escape absorption in the detector, the spectrum will show the single-line photopeak plus an "escape peak" at an energy equal to that of the photopeak less the energy carried by the escaping radiation.

At medium energies the γ ray may undergo successive Compton interactions and finally be absorbed by ejecting a photoelectron. Such interactions can also lead to partial absorption if either the original γ ray, with degraded energy, or a secondary γ ray escapes from the detector. The resulting spectrum will not appear as peaks, as with photoelectric absorption, but as a broad spectrum up to a cutoff called the Compton edge. The cutoff represents the maximum energy transfer or minimum energy [E', given by Eq. (1.1.6)] for the second γ ray. The broad Compton spectrum may also show backscatter peaks resulting from absorption in the detector of γ rays that underwent Compton interactions in the vicinity of the detector, suffered maximum energy transfer, and then were backscattered into the detector. Compton events clearly make significant contributions to the observed γ-ray spectrum. Large detectors make for proportionately fewer "escape" γ rays and enhance the single-line spectrum. The photoelectric absorption is proportional to about the fifth power of the atomic number Z of the absorber, and the Compton absorption shows a linear dependence on Z. Although such factors would suggest that γ-ray spectra are best resolved in large detectors of high Z, with present technology smaller-volume high-resolution Ge(Li) detectors are much preferred for γ-ray spectrometry.

At energies above 1.02 MeV the γ rays may interact with the detector by pair production. The positron created has little chance to escape and annihilates with the emission (usually) of two 0.511-MeV γ rays, one or both of which may escape resulting in "escape" or "double escape" peaks. These peaks are 0.511 and 1.02 MeV lower than that of the incident γ ray; however, the 0.511-MeV peak can also occur in the spectrum from interactions in the material surrounding the detector. In certain detector geometries, both of the annihilation γ rays can be scattered back into the detector, producing a 1.02-MeV sum peak. At high counting rates, two or more γ rays may coincide in a detector, resulting also in sum peaks.

The spectrum from even monoenergetic γ-emitting radioisotopes is clearly not necessarily a single photopeak, and the use of mixtures of isotopes with multiple γ-ray energies requires careful analysis. Radiotracer studies commonly use carefully chosen and well-separated single radioisotopes so that the total observed spectrum is that of the desired isotope. In such studies, integral counting ignores the complexity of the spectrum and can yield improved statistics by recording all or large segments of the spectrum. The greatly improved resolution available using Ge(Li) detectors has, however, allowed the careful analysis of γ-ray spectra and in turn allows radiotracer experiments of considerably wider scope. Not only can known mixtures of single radioisotopes be readily resolved, but use can also be made of those radioisotopes that contain unwanted and chemically inseparable radioimpurities. In order to measure the γ-ray spectra, the pulse-height discrimination techniques discussed in Section 1.4.3 are required.

1.4.4.4. Unfolding of Pulse-Height Spectra. The pulse-height spectra recorded in an MCA can be analyzed by spectrum unfolding methods. The use of the general mathematical technique of response matrices has been described and reviewed by Monahan.[21] This review contains an extensive bibliography of both response-function analysis and least-squares analysis programs used for unfolding γ-ray spectra from scintillation counting with

---

[21] J. E. Monahan, *in* "Scintillation Spectroscopy of Gamma Radiation" (S. M. Sudworth, ed.), Vol. 1, p. 371. Gordon & Breach, New York, 1967.

1.4. DETECTION AND ASSAY

NaI(Tl) detectors. Several useful hand evaluation methods that can be helpful in setting up counting experiments are described by Quittner.[22] Such methods as spectrum smoothing and spectrum stripping can provide useful qualitative information when several radioisotopes are present. However, the accuracy of most stripping techniques decreases rapidly after only a few components (about three) have been removed from the total spectrum.

High-resolution spectroscopy with solid state detectors is a particularly useful tool for identifying isotopes. Because of the order-of-magnitude improvement in energy resolution over NaI(Tl), it is almost always possible to identify several radiation lines associated with a particular isotope. Relative intensities of these lines and the energies of the lines can be used together to identify the radioisotope unambiguously. Elemental analysis is particularly simple for isotopes that decay by internal conversion, yielding characteristic x-ray spectra.

The possibility of using all the information contained in a high-resolution γ-ray spectrum has prompted the development of a variety of γ-ray analysis codes. Many programs, although lacking in sophistication, allow significant interaction with the user. Other programs use libraries of radioisotope energies and intensities. The major fault of almost all the present analysis programs is their inability to take into consideration the details of the response function energy resolution of the particular detector system. The internal description of peak shapes is often too idealized to allow for a high level of accuracy, although precision can often be obtained by taking care not to vary parameters of the counting, which could lead to substantial changes in the energy resolution of the counting system.

In applications involving radiotracer studies, the value of spectral unfolding methods is realized in the planning and design of experiments. Often, however, precision of the order of $\pm 2\%$ can be achieved, which may be adequate for some tracer experiments. Because there are many pitfalls in unfolding methods and the investment in analysis time can become very large, one is well advised (as in the case of coincidence counting) to examine the details of each counting problem so as to best decide the ultimate approach to separation of contaminating radiation. Often, the method of decay-curve analysis is faster and simpler. In fact, the chemical purification of the tracer solution may be the best route to take in many situations.

---

[22] P. Quittner, "Gamma Ray Spectroscopy." Wiley, New York, 1972.

1.4.5. Total Relative Counting

By J. N. Mundy and S. J. Rothman[*]

1.4.5.1. Criteria for Pulse Counting. The radiation detection systems for pulse counting discussed in the previous sections have a number of common features, and it is important to understand the criteria used in choosing systems.

Counting techniques may be classified as integral or differential. An example of differential counting is the determination of the pulse-height distribution originating from radiation incident on the detector. An integral experiment, as the name implies, would provide information only about the total number of counts above some minimum threshold energy. Integral techniques, when they are feasible, are preferred over differential methods in nuclear counting because they tend to be less sensitive than other methods to a variety of instabilities.

Most properties of the defect solid state that are studied by radioactivity involve the comparison of different radioactive counting rates, and the measurements can be compared directly on a relative basis. Although absolute measurements are not usually necessary, it is valuable in the planning stages to make reliable estimates of the efficiencies of detection systems. A knowledge of the efficiencies allows one to determine the quantity of radioisotope required for the particular experiment, to design equipment and choose methods of sample preparation to maximize the count rate, and to ensure that the efficiency does not change during the course of the experiment and that contamination and radiation levels consistent with radiation safety are not exceeded.

Common to all physical measurements are both systematic and random errors. An example of a systematic error in a radioactivity measurement is the incorrect determination of the detection efficiency, which in the case of relative counting is not important. A random error could arise from a surge in the line voltage while a sample is being counted. The random nature of the radioactive decay process also introduces an uncertainty into all measurements of radioactivity; thus, although the ultimate precision is determined by the statistics of the decay process, the practical limits of precision are usually related to factors ostensibly under the control of the experimenter.

Another property common to most radioactivity measurement systems is

---

[*] Materials Science and Technology Division, Argonne National Laboratory, Argonne, Illinois.

1.4. DETECTION AND ASSAY

the ability to resolve the electrical pulses from the detector with respect to both time and amplitude. The resolving or dead time is the minimum time required for two successive pulses to be detected as separate pulses. In a proportional system, the amplitude of the pulse reflects the magnitude of the energy of the incident radiation. The common features of efficiency, statistics, and dead time are discussed in the following sections.

1.4.5.2. *Counting Efficiency.* The overall efficiency of a counting system can be simply defined as the detected fraction of the disintegrations occurring in a source. The efficiency can be measured with radioisotopes with known disintegration rates. In principle, the standard sources could be made in the same irradiation facilities used by the experimenter in preparing the particular radioisotope of interest. However, to obtain accurately calibrated sources the irradiation flux must be well known and appropriate corrections made for the shielding that results from the sample and its container. Calibrated sources of many α-, β-, and γ-emitting radioisotopes are available commercially. However, not all radionuclides are available as calibrated sources; even when they are, it is not always possible to prepare the calibrated source in the same chemical and physical form as the sources for which the efficiency factor is required. In such cases it is useful to be able to calculate the efficiency. The following factors enter into the efficiency of a counting system:

(a) The geometry factor Ω, which for a point source is simply defined as the fraction of the total solid angle 4π that the sensitive volume of the detector subtends at the source. In general, Ω is calculated from the dimensions of the sources and detector and their separation. Accurate geometry factors are difficult to determine for proportional counters and Ge(Li)-drifted detectors, in which the sensitive volume is not easily defined.

(b) Absorption in the sample, sample holder, air, or the window of the detector.

(c) The intrinsic efficiency ϵ of the detector, which is defined as the fraction of the radiation incident on the surface of the detector that produces a measurable signal. The angles of incidence are defined by the same geometry that defines Ω, and the value of ϵ may be calculated from these and the absorption coefficients for the particular radiation in the material of the detector. The common use of NaI(Tl) scintillation detectors for γ-counting has led to the establishment of tables of efficiencies as a function of γ-ray energy for various sizes of cylindrical detectors.

(d) Scattering of the radiation within the source and from the source or detector holder or surrounding shielding.

(e) Dead time of the counting system.

(f) The fraction of disintegrations that provides the radiations that are detected and counted.

In the detection of γ rays, the energy of the γ ray may be absorbed in the detector by a single interaction or multiple coincident interactions that involve photoelectric, Compton, and pair-production processes. The efficiency of detecting only those γ-ray energies that lie in the photopeak is, for monochromatic radiation, a specific fraction of the efficiency of a particular detector.

Of the factors that enter into the determination of the efficiency of a counting system, some are clearly not easily calculated. If an accurate overall efficiency is required, as in absolute counting, then an experimental determination with calibrated sources is recommended. The above description of the factors influencing the efficiency illustrates the advantages of relative counting. Errors in relative counting can arise from variations in the volume, homogeneity or shape of replicate samples or in the positioning of the source relative to the detector, as well as from a variety of count-rate-dependent effects (e.g., dead time and gain shift). Careful preparation of replicative samples can ensure that the overall efficiency remains constant at better than 1%, but it is important to examine the effect of all the factors mentioned above if accurate relative counting is to be achieved.

Electronic drift can change the efficiency of the counting system. Drift can be reduced by controlling the temperature of the counting system to $\pm 0.5°C$ or better. When many samples have to be compared over a long time, a standard source should be counted periodically so that any efficiency changes can be observed and appropriate corrections made. The ideal standard sources are radioisotopes emitting the same radiation and with a similar energy spectrum. The strength of such standard sources should be adjusted to match the range of experimental samples.

1.4.5.3. Counting Statistics. The probability law that describes the statistical fluctuations of radioactive decay is the Poisson distribution, which for the extensive sampling characteristic of most radioactive counting is equivalent to the Gaussian distribution. The standard deviation (σ) of the Gaussian distribution is the square root of the true average number of counts $\sqrt{m}$, which for large count number is taken to be the square root of the measured number of counts $\sqrt{n}$. In a Gaussian distribution there is a 68.3% probability that the individual counts fall within $\pm \sigma$ of the true average number of counts. The standard deviation of a counting rate is governed by the number of counts taken, so that the relative standard deviation, $1/\sqrt{n}$, decreases with increasing total number of counts.

If repeated measurements are made, giving $n_1, n_2, \ldots, n_i$ counts, the individual standard deviation is given by

$$\sigma = \left[\sum_{i=1}^{x} (n_i - \bar{n})^2/(x-1) \right]^{1/2}, \qquad (1.4.1)$$

1.4. DETECTION AND ASSAY

where $\bar{n}$ is the mean of the x observations. The standard deviation determined in this way should agree with $\sqrt{n_i}$. If differences occur, a χ^2 test can be performed to determine whether these differences are statistically significant.

Radioactive counting usually consists of the measurement of the net counting rate r_c that results from a source counting rate r_s, counted for time t_c, in the presence of a background counting rate r_B, counted for time t_B. The rules for error propagation show that the standard deviation for the source alone is given by

$$\sigma_{r_c} = (\sigma_{r_s}^2 + \sigma_{r_B}^2)^{1/2} = \left(\frac{r_s}{t_c} + \frac{r_b}{t_B}\right)^{1/2}. \tag{1.4.2}$$

The fractional standard deviation σ_f is given by

$$\sigma_f = \frac{\sigma_{r_c}}{r_c} = \left(\frac{\alpha/t_c + 1/t_B}{r_B(\alpha - 1)^2}\right)^{1/2}, \tag{1.4.3}$$

where $\alpha = r_s/r_B$.

To realize the most efficient use of a counter for a fixed total time, $T = t_c + t_B$, σ_f must be minimized by appropriate choice of t_c/T. Alternatively, T can be minimized for a fixed value of σ_f. From Eq. (1.4.3) one can derive the optimal division of time between source and background counting,

$$t_c/t_B = (r_s/r_B)^{1/2}. \tag{1.4.4}$$

When many samples are being counted or the counting times are long, backgrounds are measured periodically, and an average value is used for the calculation of sample rates. Background counts are interspersed between several sample counts so that the overall criterion of Eq. (1.4.4) is met; the dispersion of background counts allows the experimenter to check whether any samples have contaminated the detector or instrument malfunctions have occurred.

From Eq. (1.4.3), one can also derive the minimum fractional standard deviation σ_f for a total counting time T,

$$\sigma_f = \frac{\alpha^{1/2} + 1}{(\alpha - 1)(r_B T)^{1/2}}. \tag{1.4.5}$$

Thus, when $r_s \gg r_B$, the overall standard deviation is determined by $\sqrt{r_s}$, and when $r_s \ll r_B$, one can determine a figure of merit, defined as $M = \sigma_f^{-1} T^{-1/2}$, given by

$$M = r_s/2r_B^{1/2}. \tag{1.4.6}$$

The criteria for minimizing $\sigma_f T^{1/2}$ are helpful in the choice of counting systems and their operating conditions. For example, in the measurement of diffusion, $r_s \gg r_B$ and the counting system that allows the greatest number of counts in the minimum time would be preferable.

A systematic error can occur in following the decay of a radioactive sample when the counting rate n/t is taken to represent the activity at a time halfway through the counting interval. This approximation is valid if $t \lesssim T_{1/2}$, but the correct time expressed as a fraction f of the counting interval t is given by

$$f = \frac{1}{\lambda t} \ln \left[\frac{\lambda t}{1 - \exp(-\lambda t)} \right], \qquad (1.4.7)$$

where λ is the decay constant of the radioisotope.

1.4.5.4. Dead Time. The dead time τ is the minimum time that a detection system requires after counting one pulse before it can count the next pulse. The dead time may be determined by the detector and/or by the associated electronics. Detectors have a range of dead-time values, e.g., a few hundred microseconds per count for Geiger-Müller counters and a few microseconds per count for proportional counters. For scintillation detectors, τ ranges from 10 ns per count for organic detectors to 1 μs per count for NaI(Tl) detectors. Semiconductor detectors cover a range of ten to a few hundred nanoseconds. In "paralyzable" detectors, which include most detector types, τ varies not only from one detector to another but also from one pulse to another because events occurring within the "dead" period add to the dead time despite the lack of response. The uncertainty introduced is somewhat overcome by creating a constant nonparalyzable dead time in the electronics that is a little larger than that of the detector.

If r is the input count rate and r_o the observed output count rate, then the system is dead for a fraction of time $r_o \tau$. The count loss per unit time is

$$r - r_o = r r_o \tau, \qquad (1.4.8)$$

from which

$$r = r_o / (1 - r_o \tau). \qquad (1.4.9)$$

In order to correct for the lost counts and establish the time efficiency of the counting system, τ must be measured as accurately as possible. When $\tau = 1$ μs, a count rate of 10^4 counts per second will require a 1% correction. For 0.1% accuracy, 10^6 counts must be taken, and in the present example τ would need to be known within an uncertainty of 10%. Count rates above 10^4 counts per second would have an uncertainty progressively larger than 0.1%. The loss of counts arises essentially from the random nature of the pulse arrival times, so it is important to measure τ using radioactive sources and not conventional pulse generators. One method is to allow a pulse to trigger an oscilloscope; the length of time before the counting system again accepts pulses is then readily observed. Unfortunately, the decay of the triggering pulse and the envelope of the following pulses are variable, and τ is not well defined. One may follow the decay of a short-lived isotope with an accurately known half-life. However, the most commonly adopted method is the paired-source method.

Ideally, pairs of sources are made of the same radioisotope to be used in the experiment. If this is not possible because of a short half-life, then a radioisotope with radiation of a similar energy should be chosen. τ is dependent on the electronic gain and discriminator setting, and these should be kept the same for the dead-time measurement as in the experiment. The counting rate also affects τ, so a series of paired sources should be made with count rates spanning those of the proposed experiment. The sources are counted in the sequence r_1, r_{12}, r_2, etc. When the background rate is r_B, the sum of the separate counts must be given by

$$\frac{(r_1)_o}{1 - (r_1)_o \tau} + \frac{(r_2)_o}{1 - (r_2)_o \tau} = \frac{(r_{12})_o}{1 - (r_{12})_o \tau} + r_B. \tag{1.4.10}$$

The measured rates allow Eq. (1.4.10) to be solved and τ determined. The difference between $(r_1)_o + (r_2)_o$ and $(r_{12})_o$ is small, and for $\tau = 1$ μs, 10^7 counts need to be taken to determine τ to within 10%. A special holder should be made so that each source has a fixed position and can be removed and replaced without disturbing the other source. When counting an individual source, a blank source holder should be placed in the other position so that absorption and scattering are always closely similar.

High count rates can distort the energy spectrum itself (gain shifts), and under some circumstances the dead time per pulse may be count-rate dependent. Unless special precautions are taken, dead-time losses should be kept to a few percent or less if results at the levels of parts per thousand are to be obtained. When coincidence-counting techniques are used, then the problems associated with dead-time loss and chance coincidence are highly dependent on the functional logic of the components of the counting system (see the article by Remsberg[20]). The general equation governing random coincidences when the single rates are much greater than the coincidence is

$$r_r = r_1 r_2 (\tau_1 + \tau_2), \tag{1.4.11}$$

where r_r is the random rate and the indices refer to the two counting channels. The dead time of a Wilkinson ADC is proportional to pulse height. However, most MCAs keep track of "live time," so the pulse-height-dependent dead time is not a problem.

1.4.6. Isotope Effects

*By S. J. Rothman and J. N. Mundy**

One of the more difficult counting problems in diffusion experiments is encountered in so-called isotope-effect experiments,[23] which measure the effect of the isotopic mass on the diffusion coefficient. The experiments can reveal some of the details of the diffusion mchanisms because of the relationship between the isotope effect, E_i, and the Bardeen–Herring correlation factor,[24] f:

$$E_i \equiv \frac{(D_\alpha/D_\beta) - 1}{(m_\beta/m_\alpha)^{1/2} - 1} = f\,\Delta K. \qquad (1.4.12)$$

Here D_α and D_β are the diffusion coefficients of two isotopes of the same element of masses m_α and m_β, respectively, and ΔK is a correction factor for many-body effects at the saddle point.[24,25] The experiment is difficult because the quantity to be measured, $(D_\alpha/D_\beta) - 1 \equiv \Delta D/D$, is small. (Both f and ΔK lie between 0 and 1, and $(m_\beta/m_\alpha)^{1/2} - 1$ ranges from 0.07 for ^{6}Li and ^{7}Li to, e.g., 0.022 for ^{105}Ag and ^{110m}Ag.) The diffusion of the two isotopes must therefore be measured simultaneously so that uncertainties in the temperature, diffusion time, and section thickness, which are usually larger than $\Delta D/D$, are the same for both.

The quantity $\Delta D/D$ can be obtained from the equation

$$\ln(S_\alpha/S_\beta) = \text{const} - \ln S_\alpha\,(\Delta D/D), \qquad (1.4.13)$$

which in turn is obtained from the solution of the diffusion equation for the thin-layer geometry [Eq. (1.3.1)]. (Here the S are the specific activities of the two isotopes in each section.) As the maximum practical range in S is a factor of $\sim 10^3$, $\ln(S_\alpha/S_\beta)$ will change by only ~ 0.2–0.5. Thus, if $\Delta D/D$ is to be measured to a few percent, $\ln(S_\alpha/S_\beta)$ must be measured to $\sim 0.1\%$. That is, very good counting statistics must be obtained, and it is necessary to take 10^6 counts on each section.

[23] N. L. Peterson, in "Diffusion in Solids" (A. S. Nowick and J. J. Burton, eds.), p. 116. Academic Press, New York, 1975.
[24] J. G. Mullen, *Phys. Rev.* **121**, 1649 (1961).
[25] G. H. Vineyard, *J. Phys. Chem. Solids* **3**, 121 (1957).

* Materials Science and Technology Division, Argonne National Laboratory, Argonne, Illinois.

1.4. DETECTION AND ASSAY

Obviously, an isotope experiment can be carried out only for the diffusion of an element that has two radioisotopes with masses that differ by at least several percent and with decay schemes that allow accurate separation. Four different types of separation schemes have been employed, as detailed below.

1.4.6.1. Half-Life Separation. This technique is especially useful for pairs of γ-emitting isotopes in which one isotope has a half-life of the order of tens of hours and the other is at least five times longer lived. The isotopic composition is adjusted so that the activity of the short-lived isotope is four to five times that of the long-lived one, and the total activity A of each section from the diffusion sample is counted approximately six times during a period equal to approximately five times the half-life of the short-lived isotope. Then for each section

$$A = A_\alpha \exp(-\lambda_\alpha t) + A_\beta \exp(-\lambda_\beta t); \qquad (1.4.14)$$

i.e., the total activity can be fitted to a linear equation involving the time if the half-lives are accurately known. The ratio $A_\alpha/A_\beta = S_\alpha/S_\beta$, as well as the statistical uncertainty of this ratio, can be obtained from the fit. It is best to use a single lower-level discriminator set in the valley below the lowest-energy photopeak of the isotope mixture, and to use a large well-type NaI(Tl) detector for increased counting efficiency and reproducibility in positioning the samples. The lowest practical starting activity for the last section is $\sim 10^4$ cpm; the first sections would thus run 10^7 cpm and should be diluted, maintaining the same volume, in order to reduce the uncertainty of the dead-time corrections. This technique has been successfully applied to the isotope pairs ^{22}Na – ^{24}Na,[26-28] ^{69m}Zn – ^{65}Zn,[29] ^{64}Cu – ^{67}Cu,[30] ^{55}Co – ^{60}Co,[31] ^{52}Fe – ^{59}Fe,[32] and ^{48}Cr – ^{51}Cr.[33]

1.4.6.2. Spectrometry. Half-life separation is difficult to apply if the shorter half-life is longer than a few days or if the half-lives do not differ sufficiently. If the energy of the γ rays from the two isotopes is sufficiently different, $\ln(S_\alpha/S_\beta)$ can be obtained by γ-ray spectroscopy. In some cases even scintillation spectrometry can be applied; a separation of the γ radiation from ^{105}Ag and ^{110}Ag can, for example, be made according to the

[26] L. W. Barr and A. D. LeClaire, *Proc. Br. Ceram. Soc.* **1**, 109 (1964).
[27] S. J. Rothman, N. L. Peterson, A. L. Laskar, and L. C. Robinson, *J. Phys. Chem. Solids* **33**, 1061 (1972).
[28] J. N. Mundy, *Phys. Rev. B* **3**, 2431 (1971).
[29] S. J. Rothman and N. L. Peterson, *Phys. Rev.* **154**, 552 (1967).
[30] S. J. Rothman and N. L. Peterson, *Phys. Status Solidi* **35**, 305 (1969).
[31] W. K. Chen and N. L. Peterson, *J. Phys. Chem. Solids* **41**, 647 (1980).
[32] C. M. Walter and N. L. Peterson, *Phys. Rev.* **178**, 922 (1969).
[33] J. N. Mundy, C. W. Tse, and W. D. McFall, *Phys. Rev. B* **13**, 2349 (1976).

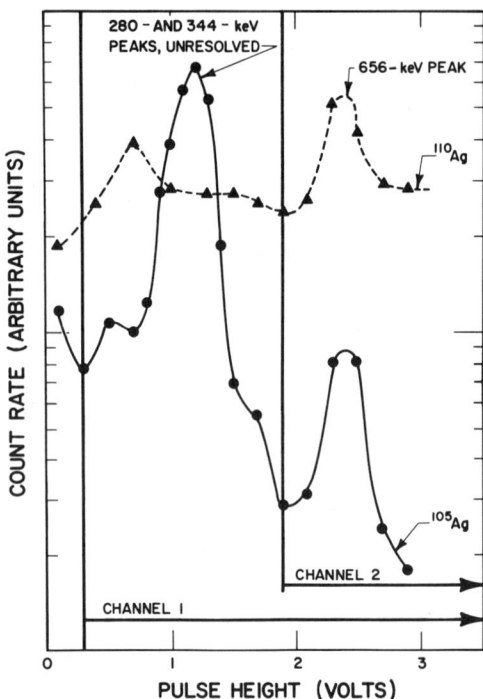

FIG. 5. Level scheme for separating radiation from ^{105}Ag and ^{110}Ag. Pure isotopes are used for calibration.

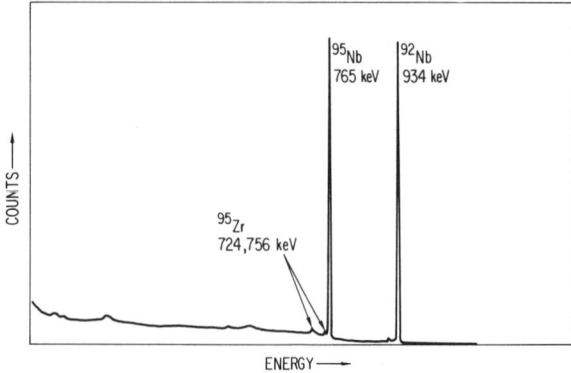

FIG. 6. Ge(Li) spectrum of the γ peaks used in the measurement of isotope effect for self-diffusion in Nb. (From Bussman et al.[36])

scheme shown in Fig. 5.[34] If better resolution is needed, for instance, to separate the γ-ray peaks from ^{92}Nb and ^{95}Nb (Fig. 6), a Ge(Li) detector must be used. The high-resolution detector also allows one to discriminate the radiation from unwanted radioisotopes, such as ^{95}Zr in Fig. 6. High-efficiency well-type Ge(Li) detectors are now available and have been used in a series of elegant experiments by Herzig and co-workers for measurements of diffusion[35,36] and of thermodynamic activity.[37] Care must be taken in such measurements because the low-energy peaks are superimposed on a mound of Compton-scattered radiation, the relative height of which varies from sample to sample.

1.4.6.3. Separation of Beta and Gamma Radiation. The separation of β and γ radiation can be done either by using different filters as Mullen[24] did with ^{55}Fe and ^{59}Fe, or by using two counters one much more sensitive to β and the other to γ radiation. Peterson et al.[38] used a thin plastic scintillator for the β and a NaI(Tl) detector for the γ in separating the radiation from ^{105}Ag and ^{111}Ag. Calibration of both counters or filters must be made with both pure radioisotopes. It should be noted that β radiation is absorbed much more strongly than γ, and thus reproducibility of the sample geometry, especially the thickness, must be controlled with extreme care.

1.4.6.4. Use of Two Beta-Emitting Isotopes. Notwithstanding problems of self-absorption, it is still possible to measure isotope effects with weak β-emitters. As an example, we mention the elegant work of Sherwood and his collaborators[39,40] on diffusion in organic crystals. They change the mass of one tracer molecule by fully deuterating it, label one tracer molecule with ^{14}C and the other with ^{3}H, and do the separation by liquid scintillation counting.

Acknowledgments

We thank B. A. Rothman for much of the preliminary typing of the first draft, H. Mirenic for her secretarial skills and diligent effort to get the manuscript ready for the editors, and E. M. Stefanski for the editing of the manuscript. This paper was prepared under the auspices of the U.S. Department of Energy, Division of Basic Energy Sciences.

---

[34] S. J. Rothman, N. L. Peterson, and J. T. Robinson, *Phys. Status Solidi* **39**, 635 (1970).
[35] C. Herzig and H. Eckseler, *Z. Metallkd.* **70**, 215 (1979).
[36] W. Bussman, C. Herzig, H. A. Hoff, and J. N. Mundy, *Phys. Rev. B* **23**, 6216 (1981).
[37] H. E. Peltner and C. Herzig, *Acta Metall.* **29**, 1107 (1981).
[38] N. L. Peterson, L. W. Barr, and A. D. LeClaire, *J. Phys. C* **6**, 2020 (1973).
[39] A. V. Chadwick and J. N. Sherwood, *J. C. S. Faraday I* **68**, Part 1, 47 (1972).
[40] R. Fox and J. N. Sherwood, *Trans. Faraday Soc.* **167**, 3364 (1971).

Bibliography

Radioactivity and Counting Statistics

D. J. Carswell, "Introduction to Nuclear Chemistry." Elsevier, New York, 1967.
R. D. Evans, "The Atomic Nucleus." McGraw-Hill, New York, 1955.
G. Friedlander, J. W. Kennedy, and J. M. Miller, "Nuclear and Radiochemistry," 2nd ed. Wiley, New York, 1964.
B. G. Harvey, "Introduction to Nuclear Physics and Chemistry," 2nd ed. Prentice-Hall, Englewood Cliffs, New Jersey, 1969.

Radioisotope Production

J. R. DeVoe, ed., "Modern Trends in Activation Analysis" NBS Spec. Publ. No. 312, 2 vols. Natl. Bur. Stand., Washington, D.C., 1969.
R. T. Overman and H. M. Clark, "Radioisotope Techniques." McGraw-Hill, New York, 1960.
L. C. L. Yuan and C.-S. Wu, eds., "Nuclear Physics," Methods of Experimental Physics, Vol. 5, Part B. Academic Press, New York, 1963.

Nuclear Data Sources

G. Erdtmann, Neutron Activation Tables. *In* "Kernchemie in Einzeldarstellungen," (K. H. Lieser, ed.) Vol. 6. Verlag Chemie, Weinheim, 1976.
General Electric Co., "Chart of the Nuclides," 11th ed. General Electric Co., Schenectady, New York, 1972.
C. M. Lederer, J. M. Hollander, and I. Perlman, "Tables of Isotopes," 7th ed. Wiley, New York, 1978.
Nuclear Data, a journal published in two sections: Section A, Nuclear Data Tables; Section B, Nuclear Data Sheets. Academic Press, New York.

Radioisotope Handling

"Radiological Health Handbook," rev. ed., U.S. Dep. Health, Educ. Welfare. U.S. Gov. Print. Off., Washington, D.C., 1970.
L. S. Taylor, *Health Phys.* **20,** 499 (1971).

Source Preparation

Subcommittee on Radiochemistry, Committee on Nuclear Science, National Academy of Sciences–National Research Council, "Source Material for Radiochemistry," Nuclear Science Series, Rep. No. 42 (1970 revision); available from NITS, U.S. Dep. Commer., Springfield, Virginia, 1970.
W. Parker and H. Slatis, *in* "Alpha, Beta, and Gamma Ray Spectroscopy" (K. Siegbahn, ed.), Vol. 1, Chap. VII. North-Holland Publ., Amsterdam, 1965.

Detection and Assay

R. D. Brooks, Organic scintillators. *Prog. Nucl. Phys.* **5,** 252 (1960).

C. E. Crouthamel, ed., "Applied Gamma Ray Spectrometry." Pergamon, New York, 1960.

G. Dearnaley and D. C. Northrup, "Semiconductor Counters for Nuclear Radiation," Spon, London, 1966.

E. Fergues and O. Haiman, "Physical Principles of Nuclear Radiation Measurements." Academic Press, New York, 1969.

G. F. Knoll, "Radiation Detection and Measurement." Wiley, New York, 1979.

P. Kruger, "Principles of Activation Analysis." Wiley (Interscience), New York, 1971.

W. E. Mott and R. B. Sutton, Scintillation and Cerenkov Counters. *In* "Handbuch der Physik," (S. Fluegge and E. Creutz, eds.) Vol. 45, p. 86. Springer-Verlag, Berlin and New York, 1958.

K. D. Neame and C. A. Homewood "Liquid Scintillation Counting." Wiley, New York, 1974.

K. Siegbahn, ed., "Alpha, Beta, and Gamma Ray Spectroscopy," Vols. 1 and 2. North-Holland Publ., Amsterdam, 1965.

L. C. L. Yuan and C.-S. Wu, eds., "Nuclear Physics," Methods of Experimental Physics, Vol. 5, Part A. Academic Press, New York, 1961.

2. EXPERIMENTAL METHODS OF POSITRON ANNIHILATION FOR THE STUDY OF DEFECTS IN METALS*

By L. C. Smedskjaer and M. J. Fluss

Materials Science and Technology Division
Argonne National Laboratory
Argonne, Illinois

2.1. Introduction

The use of the annihilation radiation from positron–electron pairs to study the nature of defects in metals has, over the past two decades, taken a definite place in the armamentarium of experimental techniques used by materials and solid state scientists. The most clear-cut successes have been in monitoring changes in defect properties as a function of temperature, pressure, or annealing treatments. New and more complex areas of investigation are also being explored, and one might reasonably expect that studies of the atomic and electronic structure of defects will be included among future experiments.

This part has two purposes:

(1) to introduce the materials scientist who is unfamiliar with the various positron techniques to the types of experiments that are possible and the required instrumentation, and

(2) to outline several special experimental topics of interest to specialists in the field, and to serve as a guide to new workers as well.

First we introduce the basic underlying theory and models for interpreting positron studies of defects in metals. Then each of the three major positron methods are discussed in turn: Doppler broadening, angular correlation, and lifetime experiments. A discussion of stabilization methods used in Doppler broadening and lifetime experiments is then presented. The goal is to provide a basic source of information for the materials scientist interested in defect studies so that he may evaluate the usefulness of this technique and begin to consider experimentation of his own.

---

* This work was supported by the U.S. Department of Energy.

2.2. Basic Theory

This chapter is intended to familiarize the reader with the basic properties and behavior of the positron. For a deeper study of the underlying theory see Hautojärvi[1] and West.[2]

2.2.1. The Positron

The positron is the antiparticle of the electron, and hence a positron–electron pair is unstable and will annihilate. In the annihilation process two γ rays are usually emitted. Annihilation resulting in three γ quanta is also possible, but the probability of this happening is reduced approximately by the fine-structure constant ($\frac{1}{137}$). Single-photon annihilation is also possible, but only if a third body is close enough to absorb the recoil. Normally this is improbable and is disregarded here.

The annihilation cross section was calculated by Dirac in 1930. Using his result in the nonrelativistic limit, one finds that the annihilation cross section between a positron with velocity v and an electron at rest is

$$\sigma = \pi r_0^2/(v/c), \quad v \ll c, \tag{2.2.1}$$

where r_0 is the classical electron radius. Consequently, for a positron embedded in a sea of "cold" electrons with a density n_e, one obtains the decay rate,

$$\lambda = \pi r_0^2 n_e c. \tag{2.2.2}$$

2.2.2. Positronium

The positron–electron pair can also form a quasi-stationary state called positronium (Ps). Apart from the effects due to annihilation, Ps is an analogue of the hydrogen atom. Positronium, therefore, is found either in the para state ($S = 0$, spins antiparallel) or in the triply degenerate ortho state ($S = 1$, spins parallel). Owing to conservation of spin momentum, the para positronium will decay by two-γ emission with a rate of $1/(125\text{ ps})$, and the ortho state will decay by three-γ emission at a rate of $1/(140\text{ ns})$. Positronium is generally not formed in metals, but is found in metal oxides, molecular solids, liquids, and gases.

---

[1] P. Hautojärvi, ed., "Positrons in Solids," Topics in Current Physics. Springer-Verlag, Berlin and New York, 1979.
[2] R. N. West, *Adv. Phys.* **22**, 263 (1973).

2.2.3. Electron Momenta

In the actual application of positron annihilation, it is the two-γ emission which is of interest. The probability of observing a photon pair of total momentum **K** is given by,

$$\rho_e(\mathbf{K}) = \frac{\pi r_0^2 c}{(2\pi)^3} \left| \int d^3x \, e^{-i\mathbf{K}\cdot\mathbf{x}} \, \psi_e(\mathbf{x})\psi_p(\mathbf{x}) \right|^2, \qquad (2.2.3)$$

where $\psi_e(\mathbf{x})$ and $\psi_p(\mathbf{x})$ are the wave functions of the electron and positron, respectively. The wave function $\psi_p(\mathbf{x})$ can usually be taken as the positron ground state in the lattice. This state can be described as a superposition of plane waves with wave vectors corresponding to thermal energies. Thus, the total photon momentum **K** is approximately equal to that of the electron since electrons have energies in the eV range. The decay rate of the positron is then given by $\lambda = \int d^3K \rho(\mathbf{K})$, where $\rho = \Sigma \, \rho_e$ is summed over all occupied states. It is seen from the above equation that the positron–electron overlap determines the annihilation properties.

Table I, along with Eq. (2.2.3), shows the ways in which positrons may be used as a probe of electron density in momentum space. Positron experiments fall into two categories: lifetime and momentum measurements. Lifetime experiments, as the name implies, involve only a determination of the lifetime of the positron in the material under study. Four different types of experiments, listed in Table I, provide momentum measurement in one, two, or three dimensions. The fundamental quantity measured in each case is presented in the last column of Table I.

It should be emphasized that **K** is never directly measured because the

TABLE I. The Positron Methods

Dimension of momentum	Experimental method	Physical quantity measured
0	Lifetime	$\lambda = \int d^3K \rho(\mathbf{K})$
1	One-dimensional angular correlation or Doppler broadening	$F(K_y) = \frac{1}{\lambda} \cdot \int dK_z dK_x \rho(\mathbf{K})$
2	Two-dimensional angular correlation	$F(K_y, K_z) = \frac{1}{\lambda} \int dK_x \rho(\mathbf{K})$
3	Three-dimensional position, and energy-sensitive γ-ray spectrometry	$F(K_x, K_y, K_z) = \frac{1}{\lambda} \rho(\mathbf{K})$

angular correlation equipment measures an angle θ_y between the γ quanta, where

$$\theta_y = \hbar K_y/mc, \qquad (2.2.4)$$

and the Doppler broadening measures an energy shift ΔE_y, where

$$\Delta E_y \simeq \hbar c K_y/2. \qquad (2.2.5)$$

For the positron to be an ideal probe, one requires that the electron states are unaffected by the presence of the positron; this is not the case for metals because the electrons screen out the positive charge. Thus the electron density is enhanced around the positron, resulting in a substantial increase in the decay rate. This problem could be of concern in the study of the details of the K space. Considerable effort has gone into the theoretical understanding of the interactions between the positron and the electrons.[3-5] Recently, positron annihilation angular correlation in mono- and divacancy defects in Al was calculated[6] using a self-consistent pseudopotential scheme based on a local-density functional formalism incorporating many-body enhancement effects. The basis for this method was the simulation of the vacancies by an unrelaxed 27-atomic-site supercell to which the band structure scheme was applied.[7-9] Corresponding experimental results of Berko and co-workers[10] have shown good agreement with these calculations.

2.2.4. Positrons in Defects

The use of the positron as a probe is based on our understanding of the positron's behavior in matter. In perfectly periodic structures, the positron and the positronium atom will be in Bloch states. Two-dimensional angular correlation experiments in quartz[3] and aluminum[3,5] demonstrated the delocalized nature of the positronium atom and the positron, respectively.

---

[3] S. Berko, M. Haghgooie, and J. J. Mader, *Phys. Lett. A* **63A**, 335 (1977).

[4] S. Wakoh, *Proc. 5th Int. Conf. Positron Annihilation, Lake Yamanaka, Japan, 1979* (R. R. Hasigawa and K. Fujiwara, eds.), p. 49. Jpn. Inst. Metals, Sendai, 1979.

[5] J. Mader, S. Berko, H. Krakauer, and A. Bansil, *Phys. Rev. Lett.* **37**, 1232 (1976).

[6] B. Chakraborty, *Phys. Rev. B* **24**, 7423 (1981).

[7] B. Chakraborty, *Proc. 6th Int. Conf. Positron Annihilation, Arlington, Texas, 1982* (P. G. Coleman, S. C. Sharma, and L. M. Diana, eds.), p. 207. North-Holland Publ., New York, 1982.

[8] B. Chakraborty, R. W. Siegel, and W. E. Pickett, *Phys. Rev. B* **24**, 5445 (1981).

[9] B. Chakraborty and R. W. Siegel, *Proc. 5th Yamada Conf. Point Defects Defect Interact. Met., Kyoto, Japan, 1981* (J. Takamura, M. Doyama, and M. Kiratani, eds.). p. 93. Univ. of Tokyo Press, Tokyo, 1982.

[10] M. J. Fluss, S. Berko, B. Chakraborty, K. Hoffmann, P. Lippel, and R. W. Siegel, *Proc. 6th Int. Conf. Positron Annihilation, Arlington, Texas, 1982* (P. G. Coleman, S. C. Sharma, and L. M. Diana, eds.) p. 454. North-Holland Publ., New York, 1982.

2.2. BASIC THEORY

In materials containing defects, the positron or Ps atom may localize in the defect. The positron will tend to localize in a defect if the density of positive ion cores (which are repulsive to the positron) is less than in the surrounding material. The Ps atom also tends to decrease its overlap with the surroundings and may localize if the space provided is sufficiently large. Since the electron-momentum distribution and density in a defect will be different from that of the perfect material, the annihilation properties will change if the positron annihilates while trapped at a defect.

In the case of Ps, a further explanation is required. In vacuum, ortho-Ps decays through three-γ emission. In dense matter, however, the Ps atom undergoes many collisions during its lifetime. Surrounding electrons compete with the original electron for the positron. This process, referred to as pickoff, has a rate $\sim 1/(1$ ns$)$, which is substantially faster than the vacuum decay rate of ortho-Ps. The pickoff process results in emission of two photons, as does the annihilation of the para-Ps. Pickoff will be less important for para-Ps because of the much higher decay rate of the para-Ps atom itself. The annihilation of the ortho state reflects the electron momenta as in the case of the positron, and the annihilation of the para-Ps mainly reflects the momentum of the Ps atom itself.

The application of positron annihilation to the study of defects in solids is thus based both on the ability of the positron to become trapped and on the changed electron density in the traps.

2.2.5. Positron Sources

Positrons are most commonly obtained from radioactive nuclei such as ^{22}Na, ^{58}Co, ^{64}Cu, and ^{68}Ge with end-point energies in the 500–2000-keV range.[11] A summary of end-point energies, half-lives, and positron yields is shown in Table II for these commonly used isotopes.[11] One can thus expect,

TABLE II. Commonly Used Positron-Emitting Isotopes

Isotope	End-point energy (MeV)	Half-life	Positrons per decay	Coincident γ yield
^{22}Na	0.54	2.6 yr	0.90	1.0
^{58}Co	0.47	71 d	0.15	1.0
^{64}Cu	0.66	12.7 h	0.19	0.0
^{68}Ge	1.9	275 d	~ 0.86	~ 0.0

[11] E. Browne, J. Dairiki, R. E. Doebler, A. A. Shihab-Eldin, L. J. Jardine, J. K. Tuli, and A. B. Buyrn, in "Table of Isotopes" (C. M. Lederer and V. S. Shirley, eds.), 7th ed., pp. 36, 170, 196, 216. Wiley, New York, 1978.

by virtue of these end-point energies, that the most energetic positron can penetrate the sample ~ 1 mm, although some positrons will stop or return to, and annihilate at, or near, the surface of the sample.

In addition to the sources mentioned in Table II, beams of slow positrons can be used for the study of metal surfaces or regions just below the surface. Positrons from a radioisotope are moderated by passage through a thin foil or metal oxide surface; if the work function of the moderator is negative, slow positrons will be emitted with an efficiency of $\sim 10^{-6}$ to 10^{-4}. These slow positrons, with energies of a few eV, are guided onto the sample by means of a magnetic or an electrostatic beam transport system. Due to their low energy, the positrons do not penetrate the sample deeply and may eventually diffuse back to the surface. At present, the intensities of such sources are so low that only the most efficient detection techniques can be used. In the future, one hopes to obtain intensities sufficient to permit experiments such as angular correlation, in which details of the surface electronic structure can be studied. Since metallurgical samples are often prepared in the form of thin foils, the use of beams would provide a means of studying the bulk of such samples without the need to stack several foils. A problem, of course, would be separating the bulk annihilation signal from the near-surface annihilation signal, the latter coming from positronium formed at or near the surface or positrons trapped at the surface.

2.2.6. Thermalization of the Positron

The positron is thermalized in a sample within a few picoseconds of its injection, according to the theoretical estimates made by Perkins and Carbotte.[12] Experimentally, Kubica and Stewart[13] demonstrated that positrons in metals reach thermal, or very near thermal, velocities prior to annihilation, even at sample temperatures as low as 4.2 K. Kubica and Stewart considered the smearing of the positron–electron momentum near the Fermi edge. Close to the Fermi edge the electron-momentum distribution is sharp, and an observed smearing of the data could be interpreted as a result of the finite momentum of the positron itself.

2.2.7. Trapping of Positrons

Defect studies utilizing the positron depend on its ability to become trapped; hence the fundamental question: How does the positron become trapped in a defect? Immediately after thermalization the positron is in the free or untrapped state. This state is commonly described as a superposition of Bloch waves such that the positron has zero probability of being near the

---

[12] A. Perkins and J. P. Carbotte, *Phys. Rev. B* **1**, 101 (1970).
[13] P. Kubica and A. T. Stewart, *Phys. Rev. Lett.* **34**, 852 (1975).

2.2. BASIC THEORY

positive ion cores and maximum probability of being in the interstices. This free or delocalized state has the character of Swiss cheese.[14,15] In the localized or trapped state, which has an energy below that of the free state, the positron probability density is nonzero only in the vicinity (within a few atomic distances) of the defect in which the positron is trapped. Examples of such localized states are a positron trapped in a vacancy or in a dislocation. In the vacancy the positron is localized mainly to the atomic volume containing the vacancy; in the dislocation the positron may be localized in the plane perpendicular to the dislocation line and delocalized in the direction along the dislocation line. Thus the trapped state of the positron depends on the defect in which the positron is localized.

In studying the transition from the delocalized to the localized state one considers only those mechanisms that conserve energy and momentum for the entire system (positron and sample). For large binding energies ($E_b \sim$ 1 eV) between the positron and the defect this is accomplished by electron–hole pair emission,[16,17] where, for example, the transition for shallow traps ($E_b < 0.1$ eV) may be accompanied by phonon emission.[18]

Once the positron is trapped one might consider the possibilities of subsequent detrapping. In many metals this can be disregarded because of the high binding energy of the positron with the vacancy,[16,17,19] with the major exceptions being the alkali metals. However, for positrons trapped in shallow traps, such as dislocations, detrapping is very important.[18] A complete discussion of detrapping has been given by Manninen and Nieminen.[20]

In addition to the question of the transition from the delocalized to the localized state, one must also deal with the problem of how the positron arrives at the trap, that is, the positron's mobility or diffusion, a process that has been considered both classically and quantum mechanically. For low transition rates (weak trapping), the presence of the defect does not affect the initial (delocalized) positron wave function, thereby permitting the "golden rule" to be used when calculating the specific trapping rate. This is often referred to as transition-limited trapping and is assumed to be valid for positron trapping in vacancies.[21–23] If, on the other hand, the transition rate is high, motion-limited trapping may occur. In this case the rate is so high

---

[14] P. Kubica and M. J. Stott, *J. Phys. F* **4**, 1969 (1974).
[15] M. J. Stott and P. Kubica, *Phys. Rev. B* **11**, 1 (1975).
[16] C. H. Hodges, *Phys. Rev. Lett.* **25**, 284 (1970).
[17] M. Manninen, R. Nieminen, P. Hautojärvi, and J. Arponen, *Phys. Rev.* **12**, 4012 (1975).
[18] L. C. Smedskjaer, M. Manninen, and M. J. Fluss, *J. Phys. F* **10**, 2237 (1980).
[19] R. P. Gupta and R. W. Siegel, *Phys. Rev. Lett.* **39**, 1212 (1977).
[20] M. Manninen and R. Nieminen, *Appl. Phys.* **26**, 93 (1981).
[21] B. Bergersen and D. W. Taylor, *Can. J. Phys.* **52**, 1594 (1974).
[22] T. McMullen and B. Hede, *J. Phys. F* **5**, 669 (1975).
[23] T. McMullen, *J. Phys. F* **7**, 2041 (1977); **8**, 87 (1978).

that a zone depleted of positrons is created around the defect, and further trapping depends on the mobility of the positron. McMullen[23] has developed a quantum-mechanical treatment of both motion- and transition-limited trapping for positrons in vacancies that predicts only a slight temperature dependence of the specific trapping rate. A classical diffusion picture of the positron-trapping process has been given by Frank and Seeger[24] and by Seeger[25] leading to a stronger temperature dependence of the specific trapping rate.

For extended defects (e.g., dislocations, voids, or surfaces), temperature-dependent specific trapping rates may be anticipated either because of a temperature dependence of the transition rate itself or because of motion-limited trapping or a combination of these effects. Although a consensus seems to have been reached as to vacancies, the temperature dependence of the specific trapping rate for positrons into extended defects is still disputed. Unfortunately, experimental results[26-29] do not lead to a firm conclusion regarding the specific trapping rate of positrons into vacancies, dislocations, or voids because questions about the nature and types of defects present in the samples can be raised. Samples prepared by electron irradiation and subsequent annealing below stage III, where vacancies are not mobile (e.g., Mantl et al.[28]) may contain dislocation loops in addition to vacancies. This may also be true for samples with quenched-in vacancies (e.g., Hall et al.[26] and McKee et al.[27]). Samples containing voids often also show evidence for the presence of dislocation loops (e.g., Pagh et al.[30]). The presence of defect types other than those intended to be studied results in competitive trapping, yielding a change in the temperature dependence of the observed trapping rate. The role of competitive trapping between voids and dislocations loops in Mo have been studied by Pagh et al.[30] It is found that the temperature dependence of the apparent positron-trapping rate into voids is dependent strongly on the presence of the dislocation loops. Furthermore, studies of the positron-trapping rate at low temperatures may be affected by prevacancy effects, which are sometimes observed even in supposedly well-annealed samples. In the presence of prevacancy effects a strong curvature in

---

[24] W. Frank and A. Seeger, *Appl. Phys.* **3**, 61 (1974).
[25] A. Seeger, *Appl. Phys.* **7**, 257 (1975).
[26] T. M. Hall, A. N. Goland, K. C. Jain, and R. W. Siegel, *Phys. Rev. B* **12**, 1613 (1975).
[27] B. T. A. McKee, H. C. Jamieson, and A. T. Stewart, *Phys. Rev. Lett.* **31**, 634 (1973).
[28] S. Mantl, W. Kesternich, and W. Triftshäuser, *J. Nucl. Mater.* **69/70**, 593 (1978).
[29] R. M. Neiminen, J. Laakkonen, P. Hautojärvi, and A. Vehanen, *Phys. Rev. B.* **19**, 1397 (1979).
[30] B. Pagh, H. E. Hansen, B. Nielsen, G. Trumpy, and K. Petersen, *Proc. 6th Int. Conf. Positron Annihilation, Arlington, Texas, 1982* (P. G. Coleman, S. C. Sharma, and L. M. Diana, eds.), p. 398. North-Holland Publ. New York, 1982.

the temperature dependence of the positron annihilation signal (e.g., line shape) is observed at low temperatures. The curvatures are too strong to be explained by the current theories (e.g., Stott and West[31] and Tam et al.[32]) taking both the static lattice expansion and positron–phonon interaction into account. The origin of the prevacancy effects is still disputed, and the reader is referred to Smedskjaer[33] for a more detailed discussion, where it is pointed out that the prevacancy effects are of an extrinsic nature and thus not characteristic for positron annihilation in a perfect lattice.

Although some of the detailed properties of positron annihilation in extended defects are not yet fully understood, the basic properties of the positron–defect interaction are known well enough that the utilization of positron annihilation as a practical tool for defect studies is presently realized.

2.2.8. The Trapping Model

The basic theoretical tool in interpreting positron data is the trapping model, which was first described by Brandt,[34] Bergersen and Stott,[35] and Conners and West.[36] In its simplest form one considers a single type of defect in an otherwise perfect lattice, and the positron then either occupies the ground state in the lattice (the free state) or it is trapped. The transition rate from the free to the trapped state is thought to be proportional to the defect concentration. This model is expressed by two coupled differential equations,

$$\frac{dn_b}{dt} = -\lambda_b n_b - \kappa n_b, \qquad (2.2.6)$$

$$\frac{dn_t}{dt} = -\lambda_t n_t + \kappa n_b, \qquad (2.2.7)$$

where n_b and n_t are the probabilities that the positron is found in the free or trapped state, respectively. The quantities λ_b and λ_t are the decay rates of the free and trapped positron, and κ signifies the concentration-dependent trapping rate. In a lifetime experiment the decay rates are not directly observed, but the quantity $-dN(t)/dt$ is (assuming infinite instrument

---

[31] M. J. Stott and R. N. West, *J. Phys. F* **8**, 635 (1978).

[32] S. W. Tam, S. K. Sinha, and R. W. Siegel, *J. Nucl. Mater.* **69/70**, 596 (1978); *J. Nucl. Mater.* **101**, 242 (1981).

[33] L. C. Smedskjaer, *Proc. Int. Sch. Phys. "Enrico Fermi,"* *1981* (to be published).

[34] W. Brandt, in "Positron Annihilation" (A. T. Stewart and L. O. Roellig, eds.), p. 155. Academic Press, New York, 1967.

[35] B. Bergersen and M. J. Stott, *Solid State Commun.* **7**, 1203 (1969).

[36] D. C. Conners and R. N. West, *Phys. Lett. A* **30A**, 24 (1969).

resolution), where $N(t)$ is the probability that a positron exists in the system at time t. The quantity $N(t)$ is given by the expression,

$$N(t) = I_1 e^{-\Lambda_1 t} + I_2 e^{-\Lambda_2 t}, \qquad (2.2.8)$$

where

$$I_1 + I_2 = 1, \qquad (2.2.9)$$

$$I_2 = \kappa/(\lambda_b + \kappa - \lambda_t), \qquad (2.2.10)$$

and

$$\Lambda_1 = \lambda_b + \kappa, \qquad \Lambda_2 = \lambda_t \qquad (2.2.11)$$

(see, e.g., Fluss et al.[37,38]). In angular correlation and Doppler broadening one observes the probability that a positron will annihilate while in a defect, which is given by

$$A_t = \kappa/(\lambda_b + \kappa). \qquad (2.2.12)$$

For angular correlation experiments this is done by observing $F(K_y = 0)$ (peak counting); in Doppler broadening by calculating the lineshape S,

$$S = \int_{-K'_y}^{K'_y} F(K_y) \, dK_y. \qquad (2.2.13)$$

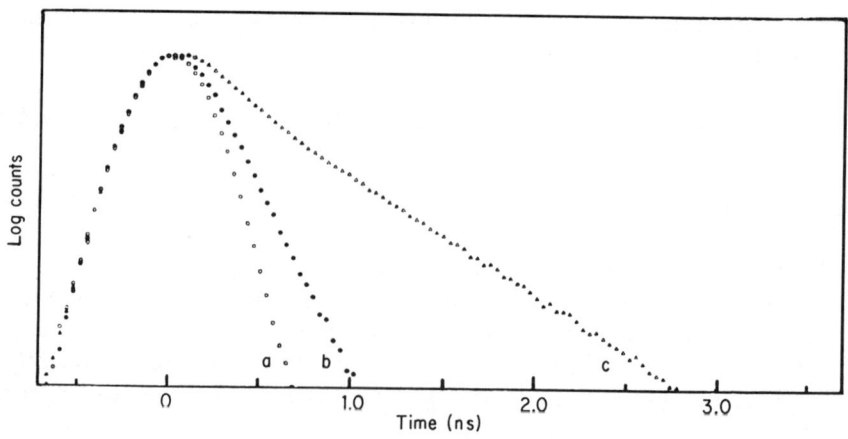

FIG. 1. Lifetime spectra obtained for Mo.[39] Curve a is the instrumental resolution function, b is obtained for an annealed sample ($\tau \sim 146$ ps), and c is a consequence of positron trapping in voids ($\tau \sim 467$ ps).

[37] M. J. Fluss, L. C. Smedskjaer, M. K. Chason, D. G. Legnini, and R. W. Siegel, *Phys. Rev. B* **17**, 3444 (1978).
[38] M. J. Fluss, R. P. Gupta, L. C. Smedskjaer, and R. W. Siegel, in "Positronium and Muonium Chemistry" (H. J. Ache, ed.), Advances in Chemistry Series, No. 175, p. 243. Am. Chem. Soc., Washington, D.C., 1979.

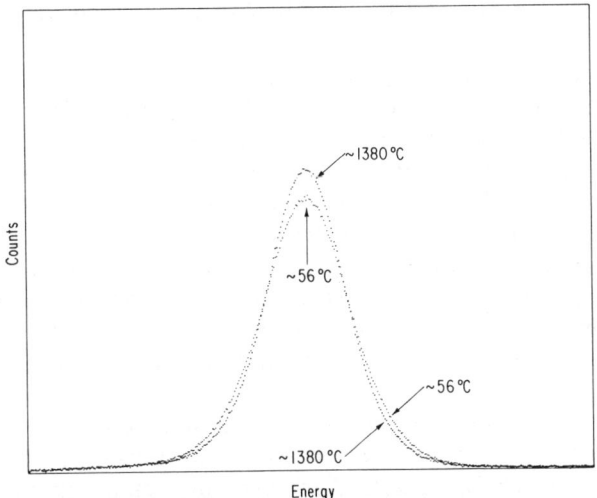

FIG. 2. Doppler-broadening spectra for positron annihilation in Ni(0.3 at. % Ge) showing annihilation in the perfect lattice (56°C) and in the vacancy (1380°C), respectively. The dispersion is ~ 55 eV/channel.

The relation between the line shapes or peak counts and the above probability is given by

$$S = (1 - A_t)S_b + A_tS_t, \qquad (2.2.14)$$

where S_b and S_t are the shapes characteristic of the free and trapped positron states. Note that Eq. (2.2.13) assumes infinitely good instrumental resolution. The effects of trapping are shown for two metals in Figs. 1 and 2. In Fig. 1 lifetime data are shown for Mo,[39] the shorter lifetime is that of the untrapped positron, and the longer lifetime is that of a positron trapped in a void. Also shown in the figure is the instrumental resolution function for comparison with the overall spectrum. In Fig. 2 Doppler-broadening data are shown for Ni(Ge), the broad (56°C) curve being that of the free positron, and the narrow (1380°C) curve is for the positron trapped in a vacancy.

2.3. Positron Studies and Metal Physics

In this section we describe the types of studies in the area of metal physics where the positron can provide information on defects. A review by Siegel[40] provides an overview of current applications of positron annihilation to the

---

[39] R. M. J. Cotterill, I. K. MacKenzie, L. Smedskjaer, G. Trumpy, and J. H. O. L. Tráff, *Nature (London)* **239**, 101 (1972).
[40] R. W. Siegel, *Annu. Rev. Mater. Sci.* **10**, 393 (1980).

study of defects in metals and includes an extensive bibliography to which we refer the reader.

There is a complex interplay between the interpretation of positron experiments and the understanding of the properties of the material involved. One often must have a priori information about the material and its history to interpret the positron experiment. Several examples of this interrelation are given here. The various uncertainties in the two-state trapping model (e.g., the uncertainty in the temperature dependence of the specific trapping cross section) yield a corresponding uncertainty in the accuracy of the materials information that can be deduced from positron experiments.[38] A discussion of this interrelation between theoretical uncertainties and the resulting accuracy of materials parameters can be found in the review by Fluss et al.,[38] which focuses on the problem of measuring the vacancy formation enthalpy in metals by positron annihilation spectroscopy.

The application of a simple two-state trapping model to complex systems of defects requires both theoretical and experimental information from the materials side of the problem, as well as from the positron side. As an example of this interplay, consider temperature-dependent lifetime experiments, interpreted under a particular trapping model, that contain experimental information about both the temperature dependence of the equilibrium trapping-site concentration and the temperature dependence of the bulk- and vacancy-trapped lifetimes.

Similarly, some nonequilibrium temperature-dependent studies using quenched-in trapping centers have been performed to measure the temperature dependence of the trapping-rate term. The ultimate interpretation, however, requires significant guidance from metal physics theory and experiment. The interpretation of nonequilibrium experiments aimed at determining the temperature dependence of the trapping rate depends on one's a priori knowledge about the type of trapping centers (defects) that are present and how they behave during the course of the experimental temperature changes.

As an additional example of the interplay of positron and metallurgical theory, consider the observed negative temperature dependence of the vacancy lifetime in Pb.[41] One might expect to have a quantitative understanding of these results if theory could predict the expected temperature-dependent relaxation of the vacancy defect with the positron present. However, this calculation is made somewhat more difficult as a consequence of the theoretical observation[42] that noted that the trapped positron exerts a significant outward pressure.

The nature of the positron to become trapped at voids (clusters of

[41] S. C. Sharma, S. Berko, and W. K. Warburton, *Phys. Lett. A* **58A**, 405 (1976).
[42] S. W. Tam and R. W. Siegel, *J. Phys. F* **7**, 877 (1977).

vacancies) provides a useful source of qualitative information that is usually interpreted by the materials scientist in conjunction with ancillary experiments, e.g., annealing studies, resistivity measurements, and electron microscopy. For those cases where it is believed that voids may be present, an increase or decrease in the positron lifetime is often attributed to an associated growth or shrinkage of the voids. The possibility has been raised that local impurities at the surfaces of the void may result in significant changes in the positron lifetime as well.[43] Thus it may be important to consider the motion of impurities as well as the interstitials and vacancies when using the positron as a probe in annealing studies.

2.3.1. Surface Studies

The study of positron annihilation in voids of large diameter can be viewed as a study of internal surfaces. Studies of external surfaces, however, can be performed using slow-positron beams of sufficient intensity first developed for studying atomic physics by Canter et al.[44] In this technique the sample is bombarded with low-energy monoenergetic ($\sim 0-30$ eV) positrons, which remain near the surface. The positrons can either annihilate in this region or form positronium at the surface. By detecting the three-quantum decay of ortho-Ps, one obtains a measure of the combined probability of positronium formation at the surface and of a positronium atom leaving the surface. Such experiments can provide information about the surface conditions of the sample, although positron behavior at surfaces is only rudimentarily understood. If the positron is implanted below the surface, the slow-beam method can also provide information about the defects in the near-surface region. In this case, the positronium formation will depend on the probability that the positron diffuses back to the surface. This probability will be reduced if positron traps are present in the near-surface region. The application of slow beams to defect studies at the surface and in the near-surface layer has been reviewed by Lynn.[45] Mills[46] has given an extensive review with respect to the more general applications of the slow-beam technique.

2.3.2. Vacancy Formation Enthalpy

By far the most active area of positron research in recent years has been the behavior of positrons in metals. Since it was first realized through the

---

[43] B. Nielsen, A. van Veen, L. M. Caspers, H. Filius, H. E. Hansen, and K. Petersen, *Proc. 6th Int. Conf. Positron Annihilation, Arlington, Texas, 1982* (P. G. Coleman, S. C. Sharma, and L. M. Diana, eds.), pp. 438. North-Holland Publ., New York, 1982.
[44] K. F. Canter, A. P. Mills, and S. Berko, *Phys. Rev. Lett.* **33**, 7 (1974).
[45] K. G. Lynn, *Proc. Int. Sch. Phys. "Enrico Fermi," 1981* (to be published).
[46] A. P. Mills, *Proc. Int. Sch. Phys. "Enrico Fermi," 1981* (to be published).

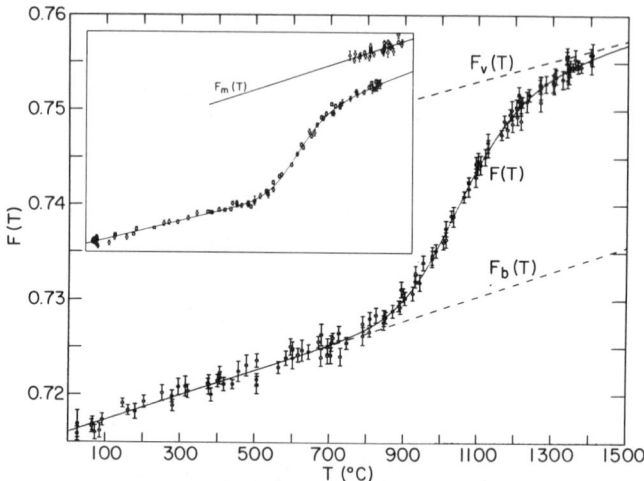

FIG. 3. A typical Doppler-broadening peak-counting parameter, F versus temperature is illustrated for the case of Ni[50]. The two-state trapping model parameters deduced from the least-squares fit to the data are indicated as described in the text. The data in the inset include annihilations from the liquid metal as well.[51]

experiments of MacKenzie et al.[47] and Berko and Erskine[48] that the positron was sensitive to vacancies, the technique has been widely applied to thermal-equilibrium measurements of the vacancy formation enthalpy in metals. It is pointed out by Triftshäuser,[49] that one of the features of the positron technique is its sensitivity over a broader temperature range as compared to the more usual thermal-equilibrium technique of differential dilatometry. Since the application of positron annihilation is of importance for thermal-equilibrium studies, a detailed discussion of how a vacancy formation enthalpy, H_v, can be deduced from Doppler-broadening data will be given. A typical line-shape parameter, $F(T)$ (defined in Section 2.4.4.2, for Ni[50,51] is shown in Fig. 3 as a function of temperature T. It will be assumed that

1. The vacancy concentration is $C_v(T) = \exp\{(TS_v - H_v)/kT\}$, where S_v is the formation entropy.
2. The trapping model describes transitions of positrons from the deloca-

---

[47] I. K. MacKenzie, T. L. Khoo, A. B. MacDonald, and B. T. A. McKee, *Phys. Rev. Lett.* **19**, 946 (1967).

[48] S. Berko and J. C. Erskine, *Phys. Rev. Lett.* **19**, 307 (1967).

[49] W. Triftshäuser, *Festkörperprobleme* **15**, 381 (1975).

[50] L. C. Smedskjaer, M. J. Fluss, D. G. Legnini, M. K. Chason, and R. W. Siegel, *J. Phys. F* **11**, 2221 (1981).

[51] M. J. Fluss, L. C. Smedskjaer, B. Chakraborty, and M. K. Chason, *J. Phys. F* in press (1983).

2.3. POSITRON STUDIES AND METAL PHYSICS

lized states (in the perfect lattice) into the localized (defect-trapped) states.

3. The trapping rate of the positron into traps other than vacancies can be disregarded.

4. The line shapes $F_v(T)$ and $F_b(T)$ characteristic of positrons annihilating in the vacancy and the bulk (perfect lattice), respectively, can be approximated by $F_v(T) = F_{v0}(1 + \beta T)$ and $F_b(T) = F_{b0}(1 + \alpha T)$, where F_{v0}, F_{b0}, β, and α are constants.

5. The temperature dependence of the specific trapping rate, μ_0, is small in comparison with that of $C_v(T)$.

The first assumption is fulfilled for well-annealed high-purity samples. Present experience shows that the second assumption is also a good approximation, and the third assumption, based upon available experimental and theoretical work, is reasonable for well-annealed high-purity samples. The validity of these assumptions is confirmed indirectly through the excellent fits between observed data, $F(T)$, and the model derived from the assumptions. The observed line shape can be written

$$F(T) = (1 - A_t)F_b(T) + A_t F_v(T),$$

where A_t is the trapping probability. By using assumption 3 and invoking the trapping model one obtains

$$A_t = \frac{\kappa}{\kappa + \lambda_b}, \quad (2.3.1)$$

where, using assumption 3,

$$\kappa = \mu_0 C_v(T) \quad (2.3.2)$$

and λ_b is the positron decay rate in the bulk. The temperature dependence of λ_b is generally known, but in the determination of H_v from Doppler-broadening data this dependence can be disregarded. Using assumption 4 the model can be written in a form ideally suited for least-squares analysis:

$$F(T) = \frac{1}{1+Q}F_b(T) + \frac{Q}{1+Q}F_v(T),$$

$$F_b(T) = F_{b0}(1 + \alpha T), \quad (2.3.3)$$

$$F_v(T) = F_{v0}\{1 + \beta(T - T')\},$$

where

$$Q = \exp\left\{\frac{-H_v(1/T - 1/T_0)}{k}\right\}, \quad T_0 = \frac{H_v}{S_v + k\ln(\mu_0/\lambda_b)}, \quad (2.3.4)$$

and T' is introduced as a constant temperature near the melting point in order to expedite the convergence of the fitting program. The parameters to be determined by the least-squares fitting procedure are F_{b0}, F_{v0}, α, β, T_0, and H_v (T_0 is the temperature at which half of the positrons become trapped prior to annihilation). Clearly, S_v, and therefore $C_v(T)$, cannot be deduced unless λ_b and μ_0 are known. The limiting factor is μ_0, which is commonly estimated to be $10^{14}-10^{15}$ s^{-1} for most metals. Thus the data shown in Fig. 3 do not contain any independent information regarding the absolute vacancy concentration, but only contain information regarding the temperature dependence of $C_v(T)$. This is not specific for positron annihilation, but it will be the case for any method where a probe–defect cross section appears in the model (e.g. resitivity measurements). The advantage of the positron method is its ability to be used on samples in thermodynamic equilibrium. When broad temperature ranges are considered, this raises the fundamental question of whether only the defect concentration depends on temperature, or whether the defect structure itself is also temperature dependent [i.e., $F_v(T)$]. Although such questions might be addressed when interpreting positron data, this additional complication is not inherent in the positron physics but rather in the defect properties themselves. One might therefore anticipate that future applications of the method can improve our understanding of the temperature dependence of defect structures. A possible example of future investigations with positron annihilation radiation is also indicated in Fig. 3. The data along the line labeled $F_m(T)$ were obtained from positron annihilation in liquid Ni using an ion-implanted ^{58}Co positron source. Comparative studies of the high-temperature defect states and those of corresponding disordered systems such as liquids and amorphous alloys promise improved insight into the structural (dynamic and static) characteristics of vacancy type defects.

2.3.3. Vacancy–Impurity Binding

Positron annihilation may also be used in the study of vacancy–impurity binding. The major effect of binding between impurities and vacancies is an increase in the vacancy concentration, and this effect can be detected using positrons. The vacancy–impurity binding energy can be deduced by applying the Lomer model,[52] as discussed by Doyama.[53] In the work by Hehenkamp and Sander[54] it was pointed out that in sufficiently concentrated alloys the positron annihilation technique might provide information with regard

---

[52] W. M. Lomer, in "Point Defects and Diffusion in Metals and Alloys," Monogr. Rep. Ser., No. 23, p. 85. Inst. Met., London, 1958.
[53] M. Doyama, *J. Nucl. Mater.* **69/70**, 350 (1978).
[54] T. Hehenkamp and L. Sander, *Z. Metallkd.* **70**, 202 (1979).

to the binding energies of vacancies bound to multiple impurity atoms when the Dorn–Mitchell model[55] is applied in the interpretation. However, when applying either the Lomer model or the more general Dorn–Mitchell model, one must make assumptions about the specific trapping rate of the positron into a free vacancy compared to a (multiple) impurity-bound vacancy. Thus in the study of complex metallurgical systems positron annihilation by itself is unlikely to provide a full theoretical understanding and should therefore be considered as a complementary tool.

2.3.4. Annealing Studies of Defects

The application of positron annihilation in annealing studies has been of great interest to metallurgists because annealing behavior is a recognized standard technique for the study of defects. When comparing positron annihilation to, for example, resistivity measurements, one notes that although electrons are scattered by both interstitial and vacancylike defects, the positron is trapped only by vacancylike defects. Thus in this sense the positron is defect specific.

As an example of an annealing experiment utilizing the defect specificity of the positron one might consider the work of Petersen et al.[56] Neutron-irradiated molybdenum was annealed and the line shape and lifetime measured as functions of annealing temperature. The results are shown in Fig. 4. The lifetime spectra were resolved into the two lifetimes τ_1 and τ_2, along with the relative intensities I_1 and I_2 ($= 1 - I_1$) of the respective lifetimes. Considering the complexity of the defect structure present in the neutron-irradiated sample, the resolved lifetimes are unlikely to be due to annihilations in a single type of defect, but should rather be interpreted as average values for the defect population present.

The general increase in τ_2 until $\sim 900°C$ may be interpreted as a result of the formation of vacancy clusters of increasing size (voids). The signature for the presence of voids, namely, the exceptionally long lifetime $\tau_2 > 400-500$ ps, was first discovered by Cotterill et al.[39] and has been extensively used in conjunction with the annealing of irradiated metal samples. A review of this has been given by Petersen.[57]

Besides illustrating an annealing experiment, Fig. 4 also illustrates an important difference in the use of the various positron annihilation techniques. The behavior of τ_2 versus annealing temperature is the key to understanding the annealing experiment shown in Fig. 4. In contrast to the lifetime, the change of the line-shape parameters exhibits less structure

---

[55] J. E. Dorn and J. B. Mitchell, *Acta Metall.* **14,** 70 (1966).
[56] K. Petersen, N. Thrane, and R. M. J. Cotterill, *Philos. Mag.* **29,** 9 (1974).
[57] K. Petersen, *Proc. Int. Sch. Phys. "Enrico Fermi," 1981* (to be published).

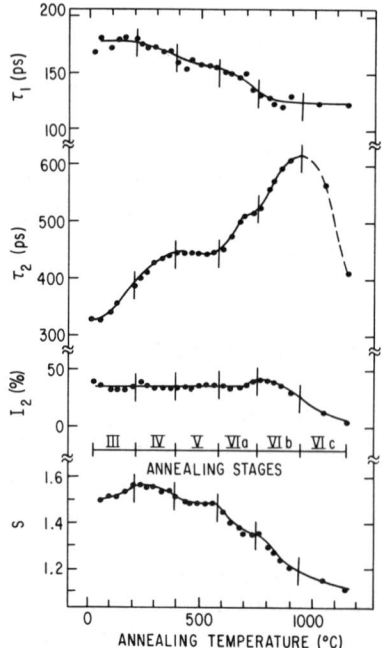

FIG. 4. Combined lifetime and Doppler-broadening annealing study of neutron-irradiated Mo.[56] The two resolved lifetimes τ_1 and τ_2, and the intensity I_2 (of τ_2) are shown along with the line-shape parameter S versus annealing temperature. Note the difference in the response to annealing between the Doppler broadening (S) and the lifetime (τ_2). The annealing stages, as given in Petersen et al.,[56] are indicated.

during the anneal. Based upon the behavior of S alone, one would probably not have reached the conclusion that voids are formed during the course of the annealing. Thus it is an advantage to measure the lifetime spectrum and the line shape since the lifetime tends to distinguish more clearly between different types of defects.

2.3.5. Electron Structure

Although the Fermi surfaces of metals have been studied for some years with existing one-dimensional techniques, the use of two-dimensional angular correlation instruments offers increased speed and sensitivity. As a technique for the study of the Fermi surface of metals and alloys, positron annihilation does not suffer from the limitations imposed on other techniques from impurity and defect scattering, such as in the case of de–Haas–Van Alphen measurements. The promise of the positron technique is in its applications to disordered systems, a topic which has been reviewed by Berko[58] and Minjarends.[59] The two-dimensional angular correlation tech-

---

[58] S. Berko, in "Electrons in Disordered Metals and at Metallic Surfaces" NATO Advanced Study Institute Series B Physics, vol. 42 (P. Phariseau, B. L. Gyoerffy, and L. Scheire, eds.), p. 239. Plenum, New York, 1979.

[59] P. E. Minjarends, in "Positrons in Solids" (P. Hautojärvi, ed.), Topics in Current Physics, p. 25. Springer-Verlag, Berlin and New York, 1979.

nique may also have important future applications in the study of defects: the details of the electron structure of the defects, their dimensions, and various properties as a function of temperature (particularly the very-high-temperature properties, that may shed new light on the processes involved in mass transport) can be studied by this technique. So far, only preliminary experiments have been reported by two groups.[10,60]

In summary, the positron technique is well established as a method for the study of vacancylike defects. It is used in a variety of defect studies, in both equilibrium and nonequilibrium situations. The usefulness of the technique derives from its defect specificity, that is, the positron's ability to localize in vacancylike defects. The eventual extension of the positron studies to more complex systems and/or more detailed studies of defect structure awaits the corresponding development of theoretical tools related to the electronic structure of defects.

2.4. Detection of Annihilation Radiation

The detection and measurement of γ radiation is in itself a complete topical area. The reader is referred to several books and monographs on γ-ray spectroscopy and counting equipment for more detailed discussions of γ-ray detection techniques.[61-63] For the particular case of positron annihilation the problem is reduced to the detection and measurement of simple γ-ray spectra from the commonly used radionuclides. Here, our purpose is to emphasize special problems attendant to the positron technique.

2.4.1. Resolution Considerations

The three basic techniques of positron annihilation studies as applied to metals are angular correlation, Doppler broadening, and lifetime measurements. Each measures a different manifestation of the positron's characteristics in the material. Thus each technique places different restrictions on the nature of the detection apparatus that is to be used. Three fundamentally different kinds of resolution are considered when choosing a detector system:

---

[60] A. Alam and R. N. West, *Proc. 5th Yamada Conf. Point Defects Defect Interact. Met., Kyoro, Japan, 1981* (J. Takamura, M. Doyama, and M. Kiratani, eds.), p. 228. Univ. of Tokyo Press, Tokyo, 1982.

[61] K. Siegbahn, ed., "Alpha-Beta-Gamma Ray Spectroscopy," Vols. 1 and 2. Am. Elsevier, New York, 1968.

[62] G. F. Knoll, "Radiation Detection and Measurement." Wiley, New York, 1979.

[63] G. G. Eichholz and J. W. Poston, "Principles of Nuclear Radiation Detection." Ann Arbor Sci. Publ., Ann Arbor, Michigan, 1979.

TABLE III. Importance of Various Resolutions

Technique	Resolution[a]		
	Energy	Timing	Spatial
Angular correlation	2	3	1
Doppler broadening	1	3	—
Lifetime	3	1	—

[a] Key: 1 = very important, 2 = moderately important, 3 = not important, and — = not applicable.

(1) timing resolution,
(2) energy resolution, and
(3) spatial resolution.

Table III indicates, although somewhat arbitrarily, the relative importance of these features for each of the three techniques.

2.4.1.1. Angular Correlation. For angular correlation experiments spatial resolution is the primary consideration. Spatial accuracy and precision are achieved by mechanical means (detectors moving through space) or by electronic means (either arrays of small detectors or by schemes for readouts of position coordinates from continuous position-sensitive detectors such as the Anger camera or the multiwire gas proportional counter). Energy resolution is also important in this technique, since it is a primary means of discriminating against the ever present scattered events that, if included, would lead to a degradation of the spatial resolution. The timing resolution, used to discriminate against random events, is usually not a principal consideration. However, at high count rates, or if one desires to study the highest momentum of the distribution where the coincident-to-singles ratio is the poorest, the timing resolution may have to be considered because of its effect on the true-to-chance-coincidence ratio.

2.4.1.2. Doppler Broadening. Doppler-broadening measurements are limited by energy resolution; the observed broadening due to the momentum of the positron–electron pair is of the order of the instrumental resolution. Because it is easy to achieve high statistics in short periods, it is possible to define spectral parameters that are very precisely known, although their sensitivity is reduced by the broad resolution function. Timing properties are considered in some cases, e.g., where a 511–511 keV coincidence is to be used to remove the Compton background of higher-lying γ-ray lines.

2.4.1.3. Lifetime. Lifetime measurements depend on a precise measurement of the time between positron birth (signaled by a coincident γ ray) and positron annihilation (signaled by the detection of one of the two 511-keV γ

2.4. DETECTION OF ANNIHILATION RADIATION

rays). Energy information is required in order to select those events that fall in the optimal operating range of the timing electronics and represent the desired combination of birth and annihilation γ rays. Thus, for a given event, the birth and annihilation γ rays are signaled by different detectors. The energy resolution, however, is limited since the spectrum obtained by use of a fast plastic scintillator such as Pilot-U consists of the Compton spectra of the incident radiation. The energy information plays an important role in controlling the timing resolution and removing events arising from scattering processes and the like.[64]

2.4.2. Counting Rate and Resolution

In all three kinds of positron experiments one often encounters a trade-off of resolution for count rate. For example, in a lifetime experiment one may obtain an excellent resolution function (e.g., full width at half maximum (FWHM) less than 200 ps) by using narrow energy windows for the pulses that are accepted by the system. However, the final count rate is proportional to the width of the energy windows, which means that the resolution is obtained at the cost of lower count rate. Conversely, broad energy windows give a better count rate but a poorer resolution function for the instrument.

The statistical uncertainty with which a given physical parameter (e.g., a lifetime) is determined will in general increase with increasing width of the resolution function; it will decrease with the square root of the total counts in the spectrum. Thus, for a given set of parameters that are to be determined, there exists an optimal balance between resolution and counting efficiency. It must be emphasized that this balance will depend strongly on the parameters that are considered, as well as their anticipated magnitudes. It is, therefore, impossible to give any general rules for determining this optimal balance, except to point out that solutions on a case-by-case basis may be obtained by computer simulation. Some approximate rules may be given for Doppler broadening that show how counts and resolution are traded off against one another without any loss of information, thereby allowing the experiments to be completed as quickly as possible.

2.4.3. Detector Systems

The detectors used in positron experiments are, in themselves, nontrivial systems. The most well known of these types of detector systems is the NaI(Tl) scintillator coupled to an appropriate phototube. Because of good efficiency and modest energy resolution, such detector systems are suitable

---

[64] L. Dorikens-Vanpraet, D. Seegers, and M. Dorikens, *Appl. Phys.* **23**, 149 (1980).

for angular correlation experiments. A more complex application of the NaI(Tl) detector and phototube is the position-sensitive Anger camera, which when used in a coincident arrangement to detect the 511–511-keV γ-ray pair, functions as a two-dimensional angular-correlation detector. In this application careful tuning of phototubes, light coupling, and resistive chains is required to produce position and energy information rather uniformly over the coincident detection field of view. Often some of the subsystem components can be replaced by digital algorithms. This is the case for the Anger camera, where a substantial part of the analog circuitry can be replaced by digital operations on the raw pulse-height information.

Often NaI(Tl) detector systems are used to define coincident events in conjunction with other detectors, such as Ge detectors and plastic scintillators, to remove Compton background and/or nonphysical events.

Doppler-broadening experiments require a γ-ray spectrometer that has sufficiently good energy resolution ($\sim 1.1-1.7$-keV FWHM) to detect the broadening of the 511-keV line due to the momentum of the positron–electron pair. The detector is a diode of either intrinsic Ge or Li-drifted germanium Ge(Li). The latter is being replaced in almost all applications by the former because the intrinsic detectors are more rugged and do not require continuous cooling by liquid nitrogen when not in use.

Reduced to their simplest concept, these detectors are ionization chambers where the electrons and holes deposited by the stopped and scattered γ rays are collected rapidly in a high field. The resulting current pulse is then amplified and constitutes the energy information obtained from these detectors. The system concept is very important for this class of detectors since their best operation depends on the matching of several key components in the first stages of amplification of the preamplifier. Often the field-effect transistor in the first stage is cooled to near-liquid-nitrogen temperatures along with the Ge detector itself.

The detectors for a lifetime system, consisting of a pair of fast scintillators and associated phototubes, represent a very specialized subsystem. Unfortunately, the matched scintillator and phototube are not available as an "off-the-shelf" system as is the germanium detector. The details of how one goes about selecting components for the lifetime detectors are discussed in Section 2.4.6.3. Another systems aspect of lifetime spectrometers is the need to tune the associated electronics carefully so as to obtain the appropriate degree of timing resolution, efficiency, and long-term system stability.

2.4.4. Doppler Broadening

The Doppler-broadening technique accounts for a large, if not major, fraction of the experiments carried out with positron annihilation to study

2.4. DETECTION OF ANNIHILATION RADIATION

defects in metals. The reason for this is, no doubt, the rapidity with which one can collect data, coupled with the relative lack of constraints related to sample geometry and detector placement. However, it should be mentioned that the major drawback of the Ge or Ge(Li) detector is its intrinsic resolution (~ 1.1 keV FWHM), which is of the order of the observed broadening from the momentum of the positron–electron pairs. At present, this resolution appears to be the limiting factor in the application of the Doppler-broadening technique.

2.4.4.1. Detectors and Electronics.

Doppler broadening is usually performed with either an intrinsic germanium (Ge) or a lithium-drifted germanium (Ge(Li)) detector with energy resolutions for the 511-keV line in the range 1.1–1.7 keV FWHM and efficiency in the range 1–20% relative to a 3×3 in. NaI(Tl) crystal. The electronics associated with the Doppler-broadening detector, although sophisticated in nature, are available "off the shelf."

Typical electronics for a Doppler-broadening system are shown in Fig. 5. The output from the preamplifier is fed to a spectroscopy amplifier that has a high energy resolution. To obtain a reasonable energy dispersion (~ 70 eV per channel) in the analog-to-digital converter (ADC), the amplifier output is passed to a gated biased amplifier (GBA), where a preset pedestal voltage is subtracted from the incoming pulse and then the difference (biased signal) is

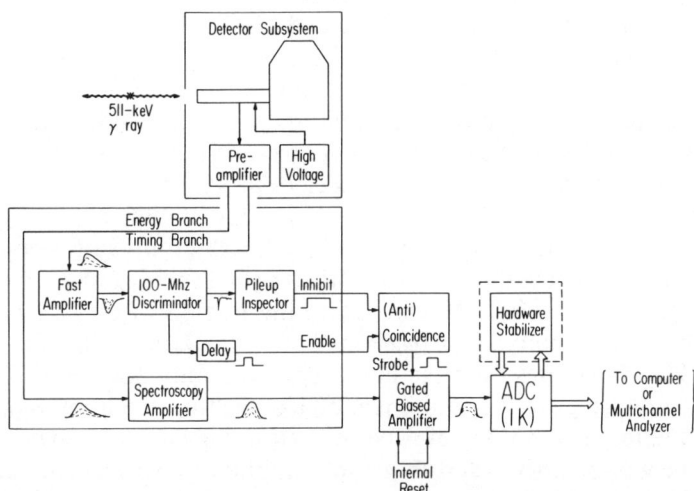

FIG. 5. Electronics for Doppler broadening. The major components of a Doppler-broadening system are indicated. The large boxes (solid lines), one around the detector subsystem, and the other around the pileup protection circuitry and the spectroscopy amplifier, indicate that these subsystems can be obtained as "single" units. The dashed line around the hardware stabilizer indicates the optional nature of this component.

amplified again. In principle, an ADC with enough channels (∼ 8000) could do the same job by use of digital biasing after the conversion, thus avoiding the inherent problems associated with analog biasing.

The requirements of good energy resolution are often in conflict with the desire to obtain high count rates. In order to obtain good energy resolution the preamplifier pulse must be "integrated" for long times, which is done in the spectroscopy amplifier. The "integration" time can be selected by varying the shaping-time constant on the amplifier. When long shaping times and/or high count rates are used, it becomes more likely that two or more γ quanta will arrive at the detector within this time interval and that they will be mixed or piled up with one another at the output of the spectroscopy amplifier. Thus the measurement of the energy of such piled-up events will be incorrect, and they should be rejected by the system. A pileup inspector, when introduced into the circuitry, helps to relieve this problem by identifying events that are too close in time to be resolved by the slower components in the spectroscopy system.

As an example of the use of a pileup inspector, let us consider a detector count rate of 30 kHz (integrated above the noise level). This corresponds to a count rate of 1.5 kHz in the 511-keV peak in an 8% efficient detector, provided ^{22}Na is used as a source of positrons. The width of the energy pulse from the spectroscopy amplifier will be ∼ 20 μs (assuming a shaping time of ∼ 3 μs); therefore, the probability that a pulse is piled up with another pulse is of the order of 45%. Some ADCs and GBAs may provide a level of protection against piled-up events. A feature of this type is called positive-on-positive pileup rejection. With this feature, a reset of the discriminator level is required prior to any subsequent peak detection, indicating that the input voltage is below the discriminator level and ready to receive the next pulse. However, this may not be sufficient at very high count rates. An alternative solution is indicated in Fig. 5. The fast amplifier strongly differentiates the incoming energy pulse, and the fast discriminator generates an output pulse signaling the presence of the original energy pulse. The pulse-pair resolution is ∼ 300–400 ns. Using the numbers from the previous example the probability of not resolving a pulse pair is now 400 ns × 30 kHz or ∼ 1.2%. The pulses from the fast discriminator are now fed to a pileup inspector, which gates the equipment off if a piled-up condition is detected within the inspection interval. The inspection interval is "updated" each time a new pulse arrives and thus triple and higher-order pileups are also rejected.

The use of pileup inspection becomes exceedingly important when the spectra to be compared are collected at different count rates. This is especially true if deconvolution techniques are to be applied to the spectra; since the effective resolution will vary in an unknown way, this could be erroneously interpreted as physical information.

A Doppler-broadening energy distribution sufficient for most applications will contain $\sim 1 \times 10^6$ counts under the 511-keV photopeak. Except for special cases where time-dependent phenomena are under consideration, a counting rate of ~ 1 kHz in the 511-keV peak will normally be more than sufficient. The practical limits on the progress of the experiment will be those imposed by sample and source preparation and by characterization.

One of the dominant limitations in Doppler-broadening measurements, apart from the resolution function, is the short- and long-term stability of the overall system. The speed with which one can collect data is not a practical protection against the effects of instrumental instabilities. In Fig. 5 it was assumed that the computer would also act as a stabilizer. This stabilization method is new to the positron field; usually such stabilization is performed with various hardware analog feedback devices. The need for such stabilization has, however, decreased as the intrinsic stability of the various electronics has improved. In addition, it is possible to stabilize the temperature of the electronics to $\sim 0.2°C$, which is sufficient to realize a significant operational stability.

2.4.4.2. Basics of Data Analysis. Only a few workers attempt to analyze the details of the 511-keV line obtained in Doppler-broadening experiments, e.g., by separating or deconvoluting the physical information from the resolution function of the system. A description of such a deconvolution procedure has been given by, for example, Dauwe et al.,[65] Jackmann et al.,[66] and Dannefaer and Kerr.[67] There are two reasons for the reluctance to deconvolute Doppler-broadening data:

1. The energy resolution of the apparatus is intrinsically poor. Much-higher-resolution experiments could be executed with the angular correlation technique.
2. In many applications the simple presence of the defects is of interest rather than the details of the electron-momentum distribution.

A practical alternative to deconvolution is to describe the shape of the observed momentum distribution in such a way as to include the resolution function. As long as the resolution function does not change owing to changes in electronics, instabilities, or pileup, and as long as direct comparisons with data collected with other instruments are not required, a shape parameter is adequate.

The 511-keV annihilation line is often described by various simple line-shape parameters, an example of which is shown in Fig. 6. The line shapes illustrated are commonly referred to as an "S" and a "W" parame-

---

[65] C. Dauwe, L. Dorikens-Vanpraet, and M. Dorikens, *Solid State Commun.* **11**, 717 (1972).
[66] T. E. Jackmann, P. C. Lichtenberger, and C. W. Schulte, *Appl. Phys.* **5**, 259 (1974).
[67] S. Dannefaer and D. P. Kerr, *Nucl. Instrum. Methods* **131**, 119 (1975).

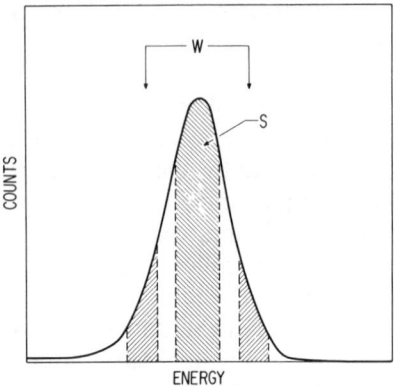

FIG. 6. Schematic illustration of the definition of a center line-shape parameter S and a wing parameter W. The line-shape S is defined as the ratio between the shaded center region and the total area; W is the ratio between the sums of the areas of the two wing regions and the total area.

ter. The definitions of these parameters are the ratio between the shaded areas and the total area under the curve. The S parameter was first suggested and applied by MacKenzie et al.[68] and has more recently been discussed by Campbell[69] in some detail. Its advantages are that it is easily calculated and affords high statistical precision. A serious disadvantage, though, is the lack of a unique physical meaning.

The statistical properties of parameters such as the S parameter can be understood in the following manner. For small variations in line shape, the S parameter will vary linearly with the second-order moment of the curve or momentum distribution. The S parameter and the second-order moment should therefore, in this sense, be equivalent. The statistical uncertainty of the second-order moment is derived principally from the fourth-order moment, which is very large for actually observed line shapes. The S parameter, however, is sensitive only to changes in a limited region of the momentum distribution and thus avoids the larger uncertainties associated with the broad distribution and the large values of the higher moments. The uncertainty δS for an S parameter is $\delta S = \sqrt{S(1-S)/N}$, where N is the total number of counts and the uncertainties associated with the background and the centroid of the energy distribution have been neglected. Thus the figure of merit for an S parameter can be expressed as $(\delta S/\Delta S)$, where ΔS is the change in the line shape under consideration.

A problem that one may encounter when calculating the line-shape parameter is the subtraction of the background counts under the 511-keV peak. The apparent background at energies below 511 keV is slightly greater than at higher energies because of incomplete charge collection. This back-

---

[68] I. K. MacKenzie, J. A. Eady, and R. R. Gingerich, *Phys. Lett. A* **33A**, 279 (1970).
[69] J. L. Campbell, *Appl. Phys.* **13**, 365 (1977).

2.4. DETECTION OF ANNIHILATION RADIATION

ground difference should be viewed as an inherent feature of the detector resolution function. However, small-angle scattering (e.g., in the sample) may cause a similar effect. A common procedure, which is appropriate for peak-count analysis, is to subtract an average of these two backgrounds.

The integration limits in the definition of a line-shape parameter can be chosen so as to give the optimal statistical significance for a particular effect. Campbell[69] has discussed the implications for different choices of parameters. One of the main conclusions that can be drawn from his work is that a line-shape parameter calculated from the center of the peak is optimized when its value is in the range 0.4–0.6, and the optimization is rather uncritical with respect to the exact choice of boundaries. Parameters derived from integrals of the high-momentum regions of the 511-keV line have a slightly lower sensitivity than the central parameters, and the boundary choice is more critical for the high-momentum parameters.

An oft sought goal of shape parameter analyses is to define defect-specific parameters that are independent of defect concentration. Mantl and Triftshäuser[70] introduced the R parameter with this idea in mind. Consider the case of a single type of defect in an otherwise perfect host material. A low-momentum (S) integral and a high-momentum (W) integral are used to define the R parameter, where

$$S = (1 - A_t)S_b + A_t S_t \qquad (2.4.1)$$

and

$$W = (1 - A_t)W_b + A_t W_t, \qquad (2.4.2)$$

$$R = \frac{S - S_b}{W - W_b} = \frac{S_t - S_b}{W_t - W_b}. \qquad (2.4.3)$$

The term A_t is the probability that the positron becomes trapped, while S_t, W_t, and S_b, W_b are the line shapes characteristic of the defect (positron-trapped) state and free (Bloch) state, respectively, in the perfect material. The principal idea is that the S parameter is dominated by low-momentum electrons, and the W parameter is to a large degree dominated by high-momentum electrons, and R describes the balance between low- and high-momentum electrons, thus giving rise to a defect specificity.

In general, the line-shape parameter is tailored to the need at hand, and although its statistical sensitivity may be of some value, it does not denote any well-defined physical concepts. Thus when comparisons are to be made of data from different laboratories, ambiguities may arise. Such ambiguities

---

[70] S. Mantl and W. Triftshäuser, *Phys. Rev. B* **17**, 1645 (1978).

can be avoided through the use of generalized line-shape parameters,[71] which, however, still do not solve the more fundamental problem of the lack of physical significance.

2.4.4.3. Positron Sources. The arrangement of the source and sample is particularly simple in the case of Doppler broadening. The source is prepared by evaporating to dryness a solution of the source material (e.g., ^{22}NaCl) in a small area of the metal sample. This source region is then covered with another piece of identical sample material and the "sandwich" is mounted in the experimental apparatus (usually a vacuum system) in front of the detector. A major disadvantage of this technique is that the properties of the source deposit may limit the applicability to low temperatures. For example, the source material (^{22}NaCl) may liquify and disperse at high temperatures. In addition, the commonly used chemical form as a salt can accelerate surface corrosion. However, Herlach and Maier[72] describe the preparation, by molecular evaporation, of a source of ^{22}NaCl, capable of withstanding high temperatures. This technique has been applied to studies of Nb, Ta, and W, in which its usefulness at even the highest temperatures was demonstrated.[73] In this technique the ^{22}NaCl is vacuum-evaporated onto a thin sample foil which is placed inside a container also prepared from the sample material. The container is sealed by electron beam welding. A possible problem with both methods outlined above is that the purity of the sample could be compromised if the source solution contained impurities other than those of the source material.

It is also common to deposit the source solution on a thin substrate other than the source material (e.g., Mylar, Kapton, or Ni), which is wrapped up to form an envelope once the source solution has evaporated. This envelope is subsequently sandwiched between two pieces of the sample to be investigated. The advantage of this method is that the source envelope can be reused for different investigations; its limitations are that at high temperature the sample package may disintegrate or the envelope material may diffuse into the sample. Further, a finite amount of positron annihilation takes place in the sample package itself, thereby giving rise to a source contribution in the observed data.

Techniques using remote sources, which are not in contact with the sample, can also be developed for use with the Doppler-broadening method. Remote-source techniques solve the problem of the introduction of impurities into the sample from the source material. Moreover, with the source

---

[71] G. Kögel, L. Smedskjaer, and W. Triftshäuser, *Proc. 6th Int. Conf. Positron Annihilation, Arlington, Texas 1982* (P. G. Coleman, S. C. Sharma, and L. M. Diana, eds.), p. 903. North-Holland Publ., New York, 1982.

[72] D. Herlach and K. Maier, *Appl. Phys.* **11,** 197 (1976).

[73] K. Maier, H. Metz, D. Herlach, and H. E. Schaefer, *J. Nucl. Mater.* **69/70,** 589 (1978).

shielded from the detectors, positron annihilation in the source is of minimal consequence. The remote sources, unlike deposited or implanted sources, can be reused. In some special circumstances, it may be possible to produce internal sources, either by nuclear transmutations of the sample material or by the technique of ion implantation.[74,75] In the latter technique, the positron-emitting nuclei are implanted into the sample with an energy of ~ 50 keV in a mass separator. Since only $\sim 10^{14}$ positron-emitting nuclei are needed to obtain a sufficient count rate in the detector, the purity of the sample is not compromised, and the source contribution is nonexistent.

The source strength used in Doppler-broadening experiments depends on the radioisotope and the detector efficiency, but typical activity levels range from 5–100 μCi. Three positron-emitting isotopes are commonly employed in Doppler-broadening experiments. Sodium-22 is deposited either from solution or by evaporation as NaCl, or in its atomic form by ion implantation. Ion-implanted ^{58}Co has been used successfully in our laboratory and is, from a metallurgical point of view, particularly well suited for experiments on the transition series of metals, Fe, Ni, and Co. The ^{22}Na source is less suitable as an implanted or internal source because of its low solubility in most metals. Germanium-68 is also commonly used and may be the best source for Doppler broadening since the 511-keV line is the dominant high energy observed, whereas ^{22}Na emits a 1.28-MeV γ simultaneously with the positron, and ^{58}Co has a low positron-branching ratio with a strong line at 810 keV. The 1.28-MeV and 810-keV lines cause a Compton background under the 511-keV line, and in both cases the detector and the electronics must bear the consequences of an increased counting load leading to pile up and deterioration of the resolution. The positron endpoint energy for ^{68}Ge is also substantially higher than that for ^{22}Na (see Table II), and therefore the positron penetrates deeper into the bulk of the sample, lessening the effects due to annihilations at the surface of the sample.

2.4.4.4. Sample and Detector Configuration. Among the three commonly used positron annihilation methods, Doppler-broadening experiments impose the fewest restrictions on the sample–detector configuration. In general, the sample (containing the source) is mounted as close to the detector as permitted by other anicillary equipment such as vacuum enclosures. The typical sample–detector distance will thus be a few centimeters. Any material through which the radiation has to penetrate in order to reach the detector will cause attenuation as well as small-angle scattering, which looks similar to incomplete charge collection and therefore degrades the

[74] M. J. Fluss, M. K. Chason, J. L. Lerner, D. G. Legnini, and L. C. Smedskjaer, *Appl. Phys.* **24**, 67 (1981).
[75] M. J. Fluss and L. C. Smedskjaer, *Appl. Phys.* **18**, 305 (1979).

apparent instrument resolution. Thus one should avoid thick equipment walls between sample and detector as well as thick (>2-3 mm) samples. When using sources like ^{22}Na and ^{58}Co, one should consider that the high-energy γ quanta may be scattered by the ancillary equipment before reaching the detector, resulting in an increased background in the 511-keV region. A more serious consideration is that it may sometimes, depending on geometry, result in a spectrum with structure in this region. Such structure may influence the interpretation of the observed data. It is also known that the observed spectra may depend on the direction of incidence of the γ quanta on the Ge detector, with regard to both centroid position (e.g., Lichtenberger and MacKenzie[76]) and resolution. One should therefore attempt to use a constant sample–detector geometry. Finally, in designing experiments in which detailed comparisons are to be made among samples of the same material, one should use samples of identical thickness (small-angle scattering), the identical ancillary equipment (scattering of high-energy quanta into the detector and small-angle scattering), and keep the detector position constant relative to the sample (possible resolution function dependence on the direction of incidence).

2.4.5. Angular Correlation

Angular correlation of annihilation radiation is used in different ways in positron annihilation studies of defects in metals. One- and two-dimensional measurements can be analogous to Doppler-broadening experiments in that shape parameters can be derived from the experimental data and studied as a function of temperature in equilibrium experiments or as a function of treatment, e.g., in annealing studies. However, the use of angular correlation extends beyond these types of experiments by virtue of significantly better momentum resolution than that available in a γ-ray spectrometer such as a Ge(Li) detector. Several schemes for collecting two-dimensional angular correlation data have been implemented.[3,77,78] The extra dimensionality of the two-dimensional apparatus opens the way for studies of defect structure and morphology. The large-area position-sensitive detectors offer a relatively full field of view of the momentum distribution, and in this sense are similar to the equivalent feature of the Doppler-broadening method. This, coupled with its resolution ($\gtrsim 0.2$ mrad) is justification for its implementation in a positron laboratory where defect studies are in progress. In this section we describe the basics for one-dimensional and two-dimensional angular correlation systems.

[76] P. C. Lichtenberger and I. K. MacKenzie, *Nucl. Instrum. Methods* **116,** 177 (1974).
[77] R. N. West, J. Mayers, and P. A. Walters, *J. Phys. F* **14,** 478 (1981).
[78] P. E. Bisson, P. Descouts, A. Dupanloup, A. A. Manuel, E. Perreard, M. Peter, and R. Sachot, *Helv. Phys. Acta* **55,** 100 (1982).

2.4.5.1. Long-Slit Angular Correlation.

We have already mentioned that the one-dimensional angular correlation equipment is used to measure the quantity $F(K_y)$ (Table I). A schematic representation of such equipment is shown in Fig. 7. The basic detectors and electronics are quite simple. The detectors consist of NaI(Tl) scintillators of sufficient length to approximate an infinite integration along their length. These in turn are coupled to suitable phototubes, either at both ends or at one end, but in either case in such a way as to ensure a relatively uniform response over the detector length.

The electronics associated with a long-slit apparatus is also shown in Fig. 7. The main design requirement is a highly efficient coincidence system that minimizes the number of accidental coincidences. By the additional use of shielding or by energy selection one avoids Compton-scattered events, which tend to broaden the effective resolution, and worse yet, tend to be interpreted as physical information. Since measurements with a long-slit apparatus usually consist of scanning the momentum space by moving one of the detectors, a spectrum can be regarded as a collection of discrete measurements; hence to achieve stability the constancy of the coincidence discriminating circuitry must be maintained. Coincidence-resolving times between 10 and 100 ns are acceptable; the lower end of this range is achieved with fast discriminators coupled to the high gain output of the phototube.

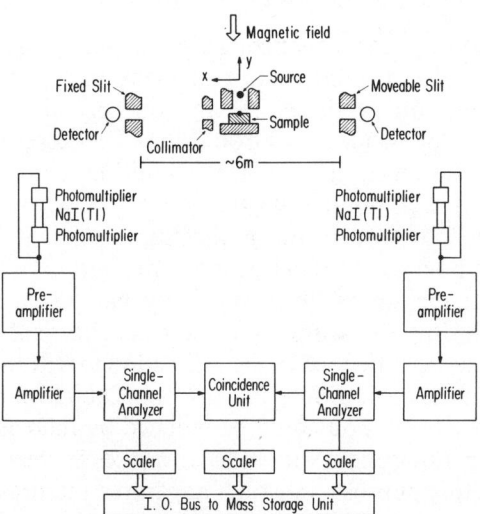

FIG. 7. A schematic illustration of a long-slit angular correlation apparatus. The upper part of the figure is a schematic of the relative positions of the major components of a long-slit system. Note the shielding of the source from the view of the detectors and the use of a magnetic field to guide the positrons onto the sample. The lower part of the figure is the electronics. Note that this particular system uses two phototubes on the scintillators to ensure uniform response.

The major fundamental problems relate to collimation and geometry rather than to difficulties in the electronics. The source is usually remote from the sample, and it is practical to guide the positrons into the sample by use of a magnetic field oriented along the axis defined by the source and sample. Although the use of a magnetic "focusing" system is not essential, it does help to overcome the inefficiency of the finite solid angle subtended by the sample, which increases with distance squared. Also, there will be an improvement in the signal-to-noise ratio because a larger fraction of the detected γ rays are from positron annihilations in the sample.

Typical distances from sample to detector are $\sim 2-5$ m. The sample length is defined by the extent of the positron beam impinging on the sample. It is important that the detector view only those positron annihilations that take place in the sample and not in surrounding or supporting structures. Such superfluous events are minimized by a combination of strategies. Careful collimation of the γ-ray detectors to ensure that the viewing area is restricted to the sample alone and/or collimation of the positron "beam," by either collimators or magnetic focusing, are usually employed.

For a given sample–detector distance the angular resolution is determined by two factors:

(1) the width of the γ-ray collimating slits, and
(2) the distribution of the positrons in the y direction (into the sample).

The y distribution is a consequence of the stopping profile of positrons by the sample material and of the positron incident-energy distribution, as well as a spatial extension produced by any off-normal angle between the sample's face and the incident positron beam. The stopping distribution of the positron in the material is often the limiting factor in obtaining the desired resolution. This point is illustrated in Fig. 8, where the "active" part of the sample is extended in the y direction so that it is seen from the detectors within an angle Ω. This gives a contribution of 2Ω to the resolution function. If the slit width is kept very narrow so as to obtain the maximum resolution, the sample–detector distance is 2 m, and the positron's position has been defined in the y direction to within ~ 0.2 mm, then the limitation on the resolution is 2×0.2 mm/2000 mm $= 0.2$ mrad. This broadening of the resolution could be reduced by increasing the sample–detector distance. Equipment with radial distances as great as 10 m has been constructed for this purpose, but such apparatus requires detectors of the order of 1 m in length to fulfill the approximation of the infinite integration in the z direction.

Two other problems associated with the sample and sample mounting should also be noted, although they can not be assessed in general terms.

2.4. DETECTION OF ANNIHILATION RADIATION

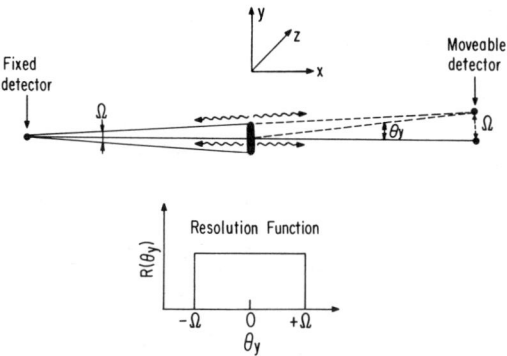

FIG. 8. Effect of positron stopping profile on the resolution function. The stopping profile extends the annihilation region in the *y* direction (heavy line). Two pairs of annihilation quanta are shown, one emerging from the middle and the other from the top of this region and in opposite directions. The top pair can be detected if $\theta_y = \Omega$, the middle pair if $\theta_y = 0$. The resulting resolution function $R(\theta_y)$ due to this extension is shown schematically as well.

1. The annihilation γ rays may scatter, either in the sample or in surrounding structures, including the defining slits themselves.
2. The attenuation may depend on angle. This is a consequence of the different possible paths for the annihilation γ rays through the sample material and is illustrated in Fig. 9. The effect can be minimized, although not completely removed, by tilting the sample slightly so that one of the two γ rays is always passing through sample material. The consequence of angle-dependent attenuation is a small asymmetry around $\theta_y = 0$ degrees.

2.4.5.2. Two-Dimensional Detector Systems. The use of two-dimensional angular correlation detector systems relieves many of the difficulties encountered with the long-slit systems and provides simultaneous information about two components of the electron momentum density. For long-slit experiments, each data point on the distribution represents a discrete measurement. Because these measurements are discrete, possible instabilities are not sampled in the same way for each of the data points unless

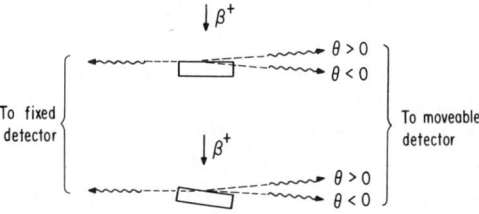

FIG. 9. Angle-dependent attenuation of annihilation quanta (upper part). The $\theta < 0$ quantum is attenuated by passing through the sample. The lower part of the figure shows how such angle-dependent attenuation may be avoided by tiltling the sample.

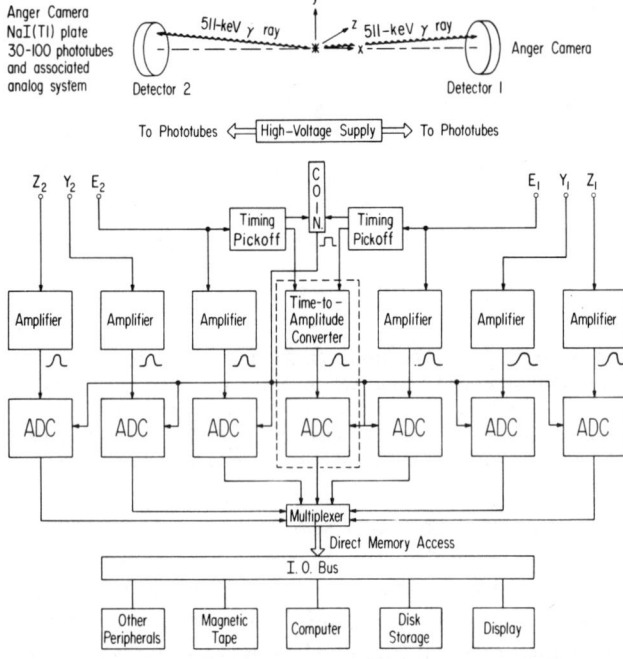

FIG. 10. A schematic diagram for an Anger camera based 2D-ACAR system. The inputs Y_i and Z_i ($i = 1,2$) carry the positional information from the detectors, while E_i carries the energy and timing information. Analog-to-digital converters are used extensively to digitize all pulse-height information for latter analysis. Note the use (optional) of a time-to-amplitude converter to measure the coincidence between events and to correct for random or accidental coincidences simultaneously.

special precautions are taken. For Doppler-broadening and two-dimensional angular correlation systems these instabilities are sampled simultaneously over the entire, or at least, most of the momentum space. The large space angle subtended by the two-dimensional apparatus means that one-dimensional experiments can be performed at least an order of magnitude faster than with a two-detector long-slit apparatus. Hence, with respect to sample variations and the time limits imposed by, for example, high temperatures, such an application becomes competitive with Doppler broadening and, of course, surpasses it in its ability to provide simultaneous information about two momentum components rather than just one.

There are at present three proven two-dimensional angular correlation of annihilation-radiation (2D-ACAR) detector systems: the discrete detector array, the Anger camera, and the Pb-converter crossed-wire chamber. A schematic of an Anger camera system is shown in Fig. 10 along with a block diagram representation of the electronics. In all three systems, one identifies

2.4. DETECTION OF ANNIHILATION RADIATION

coincident events by recording the y and z positions on each detector subsystem, evaluating the Δy and Δz,* and storing the resulting information in a two-dimensional histogram. When the detectors and sample are on a single symmetry axis, the momentum-dependent detection efficiency in the Δy, Δz histogram will be a function of the y and z momenta and will be described by a cone-shaped surface: a four-sided rectangular base pyramid for rectangular detectors and a right circular cone for circular detectors. The origin of this efficiency function is easy to understand from a comparison of the efficiency for two extreme types of events. Consider an event where the two γ rays are collinear ($p_z = 0, p_y = 0$). Such an event is identified when the pair of positions on the surfaces of the two detectors can be connected by a straight line passing through the sample. These events are detected with the highest efficiency. In contrast, high-momentum events are detected with a lower efficiency. Since the observation of most momenta are derived from events which sample large areas of the detectors' surfaces, the restrictions on the microscopic uniformity of the detectors' responses are greatly relieved, although systematic effects that correlate with momentum should be avoided.

The first successful two-dimensional angular correlation apparatus was constructed from arrays of discrete NaI(Tl) detectors and phototubes by Berko et al.[3] and by Berko and Mader.[79] The main advantage of this system is its modularity. The identification of the position of events is clear and unambiguous and is defined by the signal from a particular discrete detector. However, to achieve a resolution of the order of 0.5 mrad one must use collimators in front of each detector, and hence the intrinsic high efficiency is "sold off" to obtain resolution. This problem is improved by going to the smallest possible detectors, which at the present time would limit one to detector diameters of about 9.0 mm. Phototubes of this size could be coupled to $Bi_4Ge_3O_{12}$ scintillators to obtain maximum detection efficiency. Unfortunately the cost per detector is almost independent of size, and hence to cover a significant space angle can easily cost hundreds of thousands of dollars. If the advantages of modularity are not important, then one of the continuous detectors may be more appropriate.

The remaining two detector schemes have in common the feature of providing a continuous response in the position coordinates. The z and y positions on each detector are determined by measuring the pulse height from detector elements, either phototubes for the Anger camera or proportional crossed-wire signals for the wire chamber, and determining a center of

---

[79] S. Berko and J. J. Mader, *Appl. Phys.* **5**, 287 (1975).

* $\Delta y = y_1 + y_2$ and $\Delta z = z_1 + z_2$.

mass for the observed event in the coordinate system of the detector, which in turn is translated into the appropriate coordinate system for the sample.

The Anger camera,[80] which was originally developed for medical applications, is well suited for 2D-ACAR measurements. The Anger camera system shown in Fig. 10 represents a compromise between the high efficiency of the discrete detector system and the high resolution that has been achieved with the crossed-wire chambers. The components of an Anger camera for use in positron annihilation studies will consist of a large-area NaI(Tl) "plate" with a diameter of ~ 500 mm and a thickness of 10 mm. The thickness is a compromise between detection efficiency and position resolution. The energy and position of the incident γ ray are deduced from the response of an array of phototubes numbering from ~ 40 to as many as 100, arranged to have the characteristic of uniform linear position response. By increasing the number of phototubes and decreasing their size one can expect to achieve a more uniform detector response. The ultimate limitation of the detector is determined by the physics of the γ-ray energy loss process for producing the photoelectron, and the occurrence of Compton events in the scintillator prior to photoconversion. At present, the two positional parameters and the energy parameter are derived from simple analog circuits (capacitive networks for the positional signals) and the individual phototube signals are not usually made available. The eventual use of small rectangular phototubes and $Bi_4Ge_3O_{12}$ plates should allow one to reach better resolution and efficiency for this type of detector.

The specifications for the 12.7-mm-thick Anger-camera-based two-dimensional system operating at the University of East Anglia[77] are given in Table IV. The usable area of the detectors is significantly less than the plate size as a consequence of energy and position anomalies at the detector's edge. Such edge effects require the East Anglia group to shield this region of the detectors in order to avoid the correlation that would occur with respect to the high-momentum events. The use of larger numbers of phototubes, and hence smaller diameter tubes, is known to reduce this edge effect greatly.

The combined Pb-converter plate and proportional crossed-wire chamber of Jeavons[81] result in a detector system for positron annihilation studies that has the best intrinsic linear resolution, better than 2 mm in some cases.[78,81] The principle of operation of this detector system is the following. A converter plate, consisting of alternate layers of Pb sheets and

---

[80] H. O. Anger and A. Gottschalk, *J. Nucl. Med.* **4**, 326 (1963).

[81] A. Jeavons, *Proc. 5th Int. Conf. Positron Annihilation, Lake Yamanaka, Japan, 1979* (R. R. Hasigawa and K. Fujiwara, eds.), p. 355. Jpn. Inst. Metals, Sendai, 1979.

2.4. DETECTION OF ANNIHILATION RADIATION

TABLE IV. Comparison of Continuous Response Detector Systems for Two-Dimensional Ansular Correlation

Property	Detector	
	Anger cameras (East Anglia)	Pb-converter crossed-wire proportional gas counters[a]
Diameter	375 mm	—
Length × width	—	200 × 200 mm
Sample-to-detector distance	14.4 m	10 m
Usable solid angle	543 mrad^2	400 mrad^2
Useable fraction of coincident 511-keV γ rays	~0.17	~0.07 (0.16)
Linear resolution	~5 mm	~1.7 mm
Coincidence resolution	0.65 × 0.65 mrad^2	0.35 × 0.35 mrad^2
Coincidence resolving time	~100 ns	~300 (30–100) ns
Coincidence counts per mCi of ^{22}Na per second	~9	~1.3 (~2.6)[b]

[a] The values in parentheses are for a two-converter counter using a faster gas mixture.
[b] This value is a consequence of using a narrow sample geometry.

electrical insulating sheets of Teflon (shown in Fig. 11), is drilled with a precise array of holes. Photoelectric conversions take place in the plate, and some of the resulting electrons are guided down the holes to a multi-crossed-wire gas proportional counter. The hole spacing and the wire spacing are chosen to maximize the positional resolution. The wire spacing in the system built by Manuel and co-workers[78] is ~1 mm. Each event is identified by the simultaneous recording of the signals from the array of wires, and an appropriate center-of-mass calculation is performed to identify the position of the event. The storage of the Δy, Δz information is then much the same as for any other two-dimensional system.

One will notice on comparing the specifications in Table IV that the surface area for the crossed-wire-chamber system is significantly less than for the Anger camera; however, this is compensated by the better positional resolution. As with the Anger camera one must trade positional resolution for efficiency. The choice of converter design is at the heart of this decision. The usable detector surface area can be increased by using multiple arrays of detectors operating in crossed coincidence mode, or, as shown in Fig. 11, two converter plates can be used to achieve a greater efficiency.

The Anger camera is produced with surface diameters as large as ~50 cm, and the most modern Anger cameras have linear resolutions at

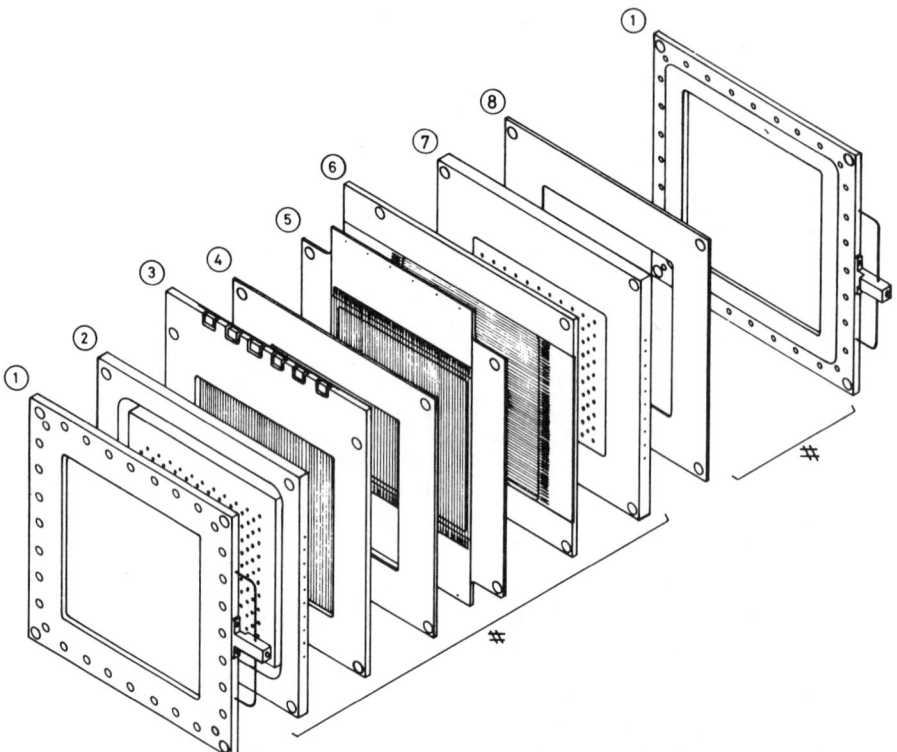

FIG. 11. A schematic of a multi-lead-converter and multiwire position-sensitive gas proportional counter system[78]. (1): window frame, (2,7): γ-ray-to-electron converters, (3,6): cathode planes, (4,8): spacer frames, (5): anode plane.

FWHM of ~2.5–3 mm. This may be compared to the best reported resolution of 1.7 mm for the wire chamber.[78] There is one feature that makes the Anger camera somewhat more attractive than the wire chamber and that is the better resolving time of ~100 ns compared to ~300 ns. The use of two converter plates in the wire-chamber system as shown in Fig. 11 makes the efficiency similar to that achieved in the Anger camera. The two systems are quite competitive, and both can be expected to find general application in the area of defect studies.

Unlike the electronics for the long-slit system, shown in Fig. 8, the two-dimensional system requires the use of some additional components to analyze the analog information related to the position of the incident γ ray. For the case of the Anger camera (Fig. 10) the detector subassembly provides three analog signals, two for position and one for energy. These analog signals must be digitized and then used in a computer to build the two-dimensional histogram. Because the Anger camera and the discrete detector systems provide sufficient energy resolution it is also possible to invoke energy-gating analysis of the data, and thus reduce the effects from scattered γ-ray events. A short coincidence resolving time is also an important criterion, particularly if one hopes to study the highest-momentum regions of the spectra where the effects from chance coincidences will be relatively large. The best coincidence resolving times are obtained with the discrete detectors, because of their small size; however, a resolving time of 100 ns is reported for the East Anglia system described in Table IV. The reported coincidence resolving time for the wire chamber is 300 ns, but it is anticipated that the use of different gas mixtures may eventually result in resolving times of 30–100 ns.

Commercial manufacturers of the Anger camera are gradually converting their systems to full digital encoding of each and every phototube, using an ADC on each tube. There are numerous advantages to this scheme, particularly for the positron experimenter, since the analog position algorithm, which is very dependent on component stabilities, can be replaced with a digital feedback system where the high voltages to the tubes are adjusted continuously to compensate for aging and where position-dependent information relating to position encoding, energy, and efficiency can be continuously updated. The scheme used by Manuel and co-workers[78] for their wire chambers is also along these lines. The output from each wire is passed to an ADC, and the data are then handled in a fully digital manner in a dedicated microprocessor.

2.4.5.3. Basic Analysis for Defect Studies. As with the Doppler-broadening technique, the long-slit angular correlation method is amenable to analysis by the use of spectral shape parameters. The most common shape parameter is the so-called peak-count parameter. It is derived experimentally from the data by placing the detectors in the collinear positions and measuring the changes in coincident count rate, e.g., as a function of temperature or annealing cycles. Experiments of this type are usually performed with resolutions of $\sim$ 1–2 mrad. Unfortunately, several practical problems arise in such experiments, and ideally one would much rather measure the entire momentum distribution and derive the shape parameter from its analysis. A frequent difficulty encountered in peak-counting experiments is undesired motion of the sample (e.g., due to thermal expansion in the sample mount), the consequence of which is that the detectors are not viewing the peak of the momentum distribution. A related concern is instabilities in the electronics, causing a time-dependent coincidence detection efficiency. The best way to reduce such count-rate effects is to measure the entire distribution. The severity of these problems increases in experiments that compare different samples over long periods of time. Of course, shape parameters other than the peak parameter are also affected by sample motion and count rate. In general, one must insist on independent verification of both angular position and normalization of the angular distribution.

The best way to obtain this independent verification is to measure the entire spectrum simultaneously. In the case of long-slit experiments, this can only be approximated by multiple scans through the spectrum, ideally in a random manner. However, the use of coincident-position-sensitive detectors with viewing areas sufficient to cover the entire momentum space greatly alleviates this problem.

The shape parameters obtained from two-dimensional experiments on single-crystal samples are potentially more defect specific (although this point remains to be proven) than those taken from one-dimensional experiments. The details of the atomic and electronic structure where the positron is localized are revealed by the changes in the local electron density. Information may also be obtained as to the way in which the positron distributes itself in the defect-trapped state. The measurement of two components of the momentum, instead of one, allows one to "see" the manifestations of these structural properties more clearly. A significant question that remains to be answered is the effect of the very presence of the positron in the defect; that is, how does the positron mask or change such structural features? The two major components of the momentum spectrum, the annihilation with valence electrons and the annihilation with localized and/or core electrons, are more easily resolved in the two-dimensional momentum space. Thus it is possible to perform radial analysis on the momentum data, e.g., for constant $p_y^2 + p_z^2$. Such an analysis is analogous to the long-slit experiment, but does a better job of separating the low-momentum from the high-momentum contributions than in a one-dimensional experiment.[10]

A major difference between positrons annihilating in the Bloch state and in the vacancy-trapped state is the almost complete disappearance of the so-called Umklapp peaks arising from the delocalized nature of the positron in the near-perfect lattice.[7] The momentum regions associated with these high-momentum components of the valence-annihilation signal can be isolated more effectively in a 2D-ACAR experiment and can thus provide a particularly sensitive shape parameter for distinguishing between trapped and delocalized positrons. To summarize, the application of large-area position-sensitive detectors greatly augments the sensitivity of the positron technique. Because of the two-dimensional resolution it is possible to separate more effectively the various momentum regions that undergo the greatest changes, either due to defect type, concentration, or possibly anisotropies arising from defect structure with respect to crystallographic orientation.

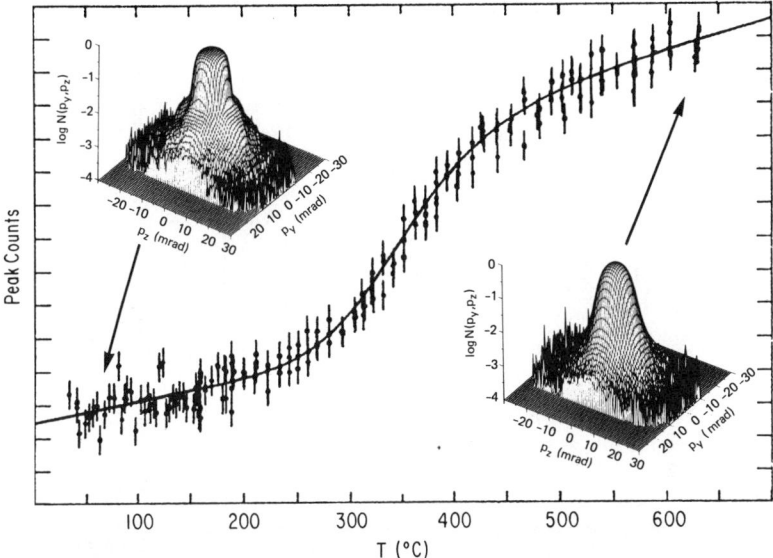

FIG. 12. Peak count and 2D-ACAR of aluminum. This figure highlights the information available in ACAR experiments. Shown are both a peak-counting experiment performed with long-slit angular correlation and, as insets, 2D-ACAR results at the highest and lowest temperatures, which emphasize the change from the delocalized Bloch state to the localized or vacancy-trapped positron state. (Data from Fluss et al.[10])

Figure 12 shows a combination of the different methods used to present and examine angular correlation data for defect studies under conditions of thermal equilibrium.[10] The use of single crystals and the choice of orientation emphasize the structural differences in the spectra from low temperatures (20°C), where the positron is annihilating in the delocalized Bloch state, to the highest temperatures, where the positron is almost exclusively annihilating while in a defect-trapped state. The data of Fig. 12 are for positron annihilation in Al. The S curve is typical of the change in the concentration of trapping sites (vacancies) in the sample as a function of temperature. The data points in the curve are those obtained for a peak-counting experiment with a resolution of 2 mrad. At selected temperatures we show the corresponding 2D-ACAR data for a single-crystal sample oriented so that the $\langle 100 \rangle$ direction is integrated out by the energy resolution of the detectors. To emphasize the structural differences in the data as the vacancy concentration and hence trapping increase we have plotted the 2D-ACAR data in a log mode. The disappearance of the Umklapp peaks in the second Brillouin zone and the narrowing of the valence momentum distribution accompanied by a smearing of the "effective" Fermi edge are clearly obvious features that can be utilized in defect studies.

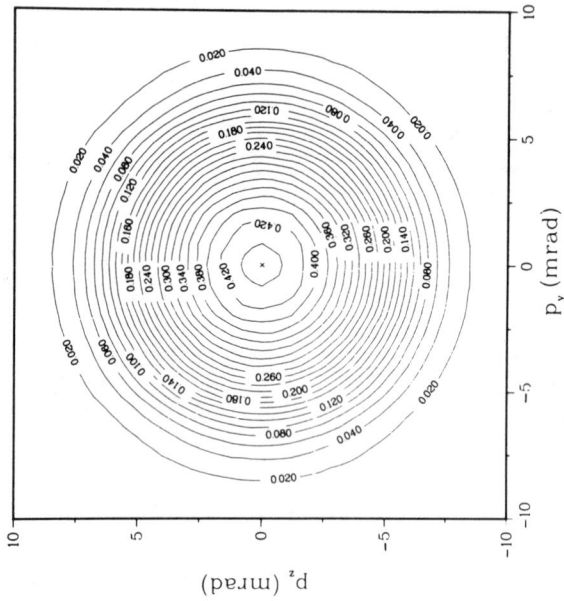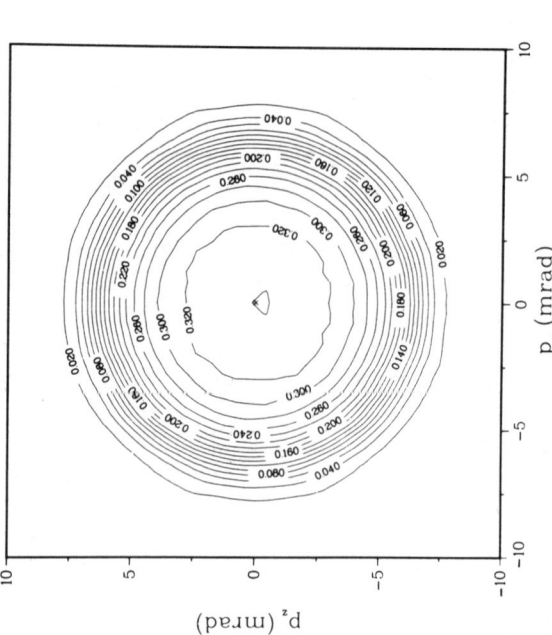

FIG. 13. 2D-ACAR contour diagrams. Contour diagrams can provide a detailed and quantitative picture of 2D-ACAR results, as illustrated here. One sees the change in the shape of the overall 2D spectrum from a rounded flat topped surface at 20°C to a peaked gumdroplike surface at 630°C. The momentum axes are both $\langle 110 \rangle$. Other plots, with different choices of contour levels, can be used to reveal changes in isotropy from state to state or from one temperature to the next. (Data from Fluss et al.[10])

2.4. DETECTION OF ANNIHILATION RADIATION

The analysis of 2D-ACAR experiments on the structure of defects is too new to permit the description of a modus operandi for such investigations. Indeed, little has been done in the area of using 2D-ACAR data in the form of shape parameters. The "separation" of high- and low-momentum components is much more easily achieved in 2D-ACAR than in 1D-ACAR experiments. The task of defining other defect-specific shape parameters could be facilitated by the use of such data.

In general, the line-shape parameters derived from angular-correlation experiments are treated in a manner similar to those obtained in Doppler-broadening experiments. Postacquisition stabilization methods can be applied, and microscopic analyses are easily accomplished. However, one expects that long-term stability of the electronics will make such stabilization procedures unnecessary. Although the exact resolution function for the angular correlation instrument is at times difficult to determine, it is sufficiently narrow that shape parameters defined by integration over several milliradians will be relatively unaffected by the resolution function itself. Thus intercomparison of results between different laboratories is more tractable with ACAR than with Doppler broadening.

Contour diagrams provide a useful graphical technique as well as a quantitative method for studying data from a two-dimensional experiment. The contour diagram of the 2D-ACAR surface illustrates the lines of equal annihilation probability in the p_y, p_z momentum space. Figure 13 compares contour diagrams for data taken in Al at 20°C and 630°C. The comparison emphasizes the dramatic changes in the momentum distributions at low momenta and the relative intensity of the anisotropy at high momenta due to high-momentum components. From such a comparison it is possible to select those regions of the momentum spectrum that best signal changes in the electron momenta. Hence one would expect the use of differences to be particularly valuable in such an application. Figure 14 shows the 2D-ACAR difference between Al at 630°C (vacancy trapped) and 20°C (Bloch state), both in perspective and contour forms. Differences such as these are basically understood in terms of the defect structure and in terms of the electron and positron wave functions. However, considerable effort is still being applied to a further understanding of the temperature dependence of the two major positron states, the Bloch and vacancy-trapped states.

Ultimately, detailed comparisons with theory will prove essential in defect studies, since the potential for identifying characteristic features of defect types through their electronic structure is contained in the information from angular correlation experiments. The comparison of experiment with theory is based on an a priori model for the defects being studied. Thus one presupposes not only a knowledge of the structure of the defect but also the nature of the population. In equilibrium studies this limits one to

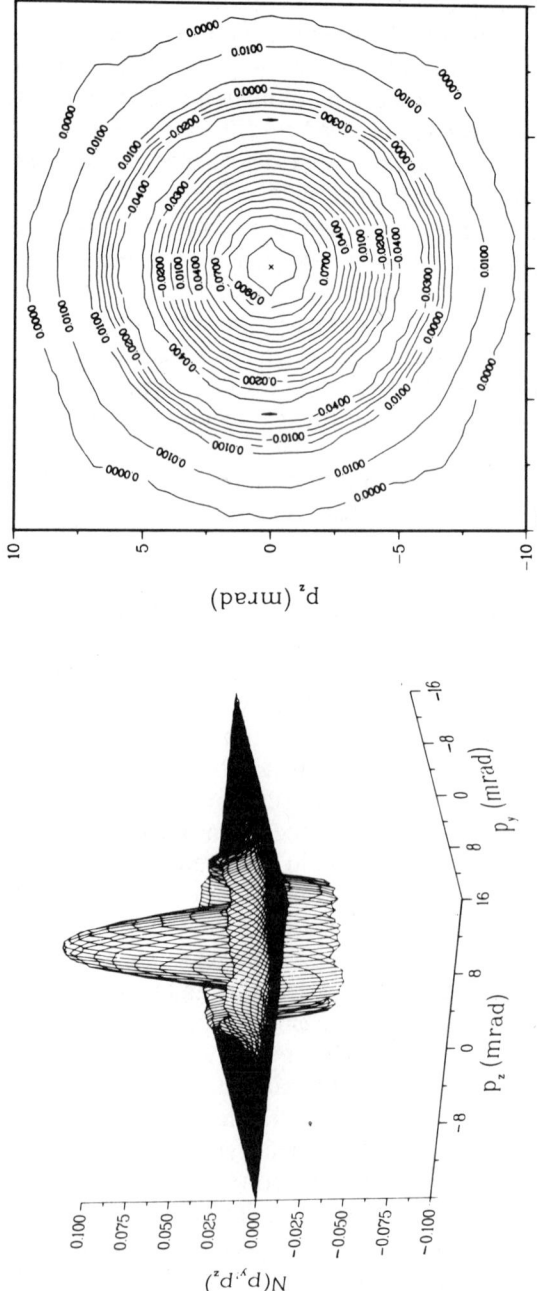

FIG. 14. 2D-ACAR difference surface. When investigating changes from one 2D-ACAR surface to another, differences can prove very useful. Here we illustrate the difference observed for positrons annihilating in the vacancy-trapped state and the Bloch state. (Data from Fluss et al.[10])

vacancies and small vacancy clusters, but for nonequilibrium studies the situation can quickly become much more complex, and such experiments are usually augmented by annealing studies in an attempt to identify the types of vacancylike defects that are present in the sample.

Eventually, one might expect to include other types of defects and parameters in comparisons of theory and experiment. As an example, phonons play an important role for positrons in the Bloch state, but their role in the case of vacancy-trapped positrons has not really been investigated. In addition to the effects from phonons, one must also include the effects from the expansion of the host lattice as a function of temperature. In some studies, pressure can be used as a variable in an attempt to separate phonon effects from lattice expansion.[82] Pressure experiments also make it possible to measure the formation volumes for vacancies in several metals (e.g., Dickmann et al.[83]).

2.4.5.4. Positron Sources. In our discussion of positron sources for angular correlation experiments we now consider the various remote sources in common use. Sources deposited directly on, or homogenized with, the sample can also be used, although to take advantage of the high resolution of the angular correlation method it is preferable to use remote sources since the source intensity must be relatively high to compensate for the small space angle that the detectors subtend. One might remark, however, that internal sources would minimize any problems associated with surface annihilations. Thus this method may be quite acceptable for experiments where high angular resolution is not paramount.

Three isotopes find common usage as remote positron sources for angular correlation experiments: ^{64}Cu, ^{58}Co, and ^{22}Na. The use of ^{64}Cu is restricted to those experimenters who have access to a reactor for producing activated Cu foil sources. The 12.7-h half-life of ^{64}Cu imposes several additional problems on the experiment: variable count rate, significant decay corrections, variable true-to-chance ratios, and the problem of continuously replacing the source itself. Sources of ^{58}Co can be produced in the form of metallic buttons on which the Co has been electroplated and subsequently diffused into the metal substrate. A careful choice of materials will allow a maximum back reflection of the positrons and hence increase the usefulness of the source. As an example, one might use a thin substrate of Cu on a Au backing as the base material for the electrodeposition. Since the size of the positron beam delivered to the sample is always a consideration, one is also faced with a practical limit to the maximum activity that can be deposited on the source when a restriction of ~ 5 mm^2 is imposed on the active region.

---

[82] I. K. MacKenzie, *Phys. Lett. A* **77A**, 476 (1980); L. C. Smedskjaer and M. J. Fluss, *Bull. Am. Phys. Soc.* **26**, 461 (1981).

[83] J. E. Dickmann, R. N. Jeffrey, and D. R. Gustafson, *J. Nucl. Mater.* **69/70**, 604 (1978).

For the case of ^{58}Co, 500 mCi seems to be the limit. The sources to be used for angular correlation experiments must be rugged enough to survive the temperature changes encountered during the course of an experiment. Moreover, no loose activity can be present, since this not only represents a health hazard but can also ruin experimental results, and cause problems of decontamination of the experimental equipment.

The use of ^{22}Na as a positron source for angular correlation experiments provides several advantages over ^{58}Co. These advantages include a half-life of 2.7 yr versus 70 d, a six times greater specific activity for positron emission, and although significantly more expensive per source, the cost per positron is nearly a factor of five less for ^{22}Na. The maximum amount of ^{22}Na that can be provided commercially as a small spot source is ~ 50 mCi; however, the deposit is much more fragile than the electrodeposited and diffused Co source. This fragility and the tendency to flake require that the source be encapsulated in a welded buttom with a thin window allowing the positrons to emerge. Because the sources will be used in the presence of magnetic fields for focusing the positrons, the materials used in the construction of the source should be nonferromagnetic; otherwise, for example, the thin window could be torn apart when the field is applied. The thin Ti windows that are presently in use absorb $\sim 15\%$ of the positrons. The window represents the limiting factor in the future design of these sources. It must be capable of surviving hundreds of pressure cycles since these sources are usually used in vacuum.

From the point of view of cost ^{22}Na is superior to either ^{64}Cu or ^{58}Co, unless one has access to large amounts of free reactor time. The radiation from a 50 mCi ^{22}Na source poses significantly less of a hazard than the corresponding 500 mCi ^{58}Co. However, the financial risk factor with the ^{22}Na source is greater since it is clearly more fragile, and precautions should be taken for recovering the ^{22}Na source material should damage result, since it can be used again in the manufacture of a new source.

2.4.6. Lifetime

Although the conceptual nature of lifetime experiments is straightforward, the associated equipment and its proper use constitute the most complex of the three general positron methods. We have included a detailed discussion of the physical properties that determine the resolution of lifetime spectrometers. This will prove of particular use to those who wish to construct their own lifetime equipment since it will enable them to calculate the expected timing resolution.

2.4.6.1. *Detectors and Electronics.* An example of the detectors and electronics is shown in Fig. 15. Two fast scintillators (e.g., Pilot-U) are

2.4. DETECTION OF ANNIHILATION RADIATION

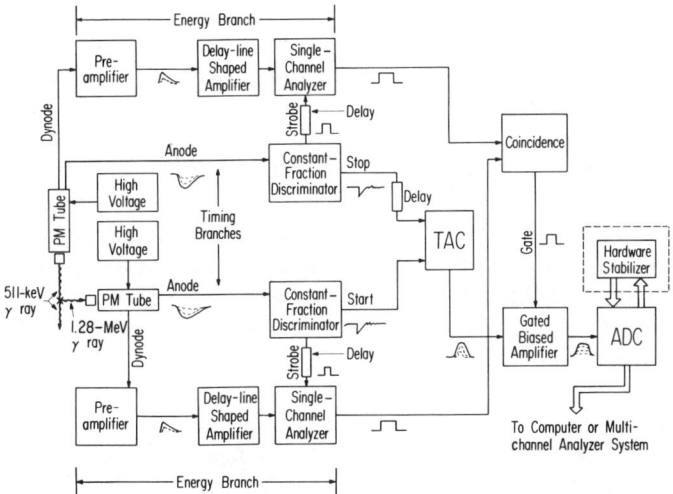

FIG. 15. Lifetime Detectors and Electronics. An illustration of a typical fast–slow lifetime system is shown. Details of its operation are covered in the text. Note that the hardware stabilizer is optional.

mounted on two fast photomultiplier tubes (e.g., RCA-8850). The anode pulse is passed directly to a constant-fraction discriminator, which produces a standard pulse timed with respect to the anode pulse. The two constant-fraction discriminators provide the start and stop signals for a time-to-amplitude converter (TAC). Either the birth or the annihilation γ ray may be used as the start signal with the other being used as the corresponding stop. Appropriate delays are used to place the event in the range of the TAC by moving the start and stop signals in time with respect to one another. It is common to optimize one detector to function as the birth γ-ray detector and the other as the annihilation γ-ray detector. The TAC then provides a voltage signal proportional to the time elapsed between the start and stop pulses. The TAC output is subsequently analyzed into a histogram by passing its output to a multichannel analyzer consisting of an ADC and the associated memory storage unit.

An energy analysis branch can also be established by taking the signal from one of the last two dynodes of each tube and then preamplifying and amplifying the signal. The correct energies are selected or gated by means of single-channel analyzers, which drive the two sides of a standard coincidence unit (typical coincidence resolution of ~ 30 ns). The presence of two acceptable coincident energy pulses is used to gate the multichannel analyzer.

The lifetime equipment thus has two branches, a timing branch, which measures the elapsed time, and the energy branch, which is used to select appropriate energies for coincidence events and thus to provide a gating condition for the storage of the timing information. This type of lifetime equipment is frequently called a fast–slow coincidence system and is distinguishable from the newer fast–fast coincidence systems.[84,85] In the latter, the energy discrimination is made in the fast timing circuitry itself, thereby avoiding the need for a slow-energy branch. The commercial availability of fast–fast constant-fraction units means a significant simplification of the equipment, although the energy selection is not quite as good as that which can be achieved with dedicated energy branches. The energies in the high-energy branch are selected to be above the Compton edge of the 511-keV quantum but below the Compton edge of the 1.28-MeV quantum (assuming ^{22}Na is used). The energies in the low-energy branch are selected below the Compton edge of the 511-keV quantum. The energy settings are, of course, independent of whether fast–fast or fast–slow or any gradation in between is used.

2.4.6.2. *Constant-Fraction Discriminators and TAC*. The heart of the electronics is the constant-fraction discriminators and the TAC. The TAC consists of a constant-current generator that begins charging a capacitor when the start pulse has been accepted and stops charging when the stop pulse arrives. The voltage over the capacitor will then be It/C, where I is the current, C the capacitance, and t the elapsed time.

The purpose of the constant-fraction units is to provide a timing pulse that is independent of the anode pulse amplitude. The anode pulse, of amplitude A, can be described as a function of time by the expression

$$V(t) = Av(t), \qquad (2.4.4)$$

indicating that all pulses have the same shape ($v(t)$ is the functional form of the pulse) but differ in amplitude. The anode pulse is now reshaped to

$$F(t) = V(t - t') - f_c V(t) = A[v(t - t') - f_c v(t)], \qquad (2.4.5)$$

where t' and f_c (the constant fraction) are constants. Examples of such pulse shaping are shown in Fig. 16 for two different amplitudes A. It can be seen that the crossover point $[F(t) = 0]$ is an amplitude-independent timing reference. The ideal situation of amplitude-independent timing is usually not realized. This dependence on amplitude is given the name of amplitude walk, or simply walk. The use of amplitude-compensated and rise-time-compensated discriminators can reduce this problem, and thus they are often employed.

[84] W. H. Hardy and K. G. Lynn, *IEEE Trans. Nucl. Sci.* **NS-23**(1), 229 (1976).

[85] M. O. Bedwell and T. J. Paulus, *IEEE Trans. Nucl. Sci.* **NS-23**(1), 234 (1976); **NS-26**(1), 422 (1979); *Proc. 5th Int. Conf. Positron Annihilation, Lake Yamanaka, Japan, 1979* (R. R. Hasigawa and K. Fujiwara, eds.), p. 375. Jpn. Inst. Metals, Sendai, 1979.

2.4. DETECTION OF ANNIHILATION RADIATION

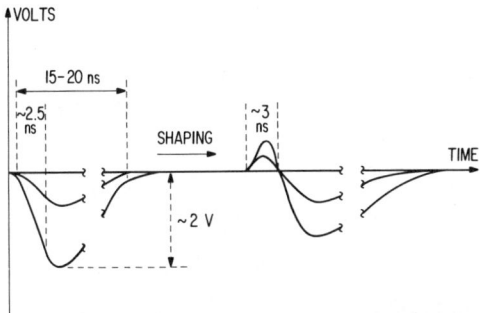

FIG. 16. Pulse shaping for fast timing. Anode output pulses from a phototube coupled to a fast scintillator (left) and constant-fraction shaped anode pulses (right). Numbers given are only approximate. It is noted that the crossover point in shaped pulses is amplitude independent.

2.4.6.3. Timing Resolution.
The most important property of a lifetime system is the timing resolution. The timing resolution depends on

(1) the scintillator and phototube subsystem, and
(2) the operating characteristics of the electronics.

The major source of timing spread is the tube–scintillator assembly, and therefore it deserves some critical attention. In Fig. 17 a typical ^{22}Na-derived spectrum obtained with Pilot-U is shown. Only the Compton edges at 341 keV and 1 MeV of the two emitted γ rays (511 keV and 1.28 MeV) are

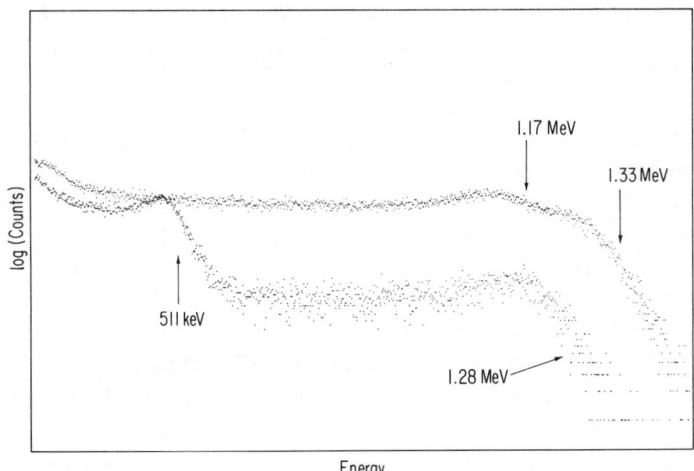

FIG. 17. Pulse-height spectra from Pilot-U. Energy spectra for ^{60}Co (upper) and ^{22}Na (lower) as measured with Pilot-U. The Compton edges are indicated (arrows) along with the total energy the incident γ ray.

observed, the photopeaks are missing. This is a consequence of the low density and low atomic number of the plastic.

After the interaction with the γ quantum the plastic emits light at a frequency suitable for the phototube cathode. The scintillator contributes to the resolution in two ways:

(1) the shape or time evolution of the emitted light pulse, and
(2) the number of light quanta reaching the photocathode.

The phototube contributes to the resolution in three ways.

(1) the photocathode efficiency (quantum efficiency),
(2) the statistical spread in the transit time of a single electron from the cathode to its arrival as an electron cloud at the anode (this spread has been measured as less than 500 ps for the RCA-8850 and RCA-C31024 tubes.[86]), and
(3) the systematic differences in the transit time from different parts of the cathode (this problem may be reduced somewhat by fully illuminating the cathode by applying a diffuse reflector to the scintillator surface).

The shape of the light pulse is determined by the scintillator material, and also by the shape and surface characteristics of the scintillators. The light pulse from Pilot-U has a decay time of $\gtrsim 1.4$ ns, which can be compared to that of NaI(Tl), ~ 230 ns. Recent work[87] has shown that the apparent decay time and the light yield depend strongly on both the dimensions and the surface treatment of the crystal; the decay time increases with scintillator dimensions, and the shortest decay times are obtained when the scintillator has a black surface. Surfaces painted with MgO give rise to diffuse light scattering and have significantly longer decay times.

The light output efficiency of the scintillator and the efficiency of the photocathode should also be considered. For Pilot-U, the output relative to anthracene is $\sim 60\%$, but the same relative yield for NaI(Tl) is 230%. The photocathode efficiency (the number of photoelectrons created per incident quantum) is important in determining the amplitude of the pulses produced and arriving at the constant-fraction discriminators. The newest generation of phototubes has quantum efficiencies in the range of $\sim 30\%$ compared to older designs with 12-15 efficiencies.

Pulse shaping and crossover triggering are required in fast-timing applications to minimize the effects of amplitude walk. The analytical representation of this technique is useful in predicting the characteristics of a particular

---

[86] B. Leskovar and C. C. Lo, *Lawrence Berkeley Lab. Rep. LBL* **LBL-3063** (1974); *Nucl. Instrum. Methods* **123**, 145 (1975).

[87] M. Moszynski, *Nucl. Instrum. Methods* **134**, 77 (1976).

2.4. DETECTION OF ANNIHILATION RADIATION

instrument. Let $f(t - t_i)$ be the crossover-shaped pulse appearing in the zero detecting circuitry inside the constant-fraction unit at time t_i in response to a single electron emitted from the photocathode. The probability density for t_i is

$$P(t_i) = \int L(\phi) T(t_i - \phi)\, d\phi, \tag{2.4.6}$$

where $L(t)$ is the probability that an electron is emitted at time t from the cathode, while $T(t)$ is the tube-transit time distribution. Strictly, one should further take into account possible variations in $f(t)$ due not only to the transit spread but also to the multiplication statistics of the tube. This will, however, be disregarded in this discussion.

The observed crossover pulse for n_e electrons emitted from the cathode and appearing at times t_i ($i = 1, \ldots, n_e$) is

$$F(t) = \sum_{1}^{n_e} f(t - t_i). \tag{2.4.7}$$

The crossover point in $f(t)$ is called t_c and the goal is to determine the crossover point t_0 in $F(t)$. The pulse $f(t)$ can be evaluated in a Taylor series:

$$\begin{aligned}F(t) &= \sum_i \sum_j \frac{1}{j!} \frac{d^j f}{dt^j}\bigg|_{t_c} (t - t_i - t_c)^j \\ &= \frac{df}{dt}\bigg|_{t_c} \sum_i (t - t_i - t_c) + \frac{1}{2} \frac{d^2 f}{dt^2}\bigg|_{t_c} \sum_i (t - t_i - t_c)^2 \\ &\quad + \cdots, \end{aligned} \tag{2.4.8}$$

where we have used $f(t_c) = 0$. The above expression should now be solved for t_0 such that $F(t_0) = 0$ to obtain t_0 as a function of the times t_i ($i = 1, \ldots, n_e$). The statistics of t_0 can then be calculated, but only if $f(t)$ and $P(t)$ are accurately known. In the following, we shall be satisfied by the approximate solution obtained from assuming $f(t)$ to be linear near the crossover point t_c. Thus one obtains

$$t_0 \sim \frac{1}{n_e} \sum_{1}^{n_e} (t_i + t_c); \tag{2.4.9}$$

and for the variance of t_0:

$$\text{var}(t_0) \sim \frac{1}{n_e^2} \sum_{1}^{n_e} \text{var}(t_i) \sim \frac{\tau^2}{n_e}, \tag{2.4.10}$$

where τ is the approximate decay time of the scintillations. Formally, the transit time spread for a single photoelectron is incorporated in $\text{var}(t_i)$, but since the spread (~ 500 ps for an 8850 tube) is small compared to that of the scintillators, it has been omitted. (One notes that for a minimal energy pulse of ~ 300 photoelectrons the observed transit spread will be ~ 30 ps.)

The expression for var(t_0) does not contain the constant fraction f_c. The reason is that the approximation corresponds to a case where the rise time of $f(t)$ is much larger than the decay time of $L(t)$. In the opposite case, where the decay time is long compared to the rise time, var(t_0) would depend on the fraction f_c. For both of these extreme cases simple analytical solutions can be obtained. The real situation is somewhere in between, since the rise time and decay time are about equal, and the expression in Eq. (2.4.10) is expected to overestimate var(t_0).

The number of electrons emitted can be written $n_e = E_\gamma A_a \epsilon_r \eta$, where E_γ is the energy deposited by the incident quantum, A_a is the number of photons generated in anthracene per eV deposited in the crystal, ϵ_r is the light yield of the scintillator relative to anthracene, and η is the quantum efficiency of the photocathode. For events in a narrow energy window around E_γ we obtain

$$\text{var}(t_0) \sim \frac{\tau^2}{E_\gamma A_a \epsilon_r \eta}. \qquad (2.4.11)$$

For actual lifetime equipment the energy windows used have a finite width. When using energies from E_1 to E_2 while assuming the energy spectrum to be constant in this region, we set

$$\text{var}(t_0) \sim \frac{\tau^2}{A_a \epsilon_r \eta} \frac{\ln(E_2/E_1)}{E_2 - E_1}. \qquad (2.4.12)$$

This expression is not exact, but it does provide the experimenter with a practical tool for selection of scintillators and energy windows, assuming the use of a good photomultiplier.

It is instructive to calculate the system resolution function. For the low-energy branch (511 keV) we take the side-channel energy window to be 100–330 keV, whereas the settings are 500–1000 keV for the high-energy branch (1.28 MeV). Using $A_a \sim 15$ photons per keV, $\eta = 0.3$, $\epsilon_r = 0.6$, and $\tau \sim 2$ ns (typical for a 1-by-1-in. crystal), we obtain for the low energy branch var(t_0) $\sim 7.7 \times 10^3$ ps^2, for the high-energy branch var(t_0) $\sim 2.1 \times 10^3$ ps^2, and therefore for the system var(t_0) $\sim 9.8 \times 10^3$ ps^2 corresponding to a FWHM of 234 ps.

The scintillator, Pilot-U, considered by itself, is capable of somewhat better resolution since $\tau \sim 1.4$–1.7 ns and $\epsilon_r \sim 0.67$ giving a FWHM of 155–188 ps. Note that the 511-keV branch is the major contributor to the resolution broadening.

To determine the final resolution of the system one must then add the effects from the electronic resolution and the walk of the system. In a

2.4. DETECTION OF ANNIHILATION RADIATION

well-adjusted system these two contributions will not be significant, but they should not be totally neglected either. In Fig. 18 the electronic resolution (in units of FWHM) of a commercially available constant-fraction discriminator is shown as a function of input amplitude. For input pulses greater than 0.3 V the resolution of the electronics is quite good and will not contribute significantly to the entire resolution of the system, which usually has values in the range 210–300 ps. A prerequisite is thus to obtain high-amplitude anode pulses so as to avoid the low-voltage region of the constant-fraction units.

Figure 19 shows as an example the FWHM, walk, and counting rates versus energy selection in the 511-keV and 1.28-MeV branches. The FWHM and walk have been measured using a ^{60}Co source, and the counting rate was determined using a ^{22}Na source. The strong energy dependence of the resolution is noteworthy. For low energies the electronic system is the dominant source of degradation in the resolution function. Plots such as Fig. 19 permit one to determine the resolution function and count rates obtainable as a function of side-channel energy selection.

The timing resolution function is experimentally determined by using the coincident γ rays from a radio isotope. The resulting resolution function is dependent on the energy spectra of the γ rays deposited in the scintillator. A commonly used γ-ray cascade is the 1.33- and 1.17-MeV γ rays from ^{60}Co, the energy spectrum of which is shown in Fig. 17 for Pilot-U. The deduced resolution function may not adequately reflect the true properties of the lifetime system when actual spectra are collected with ^{22}Na. There are a variety of reasons for such a discrepancy, but perhaps the most important is the change in the relative weighting of the different energy regions and the corresponding systematic changes in the sampling of the energy-related amplitude walk.

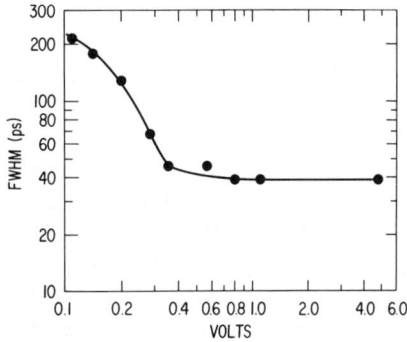

FIG. 18. Resolution versus pulse amplitude. Electronic resolution of a constant-fraction discriminator as a function of input voltage. Note that for this particular case the electronic resolution remains at ~ 40 ps for input voltages greater than 300 mV.

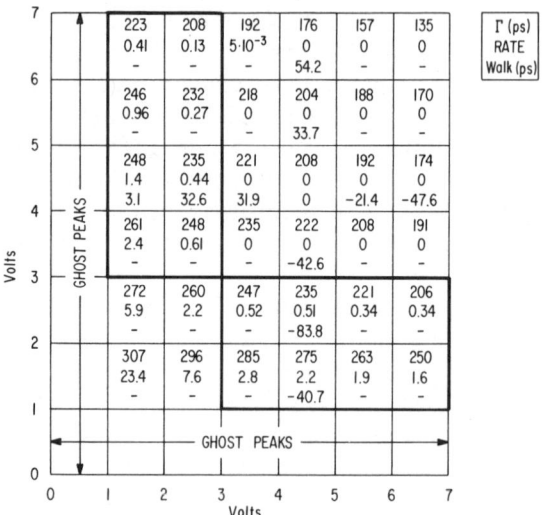

Fig. 19. Effects of side-channel energy selection. A lifetime equipment resolution function is shown versus pulse-height selection (in volts) for the two side channels. The FWHM and walk have been measured using a ^{60}Co source, and the rate (carbitrary units) was measured using a ^{22}Na source in collinear geometry. A window width of 1 V has been used. The walk has been determined only for the 4–5-V energy window in each side channel, since the remaining values can be calculated. The diagram can be used to estimate the final resolution function and positron-counting rate as a function of pulse-height selection. Pulse-height selections inside the heavily framed regions are suitable for actual experiments.

2.4.6.4. Scintillator Shapes and Assembly.

Commonly used fast scintillators are made from plastics. They are often machined into cylindrical shapes with typical dimensions (diameters and thicknesses) between 25 and 50 mm. Since the resolution function is determined by the light yield on the cathode of the phototube, along with the light scattering in the surface of the crystal, small crystals (<25 mm in diameter) have often been favored in the past in high-resolution work, although larger crystals offering higher detection efficiency but somewhat poorer resolution are used at many laboratories. The standard technique calls for coating the surfaces of the plastics with MgO powder in order to obtain diffuse light reflection at the surfaces. This technique results in a more even illumination of the cathode, thus avoiding some of the resolution broadening associated with the spread in transit time from different areas of the cathode. The MgO coating also improves the light yield, but results in a longer effective decay time of the scintillations compared to black surfaces. It is not easy to decide which is preferable, and some experimentation with different techniques should be anticipated when setting up the lifetime equipment.

The shape of the scintillator does affect the resolution, and this is particularly true for the larger-volume scintillators. Experimentation and consideration of the light scattering in the crystal has resulted in some apparent design improvements. A scintillator shaped like a truncated cone has been shown to give better resolution than a corresponding cylindrical one.[88] This work is especially significant since it allows lifetime measurements to be made with larger crystals (increased count rates) with a minimum deterioration in the resolution.

2.4.6.5. Mistriggering of Discriminator. A common and potentially serious problem in fast timing circuits arises from mistriggering of the discriminators on the leading edge of a misshaped or low-energy pulse. This problem is likely to occur if the lower-level discriminator in the low-energy branch is set too low (see Fig. 19). Often the effect of such a misadjustment is seen as ghost or side peaks in the time spectrum, although these peaks are sometimes buried beneath the main peak and are only discernable as a broadening of the resolution function. Since such misfiring is considerably more sensitive to instabilities, misadjustments must be avoided. It should be pointed out that ghost peaks can also be caused by detector–detector scattering of the quanta.[64] A practical way of separating these effects is to move the detectors and see whether the ghost peaks move as well.

2.4.6.6. Geometry and Scattering Effects. The source–sample–detector geometry has important effects on the properties of the lifetime spectrum. The collinear geometry, where detectors and source are on the same line, with the detectors separated by the source, should be avoided. This arrangement makes it possible to observe events where one detector sees a 511-keV γ ray and the other sees both the 1.28-MeV and the other 511-keV γ rays. Since the decay time of the scintillations is on the order of nanoseconds, and the elapsed time between the birth γ (e.g., 1.28 MeV from ^{22}Na) and the subsequent annihilation is ~ 100 ps, the birth γ and the 511-keV quantum can not be fully separated from one another, and thus the system resolution will be degraded. A commonly used arrangement is, therefore, to mount the detectors facing one another, with the source at the side, with an antiscatter shield placed between the detectors. This geometry tends to ensure that the scintillator pair cannot be hit by more than one 511-keV quantum per annihilation.

In experiments where the utmost precision is demanded one must consider subtle effects associated with the equipment geometry, including sample holder and vacuum chamber, and the energy-window settings. Material surrounding the source will backscatter quanta into the detectors. This makes it possible that truly coincident quanta can follow different

[88] G. Kögel, *Proc. 5th Int. Conf. Positron Annihilation, Lake Yamanaka, Japan, 1979* (R. R. Hasigawa and K. Fujiwara, eds.), p. 383. Jpn. Inst. Metals, Sendai, 1979.

routes from sample to detector, resulting in an additional time spread and degradation in timing resolution.

The detectors themselves may scatter the quanta. As an example, consider a quantum that has deposited enough energy to be within the energy window for a particular scintillator, but has backscattered out of the scintillator at an angle corresponding to the energy deposited (Compton scattering). Such backscattered low-energy quanta will be found within two cone-shaped volumes, whose position in front of the detector depends on the energy-window setting. Since the energy of these quanta is low, other detectors present in the cone volume have a good chance of detecting these scattered events because of the increase in the linear attenuation coefficient with decreasing energy. The detectors should be kept out of such cones to minimize this effect since a resolution degradation similar to that resulting from collinear geometry will otherwise occur. Fortunately, this detector–detector scattering is easily measured unlike scatter from other materials surrounding the source. For example, detector–detector scattering of the 1.28-MeV quantum can be studied by use of a NaI(Tl) detector in triple coincidence with the fast lifetime detectors. The NaI(Tl) detector serves the purpose of eliminating the possibility of scattering of the annihilation quanta in the system. Thus the energy window in the NaI(Tl) detector is set around the 511-keV photopeak, ensuring that there will be no Compton-scattered 511-keV quanta coming from this detector. Further, the NaI(Tl) detector is placed so that the other collinear 511-keV quantum cannot hit any component of the equipment. The ratio of triple coincidences to coincidences between the NaI(Tl) detector and one of the fast tubes will be the probability that a single 1.28-MeV quantum can excite both fast detectors. This procedure and the analogous procedure with respect to the 511-keV quantum can be used to determine the energy windows giving minimum detector–detector scattering. One may speculate that antiscatter shields could solve some of these problems more elegantly. However, in the authors' experience shields more often than not degrade the system resolution rather than improve it. One might suspect that scattering rather than count rates or instrumentation is the real limitation for high-precision lifetime measurements in the future.

2.4.6.7. Basics of Data Analysis. The analysis of lifetime data is unique among the various positron techniques in that the resolution of these spectra into lifetime components can be directly associated with underlying physical mechanisms or states of the positron in the sample. This can provide the experimenter with a better, or at least a more detailed, understanding of the experiment than could be achieved by Doppler broadening or angular correlation, especially when the sample contains a complex population of positron traps (see, e.g., Fig. 4).

2.4. DETECTION OF ANNIHILATION RADIATION

The least complex treatment of a lifetime curve is the determination of changes in the mean lifetime by comparing the relative positions of the 0.511–1.28-MeV timing distribution and the 511–511-keV distribution using collinear geometry. This method involves only centroid calculations and thus provides an easily employed experimental-analysis technique.

In the more detailed analysis of lifetime data, two aspects should be emphasized: first, the effect of the resolution function and, second, the source-correction effect.

The observed lifetime spectra are commonly modeled as a convolution between a sum of exponentials and the instrument resolution function. For example, Eq. (2.2.8) represents such a sum. However, it should be noted that the measured spectrum is the time derivative of the expression in Eq. (2.2.8) convoluted with the resolution function. The physical significance of the exponential terms is obtained by invoking the trapping model. Here we concentrate on how to deduce the exponential terms from the measured spectrum rather than discuss the physical meaning of these terms.

The goal of the deconvolution procedure is to estimate the intensities and the decay rates of the exponential sums. These parameters might be expressed in terms of parameters with a physical meaning. For example, compare the parameter set Λ_1, Λ_2, and I_2 to the parameter set λ_b, λ_t, and κ in Eq. (2.2.8)–(2.2.11).

Conceptually, a deconvolution procedure consists of guessing at a set of parameters, calculating the exponential sum, and convoluting it with the known resolution function. The result is then compared with the observed spectrum, iterating this procedure until agreement within statistics is achieved. Technically, the procedure is executed by a least-squares-fitting program in a computer, where the final result of this procedure depends on which resolution function is used. The resolution function is often determined by substituting a ^{60}Co source (1.17 MeV and 1.33 MeV in coincidence) for the ^{22}Na source, but not changing anything else. This procedure may result in a different resolution function than when the positron source is in place, and one might thus ask how such an error could influence the parameters to be determined.

At present, the best solution to this problem seems to be to determine the resolution function from the lifetime data themselves. Thus the parameters describing the resolution function are incorporated into the least-squares-fitting routine, which also determines the physical parameters (e.g., decay rates and intensities). For example, the resolution function may be described as a sum of two Gaussians with a relative centroid shift t_s, and standard deviations σ_1 and σ_2, and relative intensities of P_1 and P_2 ($= 1 - P_1$), respectively. Thus, in this example, the resolution function is described by four independent parameters. It is often found during the course of an experiment that fewer parameters are needed to describe variations in the resolution functions from one lifetime spectrum to the next. Such variations are due to instrument instabilities. As an example, in the work of Fluss *et al.*[37] it is concluded that all lifetime spectra reported can be fitted when only σ_1 and σ_2 are free to vary along with the physical parameters. The remaining parameters describing the resolution function could be kept fixed. Thus, although one might need several parameters to describe the resolution function, only a few of these need be taken into account in the least-squares fitting of the observed lifetime data.

To determine the actual resolution function of the instrument one might actually measure the lifetime spectrum for a sample known to exhibit only a single short lifetime, as, for example, well-annealed Cu at room temperature ($\sim$ 119 ps, Smedskjaer *et al.*[89]). The resolution function may now be determined by fitting the parameters describing the resolution function simultaneously with the decay rate in the annealed Cu sample.

In fitting both the resolution function and the physical parameters of interest simultaneously, larger statistical error bars on the physical parameters are obtained than in a fitting procedure with an a priori resolution function. It can be argued that the apparently smaller error bars obtained in the latter case are artifacts since these error bars do not include the uncertainties due to the imprecise knowledge of the instrument resolution. Fitting of the instrument resolution along with the physical parameters of interest will result in more realistic error bars and thereby help to prevent haphazard conclusions.

A variety of mathematical descriptions of the resolution function have been used with equally good results. Since the resolution function, in the context of a lifetime experiment per se, does not contain any physical parameters of interest, the mathematical description is arbitrary with the exception that the final mathematical expression fits the actual resolution function within statistics. The Gaussian sum previously mentioned is frequently used, although, for example, Hall *et al.*[90] found it useful to describe the resolution function as a convolution between a double-sided exponential and a Gaussian.

The second problem to be considered in the context of lifetime data analyses is that of the source contribution. The source contribution is not specific to the lifetime method, but will also occur in Doppler broadening and possibly in angular correlation. However, in these two latter methods it is difficult, if not impossible, to separate the term from the data; therefore,

---

[89] L. C. Smedskjaer, M. J. Fluss, D. G. Legnini, M. K. Chason, and R. W. Siegel, *J. Phys. F* **7** 1715 (1977).

[90] T. M. Hall, A. N. Goland, and C. L. Snead, Jr., *Phys. Rev. B* **10**, 3062 (1974).

2.4. DETECTION OF ANNIHILATION RADIATION

the term is usually not considered in the context of the momentum measurements.

Phenomenologically, it is often observed that a lifetime spectrum for a metal (even well-annealed metals) exhibits a long, almost voidlike lifetime (~ 400 ps or more) with a low intensity (< 5%). The origin of this lifetime is not well understood, but the name suggests that the components originate in the source material itself. Clearly, if the source is wrapped in an envelope of material other than that of the sample, a source component arising from annihilation in this envelope should be expected. If a massless source is used (see, e.g., Fluss et al.[74]), no source component should be expected, yet it is still observed. One might therefore question the origin of this long lifetime, especially since this influences the data treatment, and possibly the conclusions with respect to the experiment. Usually one determines the lifetime and intensity of the source term, either from the lifetime data themselves or from a separate experiment and then subtracts the source component term from the observed data prior to the least-squares analyses. This procedure is adequate for cases where one can be convinced that the term originates in materials extrinsic or external to the sample under study. If, on the other hand, the term originates in the sample itself (for example, due to positron trapping at internal surfaces), then the above procedure would be incorrect, and the source term should instead be treated as another lifetime associated with positron trapping in the sample. It is therefore advisable to try both procedures in the analysis of an observed lifetime spectrum.

In metal samples expected to contain only one lifetime (e.g., a well-annealed sample at low temperatures) one might analyze in terms of two exponentials rather than one, with the second exponential being associated with the source term. Such an analysis is illustrated by Smedskjaer et al.[91] In this work, lifetime spectra obtained from supposedly well-annealed Cd were analyzed in terms of two lifetimes, assuming no source correction at first. Rather than consider the long lifetime and its intensity (which are strongly correlated) separately, it is advantageous to consider their product.[89] This product depended strongly on sample temperature,[91] thereby making it unreasonable to consider it as a simple temperature-independent source correction. Since the source-term product showed a minimum near room temperature, it was decided to assume that the product at this temperature was associated with a real source term, which could be subtracted from all measured spectra. The remaining data now indicated the presence not only of positrons annihilating in the perfect lattice but also of positrons trapped in traps with vacancylike lifetimes. These observations lead to the suggestion of the unexpected presence of dislocations in the sample. Thus a combina-

---

[91] L. C. Smedskjaer, D. G. Legnini, and R. W. Siegel, *J. Phys. F* **10**, L1 (1980).

tion of the different ways of analyzing the source term may often be helpful in arriving at the final interpretations.

Computer simulations of the effects of resolution function errors and their possible interplay with the source terms by Eldrup et al.[92] provide the reader with some quantitative estimates of how these errors affect the final results. It should be emphasized that simulations of this type should be done for the real observations at hand. In this way the experimenter obtains an estimate of the intrinsic value of the data.

Observed lifetime spectra are commonly described by as many as three exponential terms, including the source term, and in exceptional cases four exponential terms are warranted. This limitation on the number of terms is due to the counting statistics and the analytical properties of the models used to describe the spectra. The exponential terms are not mutually orthogonal within the data region, meaning that a unique solution to the fitting problem might not exist. The lack of orthogonality also entails high correlation coefficients between the deduced parameters, and hence they cannot be considered independently of one another. Finally, if new quantities are to be calculated on the basis of the deduced correlated parameters, the correlation should be taken into account both when calculating the expectation values and the variances.

One often encounters the problem of too many parameters to be determined from too few data. Strictly speaking, this problem can only be solved by increasing the available information. However, many workers, including the authors, take a shortcut by assuming a priori knowledge of some of the parameters to be determined from the experimental data. This type of constrained analysis is commonly used and, when based upon reasonable assumptions, leads to approximately correct results. The statistical error bars obtained from these kinds of procedures do not include the uncertainties in the assumptions themselves (e.g., it could be assumed that a vacancy lifetime was 270 ± 30 ps, where the 30 ps plays the role of an uncertainty). There is, therefore, an additional possibility of reaching erroneous conclusions.

Lifetime spectra that are studied as a function of temperature (or any other variable for that matter) may have in common some physical model for the temperature-dependent properties of the defects themselves. An important development in the treatment of such data was originally introduced by Hall et al.[90] Since the goal is to deduce a model for the underlying physics in the observed data, one wants to compare this model with the entire collection of data at one time rather than doing it piecemeal, spectrum by spectrum. This global approach is more satisfying since assump-

---

[92] M. Eldrup, Y. M. Huang, and B. T. A. McKee, *Appl. Phys.* **15**, 65 (1978).

tions or constraints can, to some extent, be avoided owing to the more effective use of the information contained in the data. Also, after correctly accounting for correlations between the deduced parameters, the sufficiency of a model to describe all the data can be tested. The global-fitting method is not commonly used, possibly because large amounts of computer memory and computing time are required. However, with the more widespread appearance of the virtual computer concept, relieving the space and time problems, and with the availability of high-level languages aimed at manipulation of large data sets, such global analyses will be used more frequently. The computer code needed to do a global analysis using a high-level language is in fact very simple. The model (e.g., the two-state trapping model, along with the description of the concentration of vacancy traps in a well-annealed sample as a function of temperature) can be specified in a few lines of the program along with a vector pointing to the data bank. The least-squares-fitting routine, which is generalized to handle any model once the model expressions are known, is subsequently called and the iterations can start. As an example of such a high-level system, one might mention SAS (Statistical Analysis System).[93] It is stressed that the use of global fitting is not in any way a substitute for a careful investigation of the agreement between the model and the observed data, but should only be seen as a complementary tool.

Global fitting is of course not limited to lifetime data, but can be applied to angular correlation and Doppler-broadening data as well.[51,94,95]

The standard for the analysis of lifetime spectra has been set by the steadily improved versions of the POSITRONFIT programs of Kirkegaard[96] and Kirkegaard and Eldrup.[97] These programs are based upon the Gaussian description of the resolution function, and the code is flexible regarding the constraints imposed on the analyses. Similar codes like DBLCON,[98] which employ alternative descriptions of the resolution function, are also available and should be considered when designing an analysis strategy. The choice of one particular analysis system should not lead to results substantially different from those obtained with another.

In conclusion, the major problems specific to the positron-lifetime

[93] J. T. Helwig and K. A. Council, eds. "SAS Users Guide." SAS Inst., Cary, North Carolina, 1979.
[94] L. C. Smedskjaer, M. J. Fluss, D. G. Legnini, and R. W. Siegel, *Proc. 6th Int. Conf. Positron Annihilation, Arlington, Texas, 1982* (P. G. Coleman, S. C. Sharma, and L. M. Diana, eds.), p. 526. North-Holland Publ., New York, 1982.
[95] M. J. Fluss, S. Berko, B. Chakraborty, K. Hoffman, P. Lippel, and R. W. Siegel, submitted to *J. Phys. F*.
[96] P. Kirkegaard, *Comput. Phys. Commun.* **7**, 401 (1974).
[97] P. Kirkegaard and M. Eldrup, *Comput. Phys. Commun.* **3**, 240 (1972).
[98] W. K. Warburton, *Comput. Phys. Commun.* **13**, 371 (1978).

method are related to the imprecise knowledge of the actual resolution function and the unknown origin of the source term. The more general problem of obtaining a working interpretation of positron data seems to be a problem that is shared with other experimental methods.

2.4.6.8. Positron Sources. Positron sources for lifetime experiments must have characteristics such that a birth signal associated with the injection of the positron into the sample can be obtained. From the point of view of efficiency and cost, ^{22}Na has been the most useful of the positron-emitting isotopes; consequently, it accounts for most of the positron-lifetime experiments. The birth γ ray produced by the daughter of the positron emission, ^{22}Ne, is observed at 1.28 MeV, and, as already shown, the observed pulse-height spectrum in the plastic scintillator consists of the two Compton edges, one from the 1.28-MeV line and the other from the 511-keV line. The ^{58}Co isotope can also be used for lifetime work. However, the branching ratio will result in a degradation of the signal-to-noise figure, and the low energy of the birth γ may broaden the timing resolution function.

Other methods of deriving timing signals all suffer from intrinsically poorer timing resolution than that achieved by direct observation of the daughter isotope itself. Some success has been achieved in passing the positrons first through a thin scintillator and picking the resulting light pulse as the initial timing event.[99] Although this method of timing electron beams is not new to nuclear physics, its application to positron studies is quite new. Resolutions of 400–450 ps were achieved with this method, and thus it might be suitable for certain investigations where decomposition of the resulting spectra into two or more components is not a paramount consideration.

The commonly used source–sample configuration in lifetime experiments is the sandwich technique previously considered in conjunction with Doppler broadening. When preparing a lifetime source one tries to avoid geometrically extended sources, so as to minimize time-of-flight differences. The remote-source technique is generally not used in lifetime experiments, as it would result in a poor signal-to-noise ratio, posing a severe limitation on the information that could be retrieved. In addition, the scatter in the time of flight between source and sample can contribute to the resolution function.

One might speculate that the basic limitations of the sandwich technique underlie the scarcity of high-temperature lifetime experiments. The molecular evaporated source technique[72] is therefore a welcome development. The ion-implantation technique[74,75] is preferred for its cleanliness, but is limited

---

[99] K. Maier and R. Myllylä, *Proc. 5th Int. Conf. Positron Annihilation, Laka Yamanaka, Japan, 1979* (R. R. Hasigawa and K. Fujiwara, eds.), p. 829. Jpn. Inst. Metals, Sendai, 1979.

by the availability of mass separators where operations allow for radioactive contamination.

2.5. Instabilities (Detectors and Electronics)

Probably the most pervasive problem in positron annihilation studies of defects in metals is the control and monitoring of instabilities. Usually these instabilities manifest themselves in the γ-ray detectors and/or the associated electronics, although instabilities can also arise from other parts of the experiment, e.g., from sample and source geometry changes or from variations in the metallurgical properties of the sample itself. In this section we consider the detector and electronic instabilities, their monitoring and control, and methods that can be used to stabilize data in which instabilities play a significant role.

Electronic instabilities are not surprising since in both Doppler broadening and lifetime measurements the components are pressed to perform to their maximum specifications. Many instabilities are controllable simply by the application of commonsense precautions. Line-voltage stabilizers will greatly reduce possible fluctuations. Care should be taken that fast transients do not enter the voltage supplies, a somewhat tedious task because of the extensive use of silicon-controlled rectifiers. The environment of all the electronics must remain constant in both temperature and humidity. The authors' experience has led to the use of controlled-temperature ($\pm 0.1°C$) boxes for all analog electronics. Humidity levels are kept low by appropriate dehumidification to avoid any adverse effects from hygroscopic components, particularly high-voltage circuits, which are prone to surface discharge. It is always a good practice to familiarize oneself with the temperature-coefficient specifications of the electronics in use in a particular experiment and to anticipate the expected levels of instabilities in the total system.

Sometimes the environment of the detectors themselves must be carefully controlled (e.g., the phototubes and preamplifiers of a lifetime system). The temperature coefficients of some detectors can be significantly and surprisingly large. In any system using phototubes, stray magnetic fields can play havoc with both gain and timing properties and such fields must be shielded out or stringently controlled.

Even after such care is taken, instabilities will still exist. Sometimes instabilities take place at such a high frequency that they become part of the instrumental resolution function, but often they lead to systematic variations when their frequency is low and their amplitude variable. These types of instabilities require either correction or data rejection. It is emphasized that no correction in the data, or part of the data, can remove the effects of instabilities completely, and it should also be kept in mind that any instability correction has the potential of deforming the observed spectra sufficiently to produce spurious effects. It is important when using a correction or stabilization method to demonstrate which spurious effects might be present and to estimate their magnitude. The presence of such stabilization- or analysis-induced effects, along with the correction procedure itself, sets an upper limit on the information that can be retrieved from the equipment or data. This implies that there is a maximum number of counts in the spectrum beyond which no additional precision is likely. It is clear that instability corrections are an inefficient way to obtain what is sometimes considered to be stability. The more the surroundings, and hence the electronics itself, can be stabilized, the better the results.

Two methods exist by which information is obtained about the status of the equipment.

1. In the simpler method, the data themselves are used to generate information about instabilities, and the resulting correction will mainly be a dc shift. A problem inherent to a method of this kind is self-correlation between the data and the stabilization, but this is avoidable.

2. In the second method, a reference spectrum not correlated with, and distinguishable from, the data is also recorded. By using such a method it is possible to make both dc, amplification (gain), and resolution function corrections. For Doppler-broadening experiments the ^{103}Ru line at 497 keV can be recorded along with the 511-keV annihilation line. A Ru line is shown in Fig. 20 as a typical example.

The latter method opens up the possibility of accepting only those annihilation spectra which are accompanied by Ru lines having shapes within certain limits, and thereby justifies the intercomparison between spectra. Smedskjaer et al.[100] have discussed this type of stabilization more extensively. The Ru line can be used to develop stabilization feedback algorithms that go beyond the simple idea of data rejection. Valid comparisons may, however, be limited by geometric reproducibility since the line shapes will depend to some extent on the source's position relative to the detector, as previously discussed.

Hardware stabilizers are commonly used in γ-ray spectroscopy. Such devices operate in real time off the output of the ADC and monitor the count rates in preselected left-hand and right-hand regions of the peak(s) while continually adjusting the amplification and/or the dc shift until the count rates are at a predetermined ratio. A more nearly ideal stabilizer has been described by Smedskjaer et al.[100] with a computer used as the feedback

---

[100] L. C. Smedskjaer, M. J. Fluss, and D. G. Legnini, submitted to *Nucl. Inst. Meth.*

2.5. INSTABILITIES (DETECTORS AND ELECTRONICS)

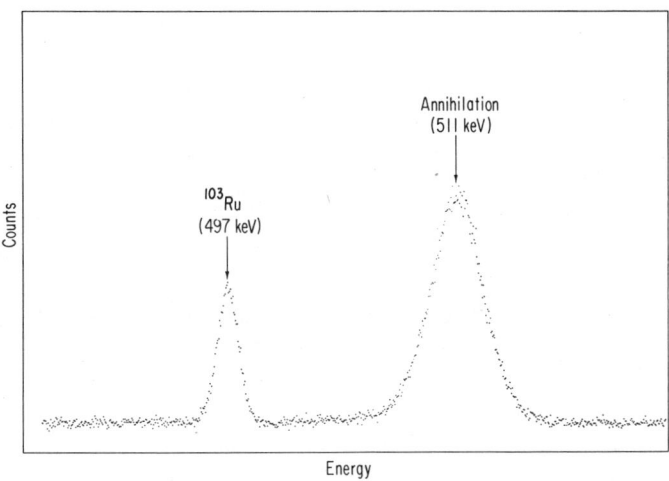

FIG. 20. The energy spectrum obtained when simultaneous monitoring of the ^{103}Ru line is used in Doppler-broadening experiments. The time interval and statistics are typical of the short experiments required for the computer-based postacquisition stabilization methods described in the text.

element. This system has as an advantage the concept of postacquisition stabilization and thus can be based on the statistical properties of the entire data set. For a Doppler-broadening experiment spectra are recorded during intervals that are short (e.g., 200 s) compared to the period of the dominant instabilities (e.g., 2000 s). The centroids of these spectra are subsequently determined, and by using the centroid as a reference, the line-shape parameter is calculated. This procedure is repeated until sufficient statistics are accumulated. The resulting list of line-shape parameters can subsequently be averaged to give the final mean line-shape parameter, and by using the central limit theorem, the uncertainty on the average is also determined. This procedure also omits the usual assumptions about Poisson statistics in the calculation of the uncertainty. The estimator of the uncertainty obtained in this way is more reliable than that based on statistics of counting alone since the scatter has been directly observed.

In lifetime experiments it is common to stabilize on the peak of the lifetime spectrum using a hardware stabilizer. Sometimes the prompt 511–511-keV peak is recorded along with the spectral data peak itself. This prompt peak is obtained by orienting the detectors opposite one another with the sample–source in between, a geometry that is normally avoided except for this purpose. Two coincidence signals are then generated, one for the 511–511 prompt events and another for the 511–1280-keV positron lifetime spectrum. The two coincidence signals act as routers for the multichannel analyzer. If it can be assumed that the 511–511-keV prompt peak and the 511–1275-keV peak are subject to the same system instabilities, then changes in the time distance between the two peaks are equal to changes in the mean life-time of the positron. This method for determining changes in the mean life-time is fast because the statistical uncertainty is related only to the second-order moments of the two peaks, whereas the uncertainty in a full deconvolution procedure is related to the fourth-order moments, a quantity that is large for a lifetime spectrum.

This method has several limitations that should be considered when evaluating the data. The two peaks often do not track completely together. This appears to be a consequence of different responses to instabilities on the part of the two different 511–511 and 1275–511 spectra. A primary reason for this might be the differences in the slopes of the crossover points (typically a factor of 3) between the anode pulses generated by the different γ-ray energies. A small change in the trigger level of the constant fraction discriminator (a not uncommon instability) could affect the timing of the 511-keV pulse substantially more than it would the timing of the 1.28-MeV anode pulse. Another limitation in this procedure is the contamination of the prompt 511–511 peak by Compton-scattered 1.28-MeV events that have energies in the 511-keV window. This means that the shape and possibly the position of the 511–511-keV prompt peak is affected by the very changes in the mean lifetime of the positron that one is attempting to measure.

When a lifetime event has been detected, it is known (from geometry arguments) that the plastic scintillator detecting the 1.28-MeV γ ray has been struck by the 511-keV quantum as well. Although the detection efficiency is low, it is not low enough to avoid summing effects. If, as an example, a 2.5-cm-thick crystal is used, the conditional probability of detecting the 511-keV γ ray is $\sim 25\%$ since the attenuation coefficients for plastics are ~ 0.1 cm^{-1}. Since the two γ rays arrive in a time that is short compared to the instrumental resolving time, a distortion in the leading edge of the anode pulse arises. Thus one might conclude by saying that the advantages of the centroid shift method are obtained at the cost of accuracy, and the final results should be evaluated with care.

Computer-based stabilization is also applicable to lifetime experiments. In those cases where mean lifetimes are to be deduced from the spectra, methods analogous to those described for Doppler broadening might be applicable. However, the deconvolution analysis of lifetime spectra requires such a large number of events that the smallest practical time interval for collecting a spectral unit would be much too long with respect to the usual type of instabilities that are encountered. It is thus more sensible to build a final histogram from all the microscopic recordings (e.g., 200 s each). This is done by calculating the centroid for each individual spectrum, then aligning or shifting the spectra according to their centroids, and finally adding all the aligned spectra together. Such a procedure becomes especially simple when only integer displacements of the microscopic data are carried out. In this case the spectra are being stabilized to ± 0.5 channels, which means that the conversion gain (time per unit channel) should be ~ 20 ps per channel or less. The average broadening of a lifetime spectrum due to this roundoff error will be ~ 0.4 ps, a value that is small enough for most realistic applications.

Any stabilization procedure, whether it takes place in real time or not, is based upon information regarding the time-dependent status of the system (e.g., resolution function, dc levels, and gain), and the procedure represents a hypothesis about how changes in system status affect the data. Therefore a statistical test of a physical model against the observations is a test not only of the model but of the stabilization hypothesis as well. From this point of view, postacquisition procedures offer clear advantages over real-time procedures since the stabilization hypothesis and the physical model can be viewed correctly as a single entity describing the observed (unstabilized) data. This continuous and direct test of the instability correction procedure cannot be achieved by real-time stabilization since the corrections are applied prior to data storage. The disadvantage of the superior postacquisition method is the observational technique (e.g., microscopic experiments) along with the large amounts of concomitant mass data storage that are required. These disadvantages can be anticipated to diminish rapidly in the future with the appearance of more advanced and less costly computer hardware. For a further discussion of the mathematical basis of various stabilization procedures see Smedskjaer et al.[100]

2.6. Some Final Remarks

It is appropriate for us to end this part by outlining some of the areas where the positron technique requires extra care while remembering that the unique nature of the technique and its general prowess in the defect area make it a useful tool in metallurgical studies.

Care must always be taken in positron experiments to ensure that the defect population being studied is indeed the dominant defect in the system with respect to positron trapping. At a low defect concentration, $\sim 10^{-6}$, which is the sensitivity limit of the positron technique for vacancy point defects, extreme care is required. It is under conditions such as this that low concentrations of additional defects (dislocations for example) have been shown to have significant or even dominant effects on the resulting data.

When one is dealing with a complex population of defects, vacancies, vacancy clusters, voids, dislocations, etc., questions will necessarily arise concerning the way in which the positron is partitioned among such a population. The partitioning of the positron among defect types present simultaneously in a system is not fully understood, and the interpretation of such data usually has a very qualitative character. Clear improvements are called for in the identification of parameters specific to different defects that can be derived from the experimental data. The number of positron states that can now be resolved is no more than 2 or 3, depending on how different the signals from these states are. The best results in separating such states are usually obtained in lifetime experiments.

To some degree one is always faced with contributions from positron annihilation events that do not take place in the part of the sample one desires to study. In experiments where bulk properties of the sample are to be investigated, the unwanted contribution to the observed data comes from annihilations in either the source or its supporting structures and/or those positrons that reach and eventually annihilate at or near the surface, or for that matter any place that can be viewed by the detection system. Currently, the best way to minimize such effects is the use of internal sources, produced by either ion implantation or activation techniques.

The positron technique has now gone through an initial period of growth with respect to its application to the study of defects in metals. This period has been successful in that increasing use is being made of the technique, not just by specialists, but by a broad group of material scientists. The technique is now entering a new stage of development where new methodologies are being explored that will uncover those properties of the annihilation signal that might be considered as defect specific. At the same time new information is being uncovered with respect to the trapping mechanism itself, particularly in traps where the binding energies of the positron are small.

The use of positron annihilation to measure vacancy formation enthalpies is well developed and is currently believed to be limited only by the variations in sample properties themselves: impurities and nonequilibrium defect concentrations arising from them.

On the instrumentation side, one can identify three general areas which hold special promise of yielding improvements in the amount and details of the data:

1. The use of large-area position-sensitive detectors operating in coincidence in ACAR experiments clearly increases data rates as well as providing considerably more flexibility in the analysis of momentum data. It is hoped that combining Doppler-broadening experiments with detailed ACAR work will lead to an efficient identification of interesting metal defects studies.

2.6. SOME FINAL REMARKS

2. Continued improvements are being made in lifetime techniques. The use of large-volume plastic scintillators is particularly noteworthy. Combined simultaneous lifetime and momentum (Doppler broadening) measurements are much more tractable, and these combined experiments yield an extra dimension of sensitivity and possible specificity. Earlier work has demonstrated that these two types of experiments are often invaluable together and provide additional insight into the underlying physical principles themselves.

3. The use of slow beams for positron sources opens the door to a whole new class of positron experiments, not only on surfaces and interfaces but also in thin foils; the latter are amenable to many other types of materials studies, from electron microscopy to resistivity, and are particularly valuable in nonequilibrium experiments such as radiation damage and quenching–annealing studies.

The techniques for analyzing experimental data are improving continually. Much more care is being taken in dealing with the statistical properties of the data and in applying stabilization methods. The relation between the physical models and the experimental uncertainties is appreciated, and more accurate results are being obtained by understanding this relation and recognizing the various limitations of the instrumental techniques.

3. NEUTRON SCATTERING STUDIES OF LATTICE DEFECTS

3.1. Static Properties of Defects

By W. Schmatz

Kernforschungszentrum Karlsruhe
Karlsruhe, Federal Republic of Germany

3.1.1. Introduction

Scattering experiments give information on the relative positions of scattering centers, i.e., on the internal structure of scattering objects. The structure can be as complicated as a biological molecule or as simple as a periodic arrangement of identical atoms. In the latter case, elastic scattering exists only for the scattering vector equal to one of the reciprocal lattice vectors, i.e., for Bragg scattering. Point defects like substitutional atoms, interstitials, and vacancies disturb the periodic arrangement. The interference pattern changes and the mutual arrangement of defects and the lattice distortion around them can therefore be studied from scattering experiments.

Regarding the scattering by point defects and small defect clusters, we distinguish between studies by Bragg scattering and those by diffuse scattering. Bragg scattering gives information by means of integrated reflectivities. there are no major difficulties in the experimental technique. The problem is mostly the so-called "primary extinction," characterized by the coefficient y_p. This coefficient is the ratio of the real integrated reflectivity to the value one would obtain for the scattering cross section calculated in the first Born approximation. However, the Born approximation holds only if the incident beam is not considerably attenuated by the Bragg reflection in question. If the coherently scattering crystallites are larger than the critical value given by the condition above, y_p is smaller than unity. The exact value of y_p depends on such large-size defects as large-and small-angle grain boundaries, dislocation networks, and others. Thus the contribution of point defects and small defect clusters to the integrated reflectivities can hardly be separated. Evidently, to obtain information on defect structure from Bragg scattering is more a problem of material preparation than of scattering technique. Therefore in this chapter we discuss only scattering from defect structures between Bragg reflections, often called diffuse scattering.

Compared to x rays, there is one essential advantage of neutrons in diffuse scattering: thermal diffuse scattering can be separated from elastic diffuse scattering. This allows, for instance, high-temperature studies with thermal diffuse scattering even a factor 30 higher than the elastic scattering. Another advantage of neutrons is the possibility of changing the coherent scattering length for a given element by use of isotopes. An excellent example is the study of clustering in CuNi alloys by de Vrijen et al.[1] Among others, they used the special alloy $^{65}Cu_{0.435}$ $^{62}Ni_{0.565}$ with the coherent scattering lengths 1.09×10^{-12} and -0.87×10^{-12} cm for the ^{65}Cu and the ^{62}Ni isotopes, respectively. For this alloy Bragg scattering vanishes and diffuse scattering is enhanced almost to its maximum value. The neutrons are complementary to x rays in all those cases in which the scattering contrast is more favorable for neutrons. The system Al–Mg is an example. Finally, only with neutrons is it possible to reveal the magnetic structure of defects. But this subject will not be treated here.

For the study of defects by diffuse scattering we are interested in the coherent scattering, i.e., the interference between the neutron wave amplitudes coming from the individual atoms. There is, however, additional incoherent nuclear scattering due to isotope and nuclear-spin disorder. In many cases this incoherent scattering is a serious limitation, and diffuse scattering can be measured only at high defect concentrations. However, there are many elements with favorably small values for the incoherent scattering cross section σ_{inc}. The σ_{inc} of C, O, Be, Bi, Pb, F, S, Al, Si, Nb, Mo, Sn, and Ta are smaller than or equal 20 mb (a typical value for the coherent scattering cross section σ_{coh} is 5 b, 1 b = 10^{-24} cm^2). Another 21 elements have σ_{inc} values below 0.5 barn. For all these 34 elements, diffuse neutron scattering is superior to or comparable with x-ray scattering. The situation may even improve: (i) Elements for which σ_{inc} values are at present badly known may well join the list of favorable elements. (ii) Isotopes with zero spin of the nucleus have $\sigma_{inc} = 0$; also, for other isotopes σ_{inc} is often very small. It may become more common to use isotopes. (iii) If in future very low temperatures (2–5 mK) can be achieved more easily, it might be possible to polarize the nuclei. Simultaneous use of polarized neutrons would also lower the σ_{inc} of isotopes to near zero. One would then have a choice between two scattering lengths for one nucleus by use of parallel or antiparallel neutron spin.

Studies of short-range order or clustering do not, in general, require small

---

[1] J. de Vrijen, C. van Dijk, and S. Radelaar, *Proc. Conf. Neutron Scattering, Gatlinburg, Tenn.* (R. M. Moon, ed.), **1**, 92 (1976).

values of σ_{inc}, at least for high defect concentrations. However, the σ_{inc} of the matrix atoms has to be small for all studies of the lattice distortion around point defects because such studies have to be performed at low defect concentrations. With special scattering techniques and for host lattices with small values of σ_{inc} (e.g., Al) considerable progress has been achieved in this field.

A scientific discipline always grows in its importance as the range of experimental possibilities increases. Because of the low intensities in diffuse scattering, the availability of high-flux reactors and cold-neutron sources was a decisive factor. Further technical improvements are neutron guide tubes, multidetectors, and on-line data reduction. Diffuse neutron scattering has now grown into maturity. In this chapter the method and the experimental techniques, in particular the specialized ones, are described. A short review on neutron small-angle scattering is included because this method proved to be successful in the study of defect clusters. The following section on theoretical background provides the nonspecialist with the general terms in the field.

There is some correlation to the chapter by N. Wakabayashi (see Chapter 3.3): elastic scattering transforms into quasi-elastic scattering for diffusing atoms. The special features of quasi-elastic scattering are discussed by Wakabayashi in Chapter 3.3. At the end of this chapter two typical examples of experimental results and the physics related to them are discussed as illustrations.

3.1.2. Theoretical Background

In the first Born approximation, the elastic scattering cross section of an ensemble of N atoms is given by

$$\left.\frac{d\sigma}{d\Omega}\right|_N = \left|\sum_{i=1}^{N} b_i \exp(-W_i) \exp(-i\mathbf{Q}\mathbf{r}_i)\right|^2 + \sum_{i=1}^{N} \frac{\sigma_{inc,i}}{4\pi} \exp(-W_i)$$

$$= \left.\frac{d\sigma}{d\Omega}\right|_{N,coh} + \left.\frac{d\sigma}{d\Omega}\right|_{N,inc}. \quad (3.1.1)$$

where $\mathbf{r}_i$ is the thermally averaged position vector of atom i and $\exp(-W_i)$ its Debye–Waller factor. $\mathbf{Q} = \mathbf{k} - \mathbf{k}'$ is the scattering vector, whereby $\mathbf{k}$ and $\mathbf{k}'$ are the wave vectors of the incident and the scattered wave, respectively. It is $|\mathbf{k}| = |\mathbf{k}'| = 2\pi/\lambda$, with λ the neutron wavelength, and thus $|Q| = 4\pi \sin(\phi/2)/\lambda$ with ϕ the scattering angle. b_i and $\sigma_{inc,i}$ are the coherent scattering length and incoherent scattering cross section of atom i, respectively. The first term in Eq. (3.1.1) is the neutron-coherent part of the scattering cross section with the coherent scattering amplitudes $b_i \exp(-W_i)$ of the individual atom at position $\mathbf{r}_i$ and the phase factor $\exp(-i\mathbf{Q}\mathbf{r}_i)$. The scattering cross section of

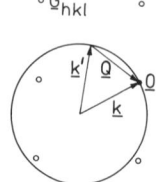

FIG. 1. Scattering experiment in reciprocal space. In this and other figures, vectors are marked with an underline.

eq. (3.1.1) is averaged with respect to nuclear spin and isotope disorder. If instead of natural elements, separated isotopes are used, the values of b_i and $\sigma_{inc,i}$ for the special isotope mixtures have to be used. Correlations between r_i, b_i, and $\exp(-W_i)$ can be of importance for light elements, especially mixtures of H and D. For molecular crystals like methane, nuclear-spin correlations have to be considered and Eq. (3.1.1) is no longer valid. For values of b, $\sigma_{coh} = 4\pi b^2$, and σ_{inc} — necessary to judge the type of possible experiments — we refer to Koester.[2] The second term in Eq. (3.1.1) is the neutron-incoherent part of the scattering cross section. It contains no positional information; however, it allows the determination of Debye–Waller factors in cases where one can neglect or correct for the coherent scattering cross section.

Before we consider special evaluations of the elastic coherent scattering cross section $(d\sigma/d\Omega)_{N,coh}$ we have to consider whether the assumption of first Born approximation is justified for the subject at hand. The criterion to be fulfilled is that the total cross section $\sigma_{tot} \simeq \int (d\sigma/d\Omega)_N \, d\Omega$ has to be smaller than the geometrical cross section σ_{geo} of the scattering object. Let us assume a crystalline spherical particle with radius R, an average atomic volume Ω_v, and an average scattering length $\bar{b}$ as the scattering object. Further, we assume the Ewald sphere touches none of the reciprocal lattice points $G_{hkl} \neq 0$ (Fig. 1), a necessary condition for a diffuse scattering experiment. Then the major contribution to σ_{tot} is given by the coherent scattering cross section in a forward direction, which, to a good approximation, is given by

$$\left.\frac{d\sigma}{d\Omega}\right|_{N,coh} = \left(\frac{4\pi}{3}\frac{R^3}{\Omega_v}\right)^2 (\bar{b})^2 \, e^{-Q^2 R^2/5}. \quad (3.1.2)$$

The criterion $\sigma_{tot} < \sigma_{geo}$ is then equivalent to

$$\sqrt{20/9} \, R\lambda \bar{b}/\Omega < 1. \quad (3.1.3)$$

For example, with $\lambda = 4$ Å, $\bar{b} = 10^{-12}$ cm, and $\Omega_v = 10^{-23}$ cm^3, Eq. (3.1.3) would require $R < 4 \times 10^{-4}$ cm. In contrast, scattering samples used

---

[2] L. Koester, *Springer Tracts Mod. Phys.* **80**, 1 (1977).

3.1. STATIC PROPERTIES OF DEFECTS

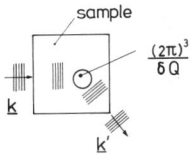

FIG. 2. Neutron-wave packets scattered in the sample.

for disorder scattering have sizes from a few millimeters up to centimeters; nevertheless we use Eq. (3.1.1). What is the argument for doing this? Experiments discussed here are mostly performed with moderate resolution δQ, say, in the range $(5 \times 10^{-3})-(1 \times 10^{-1})$ Å^{-1}. This corresponds to a real-space resolution $(2\pi/\delta Q)$ ranging from 60 to 1000 Å. This means for our purpose the effective size of the scattering object is indeed small enough to guarantee validity of first Born approximation. The correct view is to consider the scattering of the neutron wave traveling inside the large-size neutron samples (Fig. 2). The calculation of $(d\sigma/d\Omega)_{N,\mathrm{coh}}$ can be performed in principle for small volumes of $(1/\delta Q)^3$ that contain less than 10^8 atoms, but averaged for the whole sample.

Elastic neutron scattering means there is no energy transfer $\hbar\omega = E - E'$. (E and E' are the energies of the incoming and scattered neutron beams.) Experimentally, however, we always have some finite resolution $\delta\omega$, corresponding to a time element $2\pi/\delta\omega$. A photographer would say we take snapshots with a shutter speed $2\pi/\delta\omega$. The resolution $\hbar\delta\omega$ can be as small as 10 μeV. A resolution of 0.5 meV is an average value for many experiments. Thermal motions with frequencies lower than 0.5 meV are acoustic phonons with small q values ($\mathbf{q} = \mathbf{Q} - \mathbf{G}_{hkl}$), low-lying rotational modes—as they occur for large molecular units—tunneling modes, and diffusion. The experimenter must know whether his energy window includes scattering by such processes. Diffusion broadens the elastic into a Lorentzian shape described in detail by Wakabayashi in Chapter 3.3. The reciprocal of the half-width $2\omega_m(\mathbf{Q})$ is the characteristic time for diffusion out of a volume $(2\pi)^3/Q$. A simple example of the $\mathbf{Q}$ dependence of ω_m is given in Fig. 3. At small q values, $\omega_m = Dq^2$ holds with D the diffusion coefficient. Only for exceptional cases such as diffusion of hydrogen in body-centered-cubic (bcc) metals does diffusional broadening exceed our guide value of 0.5 meV. A very interesting experiment would be the determination of ω_m for vacancy diffusion, which is a factor $1/c_v$ faster than the corresponding value for

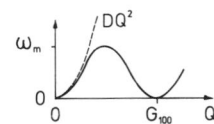

FIG. 3. Schematic sketch of $\omega_m(Q)$.

self-diffusion. (c_v is the vacancy concentration.) Such an experiment would be at the very limits of present-day possibilities.

We continue with the discussion of elastic scattering. As usual we use the scattering cross sections per atom defined by

$$\left.\frac{d\sigma}{d\Omega}\right|_{\text{coh}} = \frac{1}{N} \left.\frac{d\sigma}{d\Omega}\right|_{N,\text{coh}} \quad \text{and} \quad \left.\frac{d\sigma}{d\Omega}\right|_{\text{inc}} = \frac{1}{N} \left.\frac{d\sigma}{d\Omega}\right|_{N,\text{inc}}. \quad (3.1.4)$$

For small Q values, i.e., for $Qa \ll 1$ with a the average nearest-neighbor atomic distance, the sum in Eq. (3.1.1) is usually replaced by an integral introducing a scattering-length density $\rho(r) = \Sigma\, b_i/\Delta V$ with $\Sigma\, b_i$ the sum of coherent scattering lengths in the small volume V. We get

$$\left.\frac{d\sigma}{d\Omega}\right|_{\text{coh}} = \frac{1}{N} \left| \int \{\rho(\mathbf{r}) - \bar{\rho}\} \exp(i\mathbf{Q}\mathbf{r})\, d\mathbf{r} \right|^2 \quad (3.1.5)$$

where $\bar{\rho}$ is the average scattering length density of our N-atom sample. (The difference $\rho - \bar{\rho}$ logically has to be used if we regard the scattering of a neutron wave propagating within the sample.) For calculations, the value of $\bar{\rho}$ can be set to zero or any value between zero and $\bar{\rho}$ if the volume V occupied by the N atoms is large and the Fourier transform $\int \bar{\rho} \exp(i\mathbf{Q}\mathbf{r})\, d\mathbf{r}$ is small at the Q values of the experiment. Scattering cross sections according to Eq. (3.1.5) are derived for many large-size inhomogenities (e.g., voids, precipitates, dislocation loops, and grain boundaries), as discussed, for instance, by Schmatz et al.[3]

For Q-values higher or comparable to the reciprocal of the nearest-neighbor distance, the interference wavelength $2\pi/Q$ is comparable to or shorter than the atomic distances and details of the atomic arrangement are resolved. If we neglect lattice distortions in a first approximation, i.e., if the atoms are sitting on periodic sublattices indexed by ν, the coherent scattering cross section is given by

$$\frac{d\sigma}{d\Omega}(\mathbf{Q})_{\text{coh}} = \frac{1}{N} \left| \sum_{\alpha=1}^{\alpha_0} \sum_{\nu=1}^{\nu_0} \sum_{i=1}^{N_0} p_{i\nu\alpha}\, b_{i\nu} \exp(-W_{i\nu}) \exp[i\mathbf{Q}(\mathbf{R}_i + \boldsymbol{\rho}_\nu)] \right|^2, \quad (3.1.6)$$

with $\mathbf{R}_i$ as the periodic lattice vectors and $\boldsymbol{\rho}_\nu$ the positional vector of site ν inside the unit cell. The occupation number $p_{i\nu\alpha}$ is equal to one if an atom of species α is at the site (i,ν) and zero otherwise. It is convenient to count vacancies also as atomic species. Then we have

$$\sum_{\alpha=1}^{\alpha_0} p_{i\nu\alpha} = 1 \quad \text{for all} \quad \nu. \quad (3.1.7)$$

With the assumption that $\exp(-W_{i\nu}) = \exp(-W_{\nu\alpha})$, i.e., the Debye–Waller

[3] W. Schmatz, T. Springer, J. Schelten, and K. Ibel, *J. Appl. Crystallogr.* **7**, 96, (1974).

3.1. STATIC PROPERTIES OF DEFECTS

factors depend only on the sublattice and the atomic species, essential quantities for a disordered structure in Eq. (3.1.6) are the $v_0\alpha_0$ Fourier transform

$$\tilde{p}_{v\alpha}(\mathbf{Q}) = \sum_{i=1}^{N_0} p_{iv\alpha} \exp(i\mathbf{Q}\mathbf{R}_i). \tag{3.1.8}$$

We note that $\tilde{p}_{v\alpha}(\mathbf{Q}) = \tilde{p}_{v\alpha}(\mathbf{Q} + \mathbf{G}_{hkl})$. If we have no other restrictions than those of Eq. (3.1.7), there are $v_0(\alpha - 1)$ independent Fourier transforms $\tilde{p}_{v\alpha}$. For a binary substitutional alloy with one site per unit cell, $v_0(\alpha - 1)$ is equal to one, and we measure one physical quantity by diffuse scattering, i.e.,

$$\frac{1}{N}\left|\tilde{c}(\mathbf{Q})\right|^2 \quad \text{with} \quad \tilde{c} = \tilde{p}_1(\mathbf{Q}) = 1 - \tilde{p}_2(\mathbf{Q}), \tag{3.1.9}$$

as the reader can verify easily. For a random distribution this quantity is simply $c(1 - c)$. Short-range order or clustering gives characteristic slopes of $d\sigma/d\Omega$ as sketched in Fig. 4. For systems in thermal equilibrium $|\tilde{c}(\mathbf{Q})|^2$ is a function of temperature. It approaches the random value for infinite temperature. Evaluation with respect to $\mathbf{Q}$ and T gives insight into the effective interaction potentials and the statistical mechanics of the system; from this the great value of systematic studies of $|\tilde{c}(\mathbf{Q})|^2$ derives. For more details the reader is referred to the extensive literature.[4-6] A relation of importance in practice is

$$\int_{BZ} \frac{1}{N}|\tilde{c}(\mathbf{Q})|^2 \, d\mathbf{Q} = \frac{(2\pi)^3}{\Omega_v} c \tag{3.1.10}$$

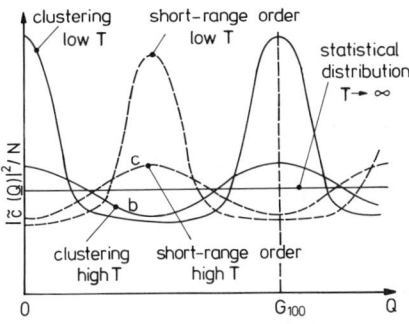

FIG. 4. The squared Fourier transform and concentration fluctuations for clustering and short-range order at various temperatures (schematic).

[4] D. de Fontaine, *Solid State Phys.* **34**, 73 (1979).
[5] M. A. Krivoglaz, "Theory of X-Ray and Thermal Neutron Scattering by Real Crystals" (transl. ed. by S. C. Moss). Plenum, New York, 1969.
[6] W. Schmatz, *Treatise Mater. Sci. Technol.* **2**, 105 (1973).

with Ω_v the volume of the unit cell. The integral is taken for one Brillouin zone (BZ). $|\tilde{c}(\mathbf{Q})|^2$ is peaked at reciprocal lattice points $\mathbf{Q}_{hkl}$. Subtraction of the peak integral leads to

$$\int_{\text{BZ(peak)}} \frac{1}{N} |\tilde{c}(\mathbf{Q})|^2 \, d\mathbf{Q} = \frac{(2\pi)^3}{\Omega_v} c(1 - c), \qquad (3.1.11)$$

an integral relation that leads to the relation

$$\int_{\text{BZ(peak)}} \frac{d\sigma}{d\Omega} (\mathbf{Q}) \, d\mathbf{Q} = \frac{(2\pi)^3}{\Omega_v} c(1 - c)(b_2 - b_1)^2 \qquad (3.1.12)$$

for the diffuse scattering.

Many substances (ternary alloys, foreign interstitials in metals, superionic conductors, and others) have more than one independent occupation number $\tilde{p}_{v\alpha}$, which makes it impossible to deduce quantities $|\tilde{p}_{v\alpha}|^2$ from one scattering cross section. Model calculations are then necessary. In some cases the use of various isotopes and/or the additional use of x-ray scattering data may solve the problem. X-ray data have to be corrected for inelastic scattering. This can be performed by calculation of the one-phonon and higher-phonon scattering cross sections or by a method developed by Borie and Sparks[7] for the correction of lattice distortions, which is analytically very similar. (Increase of the scattering amplitude is proportional to Q in contrast to the periodicity in reciprocal lattice for $\tilde{p}_{v\alpha}$.)

Rotational disorder in molecular crystals can be treated similarly (see, e.g., McKenzie and Seymour[8]). The scattering amplitude of the molecular unit is separated into an average and a fluctuating scattering amplitude; e.g., for the rotational disorder sketched in Fig. 5 the structural factor is

$$b \exp(-W)[(\cos \mathbf{Q}\rho_1 + \cos \mathbf{Q}\rho_2) + p_1(\cos \mathbf{Q}\rho_1 - \cos \mathbf{Q}\rho_2) + p_2(\cos \mathbf{Q}\rho_2 - \cos \mathbf{Q}\rho_1)] \quad (3.1.13)$$

with $p_1 + p_2 = 1$. One easily sees that diffuse scattering vanishes for $\mathbf{Q} \to 0$.

The next step is to include lattice distortions. This can be done reasonably only if a reference lattice remains, i.e., if the atoms can be indexed as in a

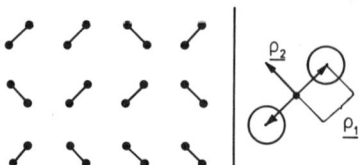

FIG. 5. Rotational disorder for dumbbells (left-hand side). Vectors used for Eq. (3.1.13) (right-hand side).

[7] B. Borie and C. J. Sparks, Jr., *Acta Crystallogr., Sect. A* **27**, 198 (1971).
[8] D. R. McKenzie and R. S. Seymour, *J. Phys. C.* **8**, 1071 (1975).

3.1. STATIC PROPERTIES OF DEFECTS

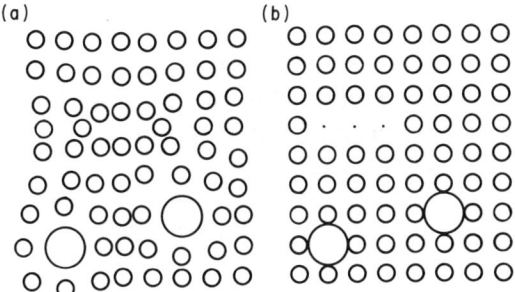

FIG. 6. (a) Real lattice after distortion by large substitutional atoms and vacancies. (b) Reference lattice.

periodic lattice. In Fig. 6a such a one-to-one correspondence between a real lattice and its reference lattice (Fig. 6b) is illustrated. The dislocation loop in the real lattice is imaged by a planar cluster of vacancies in the reference lattice. (An interstitial loop in the real lattice requires a sublattice for interstitials in the reference lattice.) With the periodic lattice vectors $\bar{\mathbf{R}}_i$ of the reference lattice, the average sublattice vector $\bar{\rho}_\nu$ and the displacements $\mathbf{u}_{i\nu} = \mathbf{r}_{i\nu} - \bar{\mathbf{R}}_i - \bar{\rho}_\nu$ the scattering cross section is given by

$$\frac{d\sigma}{d\Omega}(\mathbf{Q})_{\text{coh}} = \frac{1}{N} \left| \sum_{\alpha=1}^{\alpha_0} \sum_{\nu=1}^{\nu_0} \sum_{i=1}^{N} p_{i\nu\alpha} b_{i\nu} \exp(-W_{i\nu}) \exp(i\mathbf{Q}\mathbf{u}_{i\nu}) \exp[i\mathbf{Q}(\bar{\mathbf{R}}_i - \bar{\rho}_\nu)] \right|^2. \tag{3.1.14}$$

Evidently, the phase factors $\exp(i\mathbf{Q}\mathbf{u}_{i\nu})$ have to be very near to unity, or we need some other simplification to proceed with Eq. (3.1.14). We take the special example of a binary alloy with only one sublattice and make some simplifying assumptions: (i) the concentration of species 1 is not too high, i.e., $c \leq 0.3$, (ii) the displacements due to atoms of species 1 superimpose linearly, and (iii) for all lattice sites $\mathbf{Q}\mathbf{u}_i < 1$. We then get

$$\frac{d\sigma}{d\Omega}(\mathbf{Q})_{\text{coh}} = \frac{1}{N} |\tilde{c}(\mathbf{Q})|^2 |[b_1 \exp(-W_1) - b_2 \exp(-W_2)]$$
$$+ [cb_1 \exp(-W_1)$$
$$+ (1 - c)b_2 \exp(-W_2)]i\mathbf{Q}\,\tilde{\mathbf{s}}(\mathbf{Q})|^2, \tag{3.1.15}$$

where $\tilde{\mathbf{s}}(\mathbf{Q})$ is the Fourier transform of the displacement $\mathbf{s}(\mathbf{R}_i)$ caused by one atom of species 1 in a matrix of atoms 2. For the derivation of Eq. (3.1.15) we refer to Krivoglaz[5] or for those cases including the general case of more than one sublattice to Bauer and Kostorz.[9] An inconsistency of Eq. (3.1.15) is the use of the displacement field in a pure matrix of atoms 2. This certainly leads to errors at higher concentrations. The value of Eq. (3.1.15) is the separation of the concentration fluctuation $|\tilde{c}(Q)|^2$ and the distortion-modulated scattering cross section of a simple substitutional atom of species 1 in an average matrix. The Q dependence of

$$\tilde{s}(Q) = \sum_{i=1}^{N=1} s_i(\overline{R}_i) \exp(i Q \overline{R}_i) \tag{3.1.16}$$

can be analyzed most easily in the Kanzaki formalism. It is

$$\phi(Q) \cdot \tilde{s}(Q) = \tilde{f}(Q) \tag{3.1.17}$$

with

$$\tilde{f}(Q) = \sum_{i=1}^{n_0} f_i(\overline{R}_i) \exp(i Q \overline{R}_i). \tag{3.1.18}$$

The virtual forces produce in the harmonic approximation around a lattice site in the undisturbed lattice the same displacement fields $s_i(\overline{R}_i)$ as in the real one. They contain the real defect forces and the change of the coupling parameters. The practical advantage is the finite number n_0. The long-range elastic interaction is taken into account by the dynamical matrix $\phi(Q)$, which is known from phonon dispersion curves measured by inelastic neutron scattering. As an example, for a cubic crystal in a direction of high symmetry $\phi(Q)$ is given by

$$\phi(Q) = m \begin{bmatrix} \omega_L^2 & 0 & 0 \\ 0 & \omega_{T1}^2 & 0 \\ 0 & 0 & \omega_{T2}^2 \end{bmatrix} \tag{3.1.19}$$

with the atomic mass m and ω_L, ω_{T1}, and ω_{T2} the longitudinal and transverse phonon frequencies of the matrix.

The scattering cross section Eq. (3.1.15) has been used mainly in two approximations: (i) for negligible or small $\tilde{s}(Q)$, $\tilde{c}(Q)^2$ is determined, or (ii) for sufficiently low concentrations with $c(Q)^2$ not much different from $c(1-c)$, information on the f_i is obtained by means of Eqs. (3.1.16)–(3.1.18). Some examples are discussed at the end of this chapter.

The distortion-modulated scattering cross section has some special features for low q values: the phonon frequencies are proportional to q, and $f(Q)$ can be expressed by the elastic dipole force tensor P of the special defect,

$$f(Q) = iPq. \tag{3.1.20}$$

[9] G. S. Bauer and G. Kostorz, *Treatise Mater. Sci. Technol.* **15** (1979).

3.1. STATIC PROPERTIES OF DEFECTS

We simplify further: for symmetry directions and **q** parallel to **Q** we obtain $i\mathbf{Q}\tilde{\mathbf{s}}(\mathbf{Q}) = -QP_{11}/(v_L^2 q)$ with $v_L(Q)$ the longitudinal sound velocity. (The coordinate 1 in P_{11} is the coordinate parallel to **Q**.) This gives $i\mathbf{Q}\mathbf{s}(\mathbf{Q}) = -P_{11}/v_L^2$ for **q** around $\mathbf{G}_{000}$ and $i\mathbf{Q}\tilde{\mathbf{s}}(\mathbf{Q}) = -(P_{11}/mv_L^2) \cdot (\mathbf{G}_{hkl}/\mathbf{q})$ for **q** around $\mathbf{G}_{hkl} \neq \mathbf{0}$. The scattering amplitude in Eq. (3.1.15) is corrected by a q-independent quantity for small Qs, whereas near $\mathbf{G}_{hkl} \neq \mathbf{0}$ the scattering amplitude diverges with $1/q$. Very near to the reciprocal lattice point the diverging term dominates, and the cross section is proportional to $1/q^2$. This type of scattering has been extensively used to study defect properties by x-ray scattering. Measurements near $\mathbf{G}_{hkl} \neq \mathbf{0}$ require high resolution, and neutrons at first sight are at a disadvantage compared to x rays. However, first attempts are made to use neutrons because only with them can phonon scattering be avoided by separation of the elastic scattering. To give an impression, the cross section for three types of substitutional alloys is sketched in Fig. 7 with the simplifying assumption of a cubic lattice,

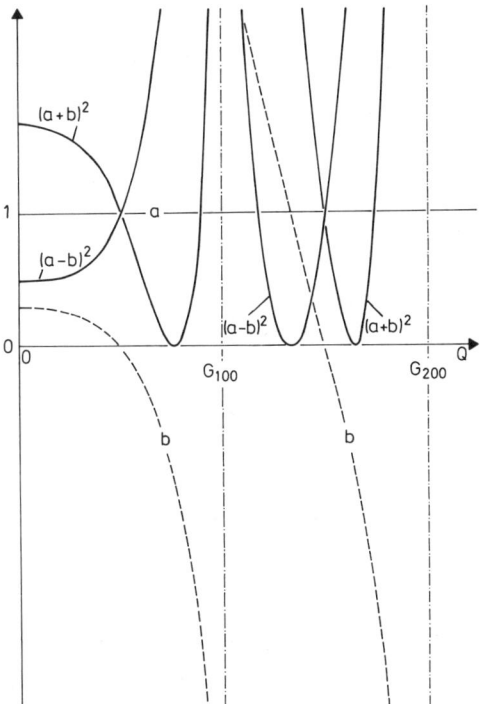

FIG. 7. Normalized defect-scattering cross sections and defect-scattering amplitudes for a substitutional atom with the most simplified I and ω_m. Curve a: scattering cross section and amplitude without lattice distortion and $\exp(-W) = 1$; curve b: distortion-dependent part of the scattering amplitude with arbitrarily fixed absolute value; defect-scattering cross sections: curves $(a + b)^2$ and $(a - b)^2$.

Q ∥ (100), central forces to nearest neighbors only and

$$\omega_L^2 = m \sin^2(2\pi Q/G_{100}) \cdot v_L^2 \cdot (G_{100}/2\pi)^2. \quad (3.1.21)$$

As the reader may realize, more complicated cases can be handled easily. Also of special interest are measurements with $\mathbf{q} \perp \mathbf{G}_{hkl}$. Combinations of the components of P can be determined by macroscopic measurements. For instance the lattice parameter change is proportional to tr P, which for isotropic defects is $3P_{11}$.

3.1.3. Experimental Techniques

From the experimenter's view there are four different regions in reciprocal space to be considered differently:

1. The reciprocal lattice points $\mathbf{G}_{hkl} \neq \mathbf{0}$ in an extension of at least their intrinsic width (a few 10^{-4} Å^{-1}) or in an extension due to radial and azimuthal mosaic spread. We discuss Bragg scattering in Section 3.1.3.1.

2. The region around $\mathbf{G}_{000} = \mathbf{0}$ with Q values ranging from a few 10^{-4} Å^{-1} up to about 0.1–0.2 Å^{-1}. This is the small-angle scattering region (see Section 3.1.3.2).

3. We speak normally about diffuse scattering for studies between reciprocal lattice points with the exception of small q values, i.e., for q smaller than approximately 0.1–0.2 Å^{-1} (see Section 3.1.3.3).

4. In regions near to reciprocal lattice points ($q \leq 0.1$–0.2 Å^{-1}) long-range fluctuations in scattering density and lattice plane bending are observed. Scattering experiments in this region require high resolution and are mainly a domain of x-ray scattering (Huang scattering, see Section 3.1.3.4).

3.1.3.1. *Bragg Scattering.* Such scattering from disordered crystals gives information on the average occupation of the sublattices and, by detailed analysis of the reciprocal lattice points, information on the large-scale imperfections. The integrated intensities yield the structure factor

$$|F_{hkl}|^2 = \left| \sum_{\nu=1}^{\nu_0} b_\nu \exp(-W_\nu) \exp(i\mathbf{G}_{hkl}\mathbf{u}_\nu) \exp(i\mathbf{G}_{hkl}\boldsymbol{\rho}_\nu) \right|^2, \quad (3.1.22)$$

where the average is taken over all unit cells and all atomic species occupying ν sites in the unit cell. The phase factors $\exp(i\mathbf{G}_{hkl}\mathbf{u}_\nu)$ enclosed in the average lead to an additional reduction of the Bragg intensities. In analogy with the factor $\exp(-W_\nu)$—the thermal Debye–Waller factor—the intensity reduction by the $\mathbf{u}_\nu$ is characterized by a *static* Debye–Waller factor $\exp(-W_{\nu,\text{st}})$. The $|F_{hkl}|^2$ are determined by conventional neutron diffractometers: powder and single-crystal diffractometers. Because mosaic crystal sizes often exceed the critical value for the first Born approximation,

3.1. STATIC PROPERTIES OF DEFECTS

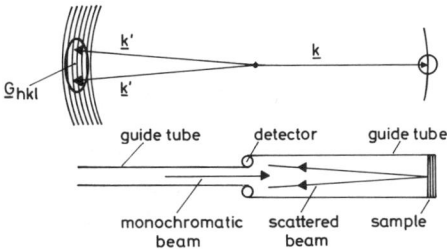

FIG. 8. Backscattering experiment in reciprocal space and a possible version in real space.

corrections for primary extinction must be applied. In addition, corrections for multiple scattering—expressed by factors $y_s \leq 1$ for secondary extinction—have to be applied. These corrections, familiar to all crystallographers in neutron diffraction, have been described by Becker.[10] Heavily distorted crystals can be produced by neutron radiation damage. Integrated Bragg intensities are reduced (e.g., in BeO, Al_2O_3, and quartz) by lattice distortions and by formation of amorphous damage zones. Separation of the two effects requires careful scans through the reciprocal lattice points with high resolution. For such scans, neutrons have not much special advantage compared to x rays. More use of neutron backscattering, however, would be desirable. By this method the intensity distribution of reciprocal lattice points for $\mathbf{q}$ parallel to $\mathbf{G}_{hkl}$ can be determined (Fig. 8).

3.1.3.2. Small-Angle Scattering of Neutrons (SANS). This method has developed within the 1970s to become a powerful technique. Among the many applications, studies of defects have been of continuous interest in the field. The initial advantage for neutrons was the possibility to use wavelengths beyond the Bragg cutoff. Thus Bragg scattering and consequently double Bragg scattering could be avoided. The latter was often a limiting parasitic scattering in x-ray work. With the availability of high-flux reactors, cold-neutron sources, multidetectors and with the consequent design of scattering devices, SANS also gained considerable intensity advantages compared to x-ray devices. The principal layout of the most productive small-angle scattering machines (in Jülich at the FRJ2 and in Grenoble at the HFR) is shown in Fig. 9. The basic underlying requirement is a sample size (say, 1 × 3 cm) smaller than the rectangular beam size of approximately 4 × 15 cm. This allows a beam collimation with an absorbing Cd aperture of about twice the dimension of the sample. The highest resolution δQ with full size of the sample is then determined by the maximum length of the collimator (20 m in Jülich and 40 m in Grenoble) and the neutron wave-

---

[10] P. Becker, *Acta Crystallogr.*, Sect. **A 33**, 1, 243 (1977).

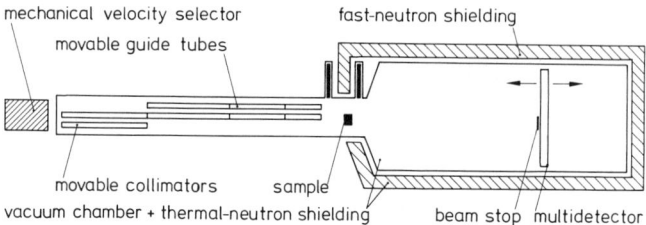

FIG. 9. Principal layout of a neutron small-angle scattering apparatus with a multidetector.

length. The multidetector is placed in a vacuum tube of about 0.8-m diameter at the same distance from the sample as the collimator length. Lower resolution can be achieved by replacing parts of the collimator by neutron guide tubes and moving the detector nearer to the sample. Neutrons are monochromated by a mechanical velocity selector in front of the collimator device. The full width at half-maximum (FWHM) is between 6% and 20%. The optimal wavelength is between 6 and 20 Å, the exact value depending on the special beam condition. In Grenoble a (64×64)-cm^2 X-Y counter with 1-cm^2 resolution is used. The multidetector in Jülich consists of six horizontal position-sensitive counters with a length of 50 cm, a diameter of 5 cm, and a resolution of 0.8×4.0 cm^2. Fast data processing and reduction with on-line computers is essential because of the high rate of data acquisition. In careful experiments, intensities of less than 10^{-5} of the primary beam can be measured immediately beside the primary beam. For this the collimator must be supplied with auxiliary apertures, scattering events like sample → vacuum wall → detector must be avoided, and the detector must be shielded against neutron background in the experimental hall. The instruments are most sensitive for particle diameters of 300 Å. Voids of this size can be measured even at a volume fraction of 10^{-4}. The sensitivity decreases for $2R > 300$ Å with $1/R$. Towards smaller values of $2R$, the sensitivity also decreases in many cases because the background due to incoherent elastic scattering increases. Further, for $2R < 100$ Å the beam collimation by the guide tubes sets a limit on the primary-beam intensity, which cannot be increased further by shortening of the collimator. Sophisticated small-angle scattering machines as described allow many experiments in a year. At the instrument D11 in Grenoble, data for a half year's work by one scientist can be produced within one week.

Developments beyond the layout sketched in Fig. 9 should be mentioned. The counter of the instrument D17 in Grenoble can be rotated around the vertical axis of the sample. The instrument D11 has additional counters at large angles that allow simultaneous measurement of diffuse scattering (D11B). Crude separation of elastic and inelastic scattering can be achieved

3.1. STATIC PROPERTIES OF DEFECTS

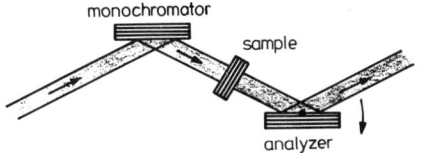

FIG. 10. Small-angle scattering apparatus with two parallel perfect crystals.

easily with a Fermi or disk chopper. At various places (Oak Ridge, Braunschweig) small-angle scattering experiments are performed with a two crystal diffractometer (Fig. 10). Two perfect crystals (silicon) are mounted parallel with the sample between them. Small-angle scattering is measured by rocking the second crystal. The instrument is very short and still has a high resolution. New developments in low-Q studies can also be expected from the use of ultracold neutrons ($\lambda = 200-500$ Å). The low Q values are obtained here by long wavelengths! Ultracold neutrons (see, e.g., Steyerl[11]) would be of special advantage for the study of surface imperfections.

3.1.3.3. Diffuse Scattering. Here by definition we are interested in $d\sigma/d\Omega$ for Q values between reciprocal lattice points with the exclusion of small Q or q values, the latter requiring special instruments. Most diffuse-scattering experiments are characterized by low Q resolution, e.g. $\delta Q \simeq 0, 1$ Å^{-1}. In many cases diffuse scattering studies are also characterized by small scattering cross sections. Let us take some examples: (i) $d\sigma/d\Omega$ for an aluminum matrix with 3% zinc in random distribution is 1.5×10^{-3} b sr^{-1}. (ii) The small-angle scattering cross section in forward direction for the same material but with all zinc atoms in spherical precipitates of 300 Å diameter would be 10^4 b sr^{-1}. (iii) Even for the favorable alloy ^{65}Cu$_{0.435}$ ^{62}Ni$_{0.565}$ with the very high value $(b_2 - b_1)^2 = 3.8$ b sr^{-1}, $d\sigma/d\Omega$ is only ~ 1 b sr^{-1}. (iv) At the limits of today's experimental possibilities would be a scattering study with 5×10^{-4} vacancies in aluminum. For this case $d\sigma/d\Omega$ would be 6×10^{-5} b sr^{-1}.

We conclude that the average diffuse scattering experiments have small scattering intensities. This disadvantage has to be compensated in most cases by high intensities of the primary beam, which can be achieved by scattering devices of low resolution δQ. In Fig. 11 two spectrometers for diffuse scattering (D7 in Grenoble and DNS in Jülich) are shown schematically. The neutron beams are monochromated by a graphite crystal and a mechanical velocity selector with a wavelength resolution of 2% and 20%, respectively. Separation of inelastic scattering is obtained with choppers and time-of-flight (TOF) operation with a duty cycle of about 1 : 15, which is a factor of 6 higher than is usual for TOF machines used in inelastic scattering

[11] A. Steyerl, *Springer Tracts Mod. Phys.* **80**, 57 (1977).

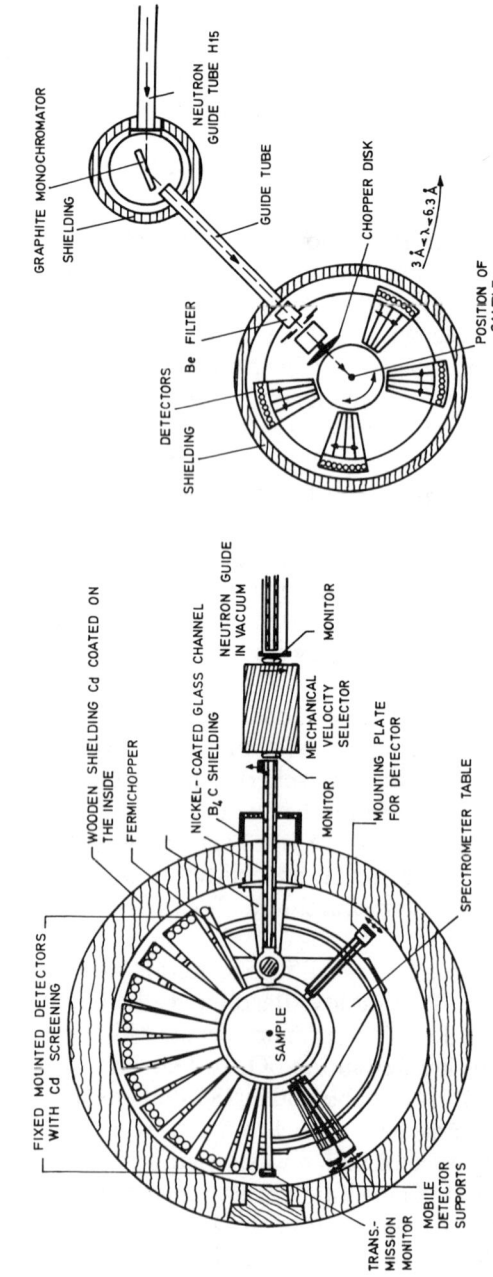

FIG. 11. Spectrometers for diffuse scattering at (a) the FRJ2 in Jülich, and (b) the HFR in Grenoble.

3.1. STATIC PROPERTIES OF DEFECTS

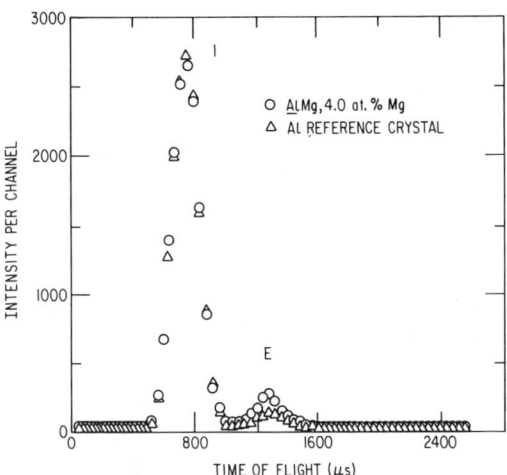

FIG. 12. Typical time-of-flight spectrum for evaluation of elastic diffuse scattering. The width of the elastic peak (E) is the experimental resolution.

work. A typical TOF spectrum (Werner[12]) obtained with the recent version of DNS is shown in Fig. 12. The resolution results from the monochromator and the chopper resolution. It is about 15% in $\Delta t/t$ at the elastic peak. Though very broad, it is still sufficient to separate the major part of the inelastic from the elastic scattering. For the instrument D7 the elastic peak is much sharper because of the better wavelength resolution. The distance of the detectors (80 – 120 cm) is in both cases well adapted to the TOF and the angular resolution at the detector site. For example, with $\lambda = 4$ Å, $\Delta\lambda/\lambda = 0.03$, a scattering angle of 90°, an effective detector width of 4 cm, and a distance of 1 m, the horizontal resolution δQ perpendicular to $\mathbf{Q}$ is about 0.08 Å^{-1} (FWHM). The vertical resolution is normally much less because of the detector height varying between 15 and 30 cm. This is justified for all experiments with a symmetry plane in the scattering plane.

Although in diffuse scattering experiments most of the background scattering is due to incoherent elastic scattering, three other major sources of background must be reduced or sufficiently controlled. (i) Counting rates with no primary neutron beam (absorber behind or in front of the monochromator) are mainly due to fast, epithermal, and thermal neutrons of the experimental hall. Against fast and epithermal neutrons the spectrometer is shielded by moderating material (wood, paraffin, or polyethylene). Thermalized and thermal neutrons are absorbed by boron and cadmium inside the moderating material. The shielding has to be closed especially carefully

[12] K. Werner, Thesis, Ruhr-Univ., Bochum, 1978.

in the direction of the neutron source. (ii) Background scattering with open beam but without the sample is produced by scattering of the primary beam at the entrance collimator or guide tube, the cryostat or furnace walls, the sample holder, the beam stop, and by the air. For counters at scattering angles between 20° and 160°, part of this background scattering can be avoided by an additional detector shielding that narrows the detector sight at the primary beam to the immediate neighborhood of the sample. This is especially effective for large-diameter (30–40 cm) vacuum vessels. On the other hand, this type of background reduction becomes more and more difficult for scattering angles toward 0 and 180°. Very good conditions can be obtained with free samples in furnaces. In experiments at low temperatures (lower than 100 K), even very thin layers of condensed oil or other hydrogeneous substrates can give considerable background scattering, which can be recognized in most cases by its continuous increase with time. The precautions needed are very good vacuum conditions, radiation shields of large diameter, and cooling of the radiation shields before cooling of the inner parts of the cryostat. (iii) The sample itself can reduce part of the background mentioned before and produce beside its own incoherent elastic and inelastic scattering, additional background scattering by double scattering processes. The process, sample $\rightarrow$ sample holder $\rightarrow$ detector is especially dangerous. In many experiments sample and standards are mounted along a vertical axis and are moved alternately into the beam. This may change the double scattering process mentioned before. The probability for double scattering processes is high if part of the primary beam is Bragg scattered by the sample. In addition, a small part of the Bragg scattered beam may come directly into the counter. Using his knowledge of wavelength, wavelength resolution, and mosaic spread of the sample, the experimenter should calculate the probability for Bragg scattering and control it by a monitor behind the sample.

In low-temperature experiments it is convenient and in some cases even unavoidable to put the sample into a (vertical) tube. This gives scattering by the tube itself, eventually by condensed layers at the tube wall and by the thin, gaseous helium layer between the sample and the tube. Sample and standard must have the same diameter in order to avoid changes of the latter type of background scattering. In summary, because of the low intensities in many defect scattering studies, careful control and reduction of the background are a must in most diffuse scattering experiments.

The instruments D7 and DNS are especially developed for diffuse scattering with cold neutrons down to wavelengths of 3 Å. The highest Q value obtainable with these instruments is $\simeq 4$ Å^{-1}, whereas 6 or even 8 Å^{-1} is desirable in many cases. This requires instruments at thermal beam holes. Sometimes, standard devices can be used. Time-of-flight spectrometers are reasonable because they have many detectors. Normally, however, their energy resolution is too good. An instrument with short sample-to-detector distance (80–150 cm) and a Fermi chopper with low duty cycle 1:30 existed at the FR2 in Karlsruhe. For diffuse scattering with high intensities, two-axis diffractometers with low resolution and without energy analysis can be used. Multicounters covering a reasonably large solid angle (≥ 0.1 sr) make such instruments rather attractive. Triple-axis spectrometers have high resolution and in most cases only one detector unit. They are of value in all cases with sharp singularities (e.g. satellite reflections) and for separation of elastic scattering.

Although considerable progress has been achieved within the 1970s, the experimental situation is not yet satisfactory. The instruments at the cold sources have an angular resolution of the primary beam of $\delta k_x \delta k_y \simeq (2 \times 10^{-2}$ Å$^{-1})^2$, whereas $\delta k_x \simeq 0.05$ Å^{-1} and $\delta k_y \simeq 0.2$ Å^{-1} would be sufficient. Because many samples are smaller in size than the primary beam, focusing the beam onto the sample with a simultaneous increase of the solid angle would be very advantageous. The beam monochromation with crystals is slightly too good, and with velocity selectors it is slightly too bad. The half-width $\Delta\lambda/\lambda$ should increase with increasing wavelength, whereas for crystals it decreases. Detectors should cover as much solid angle as possible, at least the horizontal scattering should be counted almost completely to a height of 0.5 Å^{-1}. Such steps to diminish the waste of scattered neutrons are not only expensive but are also limited by the collimators in front of the counters. Another possible refinement is the use of statistical choppers. They can have a duty cycle up to 1:2, if the elastic defect scattering exceeds 50% of the total scattering. A system with two statistical choppers in a "white" primary beam has been described by Schneider.[13] However, it has not been put in operation for routine work. Finally, it should be mentioned that fast data handling and reduction and computer-controlled operation of the essential mechanical parts are absolutely necessary for a modern instrument.

3.1.3.4. Scattering at Small q Values with $G_{hkl} \neq 0$. Such experiments can be performed with triple-axis spectrometers at thermal- and cold-neutron beams. Thus elastic scattering can be separated from inelastic scattering. The horizontal resolution normally available at such machines is well adapted for studies at small q values. The vertical resolution has to be improved in most cases, except for triple-axis spectrometers placed at thermal-neutron guide tubes, where both horizontal and vertical resolution are roughly 0.02 Å^{-1}. Indeed, the only continuous effort in elastic scattering with small q values near $G_{hkl} \neq 0$ known to the author is done by the

[13] J. Schneider, *Proc. Neutron Diffr. Conf. Petten, RCN Rep.* **RCN-234**, p. 380 (1975).

research group of Peisl at a thermal guide tube at the ILL (Institut Laue-Langevin, Grenoble).

In planning scattering studies at defects, *intensity calculations* are of considerable importance. The intensity at the detector is given by the neutron flux I at the sample position times the scattering cross section $(d\sigma/d\Omega)_{N,\text{coh}}$ of the sample times the detector solid angle $\Delta\Omega$ times various attenuation factors. As an example we calculate in the following the intensity for a Huang scattering experiment. The sample is a niobium single crystal with 5-cm length and a diameter of 1 cm. The [001] cylinder axis is vertical to the horizontal scattering plane. The crystal is loaded with 1 at. % deuterium in random distribution, which by lattice distortion gives Huang scattering at the reciprocal lattice points. We calculate the intensity at the G_{400} reciprocal lattice point for $\mathbf{q} \parallel \mathbf{G}_{400}$ and a q value of 0.05 Å^{-1}. The neutron flux I at the sample position is given by

$$I = (\phi/2\pi) \exp(-k_z^2/k_T^2) (k_z \Delta k_x \Delta k_y \Delta k_z/k_T^4) f_A \qquad (3.1.23)$$

with the z direction parallel to the neutron beam after monochromation. ϕ is the neutron flux in the reactor. $k_T \simeq 3.7$ Å^{-1} for thermal beam holes. f_A is the attenuation of the beam due to absorption, scattering, and incomplete reflectivity on the way from the reactor to the sample. We assume $f_A \simeq 0.5$, $\phi = 7 \times 10^{14}$ n s^{-1} cm^{-2}, and $k_z = 3.5$ Å^{-1}, i.e., $\lambda = 1.8$ Å. The experiment requires a resolution of $\Delta k_x = \Delta k_y = \Delta k_z = 0.02$ Å^{-1} approximately. With these numbers the directed neutron flux I at the sample is 3×10^6 n s^{-1} cm^2. The counted neutron flux in the detector is given by

$$\Delta I = I \left. \frac{d\sigma}{d\Omega} \right|_{N,\text{coh}} f_s \Delta\Omega f_D \qquad (3.1.24)$$

where the attenuation factor f_s takes account of neutron absorption and inelastic scattering by the sample, the attenuation factor f_D includes detector efficiency, analyzer crystal reflectivity, and further beam losses by scattering and absorption on the way from the sample to the detector. The detector solid angle is adapted to the resolution. We assume for our case $\Delta\Omega = \Delta k'_x \Delta k'_y/(k'_z)^2 = (0.02$ Å$^{-1}/3.5$ Å$^{-1})^2 = 4 \times 10^{-5}$ sr. Further, we assume $f_s = 0.7$ and $f_D \simeq 0.5$. The scattering cross section $(d\sigma/d\Omega)_{Ni,\text{coh}}$, for the total sample is given by Eqs. (3.1.1), (3.1.4), (3.1.14), and (3.1.20) as

$$\left. \frac{d\sigma}{d\Omega} \right|_{N,\text{coh}} = Ncb_{\text{Nb}}^2 \exp(-2W) \left| \frac{G_{400} P_{11}}{qm_{\text{Nb}} v_L^2} \right|^2 \qquad (3.1.25)$$

We insert $N = 1.7 \times 10^{23}$, $c = 0.01$, $b_{\text{Nb}} = 5 \times 10^{-13}$ cm, $\exp(-2W) = 0.7$, $G_{400} = 7$ Å^{-1}, $q = 0.05$ Å^{-1}, $P_{11} \simeq 2.8$ eV and $qm_{\text{Nb}} v_L^2 = 24$ eV in the equation above and get $(d\sigma/d\Omega)_{N,\text{coh}} = 0.08$ cm^2, which finally leads to

$$\Delta I = 4 \quad \text{counts s}^{-1}$$

This is a counting rate quite acceptable for a scattering experiment, even if the attenuation factors have been too optimistic and the final experiment has to be done at lower counting rates, possibly 1 count s^{-1}. With the numbers above we can also guess whether experiments at smaller or larger q values can be done reasonably. For $q > 0.05$ Å^{-1} the resolution elements can be increased. With the experiment at the thermal guide tube and without focusing monochromators only the elements Δk_z, $\Delta k'_x$, and $\Delta k'_y$ can be increased roughly proportional to q. This gives a final increase proportional to q because the scattering cross section is proportional to $1/q^2$. [At high q values the scattering amplitude by the deuterium atoms must be included; see Eq. (3.1.15).] Decreasing q values below 0.05 Å^{-1} requires reduction of all five resolution elements Δk_x Δk_y Δk_z $\Delta k'_x$ $\Delta k'_y$, which gives a factor proportional to q^5, which is only partly compensated by the $1/q^2$ dependence of $d\sigma/d\Omega$. The intensity drops rapidly and experiments with $q < 0.01$ Å^{-1} can hardly be performed. For small q values X rays will remain superior even with small x-ray sources and much more with the use of synchrotron radiation.

3.1.4. Typical Results

In Fig. 13 small-angle scattering data from $Al_{0.6}Zn_{0.4}$ polycrystalline alloy measured at various temperatures are shown (Schwahn[14]). Cylindrical samples with a diameter and a length of 4 cm have been used. The neutron beam was parallel to the cylinder axis. It illuminated a circle of 2-cm diameter. The data shown in Fig. 13 are already corrected for background and for inelastic scattering. The fraction of inelastic scattering has been determined at high Q values and 400°C by a separate experiment with the same sample at the diffuse neutron scattering spectrometer (DNS). Thus the scattering cross sections in Fig. 13 are purely elastic. The nearly constant value for $Q > 0.1$ Å^{-1} and for 400°C is in good agreement with the value expected for a random distribution of Al and Zn atoms with the correction for lattice distortion included. Scattering data at Q values smaller than 0.025 Å^{-1} have also been measured, but they are dominated by grain-boundary scattering and have not been used further. An analysis with Ornstein–Zernicke plots gives a critical temperature of (323.5 ± 0.5)°C. Indeed, the slopes of scattering curves measured at 320°C and 323°C are entirely different (Fig. 13). The critical temperature quoted above has to be discussed in connection with the Al–Zn phase diagram, a part of which is shown in Fig. 13 as an inset. Above the phase boundary A, the alloy is homogeneous; between the boundaries A and B, the Al–Zn alloys can decompose only in lattice-incoherent precipitates. Below B, lattice-coherent precipitates are formed. Lattice coherent

---

[14] D. Schwahn, *Ber. Kernforschungsanlage Juelich* **Juel-1379** (1977); D. Schwahn and W. Schmatz, *Acta Metall.* **26**, 1571 (1978).

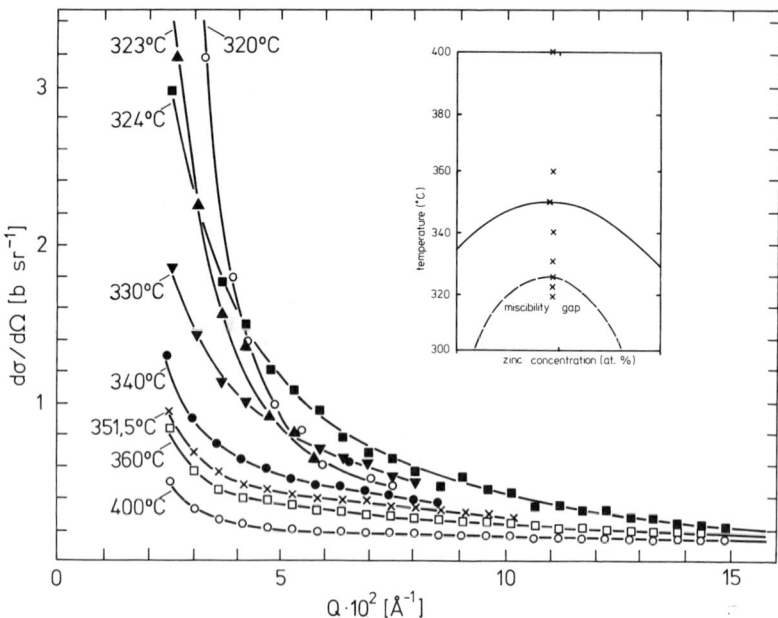

FIG. 13. Small-angle scattering data of an $Al_{0.6}Zn_{0.4}$ alloy at various temperatures.

concentration fluctuations (above T_c) and precipitates are formed rapidly, whereas the lattice-incoherent decomposition is a slow process. Therefore it is possible to measure the lattice-coherent concentration fluctuations above T_c. An example is given in Fig. 14. The scattering curves are measured at

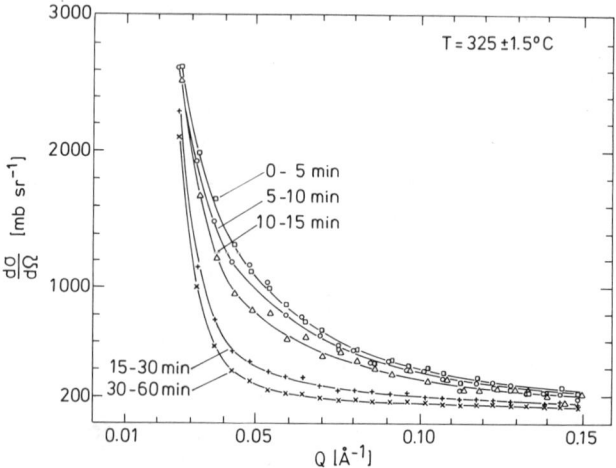

FIG. 14. Time dependence of NSAS from $Al_{0.6}Zn_{0.4}$ at a temperature between the coherent and "incoherent" critical point.

different time intervals after rapid cooling from a temperature above 351°C. Plotting the data versus time makes clear that the change of the scattering curves must be due to lattice-incoherent decomposition, which occurs with some incubation time. Thus the scattering curves at time 0 or within the first time interval are representative for lattice-coherent composition fluctuations above T_c. Only these data are shown in Fig. 13. Apparently, and in agreement with our observations, the time necessary to build up the lattice-coherent fluctuation is shorter than a few minutes. The example shows that fast data collection can be of great importance.

Lattice distortions around substitutional atoms have been studied in detail by diffuse scattering for Bi in Pb and for Cu, Mn, and Mg in Al (Werner,[12] Schumacher et al.,[15] Seitz et al.,[16] Bauer[17]). Another study was concerned with the lattice distortion around interstitial deuterium in niobium (Bauer,[17] Bauer et al.[18]). We show here for illustration the results obtained at Bi in Pb. The diffuse scattering from a cylindrical lead single crystal with 4.1% Bi has been measured at the D7 spectrometer in Grenoble at 10 K. The sample was rotated in steps of 10° around the cylinder axis, which was parallel to the [$\bar{1}$10] crystallographic direction and perpendicular to the scattering plane. For subtraction of the incoherent elastic scattering of lead, a lead single crystal with the same size and orientation has been used as reference. The diffuse scattering was measured for many $\mathbf{Q}$ values in the ($\bar{1}$10) plane. The data were reduced into a quarter of the ($\bar{1}$10) plane and have been approximated by spline functions. The result is shown in Fig. 15a as contour lines in units of mbarn per steradian and Bi-atom. The accuracy of the data is ± 2, ± 3, and ± 5 for values of 10, 20, and 40, respectively. The experiment was at the limit of present-day possibilities. A value of 8 mb sr^{-1} and Bi atom corresponds to a coherent scattering cross section of $8 \times 10^{-3} \cdot 4\pi \cdot 0.04 \simeq 4 \times 10^{-3}$ b per lead atom. This value has to be compared with the coherent scattering cross section of lead, approximately 10 b. This clearly demonstrates the sensitivity of the technique in the case of a material with very small incoherent scattering cross section ($\sigma_{\text{inc,Pb}} \simeq 2$ mb). In an earlier experiment (Schumacher et al.[15]) data from the same crystal had been obtained without energy analysis of the scattered neutrons. These data agree well with those of Fig. 15a, which demonstrates the reproducibility of such experiments. Further, at small Q values, there is good agreement of the experimental values and the values calculated from the difference in scattering lengths and the lattice parameter change per Bi atom within the experimental error. Fig. 15b shows contour lines for the scattering cross-section

[15] H. Schumacher, W. Schmatz, and W. Seitz, Phys. Status Solidi **20**, 109 (1973).
[16] E. Seitz, W. Schmatz, G. S. Bauer, and W. Just, Phys. Status Solidi **46**, 557 (1978).
[17] G. S. Bauer, Ber. Kernforschungsanlage Juelich **Juel-1158** (1975).
[18] G. S. Bauer, W. Schmatz, and W. Just, Hydrogen Met. Proc. 2nd Int. Congr., Paris, 1977, **6**, 2C15 (1978).

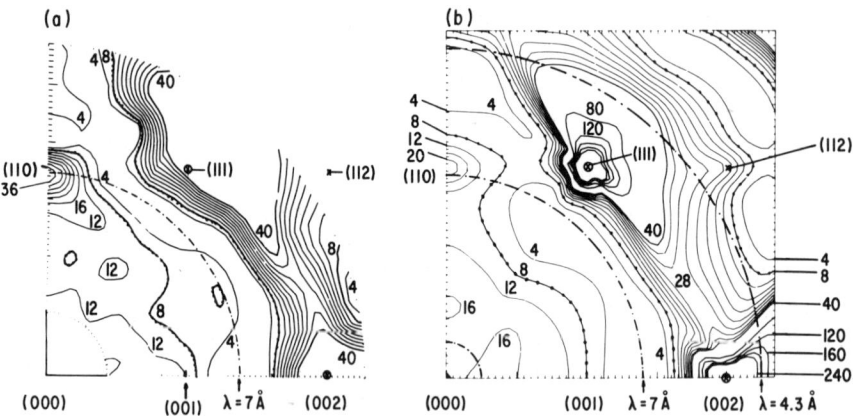

FIG. 15. Disorder scattering cross section of a Pb–Bi single crystal in units of millibarn per steradian (4 at. % for Bi). (a) experimental result, (b) calculated values for a noncentrosymmetric Kanzaki-force distribution around the Bi atom.

calculated for a model with two types of Kanzaki forces. In addition to the special assumption of the symmetry of the defect, this calculation contains only one indeterminate parameter, the ratio of the absolute values of the two types of Kanzaki forces, because one parameter is canceled by the known value of the lattice parameter change. The agreement with the experimental data is very good. It must be added that most of the structure in the contour lines is given by the **Q** dependence of the dynamical matrix $\phi(\mathbf{Q})$. There are some details—the height and width of the peak [110] and the detailed structure of the saddle between [111] and [002]—that are not satisfactorily explained by the force model. These facts may be a starting point for new experiments and theory.

In addition to the two typical examples above some investigations not included in Bauer and Kostorz[9] should be mentioned. The excellent scattering properties of the CuNi isotope alloy have been used for further studies.[19,20] A complete analysis of short-range order and displacement scattering according to the Borie–Sparks method has been achieved with thermal neutron scattering on a Ni_3Fe single crystal[21], and results concerning the same problem but with a restricted range in reciprocal space are available for the refractory compounds $TiC_{0.67}$ and $NbC_{0.73}$.[22] Diffuse scattering data

[19] J. de Vrijen, J. Aalders, and C. van Dijk, *Phys. Rev. B* **22**, 1503 (1980).
[20] W. Wagner, R. Poerschke, A. Axmann, and D. Schwahn, *Phys. Rev. B* **21**, 3087 (1980).
[21] S. Lefebvre, F. Bley, M. Bessiere, M. Fayard, M. Roth, and J. B. Cohen, *Acta Crystallogr., Sect. A* **36**, 1 (1980).
[22] V. Moisy-Maurice, C. H. de Novion, A. N. Christensen, and W. Just, *Solid State Commun.* **39**, 661 (1981).

3.1. STATIC PROPERTIES OF DEFECTS

along symmetry directions for substitutional Mg and Li in Al have been compared with theoretical results calculated with pseudopotentials.[23,24] The experimental study of Huang scattering with thermal neutron mentioned in Section 3.1.3.4 has been completed successfully for deuterium in niobium and for defects in MgO and ^{7}LiF.[25]

---

[23] G. Solt and K. Werner, *Phys. Rev. B* **24**, 817 (1981).
[24] K. Werner, *in* "Electrons in Disordered Metals and at Metallic Surfaces" (P. Phariseau *et al.*, eds.), p. 487. Plenum, New York, 1979.
[25] E. Burkel, B. von Guerard, H. Metzger, J. Peisl, and C. M. E. Zeyen, *Z. Phys. B* **35**, 227 (1979).

3.2. Dynamic Properties of Defects

By R. M. Nicklow

Solid State Division
Oak Ridge National Laboratory
Oak Ridge, Tennessee

3.2.1. Introduction

In this section we discuss the influence that defects have on the dynamical properties of a crystalline host and the extent to which these properties can be elucidated by neutron scattering experiments. The majority of experimental and theoretical investigations to date have dealt with systems in which the defect was considered to be a substitutional monatomic impurity, and our discussion centers on such defects. If an impurity atom possesses a large incoherent scattering cross section, it may be possible to study the dynamics of the impurity apart from that of the host lattice. The frequency dependence of the scattering cross section can be then directly related to the frequency distribution of the defect vibrations. Generally, however, much more detailed information can be obtained through coherent scattering measurements of the individual vibrational modes of the crystal as a whole. In this case the dynamical properties of the defects are not separated experimentally from the dynamical properties of the crystalline host. Although there may be special vibration modes in which the amplitude of the defect is particularly large, the presence of defects alters to some extent all of the modes of the host.

The range of wavelengths and energies appropriate to thermal neutrons is well matched to the interatomic distances and the energies of atomic vibrations that occur in crystal lattices. Neutron inelastic scattering measurements provide very good information about the dynamics of both perfect and imperfect crystals alike, making it possible to determine by direct measurement the influence of defects on individual modes of vibration. This possibility of comparing the dynamical properties of perfect and imperfect crystals mode by mode is a very great advantage that neutron scattering has over other experimental methods for the types of defect systems being considered here. Such experimental information can then be compared with detailed theoretical calculations of the dynamics of defects through their perturbation of the dynamics of the host lattice. Such comparisons have provided the most definitive tests available of theoretical models. Even in the absence of a detailed theoretical interpretation, neutron scattering data can contribute significantly to the understanding of those physical properties of crystals containing defects that are related to the atomic vibrations, e.g., heat capacity, thermal and electrical conductivities, the Mössbauer effect, and certain optical properties.

In the following sections we first summarize the theory for the dynamical and the neutron scattering properties of crystals without defects in order that the interpretations and the goals of the neutron scattering experiments can be more easily understood. The basic concepts that enter into the design of an experiment for a triple-axis neutron spectrometer are then discussed briefly, and finally the theory for the dynamical properties of crystals with defects and typical experimental results are presented to illustrate the practical applications of neutron scattering techniques to studies of defect dynamics. In the theoretical discussion it is convenient to express in equations the vibrational frequency in terms of the angular frequency ω, whereas in much of the published experimental work the frequency $\nu = \omega/2\pi$ is used. Throughout this chapter we use both ω and ν to denote frequency without making further distinctions between them.

3.2.2. Perfect Crystals

3.2.2.1. Lattice Dynamics. We begin with a brief description of the dynamics of a crystal having no defects in order to introduce some of the notation and the underlying physical ideas on which much of the following discussions are based. In the harmonic approximation the equation of motion for the lth atom, with position vector $\mathbf{R}(l)$, in a monatomic Bravais lattice having N atoms is[1]

$$M\ddot{u}_\alpha(l,t) = -\sum_{l'=1}^{N} \sum_{\beta=1}^{3} \phi_{\alpha\beta}(l,l') u_\beta(l',t), \quad l = 1,N, \quad \alpha = 1,2,3, \quad (3.2.1)$$

where $u_\alpha(l,t)$ is the αth Cartesian component of the displacement of the lth atom at time t, M is the mass, and $\phi_{\alpha\beta}(l,l')$ is the force in the α direction acting on atom l when atom l' is displaced a unit distance in the β direction. Because of the periodicity of the lattice the force constants $\phi_{\alpha\beta}(l,l')$ are invariant against a rigid translation of the crystal through a lattice translation vector and depend on l and l' only through their difference $l - l'$. Consequently, it is natural to choose as a solution for $u_\alpha(l,t)$

$$u_\alpha(l,t) = \sigma_\alpha(\mathbf{q}) e^{i(\mathbf{q}\cdot\mathbf{R}(l) - \omega t)}, \quad (3.2.2)$$

[1] W. Marshall and S. W. Lovesey, "Theory of Thermal Neutron Scattering." Oxford Univ. Press, London and New York, 1971.

where **q** is a wave vector (magnitude $2\pi/\lambda$), and ω is a q-dependent frequency. The polarization vectors, $\sigma(\mathbf{q})$, satisfy the equations obtained by inserting Eq. (3.2.2) into (3.2.1), namely,

$$M\omega^2(\mathbf{q})\sigma_\alpha(\mathbf{q}) = \sum_\beta D_{\alpha\beta}(\mathbf{q})\sigma_\beta(\mathbf{q}), \quad \alpha = 1,2,3, \qquad (3.2.3)$$

where

$$D_{\alpha\beta}(\mathbf{q}) = \sum_{l'} \phi_{\alpha\beta}(l,l')e^{-i\mathbf{q}\cdot[\mathbf{R}(l)-\mathbf{R}(l')]}. \qquad (3.2.4)$$

The periodicity of the lattice has enabled one to reduce the $3N$ equations of motion (3.2.1) to the problem of solving the set of only three equations given in (3.2.3). The solution is found in terms of the zeros of the secular determinant:

$$\det|M\omega^2\delta_{\alpha\beta} - D_{\alpha\beta}(\mathbf{q})| = 0. \qquad (3.2.5)$$

For each wave vector **q** there are three solutions (branches) for ω^2, which are denoted by $\omega_j^2(\mathbf{q})$ with $j = 1,2,3$, and for each $\omega_j^2(\mathbf{q})$ there corresponds a three-component eigenvector $\sigma(\mathbf{q},j)$ that satisfies certain orthonormality and closure conditions. The displacement pattern, Eq. (3.2.2), corresponding to a particular $\omega_j(\mathbf{q})$ is a normal mode of the crystal. It is described by the wave vector **q** and the vector $\sigma(\mathbf{q},j)$, and it is independent of (or is uncoupled from) all other modes. The energy quantum $\hbar\omega_j(\mathbf{q})$ is called a phonon. For **q** parallel to directions of high symmetry in the crystal lattice, the eigenvectors for the three branches of $\omega(\mathbf{q})$ correspond to displacements $\mathbf{u}(l)$ that are perpendicular (transverse branch) or parallel (longitudinal branch) to **q**.

The function $D_{\alpha\beta}(\mathbf{q})$, and hence $\omega_j^2(\mathbf{q})$, is periodic in reciprocal space, i.e.,

$$\omega_j^2(\mathbf{q} + \boldsymbol{\tau}) = \omega_j^2(\mathbf{q}), \qquad (3.2.6)$$

where $\boldsymbol{\tau}$ is a reciprocal lattice vector. The unique set of wave vectors **q**, measured relative to each $\boldsymbol{\tau}$, is contained within the first Brillouin zone with the end point of $\boldsymbol{\tau}$ as its center.

3.2.2.2. *Neutron Scattering Cross Section*. With coherent inelastic scattering experiments it is possible to measure directly the function $\omega(\mathbf{q})$ throughout the Brillouin zone. As is indicated in Eq. (3.1.1) of the preceding chapter by Schmatz, the coherent scattering of neutrons by a crystal depends on a phase factor that has the form $\exp\{-i\mathbf{Q}\cdot\mathbf{R}(l)\}$. The coherent inelastic scattering cross section is obtained by taking into account the time dependence of the atomic positions. We replace $\mathbf{R}(l)$ with $\mathbf{R}(l) + \mathbf{u}(l,t)$ and expand the resulting exponential function in powers of $\mathbf{Q}\cdot\mathbf{u}(l,t)$. The first term in the expansion that contains $\mathbf{u}(l,t)$ as a factor is the one-phonon coherent cross

3.2. DYNAMIC PROPERTIES OF DEFECTS

section, and for a monatomic Bravais lattice it has the form[1]

$$\left(\frac{d^2\sigma}{d\Omega\, dE'}\right)_{\text{coh}} = \frac{\sigma_{\text{coh}}}{4\pi} \frac{\mathbf{k}'}{\mathbf{k}} S(\mathbf{Q},\omega), \qquad (3.2.7)$$

where

$$S(\mathbf{Q},\omega) = \frac{1}{2\pi\hbar} e^{-2W} \int_{-\infty}^{\infty} dt\, e^{i\omega t} \sum_{l,l'} e^{-i\mathbf{Q}\cdot[\mathbf{R}(l)-\mathbf{R}(l')]}$$
$$\times \sum_{\alpha\beta} Q_\alpha Q_\beta \langle u_\alpha(l,t) u_\beta(l',0)\rangle, \qquad (3.2.8)$$

$\mathbf{k}$ and $\mathbf{k}'$ specify the initial and final wave vectors of the scattered neutron, E and E' are the corresponding energies, $\mathbf{Q} = \mathbf{k} - \mathbf{k}'$, $\hbar\omega = E - E'$, and Ω is a solid angle.

Evaluation of the displacement–displacement correlation function in Eq. (3.2.8) can be carried out with a variety of mathematical procedures. For our purpose it is convenient to introduce the Green-function matrix $\mathbf{P}$ defined as the inverse of $(M\omega^2 \mathbf{I} - \boldsymbol{\phi})$, where $\boldsymbol{\phi}$ is the matrix of force constants, $\phi_{\alpha\beta}(l,l')$ introduced above. The elements of $\mathbf{P}$ are[1]

$$P_{\alpha\beta}(l,l',\omega) = (NM)^{-1} \sum_{\mathbf{q}j} \sigma_\alpha(\mathbf{q},j) \sigma_\beta(\mathbf{q},j) P_j(\mathbf{q},\omega) \exp\{i\mathbf{q}\cdot[\mathbf{R}(l) - \mathbf{R}(l')]\}, \qquad (3.2.9)$$

where

$$P_j(\mathbf{q},\omega) = \{\omega^2 - \omega_j^2(\mathbf{q})\}^{-1}. \qquad (3.2.10)$$

The result for the correlation function is

$$\frac{1}{2\pi\hbar} \int_{-\infty}^{\infty} dt\, e^{i\omega t} \langle u_\alpha(l,t) u_\beta(l',0)\rangle = -\frac{1}{\pi}\{1 + n(\omega)\}\,\text{Im}\, P_{\alpha\beta}(l,l',\omega), \qquad (3.2.11)$$

where

$$n(\omega) = [\exp(\hbar\omega/kT) - 1]^{-1}. \qquad (3.2.12)$$

The coherent cross section is therefore proportional to the imaginary part of the Fourier transform of the lattice Green function. The summations over l and l' in Eq. (3.2.8) lead to delta functions that place restrictions on $\mathbf{Q}$ and $\mathbf{q}$, and Im $P_j(\mathbf{q},\omega)$ is proportional to $-\delta(\omega - \omega_j(\mathbf{q}))/\omega_j(\mathbf{q})$ to give

$$\left(\frac{d^2\sigma}{d\Omega\, dE'}\right)_{\text{coh}} = \frac{4\pi^3}{v_0} \frac{k'}{k} \frac{b^2}{M\omega_j(\mathbf{q})} \{n(\omega) + 1\} |\mathbf{Q}\cdot\boldsymbol{\sigma}_j(\mathbf{q})|^2 e^{-2W}$$
$$\times \delta\{\hbar\omega - \hbar\omega_j(\mathbf{q})\}\delta\{\mathbf{Q} - \mathbf{q} - \boldsymbol{\tau}\} \qquad (3.2.13)$$

for the coherent scattering corresponding to the creation of one phonon in the crystal. Here v_0 and W are the unit-cell volume and the Debye–Waller factor, respectively.

3.2.2.3. *Basic Experimental Design: Triple-Axis Spectrometer.* The two delta functions in Eq. (3.2.13) represent the conservation of energy and momentum and show that neutrons are scattered only when $\hbar\omega$ and $\mathbf{Q} - \boldsymbol{\tau}$ simultaneously coincide with the dispersion relation, $\omega(\mathbf{q})$, of the sample. Although many pairs of wave vectors $\mathbf{k}$ and $\mathbf{k}'$ will give the same $\mathbf{Q}$ and $\hbar\omega$, a neutron spectrometer that provides a choice of these variables provides a valuable flexibility in compromises between intensity and resolution requirements of different experiments and in the range of $\mathbf{Q}$ and $\hbar\omega$ that can be explored, e.g., $|\mathbf{Q}|_{max} = |\mathbf{k}| + |\mathbf{k}'|$. One basic spectrometer design that provides this flexibility is the triple-axis crystal spectrometer.

Triple-axis neutron spectrometers utilize coherent elastic (Bragg) scattering of neutrons from selected planes of atoms in a crystal for the selection of the energy of the neutrons incident on a sample and for the energy analysis of the scattered neutrons. A schematic illustration of one of the triple-axis spectrometers located at the Oak Ridge National Laboratory is shown in Fig. 1. A beam of thermal neutrons from the reactor having a Maxwellian energy distribution is incident on a monochromator crystal located on the first axis of the spectrometer. The wavelength of the neutrons diffracted by the monochromator is determined by Bragg's law

$$\lambda = 2d_M \sin \theta_M \qquad (3.2.14)$$

where d_M is the spacing of the diffracting planes and $2\theta_M$ is the angle of the diffracted neutrons with respect to the incident beam from the reactor. The energy of the beam incident on the sample is $E = |\mathbf{k}|^2 \hbar^2/2m$, where m is the neutron mass and the magnitude of $\mathbf{k}$ is $|\mathbf{k}| = 2\pi/\lambda$. The angular divergence of this beam, in the plane of the figure, is determined by a soller slit collimator, which is made of a series of cadmium-plated thin steel sheets separated by gaps chosen to provide divergence angles usually in the range 0.1–1°. The intensity of the beam is measured by a low efficiency ($\approx 10^{-4}$–10^{-6}) monitor detector (not shown) that consists of a thin layer of uranium inside a box with thin aluminum windows. Neutron detection is by means of the ^{235}U fission reaction. The counting rate of this detector is normally used to control the counting time of the main signal detector, which is located after the analyzer crystal. The monitor efficiency is inversely proportional to the wave vector $|\mathbf{k}|$; consequently, the counting time for a given number of neutrons incident on the sample is proportional to $|\mathbf{k}|$. The sample is mounted on a rotatable table (second axis) having an orientation with respect to the incident-beam direction, which is specified by the angle ψ. The neutrons that are scattered by the sample through the angle ϕ pass through a

3.2. DYNAMIC PROPERTIES OF DEFECTS

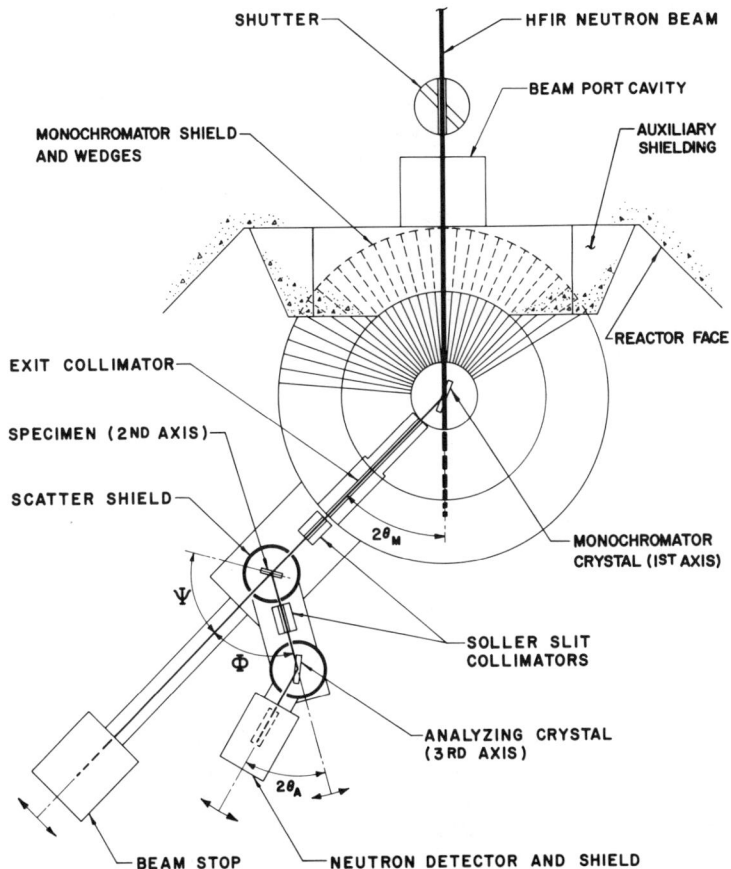

FIG. 1. Schematic illustration of a triple-axis neutron spectrometer located at the high-flux isotope reactor (HFIR) of the Oak Ridge National Laboratory.

collimator and a second monitor detector (not shown). Those neutrons having an appropriate energy may be diffracted by the analyzer on the third axis into the signal detector. The signal detector is usually a tube filled with either ^{3}He or ^{10}BF$_3$ gas. Neutron detection is by means of the charged particles produced in the (n,p) reaction in ^{3}He or the (n,α) reaction in ^{10}B.

In addition to the thermal neutrons used in the experiment, substantial fluxes of high energy neutrons (> 1 keV) and γ rays emerge from the reactor. This radiation is extremely dangerous and produces background for the experiment. Therefore, it must be prevented from escaping into the working space around the spectrometer, except along the thermal-beam path, where it is captured by the beam stop along with the thermal neutrons that pass

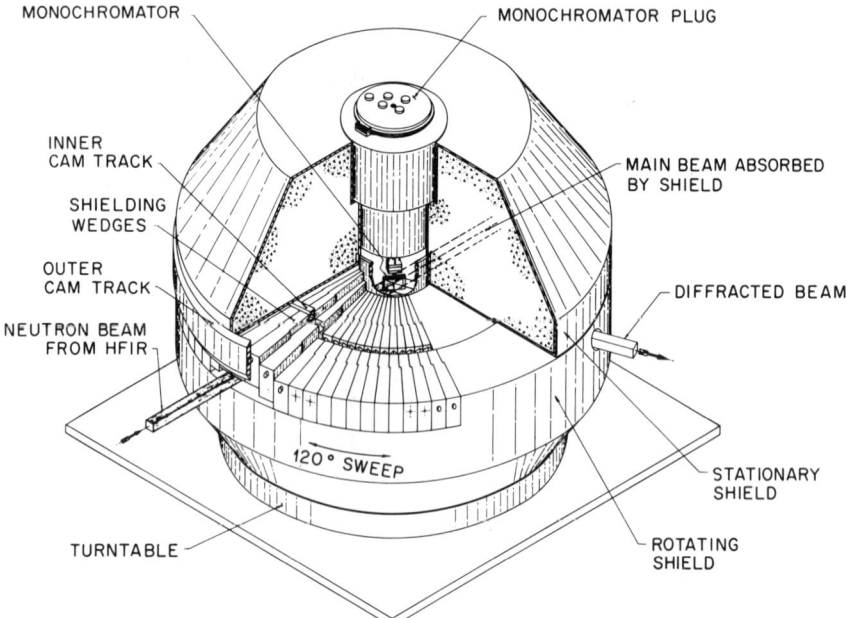

FIG. 2. Cutaway drawing of the monochromator shield in Fig. 1.

through the sample. Consequently, rather massive shielding is required around the monochromator and around the signal detector and analyzer. For example, the spectrometer illustrated in Fig. 1 utilizes a monochromator shield that is approximately 7 feet in diameter, weighs approximately 70,000 pounds, and is made of concrete, steel, and parafin. A cutaway drawing of this shield is shown in Fig. 2. It is a rather complicated instrument containing, in addition to shielding material, controls for precise alignment of the monochromator crystal and movable shielding wedges to allow continuous variation of the monochromator angle from 0° to 120° under computer control.

Typical monochromator and analyzer crystals include pyrolitic graphite, copper, beryllium, germanium, zinc, and aluminum. Such crystals can be obtained with high neutron reflectivity (25–90%) over an angular range (mosaic spread), which is well matched to typical angular divergences provided by the beam collimators. The choice of crystal to be used for an experiment obviously depends on availability and on resolution and intensity requirements. Beryllium possesses a high theoretical scattering power per unit volume, and comparatively small spacings for the first-order Bragg diffraction planes. For example, the spacing of the (002) planes in Be is 1.79 Å compared to 3.35 Å for the (002) planes of graphite or 3.27 Å for the

(111) planes of germanium. The smaller the spacing of the diffraction planes, the larger the diffraction angle for a given λ and the better the energy resolution, since from Eq. (3.2.14) the relative wavelength spread for a given angular divergence $\Delta\theta$ is

$$\Delta\lambda/\lambda = \Delta\theta \cot\theta \quad \text{or} \quad \Delta E/E = -2\,\Delta\theta \cot\theta. \quad (3.2.15)$$

Thus beryllium may be a good choice, especially if E is also large, which will make θ small for given d.

However, the availability of beryllium crystals having useful ranges of mosaic spreads is extremely limited so that most experiments are carried out with crystals of other materials. In these cases good energy resolution is achieved by making $\Delta\theta$ small with collimation. If good energy resolution is not required, a gain in intensity over that obtained with Be may be achieved with other materials simply because a larger energy slice of the reactor spectrum and of the energy distribution of the neutrons scattered by the sample is reflected by the monochromator and analyzer, respectively. A review of the properties of monochromator and analyzer crystals has been given by Schmatz.[2]

One important problem of the crystal spectrometer is that of order contamination. It arises from the fact that if, as in Eq. (3.2.14), a wavelength λ corresponding to energy E is diffracted by atomic planes of spacing d, then λ/n corresponding to an energy $n^2 E$ will also be diffracted at the same angle by planes with spacing d/n. Similarly, the analyzer may reflect neutrons with energy $m^2 E'$. Thus various scattering processes, in addition to the one we wish to measure, can take place in the sample corresponding to the conditions,

$$\hbar\omega_{nm} = n^2 E - m^2 E' \quad \text{and} \quad \mathbf{Q}_{nm} = n\mathbf{k} - m\mathbf{k}'. \quad (3.2.16)$$

Several methods for dealing with this problem, such as the use of filters and of crystals having a structure such that the scattering power of certain higher-order planes is zero, have been discussed by Dolling.[3]

Returning to the design of an inelastic scattering experiment, we see that with a crystal spectrometer the controllable variables of the experiment can be selected in such a way that both $\hbar\omega(\mathbf{q})$ and $\mathbf{q}$ for a specific phonon can be measured. Predetermined values for the neutron wave vectors, $\mathbf{k}$ and $\mathbf{k}'$, the energy transfer, $\hbar\omega = E - E'$, and $\mathbf{Q}$ are obtained for appropriate values of the monochromator and analyzer angles, $2\theta_M$ and $2\theta_A$, and the scattering angle ϕ. The angle ψ determines the orientation of $\mathbf{k}$, and hence of $\mathbf{Q}$, relative to the crystal axes to give $\mathbf{q}$ as is illustrated in Fig. 3a. In this figure a

---

[2] W. Schmatz, *Treatize Mater. Sci. Technol.* **2**, 105 (1973).
[3] G. Dolling, in "Dynamical Properties of Solids" (G. K. Horton and A. A. Maradudin, eds.), Vol. 1, p. 541. North-Holland Publ., Amsterdam, 1974.

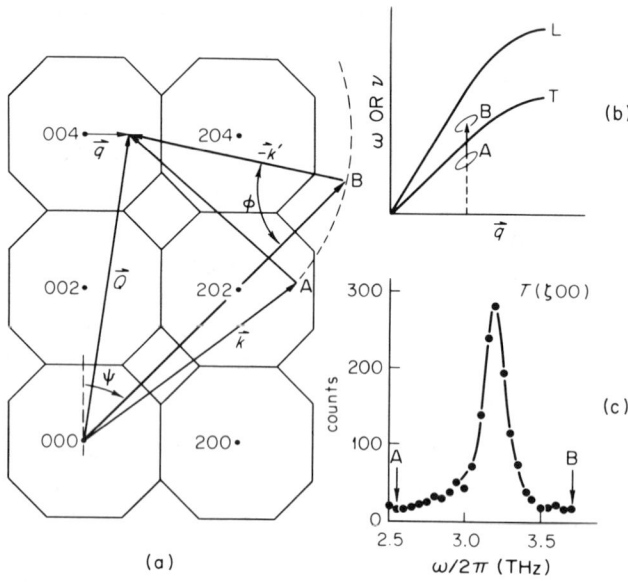

FIG. 3. (a) Schematic illustration of the relations between **k**, **k′**, **Q** and **q** (vectors marked with arrows in the figure) and the angles ψ and ϕ of Fig. 1 as viewed in the reciprocal lattice of a crystal sample having a face-centered-cubic structure. (b) Illustration of an energy scan for constant Q carried out with $|\mathbf{k'}|$ also constant as shown in (a) (L = longitudinal, T = transverse). (c) Typical intensity distribution of scattered neutrons for a constant-Q scan as shown in (b).

few reciprocal lattice points and sections of the corresponding Brillouin zones are shown for the (h0l) plane of the reciprocal lattice for a face-centered-cubic crystal structure. The neutron scattering diagram shown (i.e., the wave-vector triangle $\mathbf{k} - \mathbf{k'} = \mathbf{Q}$) is for the case of $\tau = (0,0,4)2\pi/a$ and for **q** in the [100] direction. In setting up an experiment on a crystal spectrometer, the values of **Q**, τ, $\hbar\omega$, and either E or E' are actually the quantities preselected. These are then used in a straightforward calculation to obtain the appropriate spectrometer angles. Once the spectrometer is positioned at these angles, the number of neutrons detected for that particular spectrometer setting, or equivalently for that particular $\hbar\omega$ and **Q**, is recorded, and then the spectrometer is repositioned for a new $\hbar\omega$ and **Q**. With the computer control of modern crystal spectrometers, it is possible to scan easily any line in (ω,**Q**) space. However, it is usually more convenient to interpret the results of scans carried out with either **Q** (constant **Q**) or $\hbar\omega$ (constant energy) held fixed. For example, a schematic illustration of a constant **Q** scan is shown in Fig. 3b, and the neutrons counted versus $\hbar\omega$ in an actual scan is shown in Fig. 3c. The location of the peak determines one point on the dispersion relation, which in the example illustrated in Fig. 3 corresponds to the transverse branch with **q** parallel to [100]. The branch measured in a particular scan is determined by the $\mathbf{Q} \cdot \boldsymbol{\sigma}_j(\mathbf{q})$ factor in the cross section, Eq. (3.2.13). In our example this factor is maximized for the transverse branch, $\boldsymbol{\sigma} \perp \mathbf{q}$, by a scattering geometry with **Q** nearly perpendicular to **q**. A scattering geometry with $\mathbf{Q} \parallel \mathbf{q}$ and with $\tau = (4,0,0)2\pi/a$ would be used to measure the longitudinal branch in the [100] direction.

Figure 3a also illustrates one method commonly used to perform a constant **Q** scan. The analyzer angle $2\theta_A$ is selected to give a predetermined E', and hence $|\mathbf{k}'|$, which is also held constant throughout the scan. Then, by varying $|\mathbf{k}|$, ϕ, and ψ, the energy transfer $\hbar\omega$ can be varied while keeping **Q** constant as shown in the figure. Since the magnitude of $|\mathbf{k}|$ and E increases in going from position A to B in the diagram, $\hbar\omega$ also increases as illustrated in Fig. 3b. The value of E' used for a given experiment is chosen by making various compromises between intensity and resolution considerations. The scattering intensity varies with $|\mathbf{Q}|$ as $|\mathbf{Q}|^2 e^{-2W}$. Therefore, if W is small, it may be desirable to have $|\mathbf{Q}|$ as large as possible, which requires large values for $|\mathbf{k}|$ and $|\mathbf{k}'|$. However, the larger $|\mathbf{k}'|$, the poorer the energy resolution of the analyzer for a given beam divergence $\Delta\theta$ defined by the collimators since from Eq. (3.2.15) $\Delta E' \propto |\mathbf{k}'|^3 \Delta\theta \cos\theta_A$.

The one-phonon incoherent scattering cross section for monatomic crystals is[1]

$$\left(\frac{d^2\sigma}{d\Omega\, dE'}\right)_{\text{inc}} = \frac{k'}{k} \frac{\sigma_{\text{inc}}}{M} \sum_{j,q} \frac{|\mathbf{Q} \cdot \boldsymbol{\sigma}_j(\mathbf{q})|^2}{\omega_j(\mathbf{q})}$$
$$\times \{n(\omega) + 1\}\delta\{E - E' - \hbar\omega_j(\mathbf{q})\}\exp(-2W). \quad (3.2.17)$$

This cross section is similar to that for coherent scattering except there is no condition relating **Q** and **q**. Consequently, it is not possible to measure the energies of individual phonons by incoherent scattering experiments. However, for the simple case of a cubic, Bravais crystal,

$$\left(\frac{d^2\sigma}{d\Omega\, dE'}\right)_{\text{inc}} = \frac{\sigma_{\text{inc}}}{8\pi M} \frac{k'}{k} Q^2 \exp(-2W) \frac{g(\omega)}{\omega} \{n(\omega) + 1\}, \quad (3.2.18)$$

where $g(\omega) = (\tfrac{1}{3}N)\sum_{\mathbf{q}j} \delta\{\omega - \omega_j(\mathbf{q})\}$ is the distribution of phonon energies (or density of phonon states) and the factor $|\mathbf{Q} \cdot \boldsymbol{\sigma}_j(\mathbf{q})|^2$ has been averaged over all modes to give $Q^2/3$. Therefore, $g(\omega)$ can be obtained directly from the energy distribution of neutrons incoherently scattered by the sample. Although a measurement of $g(\omega)$ is useful, it does not provide as detailed information about the lattice dynamics as does a measurement of $\omega(\mathbf{q})$. Furthermore, it is, in a sense, a more difficult measurement since it requires that intensities rather than peak positions be determined.

3.2.3. Defect Dynamics: Theory and Experiment

The main difficulty encountered in the theoretical description of the vibrational properties of a crystal containing defects is the lack of exact periodicity. Consequently, the vibrational displacements of the atoms cannot be written in terms of noninteracting normal modes (or phonons) having well-defined wave vectors and frequencies. In general, the secular determinant in this case cannot be reduced rigorously to a manageable size, 3×3, but must remain $3N \times 3N$. However, for a small concentration of defects that produce localized disturbances in the lattice, the dynamics of the lattice can be described in terms of modified phonons that are perturbed or scattered by the defects. This scattering generally leads to a change in the phonon energy and to a finite lifetime and a corresponding energy width for the phonons. The theoretical treatment of this defect–phonon scattering and the response of the lattice to an externally applied ω- and q-dependent probe, such as neutrons, are both conveniently described in terms of Green functions just as for the perfect crystal discussed above. In the following paragraphs we summarize very briefly the published theoretical work. For further details the reader may wish to consult several reviews (Marshall and Lovesey,[1] Taylor,[4] Elliott et al.[5]).

3.2.3.1. Outline of Theory.

We begin with Eq. (3.2.1), which is valid for the perfect and imperfect lattice alike, and define a mass-defect parameter ϵ and a force-constant change $\Delta\phi$ as

$$\epsilon(l) = [M - M'(l)]/M, \qquad (3.2.19)$$

$$\Delta\phi_{\alpha\beta}(l,l') = \phi_{\alpha\beta}(l,l') - \phi^0_{\alpha\beta}(l,l'), \qquad (3.2.20)$$

where M' is the mass of a substitutional impurity and $\phi^0_{\alpha\beta}(l,l')$ is the force constant appropriate to the perfect lattice. We assume a solution of the form

$$u_\alpha(l,t) = u_\alpha(l) \exp(-i\omega t). \qquad (3.2.21)$$

Substituting (3.2.19)–(3.2.21) into (3.2.1) gives

$$\sum_{\beta,l'} [M\omega^2 \delta_{\alpha\beta}\delta_{ll'} - \phi^0(l,l')]u_\beta(l') = \sum_{\beta,l'} V_{\alpha\beta}(l,l')u_\beta(l'), \qquad (3.2.22)$$

where

$$V_{\alpha\beta}(l,l') = M\omega^2 \epsilon(l)\delta_{\alpha\beta}\delta_{ll'} + \Delta\phi_{\alpha\beta}(l,l'). \qquad (3.2.23)$$

In abbreviated matrix form Eq. (3.2.22) becomes

$$(M\omega^2 \mathbf{1} - \boldsymbol{\phi})\mathbf{u} \equiv \mathbf{L}\mathbf{u} = \mathbf{V}\mathbf{u}, \qquad (3.2.24)$$

---

[4] D. S. Taylor, in "Dynamical Properties of Solids" (G. K. Horton and A. A. Maradudin, eds.), Vol. 2, p. 285. North-Holland Publ., Amsterdam, 1975.

[5] R. J. Elliott, J. A. Krumhansl, and P. L. Leath, Rev. Mod. Phys. **46**, 465 (1974).

3.2. DYNAMIC PROPERTIES OF DEFECTS

which has the formal solution

$$\mathbf{u} = \mathbf{PVu}, \quad (3.2.25)$$

where **P** is the Green function of the perfect crystal.

The Green function **G** for the lattice with defects is defined as inverse of $\mathbf{L} - \mathbf{V}$. Consequently, **G** and **P** are related by

$$\mathbf{G} = \mathbf{P} + \mathbf{PVG}. \quad (3.2.26)$$

Note that the zeroes in the denominator of **P** (i.e., the poles of **P**) occur at the normal-mode frequencies of the perfect crystal. Similarly, the poles of **G** give the vibration frequencies of the imperfect crystal. These poles will generally be complex when they occur below the maximum frequency, ω_m, of the perfect lattice, indicating that the corresponding modes are damped because of their interaction with the defects. However, for certain defects there are poles of **G** that correspond to modes having frequencies greater than ω_m. Such frequencies will be real (i.e., not damped), but they correspond to vibrations that are localized in space.

3.2.3.2. Mass Defects and Scattering Cross Sections. The dynamical properties discussed above, as well as the corresponding experimental observables, can be illustrated conveniently by means of a simple model of a lattice containing a very low concentration of impurities that differ from the host atoms they replace only in mass. Although we are primarily concerned here with describing experimental methods, it is desirable to be aware also of the limitations placed on the interpretation of the experimental results by certain inadequacies in our ability to carry to completion a detailed theoretical treatment of the real crystal. For example, the problem of a finite concentration of defects has not been solved. In addition, the introduction of impurities into a real crystal produces not only mass defects but also force-constant defects, defect clusters, nonuniform lattice relaxation, etc., and the theoretical treatment of these effects together has, to date, proved to be prohibitively complex. Thus, although the theory for mass defects can be used to illustrate general features of defect–phonon perturbations that are actually observed in neutron scattering experiments, most experiments on real crystals have shown some deviations from the theoretical predictions.

For a mass defect the form for **V** is particularly simple, and the defect-induced perturbations to the equations of motion are independent. Equation (3.2.26) can then be schematically written

$$\mathbf{G}^s = \mathbf{P} + \sum_i \mathbf{PV}_i \mathbf{G}^d, \quad (3.2.27)$$

where the summation is over the defect sites, and $\mathbf{G}^d$ is a Green function with its first lattice site label, e.g., l in $\mathbf{G}_{\alpha\beta}(l,l',\omega)$, designating a defect site. $\mathbf{G}^h$ is the corresponding Green function for a host atom at site l. For a very low concentration of widely separated defects, it is reasonable to assume that $\mathbf{G}^s$ (s = h or d) does not depend explicitly on the exact locations of all the defects but only on an average environment. With this assumption the details of the summation over defect sites in Eq. (3.2.27) can be ignored, and a simple relation between $\mathbf{G}^d$ and $\mathbf{G}^h$ for a site is obtained (Elliott and Taylor[6]), namely,

$$\mathbf{G}^h = \mathbf{G}^d[1 - M\epsilon\omega^2 P(\omega)], \quad (3.2.28)$$

where $P(\omega) = P_{\alpha\alpha}(0,0,\omega)$, $\epsilon = (M - M')/M$, and M' is the mass of a defect atom. The Green function averaged over all possible configurations of defect atoms is

$$\overline{\mathbf{G}} = c\mathbf{G}^d + (1 - c)\mathbf{G}^h. \quad (3.2.29)$$

This together with (3.2.23), (3.2.27), and (3.2.28) gives

$$\overline{\mathbf{G}} = \mathbf{P} + c\mathbf{P}T(\omega)\overline{\mathbf{G}}, \quad (3.2.30)$$

where

$$T(\omega) = \frac{M\epsilon\omega^2}{1 - (1 - c)M\epsilon\omega^2 P(\omega)}. \quad (3.2.31)$$

The configuration averaging has effectively restored the lattice periodicity. Calculation of the neutron coherent scattering cross section from the Fourier transform of $\overline{\mathbf{G}}$ is straightforward and follows the procedure described above for the perfect crystal. From Eq. (3.2.30)

$$\overline{G}_j(\mathbf{q},\omega) = \frac{P_j(\mathbf{q},\omega)}{1 - cT(\omega)P_j(\mathbf{q},\omega)} = \frac{1}{\omega^2 - \omega_j^2(\mathbf{q}) - cT(\omega)}, \quad (3.2.32)$$

and by comparison with Eq. (3.2.10) we see that the perturbations of the vibration modes caused by the defects are determined by $cT(\omega)$. The real part of $T(\omega)$ gives the shifts of the frequencies from the perfect-crystal values, and the imaginary part represents the damping of the modes in the imperfect crystal. These perturbations can be measured directly by neutron coherent scattering since,

$$S(\mathbf{Q},\omega) \propto \operatorname{Im} \overline{G}_j(\mathbf{q},\omega) = \frac{c \operatorname{Im} T(\omega)}{[\omega^2 - \omega_j^2(\mathbf{q}) - c \operatorname{Re} T(\omega)]^2 + [c \operatorname{Im} T(\omega)]^2}. \quad (3.2.33)$$

Consequently the delta-function peak in the scattering cross section for the perfect crystal, Eq. (3.2.13), is replaced by a broadened scattering

[6] R. J. Elliott and D. W. Taylor, *Proc. R. Soc. London, Ser. A* **296**, 161 (1967).

3.2. DYNAMIC PROPERTIES OF DEFECTS

function that is shifted in frequency from $\omega_j(\mathbf{q})$ and is nearly Lorentzian if the ω dependence of Re $T(\omega)$ and Im $T(\omega)$ is not strong and if the defect concentration is not large. Under these conditions the intensity maximum in the neutron scattering occurs, for a given wave vector $\mathbf{q}$, at a frequency ω that satisfies

$$\omega^2 - \omega_j^2(\mathbf{q}) - c \text{ Re } T(\omega) = 0 \qquad (3.2.34)$$

or

$$\omega = [\omega_j^2(\mathbf{q}) + c \text{ Re } T(\omega)]^{1/2}. \qquad (3.2.35)$$

For small c, ω is displaced from $\omega_j(\mathbf{q})$ by the amount Δ, where

$$\Delta(\omega) = \omega - \omega_j(\mathbf{q}) \cong c \text{ Re } T(\omega)/2\omega. \qquad (3.2.36)$$

The width of the intensity distribution is approximately

$$\Gamma(\omega) \simeq c \text{ Im } T(\omega)/\omega. \qquad (3.2.37)$$

In a typical experiment the phonon frequencies of a crystal with no defects or impurities are also determined by carrying out the measurements described previously in connection with Fig. 3. Such measurements give not only $\omega_j(\mathbf{q})$, but also the contribution of the instrumental resolution to the energy widths Γ of the measured phonon peaks.

From Eq. (3.2.31) the real and imaginary parts of $T(\omega)$ can be written explicitly to give

$$T(\omega) = \frac{M\omega^2 \epsilon [\beta(\omega) + i\gamma(\omega)]}{\beta^2(\omega) + \gamma^2(\omega)}, \qquad (3.2.38)$$

where

$$\beta(\omega) = 1 - (1 - c)M\epsilon\omega^2 \text{ Re}[P(\omega)], \qquad (3.2.39)$$

$$\gamma(\omega) = (1 - c)M\omega^2\epsilon \text{ Im}[P(\omega)]. \qquad (3.2.40)$$

For a cubic crystal the sum over all modes in Eq. (3.2.9) can be replaced by an integral over $g(\omega)$ to give

$$MP(\omega) = P \int_0^{\omega_m} \frac{g(\omega') \, d\omega'}{\omega^2 - \omega'^2}, \qquad (3.2.41)$$

where $P \int$ denotes the Cauchy principal-part integral when $\omega < \omega_m$. The modes most strongly perturbed by the defects have frequencies that make the denominator of $T(\omega)$ small, i.e., $\beta(\omega) = 0$. For $c \ll 1$ the condition $\beta(\omega) = 0$ is equivalent to

$$\epsilon^{-1} = \omega^2 P \int \frac{g(\omega') d\omega'}{\omega^2 - \omega'^2}. \qquad (3.2.42)$$

186   3. NEUTRON SCATTERING STUDIES OF LATTICE DEFECTS

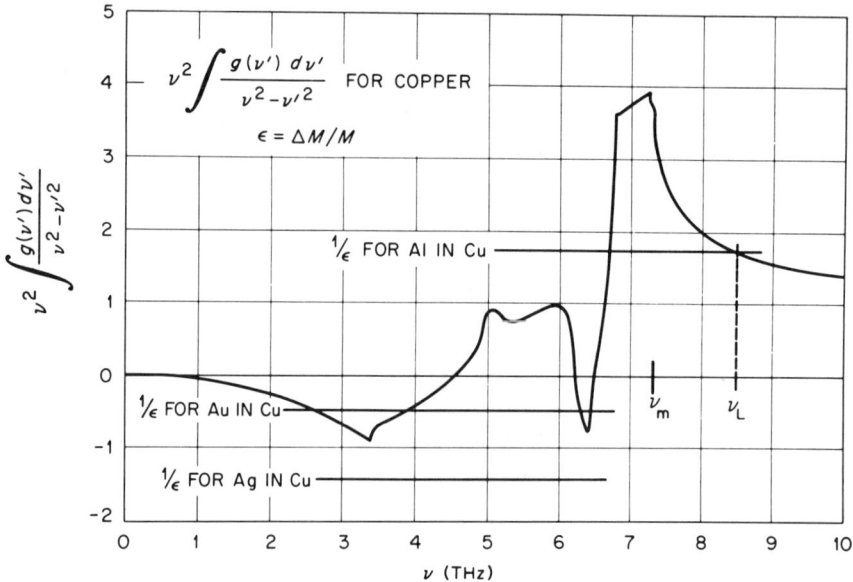

FIG. 4. The solutions to Eq. (3.2.42) shown graphically.

The solutions to Eq. (3.2.42) are illustrated graphically in Fig. 4 for a copper host (recall $\nu = \omega/2\pi$). It should be noted that for the mass-defect model $T(\omega)$ is a function of only ω. In the case of force-constant defects, T and the phonon perturbations will depend explicitly on $\mathbf{q}$ and branch index also, e.g., $T(\omega) \rightarrow T_j(\mathbf{q},\omega)$ as discussed by Lakatos and Krumhansl.[7] Although such effects are commonly observed, it is nevertheless useful to classify the neutron scattering research on defect dynamics according to the two conditions illustrated in Fig. 4: (a) light-mass impurities with $M' \le M$ for which ϵ^{-1} ranges from 1 to ∞, and (b) heavy-mass impurities with $M' > M$ for which ϵ^{-1} ranges from $-\infty$ to 0. Note that not all values of ϵ satisfy Eq. (3.2.42).

3.2.3.3. Heavy Impurities; Band Modes. For a heavy impurity all of the solutions to Eq. (3.2.42) occur for $\omega < \omega_m$, i.e., within the band of frequencies for the pure host. These frequencies correspond to resonant modes that are characterized by a large amplitude of vibration at the defect.[8] The frequencies and lifetimes of the phonons with frequencies near that of a resonant mode are considerably affected by the presence of the defects.

Typical experimental results showing these perturbations are displayed in Fig. 5 where constant-Q measurements obtained by Svensson and Brock-

[7] K. Lakatos and J. A. Krumhansl, *Phys. Rev.* **180,** 729 (1969).
[8] R. J. Elliott and P. G. Dawber, *Proc. Phys. Soc., London* **81,** 453 (1963).

3.2. DYNAMIC PROPERTIES OF DEFECTS

house[9] for copper and for copper with 3-at. % Au impurities are compared. These results show the neutron scattering by phonons with frequencies near 2.5 THz corresponding to the location of the lowest frequency resonant mode as deduced from the results shown in Fig. 4. The peaks observed for the Cu–Au sample are significantly broader than those for Cu. In addition, they are shifted in frequency from the positions measured for Cu, which are devoted by ν_0 in Fig. 5. Note that this shift is large and negative for $\nu_0 < 2.4$ THz, and it is small but positive for $\nu_0 > 2.5$ THz. This behavior is consistent with the theoretical results for Δ and Γ as is illustrated in Fig. 6, where the frequency shifts and the widths of the neutron peaks for the alloy are shown together with several curves calculated on the basis of the preceding theory [see Eqs. (3.2.36) and (3.2.37)]. In general, the agreement between theory and experiment is rather good. Svensson and Brockhouse point out that

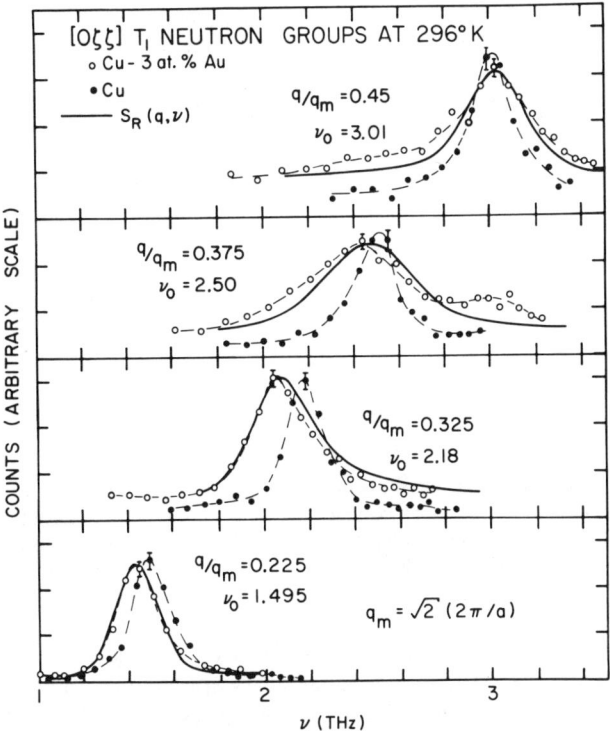

FIG. 5. Intensity distributions of neutrons scattered by Cu and Cu–3 at. % Au. The full lines are calculations obtained by folding $S(\mathbf{q},\nu)$, see Eq. (3.2.33), with the instrumental resoltuion. ν_0 is the frequency assigned to the peaks for Cu. (From Svensson and Brockhouse.[9])

[9] E. C. Svensson and B. N. Brockhouse, *Phys. Rev. Lett.* **18**, 858 (1967).

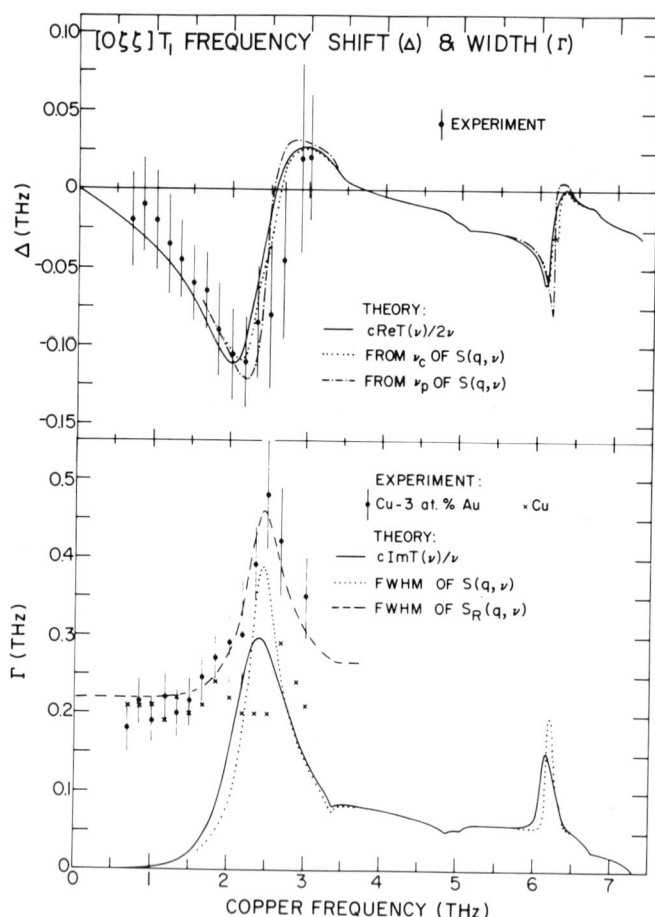

Fig. 6. Experimental frequency shifts $\Delta = \nu_{\text{alloy}} - \nu_{\text{copper}}$, neutron peak widths Γ, and the corresponding calculated curves as obtained from the resolution-broadened cross section $S_R(\mathbf{q},\nu)$. ν_c and ν_p refer to the central frequencies at half-maximum and the peak frequencies, respectively. (From Svensson and Brockhouse.[9])

even for the relatively low impurity concentration of 3 at. % in their alloy, Eqs. (3.2.36) and (3.2.37) are not satisfactory approximations in spite of the fact that the measured peak shapes appear to be nearly Lorentzian. They obtained significantly better agreement between theory and experiment for Γ on the basis of the shapes of calculated curves $S_R(\mathbf{q},\omega)$, obtained by folding $S(\mathbf{q},\omega)$ with the instrumental resolution. Several examples of these calculated peak shapes are shown in Fig. 5.

For large impurity concentrations, or if $T(\omega)$ varies strongly with ω, the peak shapes may depart significantly from a Lorentzian. In such cases the

3.2. DYNAMIC PROPERTIES OF DEFECTS

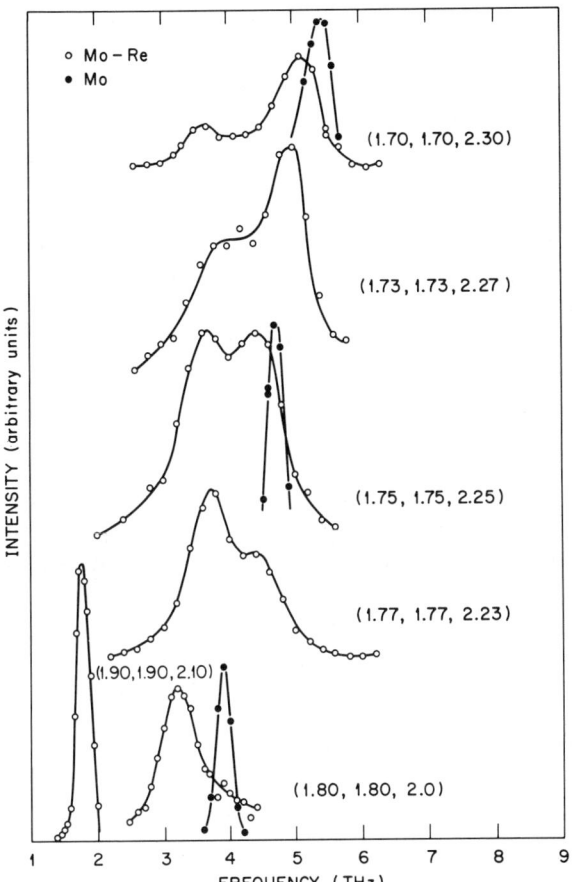

FIG. 7. Neutron-intensity distributions for transverse modes in the [111] direction for $Mo_{0.85}Re_{0.15}$ (unfilled circle) and Mo (filled circle). (From Smith et al.[10])

neutron data must be compared directly with peak-shape calculations. In fact, Im $G_j(q,\omega)$ generally possesses two peaks: one when ω is near to the shifted $\omega_j(q)$, and one near the resonant frequency where Im $T(\omega)$ can have a large peak (note the large peak in $\Gamma(\omega)$ in Fig. 6). The latter peak is usually significant only when $\omega_j(q)$ is also near the resonant frequency. These effects are clearly seen in the neutron scattering data obtained by Smith et al.[10] for the transverse phonons in the [111] direction for $Mo_{0.85}Re_{0.15}$ that are shown in Fig. 7. The measured peak locations in this series of constant-Q scans

---

[10] H. G. Smith, N. Wakabayashi, and M. Mostoller, in "Superconductivity in d- and f-Band Metals" (D. H. Douglass, ed.), p. 223. Plenum, New York, 1976.

appear to give a gap (or splitting) in the dispersion relation that is suggestive of a hybridization between the phonon and the resonant mode. A similar effect has been observed by Zinken et al.[11] for a lower impurity concentration in $Al_{0.965}Ag_{0.035}$.

3.2.3.4. Light Impurities; Localized Modes. According to the results shown in Fig. 4, when $\epsilon^{-1} < 4$ (i.e., $M'/M < 0.75$) a solution to Eq. (3.2.42) exists corresponding to a vibrational mode with a frequency, ω_L, which is greater than the maximum frequency, ω_m, of the host copper lattice. From Eqs. (3.2.3) and (3.2.4) one expects, roughly speaking, the vibrational frequency of an atom to vary with mass as $\omega^2 \alpha$ (force constant/mass). For our simplified example, therefore, a light mass will obviously vibrate at a higher frequency than a heavy mass, i.e., the host mass, for given force constants. Since the host atoms next to the impurity are directly coupled to it, they will also vibrate at the higher frequency. However, this vibrational mode does not propagate far into the host lattice because its frequency is not a normal-mode frequency of that lattice. It is therefore localized in space. A direct experimental measurement of ω_L and the degree of spatial localization of the mode can be obtained by neutron coherent scattering measurements.

When the neutron-energy change is greater than $\hbar \omega_m$, the scattered intensity corresponding to the creation of one phonon in the localized mode is determined by Eq. (3.2.33). In this case Im $T(\omega) = 0$ since $g(\omega) = 0$ for $\omega > \omega_m$ and the condition $\beta(\omega) = 0$ [see Eq. (3.2.39)] is equivalent to $\omega^2 - \omega^2(q) - c \operatorname{Re} T(\omega) = 0$ for $c \ll 1$. The calculation of Im $\mathbf{G}_j(\mathbf{q},\omega)$ from Eq. (3.2.32) therefore gives delta functions at $\omega = \omega_L$ in analogy with the delta functions obtained in Eq. (3.2.11) for Im $P_j(\mathbf{q},\omega)$ in the perfect crystal. The results is[1]

$$\left(\frac{d^2\sigma}{d\Omega \, dE'}\right)_{\mathrm{coh}} = \frac{k'}{k} \frac{b^2 c \omega_L}{2MB(\omega_L)} \{n(\omega_L) + 1\}\delta(\omega - \omega_L) \sum_j \left[\frac{\mathbf{Q} \cdot \boldsymbol{\sigma}_j(\mathbf{q})}{\omega_L^2 - \omega_j^2(\mathbf{q})}\right]^2, \quad (3.2.43)$$

where

$$B(\omega_L) = \int \frac{\omega'^2 g(\omega') \, d\omega'}{(\omega_L^2 - \omega'^2)^2} \quad (3.2.44)$$

and we have assumed for the moment that the coherent scattering length for the impurity is the same as that for the host. Therefore, the scattered neutrons will posses a peak at $\hbar\omega = \hbar\omega_L$, which is independent of $\mathbf{q}$, with an intensity which varies with $\mathbf{q}$ in a manner that depends on the difference

---

[11] A. Zinken, U. Buchenan, H. J. Fenzl, and H. R. Schofer, *Solid State Commun.* **22**, 693 (1977).

3.2. DYNAMIC PROPERTIES OF DEFECTS

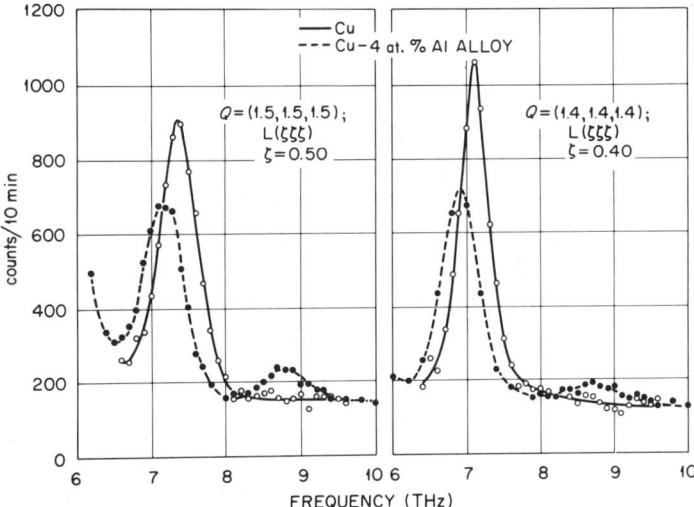

FIG. 8. Neutron-intensity distributions for constant-Q scans at the zone boundary in Cu and Cu–4 at. % Al. (From Nicklow et al.[13])

between the frequency of the local mode and the frequency of the host lattice phonon with the wave vector **q**. This particular form for the **q** dependence of the intensity is directly related to the spatial extent of the atomic displacements of the mode through Eq. (3.2.10) and Eq. (3.2.25) as discussed by Maradudin.[12]

The first studies of localized vibrational mode by coherent inelastic neutron scattering experiments were carried out on $Cu_{1-c}Al_c$ by Nicklow et al.[13] and on $Ta_{0.88}Nb_{0.12}$ by Als-Nielsen.[14] The scattering cross section takes a simple form if **Q** lies along symmetry directions because then $\mathbf{Q}\cdot\sigma_j(\mathbf{q}) = 0$ unless j refers to the longitudinal branch. The experiments on $Cu_{1-c}Al_c$ were carried out with **Q** along the [111] direction. Since the longitudinal branch in this direction reaches its highest frequency at the Brillouin-zone boundary, $\mathbf{q} = (0.5, 0.5, 0.5) 2\pi/a$, one expects the scattering by the local mode to be a maximum there according to Eq. (3.2.43).

Figure 8 shows constant-**Q** scans at and just off the [111]-zone boundary for pure copper and for a copper crystal containing 4 at. % Al. For pure copper only the expected peaks due to the longitudinal [111] phonons appear. However, for Cu–4 at. % Al an additional peak is observed near a

---

[12] A. A. Maradudin, *Solid State Phys.* **18**, 273 (1966).

[13] R. M. Nicklow, P. R. Vijayaraghavan, H. G. Smith, and M. K. Wilkinson, *Phys. Rev. Lett.* **20**, 1245 (1968).

[14] J. Als-Nielsen, *Neutron Inelastic Scattering, Proc. 4th Symp., Copenhagen*, **1**, 35 (1968).

frequency $\omega/2\pi = 8.8$ THz corresponding closely with the value expected for the local mode, 8.5 THz, on the basis of our previous discussion of Fig. 4. These results also show that the intensity of the local mode scattering decreases as **Q** is moved away from the zone boundary. The experimental results for the **Q** dependence of the local mode intensity agrees well with that predicted by Eq. (3.2.43) as is shown in Fig. 9.

One of the primary difficulties in such experiments is the low intensity of the local-mode scattering, owing to the obvious fact that for low impurity concentrations only a relatively small fraction of the sample contributes to the scattering cross section. Therefore, rather large and generally difficult to obtain single crystals (~ 5 cm^3) are required. This problem can be somewhat compensated by the factor $[\omega_L - \omega_j(\mathbf{q})]^{-2}$ in the cross section if ω_L is near $\omega_j(\mathbf{q})$, but then in order to separate the peaks in the scattering occuring at ω_L and $\omega_j(\mathbf{q})$, rather good energy resolution is required, which is also costly in intensity. Consequently, few studies of localized defect modes have been carried out by coherent scattering measurements. Of course, the scattering intensity can be increased by increasing the defect concentration. However,

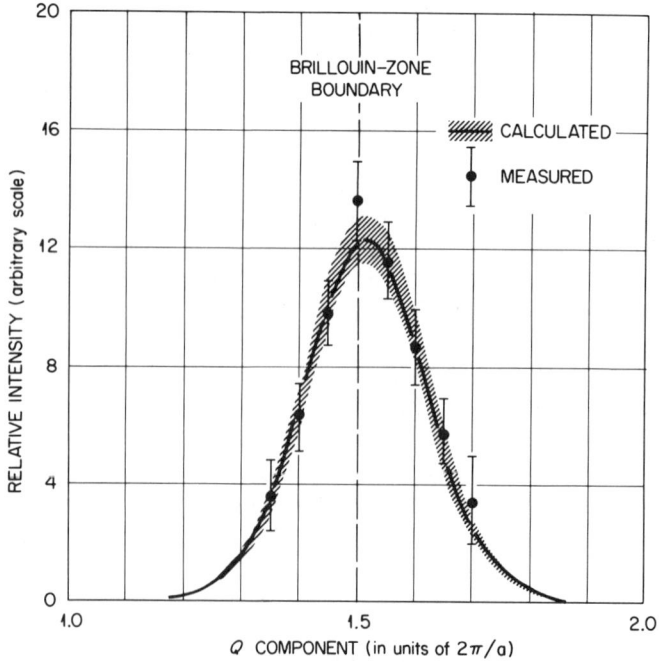

FIG. 9. Comparison of the observed Q dependence of the local-mode intensity with theory, Eq. (3.2.43). (From Nicklow et al.[13])

3.2. DYNAMIC PROPERTIES OF DEFECTS

then defect-defect interactions can become important, and the isolated defect theory on which Eq. (3.2.43) is based is no longer applicable.

Studies of localized modes may not always require single-crystal samples. Occasionally, it may be of interest to measure only the local-mode frequency and not the **q** dependence of the scattering in situations where the interpretation of other properties, e.g., the temperature dependence of the Mössbauer effect, depends primarily on knowing ω_L only. In such cases neutron scattering measurements on polycrystalline samples would be useful. A similar situation exists in studies of the localized vibrations of hydrogen in metals. The large incoherent scattering cross section of hydrogen makes the study of coherent scattering impossible.

To date nearly all coherent neutron scattering studies of localized modes have been carried out on systems with defect concentrations in the 4%–30% range. Even at the lower end of this range the applicability of the isolated defect theory is questionable as discussed by Kaplan and Mostoller[15] for $Cu_{0.9}Al_{0.1}$. At the higher-concentration range one is dealing with an alloy in which even the specification of "host" and "defect" becomes obscured. The theoretical treatment of high-concentration systems is exceedingly difficult and challenging and is the area of most current research on defect systems.

---

[15] T. Kaplan and M. Mostoller, *Phys. Rev. B* **9**, 353 (1974).

3.3. Diffusion Studies

By N. Wakabayashi*

Solid State Division
Oak Ridge National Laboratory
Oak Ridge, Tennessee

3.3.1. Introduction

The study of diffusion by means of slow-neutron scattering techniques may be classified into two general categories. One is to focus one's attention on the motion of a single diffusing atom and to follow its motion as time evolves. The other is to observe the motion of a diffusing atom with respect to other diffusing or stationary atoms. In the former type of study the self-correlation function of the migrating atom plays an essential role. In the latter, on the other hand, pair-correlation functions determine the neutron scattering cross section. As is discussed in the following, the self-correlation function determines the cross sections for *incoherent* scattering processes, and the pair-correlation function determines the *coherent* scattering cross sections. The most favorable incoherent scatterer is the proton for which the total incoherent scattering cross section is about 80 b, extremely large compared with the coherent scattering cross section (1.8 b). Niobium, for example, has the incoherent and coherent scattering cross sections of 0.3 b and 6.3 b, respectively. Thus the self-correlation of a diffusing proton in a crystal can be studied most easily, and the majority of measurements on diffusion has to date been on hydrogen in transition metals. Possibilities for studying other types of diffusion such as those of solid electrolytes have been explored, but only a limited amount of experimental information has been accumulated at present.

There are several review articles and books on slow-neutron scattering studies of diffusion. The book by Springer[1] gives a detailed discussion on the subject as well as references to the experiments reported up to 1972. The

---

[1] T. Springer, "Quasielastic Neutron Scattering for the Investigation of Diffusive Motions in Solids and Liquids." Springer-Verlag, Berlin and New York, 1972.

* Present address: Department of Physics, Keio University, Yokohama, Japan.

review article of Rowe[2] summarizes experiments performed on hydrogen-metal systems. A more recent article by Springer[3] also describes the experimental results on the subject as well as on rotational diffusion. Shapiro and Reidinger[4] reviewed the experiments on solid electrolytes.

One of the main advantages of the slow-neutron-scattering techniques is that the wavelengths of thermal neutrons are of the same order of magnitude as the interatomic distances in condensed matter. Thus studies can be made on atomic motions in the microscopic spatial scale; in the Fourier-transformed space, **Q** space, measurements may yield the **Q** dependence of various physical quantities such as the line broadening due to diffusion. In principle, therefore, slow-neutron scattering measurements yield results that can be obtained only partially or not at all by combinations of other techniques such as those of NMR, x-ray diffraction, the Mössbauer effect, and macroscopic measurements. However, in contrast to the well-developed applications of this technique to the study on dynamics of solids in thermal equilibrium, such as measurements of phonon or magnon dispersion curves, the technique as applied to the study of diffusion is not yet well established. Therefore, one must always compare the values of the physical quantities obtained by this technique with those that may be obtained by other methods. For example, the data, when extrapolated to the long wavelength limit, or $Q \to 0$, must in general be consistent with the results obtained by macroscopic or optical measurements.

3.3.2. Neutron Scattering Cross Sections and Correlation Functions

The differential scattering cross section is given by

$$\frac{d^2\sigma}{d\Omega \, dE'} = \frac{k'}{k}\left(\frac{m}{2\pi\hbar}\right)^2 \sum_{nn'} P_n |\langle n| \sum_l V_l(\mathbf{Q}) e^{i\mathbf{Q}\cdot\mathbf{R}_l} |n'\rangle|^2 \delta(\hbar\omega + E_n - E_{n'}),$$
(3.3.1)

where l specifies the scatterer and $|n\rangle$ and $|n'\rangle$ represent the initial and the final states of the entire system of the scatterers, P_n is the population factor of $|n\rangle$, **k** and **k'** specify the initial and final wave vectors of the scattered neutron, E_0 and E' are the corresponding energies, $\mathbf{Q} = \mathbf{k} - \mathbf{k'}$, $\omega = (E_0 - E')/\hbar$, Ω is the solid angle, $V_l(\mathbf{Q})$ is the Fourier transform of the interaction potential $V_l(\mathbf{r})$ between the neutrons and the scattering atom l. For nuclear

---

[2] J. M. Rowe, *Proc. Conf. Neutron Scattering, Gatlinburg, Tenn.*, p. 491 (1976).
[3] T. Springer, *in* "Dynamics of Solids and Liquids by Neutron Scattering" (S. W. Lovsey and T. Springer, eds.), p. 255. Springer-Verlag, Berlin and New York, 1977.
[4] S. M. Shapiro and F. Reidinger, *in* "Physics of Superionic Conductors" (M. B. Salamon, ed.), p. 45. Springer-Verlag, Berlin and New York 1979.

scattering in the energy range of slow neutrons, $V_l(\mathbf{Q})$ is given by the Fermi pseudopotential as

$$V_l(\mathbf{r}) = (2\pi\hbar^2/m)b_l\delta(\mathbf{r}) \qquad (3.3.2)$$

or

$$V_l(\mathbf{Q}) = \frac{2\pi\hbar^2}{m} b_l. \qquad (3.3.3)$$

The expression (3.3.1) is quite convenient to evaluate if the initial and the final states are well characterized, as for example in the case of an isolated harmonic oscillator. However, in many cases it is more advantageous to write the cross section in terms of the thermal average of physical quantities associated with the system of scatterers. Such a formulation, which is based on statistical mechanics, is essential in the discussion of the diffusion problem, and the cross sections must be expressed in terms of various correlation functions instead of the eigenstates of the total system. Detailed mathematical steps are discussed by van Hove[5] in his original paper and by Marshall and Lovesey.[6] It can be shown that the scattering cross section is made up of two parts, the coherent and the incoherent parts. In the former type of scattering, the intensity is determined by the interference between the neutron waves scattered by different nuclei. On the other hand, the incoherent scattering is given solely by the static and dynamic states of a single nucleus. The differential cross sections are written

$$\left(\frac{d^2\sigma}{d\Omega\,dE'}\right)_{\text{coh}} = N\frac{k'}{k}\frac{\sigma_{\text{coh}}}{4\pi} S(\mathbf{Q},\omega) \qquad (3.3.4)$$

and

$$\left(\frac{d^2\sigma}{d\Omega\,dE'}\right)_{\text{inc}} = N\frac{k'}{k}\frac{\sigma_{\text{inc}}}{4\pi} S_i(\mathbf{Q},\omega), \qquad (3.3.5)$$

where σ_{coh} and σ_{inc} are the total coherent and incoherent cross sections, and the scattering law is

$$S(\mathbf{Q},\omega) = \frac{1}{2\pi\hbar} \int_{-\infty}^{\infty} dt\, e^{-i\omega t} \int d^3r\, e^{i\mathbf{Q}\cdot\mathbf{r}} G(\mathbf{r},t) \qquad (3.3.6)$$

and

$$S_i(\mathbf{Q},\omega) = \frac{1}{2\pi\hbar} \int_{-\infty}^{\infty} dt\, e^{-i\omega t} \int d^3r\, e^{i\mathbf{Q}\cdot\mathbf{r}} G_s(\mathbf{r},t). \qquad (3.3.7)$$

[5] L. van Hove, *Phys. Rev.* **95**, 249 (1954).
[6] W. Marshall and S. W. Lovesey, "Theory of Thermal Neutron Scattering." Oxford Univ. Press, London and New York, 1971.

3.3. DIFFUSION STUDIES

The pair- and self-correlation functions $G(\mathbf{r},t)$ and $G_s(\mathbf{r},t)$ are given by

$$G(\mathbf{r},t) = \frac{1}{N} \sum_{ll'} \int d^3r' \langle \delta(\mathbf{r} - \mathbf{r}' + \mathbf{R}_l(0))\delta(\mathbf{r}' - \mathbf{R}_{l'}(t)) \rangle \quad (3.3.8)$$

and

$$G_s(\mathbf{r},t) = \frac{1}{N} \sum_{l} \int d^3r' \langle \delta(\mathbf{r} - \mathbf{r}' + \mathbf{R}_l(0))\delta(\mathbf{r}' - \mathbf{R}_{l'}(t)) \rangle \quad (3.3.9)$$

respectively, where $\langle A \rangle$ denotes thermal averaging of the operator A. It should be pointed out that since the **R**s are quantum-mechanical operators, $\mathbf{R}(0)$ and $\mathbf{R}(t)$ commute, in general, only at $t = 0$. However, as discussed by Springer,[1] in many cases related to diffusion studies the correlation functions may be approximated by the classical forms:

$$G^{Cl}(\mathbf{r},t) = \sum_{l} \langle \delta(\mathbf{r} - \mathbf{R}_l(t) + \mathbf{R}_0(0)) \rangle \quad (3.3.10)$$

and

$$G_s^{Cl}(\mathbf{r},t) = \langle \delta(\mathbf{r} - \mathbf{R}_0(t) + \mathbf{R}_0(0)) \rangle. \quad (3.3.11)$$

$G^{Cl}(\mathbf{r},t)$ is the probability of finding any particle at r at time t when a particle existed at $\mathbf{r} = 0$ at $t = 0$. $G_s^{Cl}(\mathbf{r},t)$ represents a similar probability of finding the same particle at $\mathbf{r}$ at t. For example, the self-correlation function for a particle the motion of which obeys the diffusion equation,

$$D\nabla^2 G_s(\mathbf{r},t) = \partial G_s(\mathbf{r},t)/\partial t, \quad (3.3.12)$$

is given by

$$G_s(\mathbf{r},t) = (4\pi Dt)^{-3/2} \exp(-r^2/4Dt), \quad (3.3.13)$$

and the scattering law $S(Q,\omega)$ is a Lorentzian in ω:

$$S(Q,\omega) \propto \frac{2DQ^2}{\omega^2 + (Dq^2)^2}, \quad (3.3.14)$$

where D is the diffusion coefficient. The width of the Lorentzian depends on Q quadratically as

$$W(\text{meV}) = 0.132 \times D(10^{-5}\,\text{cm}^2\,\text{s}^{-1}) \times [Q(A^{-1})]^2. \quad (3.3.15)$$

The pair-correlation function can be separated into two terms

$$G(\mathbf{r},t) = G(\mathbf{r},\infty) + G'(\mathbf{r},t), \quad (3.3.16)$$

and, accordingly the cross section may be divided into the elastic and the

3. NEUTRON SCATTERING STUDIES OF LATTICE DEFECTS

inelastic parts given, respectively, by

$$\left(\frac{d^2\sigma}{d\Omega\, dE'}\right)_{\text{coh}}^{\text{el}} = \delta(\hbar\omega)Nb^2 \int d^3r\, e^{i\mathbf{Q}\cdot\mathbf{r}}G(\mathbf{r},\infty) \qquad (3.3.17)$$

and

$$\left(\frac{d^2\sigma}{d\Omega\, dE'}\right)_{\text{coh}}^{\text{inel}} = Nb^2 \frac{k'}{k}\frac{1}{2\pi\hbar} \int dt\, e^{-i\omega t} \int d^3r\, e^{i\mathbf{Q}\cdot\mathbf{r}}G'(\mathbf{r},t). \qquad (3.3.18)$$

The coherent scattering length b is given by $4\pi b^2 = \sigma_{\text{coh}}$. G_s can be divided in a similar fashion.

3.3.3. Incoherent Scattering and the Jump Model of Diffusion

The theory of diffusion can be formulated from a point of view based on either a phenomenological description or more fundamental processes. Simple theories of the latter kind have been employed to calculate the diffusion constants of light atoms in metals,[7,8] but little effort has been made along this line to obtain a formula for the pair-correlation function that can be compared with the results of neutron scattering experiments. Thus only a simple phenomenological model, the jump diffusion model,[9] has been used to analyze the data from slow-neutron scattering measurements. In the following the neutron scattering cross section is derived generally from the jump model of diffusion.

In this model the diffusing atoms are assumed to occupy only certain positions that form a periodic lattice and the jump time is neglected. These equilibrium positions are designated by $\mathbf{R}(^l_\kappa) = \mathbf{R}(l) + \mathbf{R}(\kappa)$, where $\mathbf{R}(l)$ represents the origin of the lth unit cell and $\mathbf{R}(\kappa)$ the position of the κth site in the unit cell ($\kappa = 1, \ldots, n$). The rate equation for the probability, $P(^l_\kappa;t)$, of finding an atom at $\mathbf{R}(^l_\kappa)$ at the time t is given by

$$\frac{dP(^l_\kappa;t)}{dt} = -\sum_{l'\kappa'} T(^{l\,l'}_{\kappa\kappa'})P(^{l'}_{\kappa'};t), \qquad (3.3.19)$$

where $T(^{l\,l'}_{\kappa\kappa'})$ is the probability of the jump from $\mathbf{R}(^{l'}_{\kappa'})$ to $\mathbf{R}(^l_\kappa)$ in unit time. It is assumed that the presence of other diffusing atoms in the crystal does not affect the jump probability. Also, the time an atom takes to jump from one site to another, the *traveling time*, is neglected. (A formula including the traveling time has been given by Gissler and Stump.[10]) $T(^{l\,l'}_{\kappa\kappa'})$ is translation-

---

[7] C. P. Flynn and A. M. Stoneham, *Phys. Rev. B* **1**, 3966 (1970).
[8] E. Gorham-Bergeron, *Phys. Rev. Lett.* **37**, 146 (1976).
[9] C. T. Chudley and R. J. Elliott, *Proc. Phys. Soc., London* **77**, 353 (1961).
[10] W. Gissler and N. Stump, *Physica (Amsterdam)* **65**, 109 (1973).

3.3. DIFFUSION STUDIES

ally invariant; namely, T depends on l and l' only through $\mathbf{R}(l) - \mathbf{R}(l')$. Thus $T(^{l\,l'}_{\kappa\kappa'}) = T(^{l-l'}_{\kappa\;\kappa'})$. Also, for $t \to \infty$, the system is in the equilibrium state and $P(^l_\kappa;\infty)$ is given by the equilibrium distribution $P^0(\kappa)$, independently of l. Hence

$$\sum_{l'\kappa'} T(^{l-l'}_{\kappa\;\kappa'})P^0(\kappa') = 0. \tag{3.3.20}$$

If the sites of occupation form a Bravais lattice ($n = 1$) and the jump occurs only to the nearest-neighbor sites, Eq. (3.3.19) simplifies to

$$\frac{dP(l;t)}{dt} = \frac{1}{z\tau} \sum_{l'} \{P(l';t) - P(l;t)\}, \tag{3.3.21}$$

where the summation is over the nearest-neighbor sites of the site l and z is the total number of the nearest-neighbor sites. τ is called the *residence time* and related to T by $T(l - l') = (z\tau)^{-1}$ for the nearest neighbors. From Eq. (3.3.20), $\tau^{-1} = -T(l = l') = \sum_{l' \neq l} T(l - l')$.

Writing

$$P(^l_\kappa;t) = P_\kappa(\mathbf{Q}) \exp[i\mathbf{Q} \cdot \mathbf{R}(^l_\kappa) - \lambda t], \tag{3.3.22}$$

one obtains from Eq. (3.3.19)

$$\lambda P_\kappa(\mathbf{Q}) = \sum_{\kappa'} T(^{\mathbf{Q}}_{\kappa\kappa'}) P_{\kappa'}(\mathbf{Q}), \tag{3.3.23}$$

where

$$T(^{\mathbf{Q}}_{\kappa\kappa'}) = \sum_{l'} T(^{l-l'}_{\kappa\;\kappa'}) \exp\{-i\mathbf{Q} \cdot [\mathbf{R}(^l_\kappa) - \mathbf{R}(^{l'}_{\kappa'})]\}. \tag{3.3.24}$$

It should be noted that because of the periodicity of the lattice,

$$T(^{\mathbf{Q}}_{\kappa\kappa'}) = e^{-i\boldsymbol{\tau}\cdot\mathbf{R}(\kappa)} T(^{\mathbf{q}}_{\kappa\kappa'}) e^{i\boldsymbol{\tau}\cdot\mathbf{R}(\kappa')}, \tag{3.3.25}$$

where $\mathbf{q} = \mathbf{Q} - \boldsymbol{\tau}$ and $\boldsymbol{\tau}$ is a reciprocal lattice vector and $\mathbf{q}$ is confined in the first Brillouin zone. In many instances, the lattice of the equilibrium sites has the same space group as that of the host lattice and $\boldsymbol{\tau}$ is identical to the reciprocal lattice vector of the host lattice.

The secular equation

$$\det|\lambda\delta_{\kappa\kappa'} - T(^{\mathbf{Q}}_{\kappa\kappa'})| = 0 \tag{3.3.26}$$

gives eigenvalues $\lambda_j(\mathbf{Q})$ and the eigenviectors $P^j(\mathbf{q})$ with $j = 1, \ldots, n$. From the property of T matrix given by Eq. (3.3.25), it can be shown that

$$\lambda_j(\mathbf{Q}) = \lambda_j(\mathbf{q}) \tag{3.3.27}$$

and

$$P^j(\mathbf{Q}) = P^j(\mathbf{q}) e^{-i\boldsymbol{\tau}\cdot\mathbf{R}(\kappa)} \tag{3.3.28}$$

For example, $\lambda(\mathbf{Q})$ for the face-centered-cubic (fcc) Bravais lattice ($n = 1$) with the nearest-neighbor jump can be obtained by deriving Eq. (3.3.26) from Eq. (3.3.21). The result is

$$\lambda(\mathbf{Q}) = \tau\left[1 - \frac{1}{3}\left\{\cos\left(Q_x\frac{a}{2}\right)\cos\left(Q_y\frac{a}{2}\right)\right.\right.$$
$$\left.\left.+ \cos\left(Q_y\frac{a}{2}\right)\cos\left(Q_z\frac{a}{2}\right) + \cos\left(Q_z\frac{a}{2}\right)\cos\left(Q_x\frac{a}{2}\right)\right\}\right], \quad (3.3.29)$$

where a is the lattice constant. Figure 1 shows $\lambda(\mathbf{Q})$ for three principal symmetry directions.

The probability $P(^l_\kappa;t)$ can be expressed generally as superpositions,

$$P(^l_\kappa;t) = \sum_j \int d^3q\, A_j(\mathbf{q})P^j_\kappa(\mathbf{q})\exp[i\mathbf{q}\cdot\mathbf{R}(^l_\kappa) - \lambda_j t]. \quad (3.3.30)$$

In order to obtain the neutron-scattering cross section, one needs the expression for the correlation function, which, in turn, is related to the conditional probability. It is the probability of finding an atom at $\mathbf{R}(^l_\kappa)$ at time $t = t$ when the same atom was located at $\mathbf{R}(^{l_0}_{\kappa_0})$ at $t = 0$. Thus this conditional probability is the self-correlation function mentioned in Section 3.3.2, and it can be determined by Eq. (3.3.30) with the initial condition

$$1 = P(^{l_0}_{\kappa_0};0) = \sum_j \int d^3q\, A_j(\mathbf{q})P^j(\mathbf{q})\exp[i\mathbf{q}\cdot\mathbf{R}(^{l_0}_{\kappa_0})] \quad (3.3.31)$$

and

$$0 = P(^l_\kappa;0) = \sum_j \int d^3q\, A_j(\mathbf{q})P^j_\kappa(q)\exp[i\mathbf{q}\cdot\mathbf{R}(^l_\kappa)] \quad \text{otherwise.} \quad (3.3.32)$$

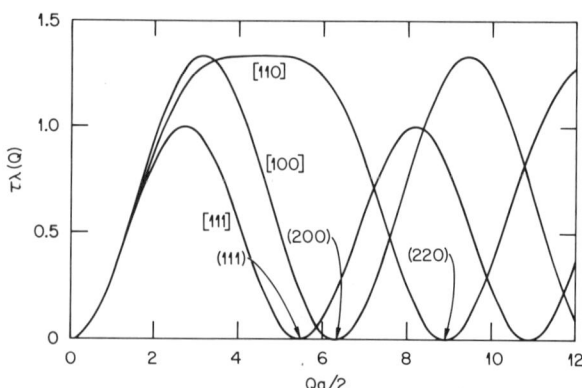

FIG. 1. $\lambda(\mathbf{Q})$ of the nearest-neighbor jump model for the fcc structure.

3.3. DIFFUSION STUDIES

The solution is given by

$$P(^{l}_{\kappa};t|^{l_0}_{\kappa_0};0) = \sum_j \int d^3q \, P^j_\kappa(\mathbf{q}) \exp[i\mathbf{q}\cdot\mathbf{R}(^{l}_{\kappa})]P^{j*}_{\kappa_0}(\mathbf{q})$$
$$\times \exp[-i\mathbf{q}\cdot\mathbf{R}(^{l_0}_{\kappa_0})] \exp[-\lambda_j(\mathbf{q})t]. \quad (3.3.33)$$

Averaged over the initial positions, this leads to the self-correlation function G_s and to the incoherent scattering cross section,

$$\left(\frac{d^2\sigma}{d\Omega \, dE'}\right)_{\text{inc}} = N \frac{k'}{k} \frac{\sigma_i}{4\pi\hbar} \sum_{\substack{ll' \\ \kappa\kappa'}} \int d^3r \, dt \, e^{-i(\mathbf{Q}\cdot\mathbf{r}-\omega t)} P(^{l}_{\kappa};t|^{l'}_{\kappa'};0)\delta(\mathbf{r} - \mathbf{R}(^{l}_{\kappa})$$
$$+ \mathbf{R}(^{l'}_{\kappa'}))$$

$$= N \frac{k'}{k} \frac{\sigma_i}{4\pi\hbar} \sum_{\substack{ll' \\ \kappa\kappa'}} \sum_j \int d^3q \, e^{i(-\mathbf{Q}+\mathbf{q})\cdot(\mathbf{R}(l)-\mathbf{R}(l'))} P^j_\kappa(\mathbf{q})P^{j*}_{\kappa'}(\mathbf{q})$$

$$\times e^{i(-\mathbf{Q}+\mathbf{q})\cdot(\mathbf{R}(\kappa)-\mathbf{R}(\kappa'))} \int dt \, e^{(i\omega - \lambda_j(\mathbf{q}))t}$$

$$= N \frac{k'}{k} \frac{\sigma_i}{4\pi\hbar} \sum_j \left|\sum_\kappa P^j_\kappa(\mathbf{q})e^{-i\boldsymbol{\tau}\cdot\mathbf{R}(\kappa)}\right|^2 \frac{1}{\pi} \frac{\lambda_j(\mathbf{q})}{\omega^2 + \lambda_j^2(\mathbf{q})}. \quad (3.3.34)$$

For a fixed value of $\mathbf{Q}$, the cross section is a superposition of several Lorentzians as a function of the energy transfer ω with the center at $\omega = 0$. The width of a Lorentzian is $\mathbf{q}$ dependent and given by $\lambda_j(\mathbf{q})$.

The incoherent scattering process represented by the cross section Eq. (3.3.34) having a peak at $\omega = 0$ is called the quasi-elastic scattering. The coherent scattering may also produce peaks at $\omega = 0$. They are often observed near phase transitions and related to the damping of collective excitations of atomic motions. In such cases, they are usually called the central peak. However, in some instances, the coherent scattering due to diffusing atoms produce a peak at $\omega = 0$, as discussed in Section 3.3.4, and it may also be called the quasi-elastic peak. The cross section can be interpreted as consisting of contributions from various diffusion modes specified by j and the weight of each mode is given by the factor

$$|K_j(\mathbf{Q})|^2 \equiv \left|\sum_\kappa P^j_\kappa(\mathbf{q})e^{-i\boldsymbol{\tau}\cdot\mathbf{R}(\kappa)}\right|^2. \quad (3.3.35)$$

In analogy with the structure factor for one-phonon scattering process,[6] this quantity may be called the quasi-elastic structure factor.

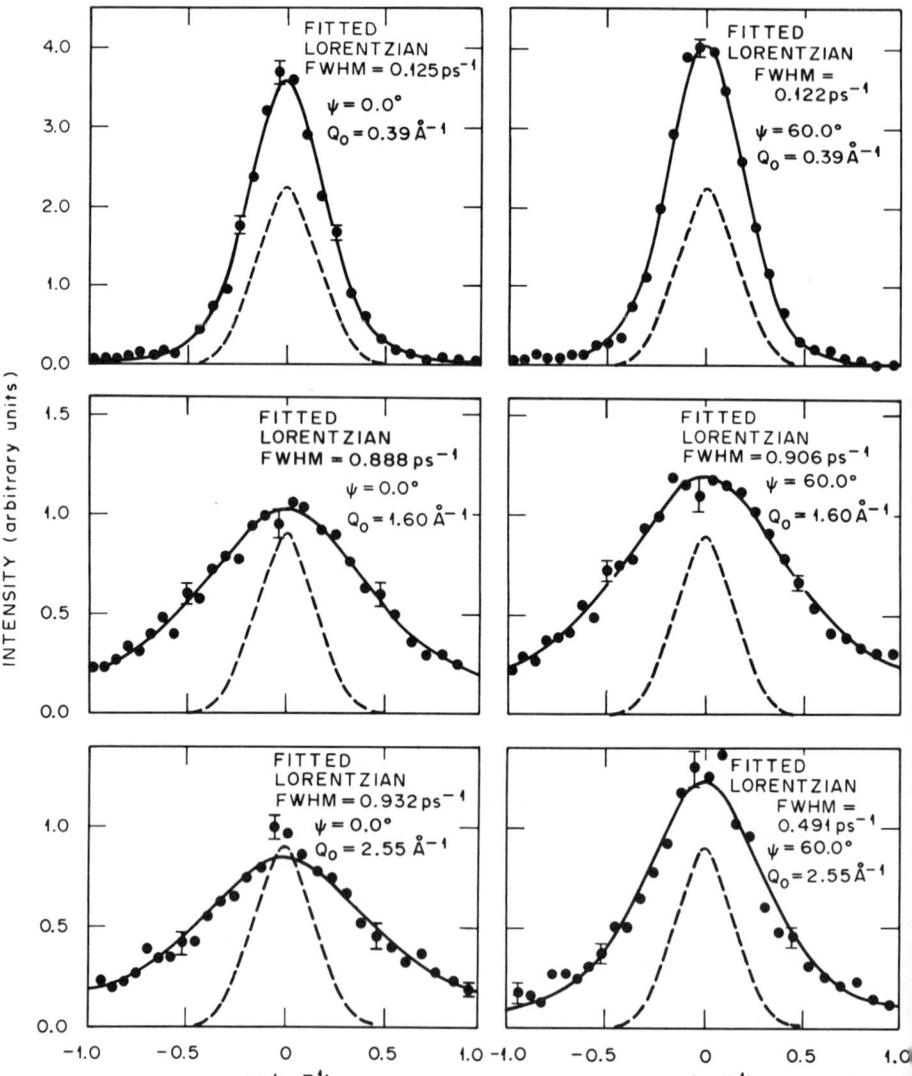

FIG. 2. Quasi-elastic lines for Pd(3 at. % H) obtained with a TOF spectrometer. Corrections have been made for sample container and palladium scattering as well as one-phonon incoherent scattering. The solid lines represent the results of fitting a Lorentzian broadened by the instrumental resolution, which is shown by the dashed lines. (From Rowe et al.[11])

3.3. DIFFUSION STUDIES

The quasi-elastic lines obtained by time-of-flight (TOF) measurement[11] on a single-crystal $PdH_{0.03}$ sample are shown in Fig. 2. The hydrogen atoms in the Pd lattice are considered to occupy so-called octahedral sites that form the fcc Bravais lattice. The diffusion constant is of the order of 10^{-5} cm^2 s^{-1} at 500 K and the corresponding energy width of a quasi-elastic line at $Q = 1$ Å^{-1} is expected to be of the order of 0.1 meV [see Eq. (3.3.15)]. The angle ψ in Fig. 2 specifies the sample orientation with respect to the direction of the incident neutron beam, which had the energy of 4.96 meV. The energy resolution of the spectrometer was 0.25 meV (full width at half-maximum, FWHM). The scattering angle ϕ determines the momentun transfer $\mathbf{Q}$. (See Sections 3.3.2 and 3.3.5.2). The observed TOF spectra were corrected for the scattering from the sample container and from the host Pd as well as the one-phonon incoherent scattering. Figure 2 shows the resulting quasi-elastic line plotted against ω, which is equal to the neutron-energy transfer divided by $\hbar$. Thus, $\omega = 1$ ps^{-1} corresponds to $\Delta E = 0.658$ meV. The solid lines represent the results of fitting the data with Lorentzian functions folded by the instrumental resolution function (dashed lines). The wave vectors, $\mathbf{Q}$s, were not, in general, along the symmetry directions in the reciprocal space, and the widths for $\mathbf{Q}$s along the [100] and [110] symmetry directions were determined by interpolations. The result (Fig. 3) is compared with the prediction of a simple jump model for the octahedral interstital sites (Fig. 1). This may be an example (and perhaps the only definite example at present) that shows the validity of a simple jump model for diffusion. The results for the body-centered-cubic (bcc) metal–hydrogen systems (Nb–H, V–H) cannot be understood on the basis of such a simple mode.

The incoherent scattering cross section

$$\left(\frac{d^2\sigma}{d\Omega\,dE'}\right)_{\text{inc}} \propto \sum_j |K_j(\mathbf{Q})|^2 \frac{1}{\pi} \frac{\lambda_j(\mathbf{q})}{\omega^2 + \lambda_j^2(\mathbf{q})} \qquad (3.3.36)$$

may be compared with the coherent scattering cross section for a one-phonon process,

$$\left(\frac{d^2\sigma}{d\Omega\,dE'}\right)_{1-\text{ph}} \propto \sum_j |F_j(\mathbf{Q})|^2 \delta(\omega - \omega_j(\mathbf{q})), \qquad (3.3.37)$$

where

$$F_j(\mathbf{Q}) = \sum_\kappa \mathbf{Q}\cdot\mathbf{e}_j(q|\kappa)e^{i\boldsymbol{\tau}\cdot\mathbf{R}(\kappa)} \qquad (3.3.38)$$

is the one-phonon inelastic structure factor and $e_j(q|\kappa)$ represents the eigenvector associated with the phonon having the wave vector $\mathbf{q}$ and the branch

---

[11] J. M. Rowe, J. J. Rush, L. A. deGraaf, and G. A. Furgeson, *Phys. Rev. Lett.* **29**, 1250 (1972).

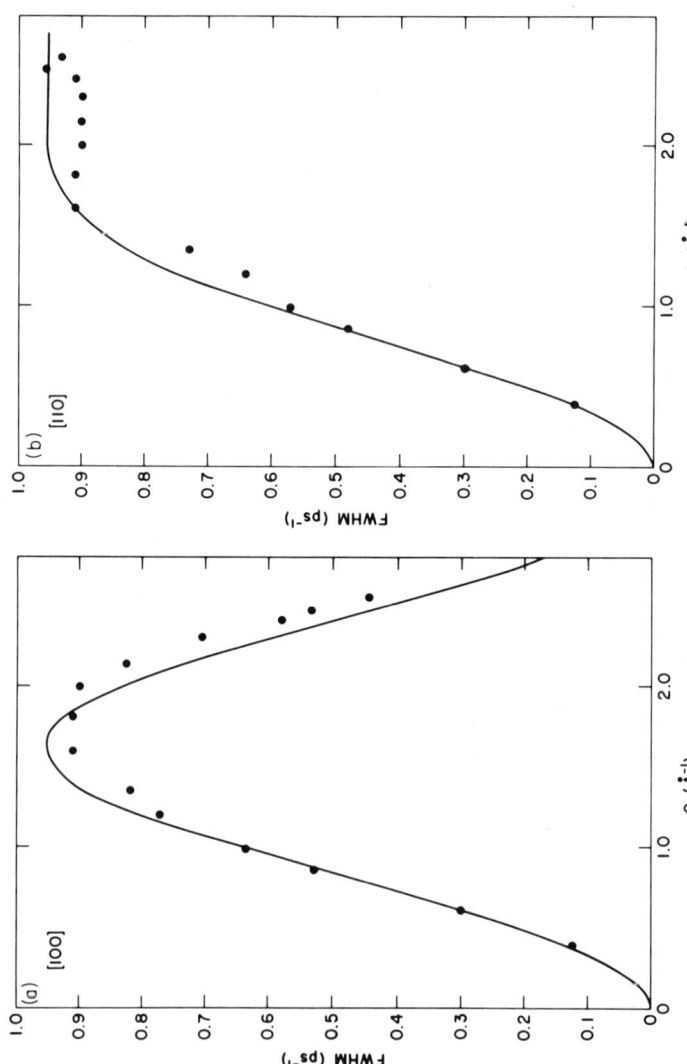

FIG. 3. FWHM of the quasielastic line for Pd(3%H) along (a) the [100] and (b) the [110] directions. Solid lines are the predictions of a jump model based on the nearest-neighbor octahedral to octahedral jumps (Fig. 1) with the residence time $\tau = 2.8$ ps. (From Rowe et al.[11])

3.3. DIFFUSION STUDIES

index j. Equation (3.3.37) for the diffusion contains Lorentzian functions as a function of ω, whereas Eq. (3.3.38) for the one-phonon scattering contains the delta functions. These functions can be derived from a common expression

$$I = \text{Im}\left[\frac{1/\pi}{\omega - \zeta}\right]. \quad (3.3.39)$$

By substituting $\zeta = i\lambda$, one obtains

$$I = \pi^{-1}\lambda/(\omega^2 + \lambda^2) \quad (3.3.40)$$

and by substituting $\zeta = \omega_0$, $I = \delta(\omega - \omega_0)$. Of course, the delta functions in Eq. (3.3.38) become Lorentzians if the phonons are damped with the width $\Gamma_j(\mathbf{q})$ as one can see by substituting $\zeta = \omega_j(q) + i\Gamma_j(\mathbf{q})$ into Eq. (3.3.39). However, it is a higher-order effect, and, in many cases, the main object of measuring the one-phonon scattering is to obtain the dispersion relation, namely, $\omega_j(q)$. Thus the determination of the peak position in the scattering cross section is generally sufficient. Furthermore, there is only one Lorentzian for one phonon and hence only one $\omega_j(q)$ for each q and j, and one can determine the width $\Gamma_j(q)$ uniquely within the approximation that Γ does not depend on ω. On the other hand, in Eq. (3.3.36), there are in general many Lorentzians centered around the same value $\omega = 0$. When there is no knowledge of the number of Lorentzians involved, it is impossible experimentally to determine uniquely all the $\lambda_j(Q)$s. Since the octahedral sites in the Pd–H system form a Bravais lattice, the jump involves only one λ for each Q and the quasi-elastic line consists of a single Lorentzian. This fact made it possible to interpret unambiguously the experimental data on this system.

The above formalism is based on the assumption that the diffusing atoms are stationary during the time of their residence at the equilibrium site $\mathbf{R}(^l_k)$. However, in general they vibrate around the equilibrium positions with a time scale much shorter than that for diffusion, the latter being given approximately by $1/\lambda_j(\mathbf{q})$. Then, the two types of motion may be separated, and the correlation function is given by a convolution of the correlation functions associated with these motions. More generally, the motion at an equilibrium site may consist of types of motion other than that of simple harmonic oscillations. Thus the convolution is

$$G_s(\mathbf{r},t) \propto \int G_s^{\text{eq}}(\mathbf{r}' - \mathbf{r},t) G_s^{\text{diff}}(\mathbf{r}',t) d^3r', \quad (3.3.41)$$

where G_s^{eq} is the self-correlation function associated with the motion at the equilibrium site. The corresponding scattering law is given by

$$S_{\text{inc}}(\mathbf{Q},\omega) \propto \int S_{\text{inc}}^{\text{eq}}(\mathbf{Q},\omega') S_{\text{inc}}^{\text{diff}}(\mathbf{Q},\omega - \omega') d\omega'. \quad (3.3.42)$$

Since the dominant term in $S_{\text{inc}}^{\text{eq}}$ is the elastic scattering contribution, the inelastic part of $S_{\text{inc}}^{\text{eq}}$ may be neglected, and

$$S_{\text{inc}}(\mathbf{Q},\omega) \sim S_{\text{inc}}^{\text{eq}}(\mathbf{Q},0)S_{\text{inc}}^{\text{diff}}(\mathbf{Q},\omega). \tag{3.3.43}$$

In the case of harmonic oscillations, $S_{\text{inc}}^{\text{eq}}(\mathbf{Q},0)$ is expressed in terms of the Debye–Waller factor and

$$S_{\text{inc}}(\mathbf{Q},\omega) \sim I_0 e^{-\langle u^2 \rangle Q^2} S_{\text{inc}}^{\text{diff}}(\mathbf{Q},\omega), \tag{3.3.44}$$

where $\langle u^2 \rangle$ is the mean-squared amplitude of the harmonic oscillation. The total intensity of the quasi-elastic line is, then,

$$I(Q) = I_0 e^{-\langle u^2 \rangle Q^2} \int S_{\text{inc}}^{\text{diff}}(\mathbf{Q},\omega)d\omega = I_0 e^{-\langle u^2 \rangle Q^2} \tag{3.3.45}$$

The plot of $\log\{I(Q)\}$ versus Q^2 is a straight line with the slope given by $\langle u^2 \rangle$. In V(H) as well as in some other metal–hydrogen systems at high temperatures, $\langle u^2 \rangle$ was found to be dependent on the range of Q in which $\langle u^2 \rangle$ was determined.[12] This indicates that $\log\{I(Q)\}$ is not a linear function of Q^2 in these systems. There are reports[13,14] in which observations and interpretations of large deviations from a linear behavior in Nb(H) were discussed. Although $\log(I)$ versus Q^2 is linear at room temperature, $\log(I)$ seems to consist of two straight lines at higher temperatures. The slope in the higher Q region is nearly equal to the slope at room temperature but that in the lower Q region is much steeper corresponding to the spatial extent of proton much larger than that at room temperature. However, more recent measurements do not seem to reproduce these results. The apparent large value for $\langle u^2 \rangle$ may be due to the failure in carrying out the integration in Eq. (3.3.45) over sufficiently large ω range to include the entire quasi-elastic scattering intensity. The difficulty associated with the experimental determination of the integrated intensity of a quasi-elastic line is discussed in Section 3.3.6.

3.3.4. Information Obtained from Coherent Scattering Processes

For simplicity, the coherent scattering due to atoms A and B in a compound AB is discussed in this section. The generalization to a compound consisting of a large number of atoms is trivial. A is assumed to be stationary and B mobile. The atom B jumps among sites that form a regular

---

[12] L. A. deGraaf, J. J. Rush, H. E. Flotow, and J. M. Rowe, *J. Chem. Phys.* **56**, 4574 (1972).
[13] W. Gissler, B. Jay, R. Rubin, and L. A. Vinhas, *Phys. Lett. A* **43A**, 279 (1973).
[14] N. Wakabayashi, B. Alefeld, K. W. Kehr, and T. Springer, *Solid State Commun.* **15**, 503 (1974).

3.3. DIFFUSION STUDIES

periodic structure. During the course of diffusion, B atoms may occupy certain interstitial sites that also form a periodic structure having the symmetry consistent with the space group of the original lattice. The position vector $\mathbf{R}(^l_\kappa) = \mathbf{R}(l) + \mathbf{R}(\kappa)$ specifies the κth site in the lth unit cell. $\kappa = 0$ refers to the site occupied by A and $\kappa = 1, 2, \ldots$, the possible sites occupied by B. $\mathbf{R}(\kappa = 0)$ is taken at the origin in the unit cell. The time the atom B takes to move from one site to another may be assumed to be negligible compared with the residence time.

The coherent scattering intensity is determined by the correlation functions G^{AA}, G^{AB}, G^{BA}, and G^{BB}. As before, the vibrational parts of the correlation functions are neglected for the moment and included in the form of Debye–Waller factors at a later stage. Thus the A atoms are rigidly bound at $\mathbf{R}(^l_0)$, and the density function is given by

$$\rho^A(\mathbf{r},t) = \sum_l \delta(\mathbf{r} - \mathbf{R}(^l_0)) \tag{3.3.46}$$

independent of t. Then,

$$G^{AA}(\mathbf{r},t) = \frac{1}{N} \int d\mathbf{r}' \langle \rho^A(\mathbf{r}' - \mathbf{r},0) \rho^A(\mathbf{r}',t) \rangle$$

$$= \frac{1}{N} \int d\mathbf{r}' \langle \rho^A(\mathbf{r}' - \mathbf{r},0) \rangle \langle \rho^A(\mathbf{r}',\mathbf{r}) \rangle$$

$$= \frac{1}{N} \sum_{l,l'} \delta(\mathbf{r} + \mathbf{R}(^l_0) - \mathbf{R}(^{l'}_0)). \tag{3.3.47}$$

Also,

$$G^{AB}(\mathbf{r},t) = \frac{1}{N} \sum_l \langle \rho^B(\mathbf{r} + \mathbf{R}(^l_0),t) \rangle = \frac{1}{N} {\sum_{l l' \atop \kappa'}}' P_\kappa' \delta(\mathbf{r} - \mathbf{R}(^{l'}_{\kappa'}) + \mathbf{R}(^l_0)) \tag{3.3.48}$$

and

$$G^{BA}(\mathbf{r},t) = G^{AB}(\mathbf{r},t), \tag{3.3.49}$$

where P_κ is the fraction of the B atoms located at the κth site. Furthermore,

$$G^{BB}(\mathbf{r},\infty) = \frac{1}{N} \int d^3\mathbf{r}' \langle \rho^B(\mathbf{r}' - \mathbf{r},0) \rangle \langle \rho^B(\mathbf{r}',\infty) \rangle$$

$$= \frac{1}{N} {\sum_{l l' \atop \kappa\kappa'}}' P_\kappa P_{\kappa'} \delta(\mathbf{r} - \mathbf{R}(^{l'}_{\kappa'}) + \mathbf{R}(^l_\kappa)). \tag{3.3.50}$$

Thus the coherent elastic scattering cross section (Eq. (3.3.17)) is given by

$$\left(\frac{d^2\sigma}{d\Omega\, dE'}\right)^{el}_{coh} = \delta(\hbar\omega) \int d^3r\, e^{i\mathbf{Q}\cdot\mathbf{r}} \sum_{ll'} \{b_A^2 \delta(\mathbf{r} + \mathbf{R}(^l_0) - \mathbf{R}(^{l'}_0))$$

$$+ 2b_A b_B \sum_{\kappa}{}' P_\kappa \delta(\mathbf{r} + \mathbf{R}(^l_0) - \mathbf{R}(^{l'}_{\kappa'}))$$

$$+ b_B^2 \sum_{\kappa\kappa'}{}' P_\kappa P_{\kappa'} \delta(\mathbf{r} + \mathbf{R}(^l_\kappa) - \mathbf{R}(^{l'}_{\kappa'}))\}$$

$$= \delta(\hbar\omega) \sum_{ll'} e^{i\mathbf{Q}\cdot(\mathbf{R}(l) - \mathbf{R}(l'))} |b_A + b_B \sum_{\kappa}{}' P_\kappa e^{i\mathbf{Q}\cdot\mathbf{R}(\kappa)}|^2. \quad (3.3.51)$$

The summations over l and l' give $[N(2\pi)/v_0] \Sigma_\tau \delta(\mathbf{Q} - \tau)$, where v_0 is the volume of the unit cell and the τs are the reciprocal lattice vectors. Thus

$$\left(\frac{d^2\sigma}{d\Omega\, dE'}\right)^{el}_{coh} = \delta(\hbar\omega) N \frac{(2\pi)^3}{v_0} \sum_\tau \delta(\mathbf{Q} - \tau) |F(\tau)|^2, \quad (3.3.52)$$

where the elastic structure factor $F(\tau)$ is defined by

$$F(\tau) = b_A + b_B \sum_\kappa{}' P_\kappa e^{i\tau\cdot\mathbf{R}(\kappa)} \quad (3.3.53)$$

The effect of the vibrational motions around the equilibrium positions can be introduced by replacing b_A and b_B by $b_A e^{-W_A(Q)}$ and $b_B e^{-W_B(Q)}$, respectively, where W is the Debye–Waller factor.

The Fourier inversion of $F(\tau)$ gives the density function weighted by bs. Thus one may be able to obtain $\mathbf{R}(\kappa)$ and the population factors P_κ that are closely related to the diffusion process of the B atoms. As is well known, the scattering intensity does not give the phase factor of $F(\tau)$, and, in general, it is rather difficult to obtain a unique set of $\mathbf{R}(\kappa)$ and P_κ for $\kappa = 1, 2, \ldots$, by this method. However, in some cases $F(\tau)$ is real for all τs, and the signs in such cases may be determined uniquely. Elastic scattering measurements on polycrystalline BaF_2 and PbF_2 at temperatures ranging from room temperature to the melting points were analyzed in this fashion.[15] The signs of $F(\tau)$s at room temperature are well known from the fluorite structure. As the temperature is raised, fluorine atoms move from their regular sites (F(I)), and the Bragg intensities change with temperature. However, the intensities studied in the experiment did not vanish in this temperature range, which indicates that the signs of $F(\tau)$ for these Bragg reflections remain the same in the entire range of temperature. Thus the Fourier inversion could be carried out uniquely. Figure 4 shows the difference plot of the weighted densities

---

[15] S. M. Shapiro, in "Superionic Conductors" (G. D. Mahan and W. L. Roth, eds.), p. 261. Plenum, New York, 1976.

3.3. DIFFUSION STUDIES

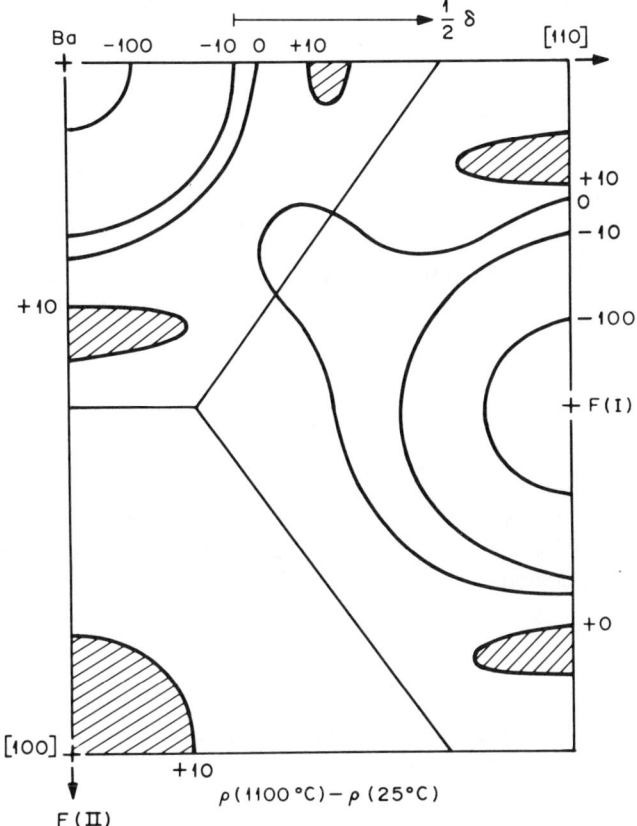

FIG. 4. Difference plot of the weighted nuclear densities between 25°C and 1100°C for BaF_2 in the (110) plane. (From Shapiro.[15])

between 25°C and 1100°C in the (110) plane. Additional fluorine density appears around F(II) sites and nearly 40% of F(II) are occupied by fluorine atoms just below the melting point (66.7% for the completely disordered state). Similar results were obtained for PbF_2. Obviously, the diffusion mechanism of fluorine atoms in the ionic conductors having the fluorite structure involves this interstitial site and associated vacancies. A later analysis[4] of the same data seems to indicate that the interstitial sites are not located exactly at the F(II) sites.

The quasi-elastic part of the coherent scattering cross section has not been formulated accurately for diffusing particles. In the cases that involve discrete atomic sites as discussed above, the coherent quasi-elastic scattering intensity is expected to be significant only near the Bragg reflections corresponding to the lattice of the atomic sites. The point may be made somewhat clearer by means of the approximate expression derived by Vineyard[16] for the coherent cross section. The correlation function G is written

$$G^{BB}(\mathbf{r},t) = G_s^B(\mathbf{r},t) + \int \{G^{BB}(\mathbf{r}',0) - G_s^B(\mathbf{r}',0)\} H_0(\mathbf{r},\mathbf{r}',t)\, d^3\mathbf{r}, \quad (3.3.54)$$

where $H_0(\mathbf{r},\mathbf{r}',t)$ is the probability that an atom is found at $\mathbf{r}$ at t when the same atom was at $\mathbf{r}'$ at $t = 0$ and another atom was at the origin simultaneously. The convolution approximation of Vineyard is

$$G^{BB}(\mathbf{r},t) \simeq G_s^B(\mathbf{r},t) + \int \{g^B(\mathbf{r}') - \delta(\mathbf{r}')\} G_s^B(\mathbf{r} - \mathbf{r}',t)\, d^3\mathbf{r}'. \quad (3.3.55)$$

where $g^B(\mathbf{r}') = G^{BB}(\mathbf{r},0)$. Thus

$$G^{BB}(\mathbf{Q},\omega) \simeq S_i^B(\mathbf{Q},\omega) + (\gamma(\mathbf{Q}) - 1) S_i^B(\mathbf{Q},\omega) = \gamma(\mathbf{Q}) S_i^B(\mathbf{Q},\omega), \quad (3.3.56)$$

where

$$\gamma(\mathbf{Q}) = \int S^{BB}(\mathbf{Q},\omega)\, d\omega = \int G^{BB}(\mathbf{r},0) e^{i\mathbf{Q}\cdot\mathbf{r}}\, d^3\mathbf{r}. \quad (3.3.57)$$

The coherent cross section is now expressed in terms of the incoherent cross section and the structure factor $\gamma(\mathbf{Q})$. However, if the flight time for the diffusion is neglected,

$$G^{BB}(r,0) = \sum_{\substack{ll' \\ \kappa\kappa'}}{}' P_\kappa P_{\kappa'}\, \delta(\mathbf{r} - \mathbf{R}(^l_\kappa) + \mathbf{R}(^{l'}_{\kappa'})) \quad (3.3.58)$$

and

$$\gamma(\mathbf{Q}) = \sum_{\substack{ll' \\ \kappa\kappa'}}{}' P_\kappa P_{\kappa'}\, \exp[i\mathbf{Q}\cdot(\mathbf{R}(^l_\kappa) - \mathbf{R}(^{l'}_{\kappa'}))]$$

$$= \left| \sum_l e^{i\mathbf{Q}\cdot\mathbf{R}(l)} \right|^2 \left| \sum_\kappa{}' P_\kappa e^{i\mathbf{Q}\cdot\mathbf{R}(\kappa)} \right|^2$$

$$= \sum_\tau \delta(\mathbf{Q} - \tau) \left| \sum_\kappa{}' P_\kappa e^{i\mathbf{Q}\cdot\mathbf{R}(\kappa)} \right|^2, \quad (3.3.59)$$

where τ signifies a reciprocal lattice vector. Thus the coherent cross section is finite only at Bragg reflections if the flight time is neglected. For certain cases, the sites for B atoms are not well defined and do not form a regular lattice. Furthermore the flight time may not be negligible in such cases. Then $\gamma(\mathbf{Q})$ given by Eq. (3.3.57) may be similar to that for a liquid and the coherent quasi-elastic cross section would be significant for general Q values. Eckold et al. interpreted their data on α-AgI on the basis of this formalism.[17] α-AgI is an example of the solid electrolytes, and because of the

---

[16] G. H. Vineyard, *Phys. Rev.* **110**, 999 (1958).
[17] G. Eckold, K. Funke, J. Kalus, and R. E. Lechner, *J. Phys. Chem. Solids* **37**, 1097 (1976).

high mobility of Ag ions the broadening of quasi-elastic lines can be observed easily by using conventional high-resolution spectrometers. Since the scattering cross section of Ag is dominated by that for coherent scattering, the above formalism may be applied to the interpretation of the observed quasi-elastic line shapes, but the conclusion reached is rather ambiguous because of the uncertainties in the theory. The quasi-elastic line measured by means of a TOF spectrometer seems to consist of more than one Lorentzian, and the results for various Q values were interpreted on the basis of the combination of a jump motion and a local random motion. The coherent scattering cross section of the Vineyard form modified by Sköld[18] was used together with the jump model of Gissler and Stump,[10] which includes the effect of the large jump time τ_1. The model parameters, τ_0, the mean residence time, the jump distance l and the parameters associated with the local motion, i.e., the spatial extent of the local motion and the local diffusion constant, were determined by fitting to the data. The model reproduces experimental data quite well, but since there are uncertainties in the validity of the theory, the values of the model parameters may be considered tentative.

3.3.5. Samples and Instruments

3.3.5.1. Samples. As discussed in the previous sections, the diffusive motions of atoms can be studied by measuring either coherent or incoherent scattering processes. However, since the motions of different diffusing atoms are not correlated strongly in most cases in which the concentration of the diffusing atoms is low, one can obtain the most direct information from the self-correlation function. The incoherent scattering intensity is proportional to the Fourier transform of the self-correlation function with the nuclear incoherent scattering cross section σ_i as the proportionality constant. Generally speaking, it is desirable that the incoherent scattering cross section for the diffusing atom be much larger than those of other constituent atoms in the solid. The values for the scattering cross section of various nuclei can be found in several books.[19,20] Table I lists those for some of the nuclei commonly found as constitutents of hydrogen storage systems or solid electrolyte materials. It can easily be seen that the proton has a unique property of being a predominantly incoherent scatterer with an extremely high incoherent scattering cross section. Vanadium is an almost perfect

---

[18] K. Sköld, *Phys. Rev. Lett.* **19**, 1023 (1967).
[19] G. E. Bacon, "Neutron Diffraction," Oxford Univ. Press (Clarendon), London and New York, 1975.
[20] L. Koester and A. Steyerl, "Neutron Physics." Springer-Verlag, Berlin and New York, 1977.

3. NEUTRON SCATTERING STUDIES OF LATTICE DEFECTS

TABLE I. Scattering Lengths and Cross Sections[a]

	$b(\sigma_{coh} = 4\pi|b|^2)$ (10^{-12} cm)	σ_{inc} (barns)[b]	σ_{abs} for $\lambda = 1.08$ Å (barns)
H	−0.374	80	0.19
D	0.673	2.0	0.0005
Al	0.345	≤0.01	0.13
V	−0.041	5.0	2.8
Fe	0.95	~0.3	1.4
Ni	1.03	4.9	2.7
Zr	0.70	~0.1	0.1
Nb	0.71	~0	0.63
Mo	0.69	~0.3	1.4
Cd	0.38 + 0.12i	—	2650
Gd	1.5	—	20,000
^{160}Gd	0.91	—	1

[a] Data from Bacon[19] and Koester.[20]
[b] 1 barn = 10^{-24} cm².

incoherent scatterer, but the value of σ_{inc} is not much larger than those of other transition metal nuclei such as Ni and Ti. The absorption cross section is also an important factor in determining the feasibility of an experiment. The absorption problem becomes more serious for longer-wavelength neutrons since the absorption cross section is proportional to the inverse of neutron velocity. Highly absorbing nuclei such as Cd, B, and Gd, cannot be host or diffusing atoms, but in some cases an isotope may exist that has a much smaller absorption cross section compared with other isotopes (e.g., ^{160}Gd). In order to obtain strong scattering intensity it is generally advantageous to use a large sample. However, the larger the sample is, the higher is the probability for multiple scattering processes. Consequently, the total scattering should be less than about 10% in order to avoid such processes the corrections for which are rather difficult. The shape of the sample is also important since various corrections are easier for regularly shaped samples.

Since hydrogen has a large incoherent cross section, the sample and the sample holder should not be contaminated by substances containing hydrogen. In particular, glues containing hydrogen cannot be used to hold samples.

3.3.5.2. Instruments. Early measurements were performed on hydrogen–metal systems by means of conventional spectrometers, such as triple-axis spectrometers and TOF spectrometers. More recently high-resolution spectrometers have been developed, and many significant studies have been carried out by these new types of spectrometers.

Triple-Axis Spectrometer. A detailed description of this type of spectrometer is given in Chapter 3.2. With this type of spectrometer the energy transfer due to scattering processes can be measured at a fixed **Q**, the momentum-transfer vector. As shown in Fig. 3 of Chapter 3.2, the momentum transfer **Q** can be held constant by varying the angles ϕ and ψ while the energy scan is performed. Thus it is very simple to carry out the measurements of quasi-elastic line shape or intensity as a function of **Q** and, in particular, along various symmetry directions of the reciprocal space. The spectrometer is most suitable for measurements on single-crystal samples, although more and more measurements with fixed values of the magnitude of the momentum transfer $|\mathbf{Q}|$ performed on polycrystal samples have been carried out by means of triple-axis spectrometers. The overall energy resolution of such spectrometers is typically more than 1%, and the **Q** resolution is about 0.01 Å^{-1}.

Time-of-Flight Spectrometer. Detailed descriptions of various TOF spectrometers can be found in the article by Dolling.[21] A TOF spectrometer utilizes a pulsed beam of monoenergetic neutrons impinging on the sample. The scattered neutrons are detected by counters (^{3}He counters are most commonly used because of the small size) placed at various angular positions. The energy of the scattered neutron is determined by measuring the interval between the time of its arrival at the sample and the time when it is detected by one of the counters (flight time). Scattered neutrons within a pulse are recorded in a multichannel time analyzer according to the flight time, and this procedure is repeated after a certain time interval. The monochromatic pulsed neutron beam is commonly produced by a mechanical chopper or a rotating monochromating crystal with a regular periodicity. Since the reactor is a continuous source of neutrons, conventional TOF spectrometers make use of only a small fraction of available neutrons with the duty cycle, as low as 1%. The duty cycle can be improved dramatically by producing monochromatic neutron beam pulses in a random pattern in time and by correlating the time distribution of the scattered neutrons with this random sequence of the incident neutron beam. This type of spectrometer is called the correlation spectrometer, and one of the most versatile correlation spectrometers[22] is illustrated in Fig. 5. The monochromatic pulsed neutron can be produced by changing the direction of the magnetic moments in the ferrite $Li_{0.5}Fe_{2.5}O_4$ monochromating crystal. The nuclear part of the (111) Bragg reflection is nearly absent in this ferrite, and the magnetic reflection can be turned on and off by changing the direction of the magnetic moments. Thus a random sequence of monochromatic neutron

---

[21] G. Dolling, *in* "Dynamical Properties of Solids" (G. K. Horton and A. A. Maradudin, eds.), Vol. 1, p. 541. North-Holland Publ., Amsterdam, 1974.

[22] H. A. Mook, F. W. Snodgrass, and D. D. Bates, *Nucl. Instrum. Methods* **116**, 205 (1974).

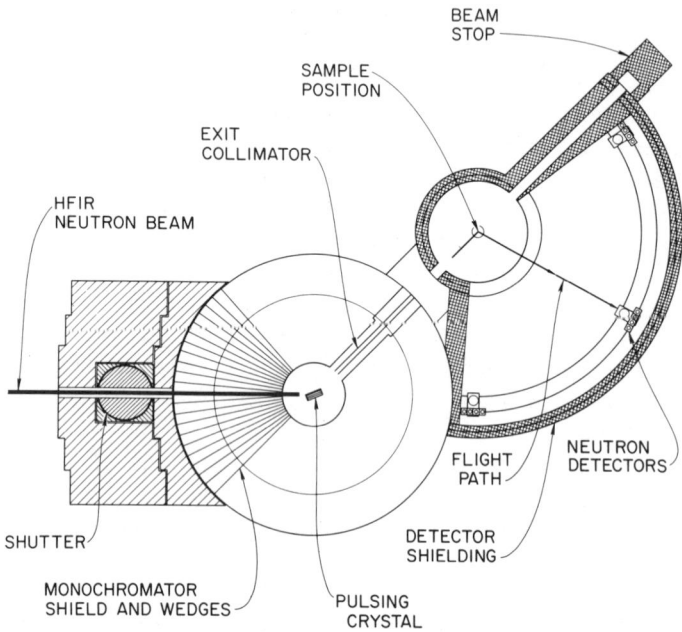

Fig. 5. Schematic diagram of the correlation TOF spectrometer with magnetic chopper. (From Mook et al.[22])

beam pulses is obtained by applying the magnetic field perpendicular or parallel to the scattering vector in a desired time sequence. The duty cycle can be as high as 25% in this spectrometer. The greatest advantage of the TOF spectrometer in general is its capability of collecting data with a multidetector system. Thus, in principle, measurements for various Qs may be performed simultaneously. On the other hand, two disadvantages arise. (1) Since the detectors are placed at fixed scattering angles ϕ, the TOF energy scan for each detector is not at constant Q. (2) The Q corresponding to the ϕ angles of these detectors do not lie along directions of high symmetry in the reciprocal space. The contour of the constant-ϕ scan in Q–E space is illustrated in Fig. 6. If the incident neutron energy is much higher than the width of the quasi-elastic line, the variation of the momentum transfer Q corresponding to the energy change may be negligible, and a constant-ϕ scan can be nearly equivalent to a constant-Q scan. However, the energy resolution may become too low for the linewidth measurements, and it is generally necessary to use a low incident energy. It is also possible to perform TOF measurements with Q along a symmetry direction by changing the sample orientation. However, the advantage of the multidetector system is not fully utilized in such a mode of operation. Of course, in the measure-

3.3. DIFFUSION STUDIES

ment on a polycrystalline sample, only $|Q|$ is of interest, and the difficulties mentioned above do not arise.

In many cases the diffusion constant (or the jump rate $1/\tau$) is so small that the quasi-elastic linewidth is extremely narrow. Thus the instrumental energy resolution must be very high. For example, for a jump rate of 10^{-9} s^{-1} the linewidth is of the order of 10^{-6} eV. A new type of spectrometer that is capable of such a high resolution is the *backscattering spectrometer*.[23] In crystal monochromators, the wavelength is determined by the Bragg relation, and the wave-length spread due to the beam divergence is given by

$$\Delta\lambda/\lambda = \cot\theta \, \Delta\theta \quad \text{or} \quad \Delta E/E = -2\cot\theta \, \Delta\theta. \quad (3.3.60)$$

Thus for $\theta = 90°$ (or $2\theta = 180°$, backscattering), $\Delta\lambda = 0$ and hence there is no energy spread. The backscattering spectrometer uses this principle both for the monochromator and the analyzer along with the beam deflector and the chopper that stops the beam periodically to reduce background while the counter is activated (Fig. 7). Both the monochromator and analyzer of the existing backscattering spectrometers consist of silicon crystals. The incident energy is varied by the Doppler effect of the moving monochromators while maintaining the backscattering configuration. An energy resolution of higher than 10^{-6} eV has been achieved by such a spectrometer. In order to increase the overall counting efficiency, the analyzer size is quite large, which increases the solid angles of the analyzer system; consequently, the Q resolution is much worse than those of conventional spectrometers. But the incoherent scattering, in general, varies smoothly with Q and the poor Q

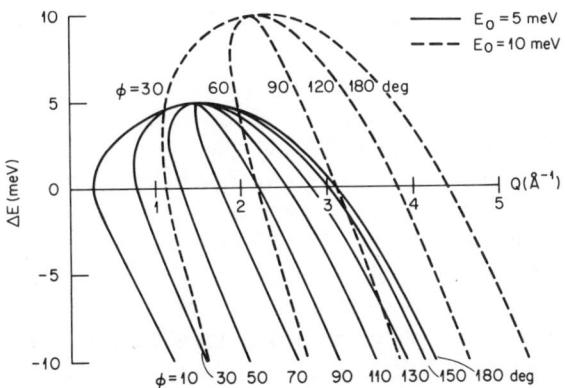

FIG. 6. TOF contour in the Q-E space at various scattering angles ϕ and incident energies E_0.

[23] M. Birr, A. Heidemann, and B. Alefeld, *Nucl. Instrum. Methods* **95**, 435 (1971).

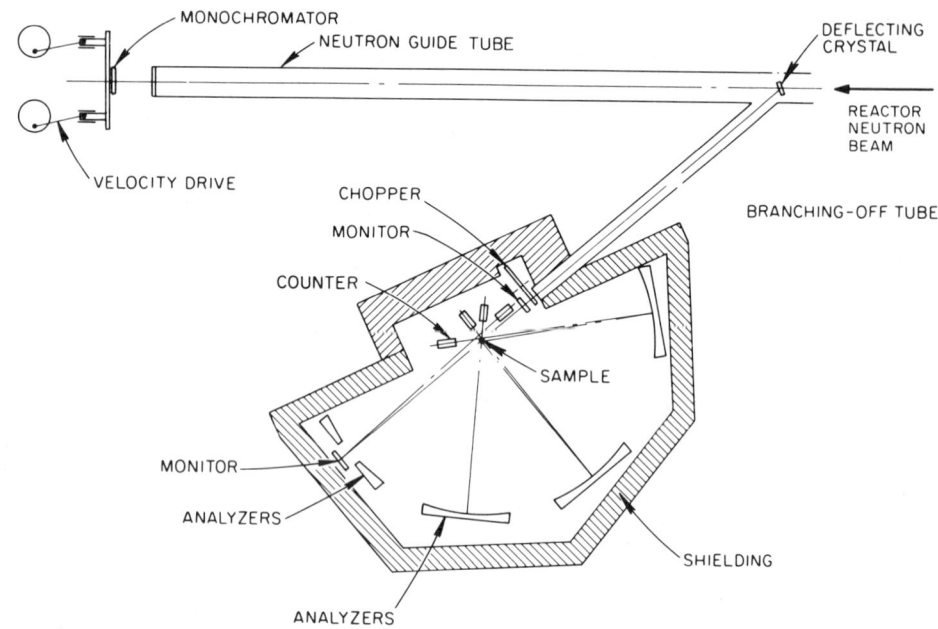

Fig. 7. Schematic diagram of backscattering spectrometer.

resolution is seldom a serious problem. The low intensity of the incident neutron beam as a consequence of the extremely small energy spread is the largest restriction in the application of this type of spectrometer. However, there are numerous problems that cannot be solved without the resolution it offers at present, and the spectrometer has been used more and more widely in recent years.

Of course, the energy uncertainty of a neutron beam can be reduced simply by using neutrons having a long wavelength. However, this poses a severe restriction on the range of the **Q** space that can be studied since the ultimate maximum of Q is given by $4\pi/\lambda$. Furthermore, the energy spectrum of a reactor has a maximum at an energy corresponding to a neutron wavelength of about 1 Å, and the flux of neutrons having much longer wavelengths is extremely low. This difficulty may be remedied to some extent by installing a cold-neutron source consisting of liquid deuterium, which moderates neutrons even further than does the moderator of the reactor. Such a cold source can increase the flux of long-wavelength neutrons by a large factor (for example, a factor 20 for 6-Å neutrons). Even then the resulting flux is still quite low, and the time required to accummulate accurate data on quasi-elastic lines is much longer than that required to perform, for example, measurements of phonon dispersion curves. An additional difficulty is that the spectrometers used at present for ultrahigh resolution measurements are rather limited in number since a cold source cannot be installed easily.

3.3.6. Data Reduction

As has been discussed above, the measurements of the response of a system with relaxation involve accurate determination of the frequency-response function, i.e., line shapes instead of simply determining the peak positions that are the positions of the poles in the response function. There are various scattering processes that distort the quasi-elastic line shapes as well as the intensities. One would like to use large samples in order to obtain large intensities, but probability of the multiple scattering becomes accordingly greater. The effect of multiple scattering is to give rise to uncertainties in the energy and momentum transfer, ω and **Q**. The measured intensity may contain not only the contribution from the scattering corresponding to the nominal ω and **Q** but also those from several consecutive scattering processes which, combined, result in the same ω and **Q**. Attempts have been made to carry out corrections for such processes numerically.[24] Without the knowledge of the response function, which is to be determined by the experiment, it is difficult to make accurate corrections to the measured intensities. Therefore, it is a standard practice to attempt to reduce the total scattering intensity to a value as low as tolerable. One can eliminate the multiple scattering involving Bragg scattering due to the host lattice by using neutrons with very long wavelength so that no Bragg reflections are possible. But, then, the range of **Q** that can be studied by such an arrangement is very limited. Also, since the Bragg scattering is discrete in **Q**, it is generally possible to predict the existence of such a multiple scattering for any orientation of a single crystal sample, and thus one may be able to avoid it.

At high temperatures the diffusion processes become faster and the quasi-elastic peak widths become larger. At the same time the phonon scattering intensity increases due to the increase in the population factor. For small ω, the incoherent phonon scattering intensity is constant with respect to ω and proportional to the temperature. Thus it is important to distinguish experimentally the constant background due to phonon scattering from the broad quasi-elastic peaks. When the instrument used has a very limited range of energy that can be scanned, a complication in the interpretation of data may arise. This is one of the difficulties in the quasi-elastic line measurements especially when the number of the Lorentzian functions involved is not known. If there is one Lorentzian whose width is so large that one cannot recognize it as a Lorentzian from the data obtained within a

[24] J. R. D. Copley, *Comput. Phys. Commun.* **7**, 289 (1974).

small energy range covered by the spectrometer, it may be mistaken as a small constant background or, worse still, may not be detected at all. If one attempts to determine the Debye–Waller factor by integrating the quasi-elastic line intensity in such a case, one may fail to include a part of or the entire contribution from the extremely broad Lorentzian. The partial intensity then corresponds to

$$I = I_0 \int_{-\Delta E/2}^{\Delta E/2} \frac{1}{\pi} \frac{\Gamma}{E^2 + \Gamma^2} \, dE = I_0 \frac{2}{\pi} \tan^{-1}\left(\frac{\Delta E}{2\Gamma}\right), \quad (3.3.61)$$

where ΔE is the range of the energy transfer covered by the spectrometer, and the apparent Debye–Waller factor may become very large as determined from the data in the range of small Q values (Fig. 8).

The resolution function of the spectrometer depends on the collimators, the monochromator and analyzer crystals, and the flight-path length (in the case of a TOF spectrometer). It can be calculated and checked against

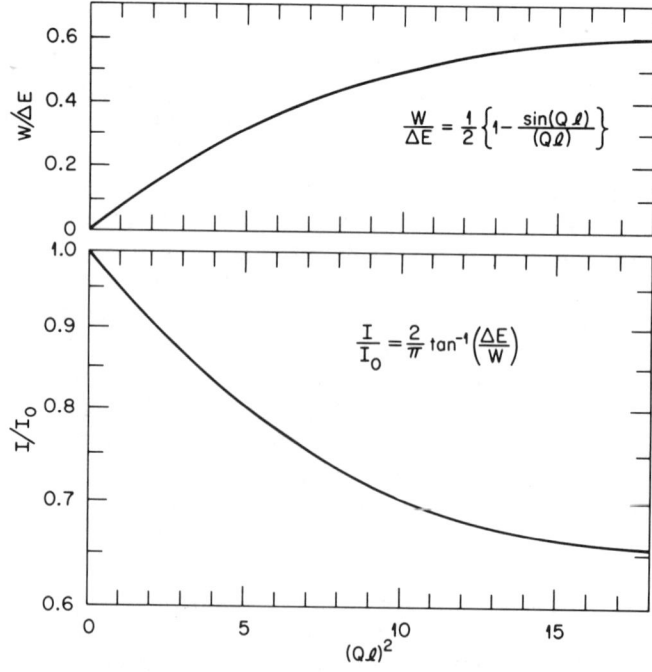

FIG. 8. Partially integrated intensity of quasi-elastic line corresponding to the isotropic jump model. The total linewidth $W = 2\Gamma$.

certain experimental information. For example, in the case of a triple-axis spectrometer, the resolution function (the socalled resolution ellipsoid) near Bragg reflections in the $\mathbf{Q}-\omega$ space can be measured by scanning the $\mathbf{Q}$ and ω around the Bragg peaks. Since the $\mathbf{Q}$ dependence of the quasi-elastic line shape is not very strong, only the energy spread needs to be taken into consideration for analyzing experimental data in many cases, and the resolution function is taken to be a Gaussian as a function of the energy deviation.

The quasi-elastic lines to be analyzed generally consist of linear combinations of Lorentzians convoluted by the resolution function, but there is no a priori way to determining the number of Lorentzians involved in the quasi-elastic line. Thus it is impractical to attempt to determine uniquely $\lambda_j(\mathbf{q})$ in Eq. (3.3.36) for each j as a function of $\mathbf{q}$ and subsequently to compare it with predictions of various models for diffusion. Instead, a certain diffusion model with a few parameters is assumed and the expression for the cross section derived from the model is convoluted with the resolution function and the model parameters are determined by fitting the resulting expression to the experimental data.

3.3.7. Future Directions

One cannot emphasize too strongly the fact that the neutron scattering technique as a means of studying diffusion process has not been established to date. Even the result of the simple measurement of the quasi-elastic line intensity is not without ambiguity as mentioned in Sections 3.3.3 and 3.3.6. As for the theoretical aspect of the problem, a more microscopic approach is desirable in order to express the neutron scattering cross sections in terms of the fundamental properties of solids. In particular, the diffusive motion is coupled to lattice vibrations of the crystal in such a specific fashion that one cannot treat the phonons as a simple heat bath. This also means that, for certain low-frequency modes of vibrations, the separation of the correlation function into that of the local vibration and that associated with the diffusive motion as described in Section 3.3.3 is not possible. Instead, the correlation function of the entire Hamiltonian must be evaluated in order to obtain the scattering law. This procedure has not yet been followed rigorously, and one does not know exactly what effects the diffusion has on the phonon dispersion relation and phonon lifetimes. The changes in the interatomic forces due to the positional disorder, rather than the diffusive motion itself, may have the most significant effects on phonons.[25]

[25] R. J. Elliott, W. Hayes, W. G. Kleppmann, A. J. Rushworth, and J. F. Ryan, *Proc. R. Soc. London, Ser. A* **360**, 317 (1978).

With further developments in high-resolution spectrometers and their increased availability, measurements on the slow-diffusion phenomena such as self-diffusion in simple metals,[26] as well as impurity effects,[27,28] would become easier.

[26] M. Ait-Salem, T. Springer, A. Heidemann, and B. Alefeld, *Philos. Mag., Part A* **39**, 797 (1979).
[27] D. Richter, B. Alefeld, A. Heidemann, and N. Wakabayashi, *J. Phys. F.* **7**, 569 (1977).
[28] D. Richter and T. Springer, *Phys. Rev. B* **18**, 126 (1978).

4. ION BEAM INTERACTIONS WITH SOLIDS

4.0. Introduction

By L. M. Howe, M. L. Swanson, and J. A. Davies

Atomic Energy of Canada Limited Research Company
Chalk River Nuclear Laboratories
Chalk River, Ontario, Canada

The interactions of light ions in the low to medium energy range (a few keV to a few MeV) with solids fall into the categories of elastic scattering, electronic excitation, and nuclear reactions. These interactions are generally well understood in terms of two-body collisions and are often independent of the chemical bonding in the solids. Thus the yields from these interactions provide an accurate analytical method for compositional studies.

In Chapter 4.1 the use of measurements of the yields of (primary and secondary) scattered particles, x rays, and nuclear reaction products for the determination of the composition of the near-surface regions of solids (0–1000 nm) is discussed.

The effect of an ordered crystal lattice on ion–solid interactions causes the phenomenon of ion channeling. In Chapter 4.2 the study of a variety of lattice defects, using yields from ion beam interactions under channeling conditions, is discussed.

4.1. Compositional Studies

By N. Cue

Department of Physics
State University of New York at Albany
Albany, New York

The richness of collisional phenomena associated with ion beam interactions with solids gives rise to a variety of techniques for compositional studies. These microanalytic techniques, which rely on the detection of characteristic signals generated in the collisions, may be classified broadly as either "destructive" or "nondestructive," according to whether or not material is eroded from the sample to be analyzed. Of course the solid is altered to some degree even if no erosion occurs since the mere passage of the ions creates lattice vacancies and interstitials along their paths and these can later agglomerate to form extended defects. Moreover, ions that differ from the host species and are stopped inside the solid constitute impurities. Thus the term "nondestructive" as used here is meant to imply that the level of impurities or defects produced either is inconsequential or can be controlled to a tolerable degree.

Because an ion beam has a finite range, the techniques to be described are most appropriate for the analysis of layered structures such as those encountered in thin films and thick, solid surfaces. To the extent that the surface is flat and laterally uniform in composition across the area sampled by the probing beam, many of these techniques have the ability to determine the concentrations of atoms with high sensitivity as a function of depth and with a resolution approaching tens of angstroms. Consequently, they provide invaluable tools for the investigation of solid state phenomena such as

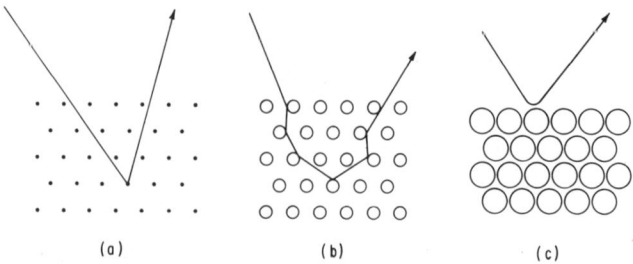

Fig. 1. Schematic of trajectories for ions backscattered from a solid target at (a) "high" energy, (b) "medium" energy, and (c) "low" energy.

4.1. COMPOSITIONAL STUDIES

thin-film reactions and impurity solubility and diffusion. In subsequent sections, the more established techniques are described, beginning with the "nondestructive" types, which include ion backscattering spectrometry, nuclear reaction analysis, and ion-induced x rays. Next the analysis of the sputtered ions and their light emission, which are the "destructive" types, are described. Complications due to channeling in crystallographic axes and planes are avoided here by focusing on amorphous or polycrystalline targets. When single-crystal targets are discussed, the beam is assumed to travel in a direction other than these symmetry axes or planes.

4.1.1. Ion Backscattering Spectrometry

Large-backward-angle elastic scattering is widely used particularly in the regime where the well-known Rutherford cross section[1] is applicable. For a given projectile incident on a target, this regime corresponds to a bombarding-energy range such that at the distance of closest approach in the collision the screening of the atomic electrons is ineffective on the one hand and the short-range nuclear interactions are not significant on the other. In the case of light ions (atomic number $Z_1 \leq 3$) incident on a target at rest, the bombarding energy range 100 keV amu$^{-1} \lesssim E_0 \lesssim$ 1 MeV amu^{-1} generally fits into this regime for not too light a target ($Z_2 \geq 20$) and is accessible by many small accelerators. The important point here is that, at these energies, an incident ion penetrating a solid loses energy quasi-continuously through the predominant interactions with the sea of valence or conduction electrons[2] and is thereby hardly deflected until the comparatively rare event of close-impact encounter occurs, causing the ion to backscatter and reemerge in a straight-line path. This is illustrated in Fig. 1a. The backscattered particles can thus be interpreted as single-collision events, and the simplicity of the two-body kinematics (with appropriate consideration of the energy losses) can be exploited. This analysis technique is known as Rutherford backscattering (RBS).

Similar considerations apply to higher bombarding energies except that the elastic scattering cross section now includes the lesser-known contributions from the nuclear interactions. This then restricts the method to special cases. On the other hand, those cases with large cross sections can be used to enhance the sensitivity for a particular impurity species, and the higher energy can be used to probe deeper layers. Further considerations of such cases and others involving the detection of nuclear reaction products are discussed in Section 4.1.2.

As the bombarding energy is lowered, the screening of the atomic electrons assumes increasing importance, and the elastic cross section begins to deviate from the Rutherford value. At the same time the "nuclear" contri-

---

[1] E. Rutherford, *Philos. Mag.* **21**, 669 (1911).

bution[2] to the energy loss, which tends to deflect the incident particles from a straight-line path, assumes increasing significance. As a consequence, the backscattered particles can undergo an increasing number of successive hard collisions as illustrated in Fig. 1b. The interpretation of the data becomes more involved, and the deduction of the number of scatterers becomes difficult.

The single-collision condition is again approached as the bombarding energy is lowered further (<10 keV) because the incident particle now penetrates no more than a few surface monolayers as illustrated in Fig. 1c. The elastic cross sections are not accurately known in this case although the characterization with some sort of screened Coulomb potential may form a good approximation in specific instances. Further complication arises from the charge-changing processes when only backscattered charged particles are detected. For singly charged incident ions there is a large probability for neutralization. Nevertheless, low-energy scattering is truly a surface-layer technique and can be made quantitative by comparison with standards of known surface composition, without reference to the scattering cross section and neutralization probability. This technique is known as low-energy ion scattering (LEIS) or ion scattering spectrometry (ISS).

In the following sections the detailed features of RBS and LEIS are amplified. Analysis using the intermediate-energy regime offers no distinctive advantage in compositional study and therefore is not considered here.

4.1.1.1. Rutherford Backscattering Spectrometry (RBS). This topic has been covered extensively in the literature and has been reviewed by several authors.[3,4]

A schematic diagram of a typical experimental setup is shown in Fig. 2. Ion beams of 0.1–1 MeV H and 1–4 MeV He are commonly used because

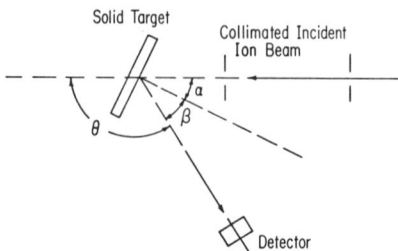

FIG. 2. Schematic of experimental geometry for backscattering measurements.

[2] P. Sigmund, in "Radiation Damage Processes in Materials" (C. H. S. Dupuy, ed.), p. 3. Noordhoff, Leyden, 1975.
[3] J. F. Ziegler, J. W. Mayer, B. M. Ullrich, and W. K. Chu, in "New Uses of Ion Accelerators" (J. F. Ziegler, ed.), p. 75. Plenum, New York, 1975.
[4] W. K. Chu, J. W. Mayer, and Nicolet, M.-A., "Backscattering Spectrometry." Academic Press, New York, 1978.

they are readily available, incur the least damage among ions of comparable velocity, and their stopping power (energy loss per path length) has been extensively investigated.[5-7] Also, the regime for which the Rutherford cross section is applicable is better known for these ions. The momentum-analyzed and collimated beam from an accelerator is directed toward the target at angle α relative to the surface normal and the elastically scattered yield coming off at an angle β corresponding to a scattering angle θ is recorded by a charged-particle detector. Solid state detectors, which are relatively inexpensive and convenient to use because of their energy-dispersive nature and reasonable energy resolution ($\delta E \simeq 15$ keV) are commonly used. Improved energy resolution can be achieved by using the bulkier electrostatic[8] or magnetic[9] spectrometers, but this must be weighed against the lengthier data-accumulation time since each point in the energy spectrum must be recorded sequentially. Also, the larger accumulated beam dose contributes to a larger radiation damage in the crystal. One other advantage of the momentum-dispersive devices is the absence of electronic pileup distortion in the backscattering spectra, which is often encountered with solid state detectors. However, such distortions in the latter devices can be minimized with a lower counting rate or the use of standard pileup rejection circuitry. Base pressure in the scattering chamber of 10^{-6} Torr is adequate for most applications except in instances where the cleanliness of the sample surface needs to be preserved for other purposes. Surrounding the sample with a cold surface (cold can) at liquid-nitrogen temperature is easily implemented and helps to slow down the buildup of surface impurities on the sample. Biasing the cold can to -300 V relative to the target is a convenient means of suppressing secondary electons if the true beam current on a thick target is to be measured. A more accurate method for beam monitoring, which is applicable to many circumstances, has been described by Mitchell *et al.*[10] For targets that are insulators, voltage buildup due to the poor conduction of accumulated charges can be alleviated by depositing a thin, conducting masking layer on the sample.

[5] J. F. Ziegler and W. K. Chu, *At. Data Nucl. Data Tables* **13**, 463 (1974).

[6] H. H. Andersen and J. F. Ziegler, "Hydrogen Stopping Powers and Ranges in All Elements." Pergamon, Oxford, 1977; J. F. Ziegler, "Helium: Stopping Powers and Ranges in All Elemental Matter." Pergamon, Oxford, 1977.

[7] P. W. Keaton, P. S. Peercy, B. L. Doyle, and C. J. Maggiore, *Nucl. Instrum. Methods* **168**, 187 (1980).

[8] A. Feuerstein, H. Grahmann, S. Kalbitzer, and H. Oetzmann, in "Ion Beam Surface Layer Analysis" (O. Meyer, G. Linker, and F. Käppeler, eds.), Vol. 1, p. 471. Plenum, New York, 1976.

[9] J. K. Hirvonen and G. K. Hubler, in "Ion Beam Surface Layer Analysis" (O. Meyer, G. Linker, and F. Käppeler, eds.), Vol. 1, p. 457. Plenum, New York, 1976.

[10] I. V. Mitchell, K. M. Barfoot, and H. L. Eschbach, *Nuc. Instrum. Methods* **168**, 233 (1980).

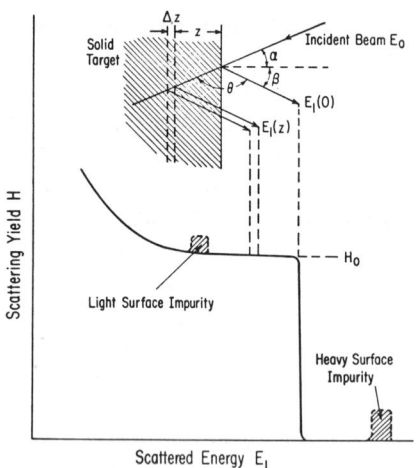

Fig. 3. Schematic of RBS yield from a thick target with surface impurities.

In order to illustrate the principles and limitations of the technique, consider first the case of a homogenous monisotopic solid target. For a fixed bombarding energy and geometry, the yield as a function of the energy of the scattered particle is depicted in Fig. 3. The edge occuring at an energy $E_1(0)$ corresponds to the scattering from the atoms at the surface layer and is governed by the two-body kinematics of elastic scattering.[11] For the energies of interest here, relativistic effects may be neglected. In terms of the incident particle mass M_0 and energy E_0, target mass M, and laboratory scattering angle θ,

$$E_1(0) = K^2 E_0, \qquad (4.1.1)$$

where

$$K = [M_0 \cos \theta \pm (M^2 - M_0^2 \sin^2 \theta)^{1/2}]/(M_0 + M). \qquad (4.1.2)$$

Here the plus sign is applicable for $M > M_0$. The yield at energies below the step therefore corresponds to the scattering from deeper layers because the scattered particle suffers energy loss both in the inward and outward paths. It can also be seen that the presence of surface impurities will show up as a peak at an energy above the edge if their mass is heavier and below if lighter. At a depth z, the incident particle energy is reduced to a value $E(z)$ before a large-angle deflection occurs and which, according to Eq. (4.1.1), will result in a scattered energy of $K^2 E(z)$. This energy is further reduced to $E_1(z)$ as the scattered particle moves to the surface. In the inward and outward paths, the

---

[11] J. B. Marion and F. C. Young, "Nuclear Reaction Analysis," p. 163. North-Holland Publ., Amsterdam, 1968.

energy losses are related to the depth z, respectively, according to

$$\frac{z}{\cos \alpha} = -\int_{E_0}^{E(z)} \frac{dE}{S}, \quad (4.1.3)$$

$$\frac{z}{\cos \beta} = -\int_{K^2 E(z)}^{E_1(z)} \frac{dE}{S}, \quad (4.1.4)$$

where

$$S = \frac{dE}{dz} \quad (4.1.5)$$

is known as the stopping power. Using tabulated values for S or their analytic approximations,[6] the integrations of the above equations allow one to convert the measured energy E_1 into a depth scale z with the elimination of $E(z)$.[7]

For a thin surface layer in which the range of the incident beam far exceeds the layer thickness, the variation of S is small, and the use of an average value is a reasonably good approximation. The integrations of (4.1.3) and (4.1.4) are then simple and result in the relation

$$E_1(0) - E_1(z) = [S]z, \quad (4.1.6)$$

where

$$[S] = \frac{K^2}{\cos \alpha} S_{\text{in}} + \frac{1}{\cos \beta} S_{\text{out}}. \quad (4.1.7)$$

The $[S]$ is commonly called the energy-loss factor. From these equations, the overall energy resolution δE_1 with which the particles are detected is seen to determine the depth resolution for a fixed geometry. This δE_1 is usually governed by the detector resolution since the energy straggling is negligible and the kinematical energy spread due to the finite acceptance angle of the detector can be arranged to be small. For a given δE_1, the depth sensitivity is seen to increase as either α or β approaches 90°. In practice the maximum values for α and β are restricted by the experimental geometry and by the flatness of the scattering surface.[12] Moreover, if the surface layer contains scatterers of closely spaced masses, the scattering angle $\theta = \pi - (\alpha + \beta)$, according to Eqs. (4.1.1) and (4.1.2), must be chosen to be sufficiently large in order to separate their respective contributions. The angles must thus be optimized by taking into account these experimental factors.

The above results can be applied to the case of a thick slab by considering it to be made up of many thin layers of thickness Δz. Using the notation

---

[12] J. S. Williams, *Nucl. Instrum. Methods* **126**, 205 (1975).

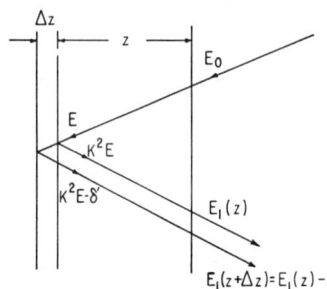

FIG. 4. Energies associated with a particle scattered from a thin layer Δz at a depth z in a solid.

given in Fig. 4, Eq. (4.1.6) becomes

$$\delta' = [S(E)] \Delta z, \qquad (4.1.8)$$

where E is the reduced incident energy at z and $[S(E)]$, as given in Eq. (4.1.7), is to be evaluated with the corresponding energy E. Experimental spectra are usually recorded as counts per channel in which each channel corresponds to a fixed E_1 increment δ. The relation between δ and δ' can be obtained by noting that the outward path lengths of K^2E and $K^2E - \delta'$ are the same (see Fig. 4), and thus

$$-\int_{K^2E}^{E_1} \frac{dE}{S} = -\int_{K^2E-\delta'}^{E_1-\delta} \frac{dE}{S}. \qquad (4.1.9)$$

Since both δ and δ' are small compared to K^2E and E_1, they can be treated as differentials and Eq. (4.1.9) reduces to

$$\delta'/\delta = S(K^2E)/S(E_1). \qquad (4.1.10)$$

The correspondence between E_1 and z may thus be established by an iterative procedure[13] starting from the surface. With increasing depth, energy straggling increases, and therefore the depth resolution becomes progressively worse.

Turning to the backscattered yield that corresponds to the height of the spectrum, the contribution of a layer Δz at a depth z can be expressed as

$$H(z) = \sigma \Omega N n \, \Delta z, \qquad (4.1.11)$$

where σ is the differential scattering cross section evaluated at $E(z)$ and averaged over the solid angle Ω subtended by the detector and N is the total number of particles incident on the sample having an atomic volume density n for that layer. Note that the product $n \, \Delta z$ is the number of atoms per unit area, the areal density, and the yield is proportional to this. To make this relation explicit in the expression for H and therefore more convenient

---

[13] J. F. Ziegler, R. F. Lever, and J. K. Hirvonen, *in* "Ion Beam Layer Analysis" (O. Meyer, G. Linker, and F. Käppeler, eds.), Vol. 1, p. 163. Plenum, New York, 1976.

4.1. COMPOSITIONAL STUDIES

for comparison with measured backscattered spectra, the stopping cross section ϵ, defined as

$$\epsilon = n^{-1}S, \qquad (4.1.12)$$

can be used in Eqs. (4.1.8), (4.1.10), and (4.1.11) to obtain

$$H(E_1) = \sigma(E)\Omega N \frac{\delta}{[\epsilon(E)]} \frac{\epsilon(K^2 E)}{\epsilon(E_1)}. \qquad (4.1.13)$$

For the surface layer, $E = E_0$, and $E_1 = KE_0$, the equation simplifies to

$$H_0 = \sigma_0 \Omega N \delta / [\epsilon_0]. \qquad (4.1.14)$$

As indicated in Fig. 3, this is the height of the edge.

The case of a thin, heavy impurity layer on top of a light substrate contains illuminating features. The heavy impurity would correspond to the peak above the edge in Fig. 3 for which Eq. (4.1.14) is applicable. If the overall energy resolution is small compared to the width of peak, this width is a direct measure of the film thickness if the stopping power is known. On the other hand, the yield represented by the total area under the peak is also a direct measure of the film thickness as implied, for example, by Eq. (4.1.14) if δ is interpreted as the total energy loss in the thin film. This then offers a self-consistency check on the analysis. Because the Rutherford cross section for a fixed scattering angle is proportional to $[Z_1 Z_2 / E]^2$, where Z_1 and Z_2 are the atomic numbers of the incident and target nuclei, respectively, and E is the projectile energy, the detection sensitivity increases with decreasing E and, for impurities, with increasing Z_2. Effects of the finite energy resolution are also manifested in the peak shape. The rounded edge and the corresponding knee on the high-energy side reflect the detector resolution and finite acceptance angle, but on the low-energy side the added contribution of straggling effects resulting from the penetration of and emergence from the film makes the rounding and knee more pronounced.[14]

Areal density, not the volume density, is also the determining factor in the energy loss since collisional (soft) processes are involved. Thus if a particle traverses two thin samples of the same material, one of which is porous and the other not, a thicker layer must be crossed in the porous sample in order to produce the same energy loss. If the volume densities are denoted n' and n for the porous and nonporous samples, respectively, the condition $n' \Delta z' = n \Delta z$ implies $\Delta z' > \Delta z$ for $n' < n$. Thus the depth scale derived from the energy loss is a measure of the areal density; only when the volume density of the sample is given can a depth scale be established.

For homogeneous solids consisting of multiple elements such as chemical compounds and solutions (alloys), the backscattered spectra can be ana-

---

[14] D. K. Brice, *Thin Solid Films* **19**, 121 (1973).

lyzed in a similar fashion. The essential difference is that the energy loss in both the inward and outward paths will now be governed by the S or ϵ of the compound medium. For a two-element composite of the form A^aB^b, where a/b is the atomic ratio, the assumption of linear additivity of the elemental stopping cross sections results in

$$\epsilon^{AB} = a\epsilon^A + b\epsilon^B. \tag{4.1.15}$$

This is the so-called Bragg rule,[15] which has been widely used. The extension to more than two elements is straightforward. In addition, the backscattering kinematic factor and cross section will be different for the different elements, and due account of these must be kept in reducing the spectra to depth profiles.

In deciding on what are the most suitable applications for this technique, it is worthwhile to describe briefly its general limitations. First, it is a mass-sensitive technique, and according to Eq. (4.1.1) the scattering from neighboring masses becomes increasingly difficult to distinguish with increasing mass. Thus depth profiling is simple only if the sample contains a few elements of highly different masses. For such cases, the atomic concentrations near the surface can be measured with typical accuracies of a few percent. In the case of impurities, a high detection sensitivity can be achieved only for those more massive than the substrate atoms. If these heavy impurity atoms lie on the surface, even a monolayer or less can be detected. Second, the depth scale established on the basis of energy loss phenomena is meaningful only if the surface is flat and laterally uniform. Although the degree of flatness can be checked by analyzing the backscattering yield as a function of the target tilt angle α or detector angle β or both, the observed deviation from a flat surface can not be directly related to the nature of the irregularity. Other techniques, such as x-ray diffraction and electron microscopy, must be used to obtain the details of surface topography. It is also clear that the effects of any surface irregularity will be magnified in a glancing geometry (large α or β) that enhances the depth resolution. Third, the accuracy of the depth scale is governed by our knowledge of the stopping cross section ϵ. For H and He ions, ϵ has been extensively investigated both experimentally and theoretically, and the values for nearly all elements are known with an accuracy of a few percent in the energy region of interest here. However, several observations, which are not completely understood, must be taken into consideration. It has been reported that the ϵ for He ions stopping in elemental-gas targets appear consistently higher than those for elemental solids[16] and that bonding effects

---

[15] W. H. Bragg and R. Kleeman, *Philos. Mag.* **10**, 318 (1905).

[16] J. F. Ziegler, W. K. Chu, and J. S. Y. Feng, *in* "Ion Beam Surface Layer Analysis" (O. Meyer, G. Linker, and F. Käppeler, eds.), Vol. 1, p. 15. Plenum, New York, 1976.

4.1. COMPOSITIONAL STUDIES

in molecular gases not considered in the Bragg rule may be important.[17] In any case the RBS itself can provide information on the stopping cross sections.[18] Finally, the specific validity of the Rutherford cross section must be independently verified.

The limitations described above have not been restrictive as evidenced by the widespread applications of the technique to a large variety of problems, some of which are described in the following section. As can be surmised, the main advantage of the technique is that the analysis is fast, simple, and quantitative.

4.1.1.2. Application of RBS. Many technologically important aspects of material sciences involve phenomena occuring at or near the solid surface. It is thus not surprising that the RBS "nondestructive" technique has been extensively applied to a wide variety of problems since the composition in depth of the material is basic to the understanding of the underlying phenomena. The applications range from routine ones, such as quality control in the fabrication of semiconductor devices, to specialized investigation, such as interfacial reactions in thin films. Their number and diversity can be seen in the proceedings of various conferences[19-22] on ion beam analysis. A description of a few representative ones will suffice to illustrate some of the more salient features of the technique.

In the modification of materials properties by ion implantation, the intent is to achieve a certain spatial distribution for the implanted species. Although general theories for the slowing down of ions in amorphous solids[23,24] give predictions on the specific range distributions, deviations from these have been observed that can be attributed to a number of factors. At high dose, target sputtering, changes in the stopping power of the target as the foreign species builds up, and radiation-enhanced diffusion can contribute to the final ion distribution. Accurate final profiles must thus be mea-

---

[17] A. S. Lodhi and D. Powers, *Phys. Rev. A* **10**, 2131 (1974).

[18] B. M. U. Scherzer, P. Børgesen, M.-A. Nicolet, and J. M. Mayer, *in* "Ion Beam Surface Layer Analysis" (O. Meyer, G. Linker, and F. Käppeler, eds.), Vol. 1, p. 33. Plenum, New York, 1976.

[19] *Proc. Conf. Ion Beam Surf. Layer Anal., Yorktown Heights, New York, 1973*, published in *Thin Solid Films* **19**, 1 (1973).

[20] O. Meyers, G. Linker, and F. Käppeler, eds., "Ion Beam Surface Layer Analysis," Vols. 1 and 2. Plenum, New York, 1976.

[21] E. A. Wolicki, J. W. Butler, and P. A. Treado, eds., *Proc. 3rd Int. Conf. Ion Beam Anal.* North-Holland Publ., Amsterdam, 1978.

[22] H. H. Andersen, J. Bøttiger, and H. Knudsen, eds., *Proc. 4th Int. Conf. Ion Beam Anal.* North-Holland Publ., Amsterdam, 1980.

[23] J. Lindhard, M. Scharff, and H. E. Schiøtt, *Mat.-Fys. Medd.—K. Dan. Vidensk. Selsk.* **33**, No. 14 (1963).

[24] O. B. Firsov, *Sov. Phys.—JETP (Engl. Transl.)* **5**, 1192. (1957); **6**, 534 (1958).

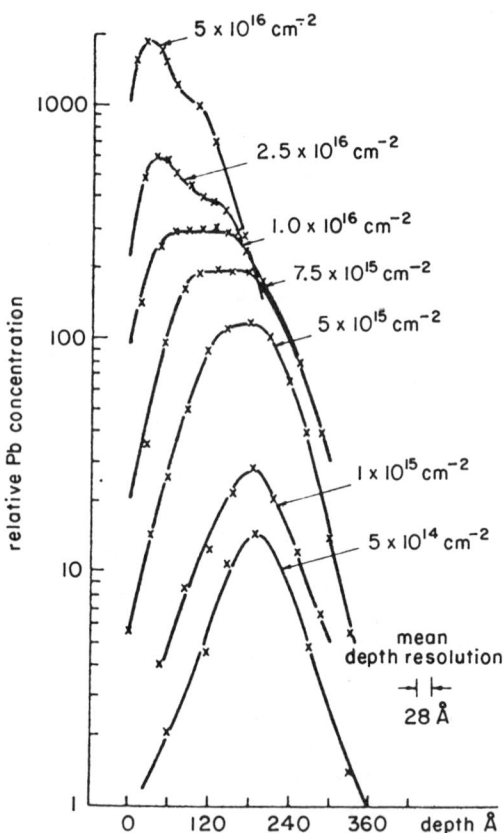

Fig. 5. Range distributions derived from 2-MeV ^{4}He$^+$ RBS measurements for 20 keV Pb$^+$ ions implanted into Si as a function of ion doses. (From Williams.[25])

sured, and the RBS technique is particularly suited to analyze the cases of implants more massive than that of the substrate species. Figure 5, taken from Williams,[25] shows such depth profiles of Pb$^+$ implanted at 20 keV at an off-axis direction into slices of $\langle 111 \rangle$ Si single crystal at room temperature for several ion doses. In the RBS analysis, where 2-MeV He$^+$ was used at an off-axis glancing incidence ($\alpha = 85°$) and with a scattering angle of $\theta = 168°$, the Pb elastic peak is well separated from the Si thick target edge. The normalized Pb yield has been taken directly as a measure of the Pb concentration in the Si target, and the depth scale in each spectrum was established by taking into consideration the effect of the local Pb concentration on the stopping power. This was achieved by an iterative procedure using the

[25] J. S. Williams, *Phys. Lett. A* **51A**, 85 (1975).

relative heights of the backscattered Si and Pb yields to construct the compound stopping power for each step of the iterations. As indicated in the figure the mean depth resolution is 28 Å. At low dose ($<10^{15}$ ions cm^{-2}), the expected near-Gaussian profile is obtained. A flat top profile develops and extends towards the surface as the dose is increased, consistent with the effect of target sputtering during ion implantation. At doses above 10^{16} ions cm^{-2}, the Pb concentration exhibits a double-hump profile with the higher one at the surface. Williams[25] had suggested that this "surface" enhancement may be the consequence of Pb or Pb–Si alloy precipitating on the surface, analogous to the effect observed in Al implanted at 400°C with 70 keV Pb$^+$ ions[26] and Pb films deposited and heated on silicon substrates.[27] Since the dimension of the probing ion beam is much larger than those of the precipitates, other techniques, such as transmission or scanning electron microscopy (TEM and SEM), must be used to observe these directly. The effects of laterally nonuniform samples on the RBS spectra have been investigated by a number of authors.[28-30]

Another area where the RBS technique has been extensively applied is in the study of thin-film interdiffusion.[31] The large surface-to-volume ratio of materials in film form gives rise to many properties that differ markedly from those in the corresponding bulk form. Again, its nondestructive nature, relatively direct reflection of the concentration profiles, and short data-accumulation time make the RBS technique ideally suited for the study of diffusion in thin films. Figure 6 shows the data of Poate and co-workers[32] on Ti–Au–Pd films deposited in the order written on a sapphire substrate with respective thicknesses of 750, 600, and 850 Å. In this case, a 1.8-MeV ^{4}He$^+$ beam was used, and the experimental parameters were chosen so that the front edge of the backscattered peaks of Au and Pd lay on top of each other. The expected backscattered energies for atoms on the surface are indicated below the graph. For the as-deposited profile (solid curve), the Au peak lies below the Au surface-scattered energy and sits on top of the Pd peak because the analyzing beam had to pass through the Pd layer. The slight curvature on the low-energy side of the peak is due to the larger thickness of Pd. Similarly, the Ti peak is moved to a lower energy due

[26] P. A. Thackery and R. S. Nelson, *Philos. Mag.* **19**, 169 (1969).
[27] S. U. Campisano, G. Foti, F. Grasso, and E. Rimini, *Thin Solid Films* **25**, 431 (1975).
[28] G. Ottaviani, D. Sigurd, V. Marrello, J. W. Mayer, and J. O. McCaldin, *J. Appl. Phys.* **45**, 1730 (1974).
[29] K. Nakamura, M.-A. Nicolet, J. W. Mayer, R. J. Blattner, and C. A. Evans, Jr., *J. Appl. Phys.* **46**, 4678 (1975).
[30] G. W. Arnold and J. A. Borders, *J. Appl. Phys.* **48**, 1488 (1977).
[31] J. M. Poate, *in* "Ion Beam Surface Layer Analysis" (O. Meyer, G. Linker, and F. Käppeler, eds.), Vol. 1, p. 317. Plenum, New York, 1976.
[32] J. M. Poate, P. A. Turner, W. J. DeBonte, and J. Yahalom, *J. Appl. Phys.* **46**, 4275 (1975).

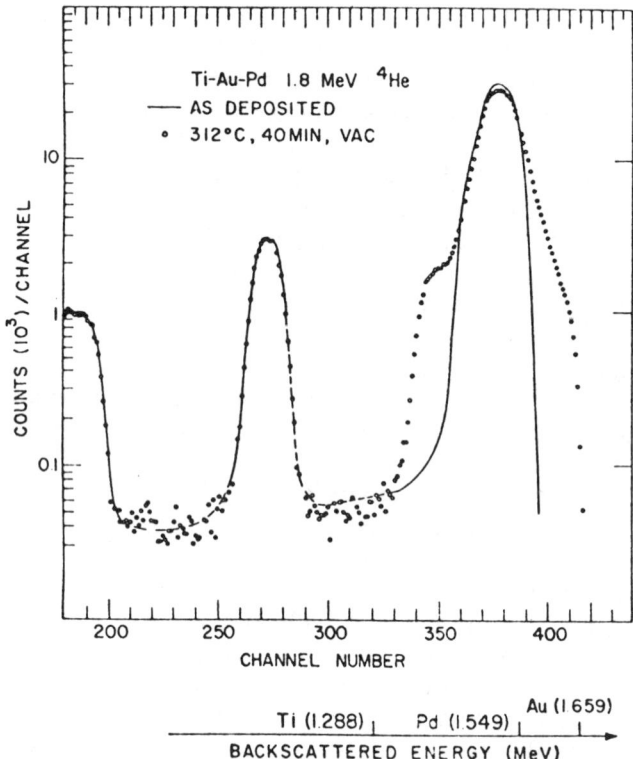

FIG. 6. RBS spectra from 1.8-MeV ^{4}He$^+$ on Ti–Au–Pd deposited on a sapphire backing. (From Poate et al.[32])

to the energy loss in the Pd and Au films. The step on the left is the higher-energy portion of the thick-target yield from the Al in the sapphire substrate. Upon annealing the sample in vacuum for 40 min at 312°C, the RBS spectrum (dotted curve) shows the results of interdiffusion: the ledge at the high-energy side of the Pd + Au peak reflects the migration of Au into Pd and toward the surface as the scattering from Au indicates a lesser energy loss, and the low-energy ledge shows the migration of Pd into Au as more energy loss is involved. Both the Ti and substrate profiles remain essentially unchanged. By investigating the annealing effects in vacuum and in air of various binary and ternary combinations of these three elements, Poate et al.[32] were able to delineate the factors controlling the interdiffusion. Although RBS studies provides valuable information, other techniques are often required to supply additional data in the detailed understanding of the processes. For example, the grain growths upon annealing, studied by TEM, were correlated with the RBS observations in order to establish that, at

4.1. COMPOSITIONAL STUDIES

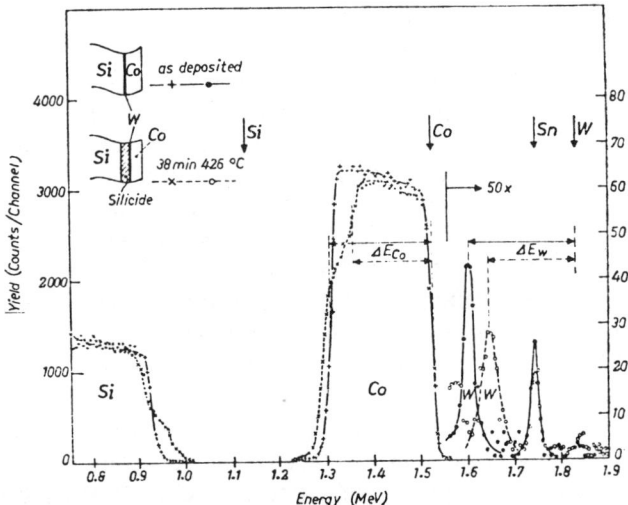

FIG. 7. RBS spectra from 2-MeV ^{4}He on a Co–Si couple with a W marker. The effects of silicide formation are manifested by the upward shift in the W peak position and the knee structure in the scattering from Co and Si. Energies corresponding to scattering from surface atoms are indicated by arrows. The Sn surface peak is due to impurities introduced during the deposition of W marker. (From van Gurp et al.[37])

temperatures ≲400°C, grain boundary diffusion dominates the interdiffusion of Pd–Au and Au–Pd couples.

In the thin-film interdiffusion studies, the unambiguous determination of the diffusing species has been made in a number of cases using an inert marker analogous to the Kirkendall marker[33] for diffusion in bulk materials. The ion implantation of inert gases as markers was first used by Brown and Mackintosh[34] and applied by other investigators.[35,36] Other kinds of thin, inert markers have also been used successfully.[37,38] The basic idea can be seen from the RBS spectra in Fig. 7, taken from the work of van Gurp et al.[37] on cobalt silicide formation starting from a cobalt film on a silicon substrate. In this case the marker consists of islands of tungsten about 30 Å thick deposited at the Si–Co interface. After annealing, the backscattered tungsten peak has increased in energy indicating that Co is the faster diffusing

[33] E. O. Kirkendall, *Trans. Am. Inst. Min. Metall. Eng.* **147**, 104 (1942).
[34] F. Brown and W. D. Mackintosh, *J. Electrochem. Soc.* **120**, 1096 (1973).
[35] W. K. Chu, H. Kraütle, J. W. Mayer, H. Müller, M.-A. Nicolet, and K. N. Tu, *Appl. Phys. Lett.* **25**, 454 (1974).
[36] W. K. Chu, S. S. Lau, J. W. Mayer, H. Müller, and K. N. Tu, *Thin Solid Films* **25**, 393 (1975).
[37] G. J. van Gurp, D. Sigurd, and W. F. van de Weg, *Appl. Phys. Lett.* **29**, 159 (1976).
[38] R. Pretorius, Z. L. Liau, S. S. Lau, and M.-A. Nicolet, *Appl. Phys. Lett.* **29**, 598 (1976).

species. Shape change in the backscattering from Si and Co has been attributed to the formation of mainly Co_2Si.

Many other phenomena in layered structure have been investigated by RBS techniques, including thin-film reactions, impurity solubility and diffusion in bulk material, epitaxial layers, oxidation and corrosion, and superconducting or magnetic thin films. As discussed in Chapter 4.2, backscattering measurements in a channeling geometry also give crystallographic information such as lattice site location of impurities or lattice disorder.

4.1.1.3. Low-Energy Ion Scattering (LEIS). With low enough energy (0–10 keV) ion backscattering can be arranged to be sensitive only to the outermost surface atoms. Reviews on the use of low-energy ions for compositional studies have been given by Buck,[39] Taglauer and Heiland[40] and Brongersma et al.[41]

The usual experimental setup is similar to that indicated in Fig. 2 except that an accelerator is no longer needed. The ion source must be well stabilized in order to get a well-defined beam energy. In instances where a high purity beam is desired, a mass filter can be positioned between the source and the target. Since the outermost surface layer is of primary interest, the vacuum in the target chamber must be sufficiently low ($< 10^{-9}$ Torr) to prevent surface alteration by absorption of residual reactive gases within the measuring time interval. The scattered ions are usually detected with an electrostatic cylindrical or spherical energy analyzer (ESA) coupled with some type of electron multiplier device such as channeltron. Niehus and Bauer[42] have described an ingenious detection scheme in which the voltages of a cylindrical mirror analyzer normally used for Auger electron spectroscopy are inverted. The NODUS scheme developed[43] at the Philips Laboratories is particularly noteworthy in achieving a substantial increase in sensitivity by arranging the setup in such a way that the beam enters the axis of a cylindrical ESA and the backscattered ions are accepted by a large solid but small divergence angle. This is illustrated schematically in Fig. 8. The multiplier in this case has a ring geometry. This increase in sensitivity made possible the investigations of more easily sputtered samples. The time-of-flight (TOF) technique has also been used for energy analysis.[44,45] Because of

[39] T. M. Buck, in "Methods of Surface Analysis" (S. P. Wolsky and A. W. Czanderna, eds.), p. 75. Elsevier, Amsterdam, 1975.
[40] E. Taglauer and W. Heiland, *Appl. Phys.* **9,** 261 (1976).
[41] H. H. Brongersma, L. C. M. Beirens, and G. C. J. van der Ligt, in "Materials Characterization Using Ion Beams" (J. P. Thomas and A. Cachard, eds.), p. 65. Plenum, New York, 1978.
[42] H. Niehus and E. Bauer, *Surf. Sci.* **47,** 222 (1975).
[43] H. H. Brongersma, F. Meijer, and H. W. Werner, *Philips Tech. Rev.* **34,** 362 (1974).
[44] T. M. Buck, Y. S. Chen, G. H. Wheatley, and W. F. van der Weg, *Surf. Sci.* **47,** 244 (1975).
[45] T. M. Buck, G. H. Wheatley, G. L. Miller, D. A. H. Robinson, and Y. S. Chen, *Proc. 3rd Int. Conf. Ion Beam Anal.*, p. 591. North-Holland Publ., Amsterdam, 1978.

4.1. COMPOSITIONAL STUDIES

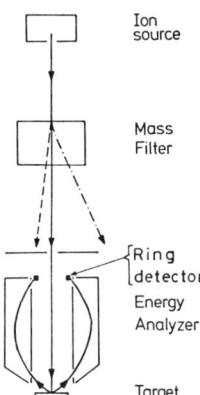

FIG. 8. Schematic diagram of the NODUS set up for low-energy ion scattering measurements. (From Brongersma et al.[41])

its energy-dispersive nature, a significantly less ion dose is required compared to the analysis by the stepwise ESA, and thus the sample will be subjected to less damage. The TOF technique being more flexible in geometry than the NODUS scheme also has the added advantage that neutral particles can be analyzed. However, there is a steep decrease in the detection efficiency of multipliers for neutrals below a few keV. An alternative technique for detecting neutral particles by first stripping their electrons in a gas cell is described by Verbeck et al.[46] Mashkova and Molchanov[47] in their review of ion scattering at somewhat higher energies also gave a description of several experimental arrangements.

A spectrum of 300-eV Ne^+ scattered at an angle $\theta = 90°$ from a nickel crystal on which Cl, Br, and I were absorbed is shown in Fig. 9. Distinct peaks corresponding to the expected surface composition are observed as shown in the figure. The correspondence is established by the good agreement between the observed peak energies and the expected ones calculated on the basis of a single binary collision as given by Eq. (4.1.1). That the elastic scattering behaves like an isolated ion–atom collision can be attributed to the fact that the collision time ($\sim 10^{-15}$ s) is much shorter than the characteristic vibration time ($\sim 10^{-13}$ s) of a surface atom.

Sharp, single-binary-collision (SBC) peaks are generally observed[48] for noble gas ions in the energy range of about $E_0 = 100$–2000 eV. With reactive gas ions, and noble gas ions at higher energies, the broad structure at the low-energy side of the SBC peak is generally present, as shown in Fig. 10. Structure at the high-energy side of the SBC peak also appears at the smaller

---

[46] H. Verbeek, W. Eckstein, and F. E. P. Matschke, *J. Phys. E* **10**, 944 (1977).
[47] E. S. Mashkova and V. A. Molchanov, *Radiat. Eff.* **16**, 143 (1972).
[48] D. P. Smith, *J. Appl. Phys.* **38**, 340 (1967).

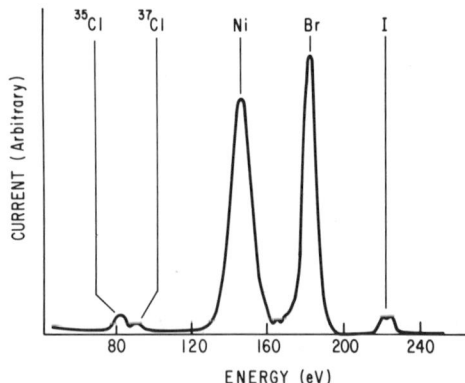

FIG. 9. Detected ion current for a 300-eV Ne⁺ beam scattered at 90° by a nickel surface to which Cl, Br, and I are absorbed. (From Brongersma et al.[43])

scattering angles θ and for low mass ratios M/M_0.[49,50] The depth of ion penetration cannot be so sharply defined and also the known electron exchange processes lead to ion neutralization[51]; thus the sharp SBC peaks are really due to a special set of circumstances. In the case of noble gas ions scattered at large angles θ, the surface electrons see a potential that is more attractive than the work function. Therefore, there is a high probability for electron tunneling in the ions' close encounter with a surface atom, and this leads to the neutralization of the ion. The probability of the ion escaping unneutralized after two or more collisions must therefore be negligibly small. This then accounts for the sharp SBC peaks because only in collisions with the outermost surface atoms will the ion have a finite probability to survive as an ion.

Surface compositional studies using low-energy ion scattering (LEIS) would clearly be facilitated in the SBC regime. Even here the scattered ion intensity from a particular surface species will depend on the probability for escaping neutralization P_i, in addition to the usual factors such as the species' areal density and scattering cross section σ. Reasonable estimates of σ have been obtained based on the use of screened Coulomb potentials,[50] but the same can not be said for the P_i.

Direct information on the P_i could, in principle, be obtained if the neutral yield were also measured, but difficulties are encountered in the detection of low-energy neutral particles. The study of photon emission could help, but

---

[49] D. J. Ball, T. M. Buck, D. McNair, and G. H. Wheatley, *Surf. Sci.* **30**, 69 (1972).
[50] W. Heiland and E. Taglauer, *Nucl. Instrum. Methods* **132**, 535 (1976).
[51] N. H. Tolk, J. C. Tully, W. Heiland, and C. W. White, eds., "Inelastic Ion–Surface Collisions." Academic Press, New York, 1977.

4.1. COMPOSITIONAL STUDIES

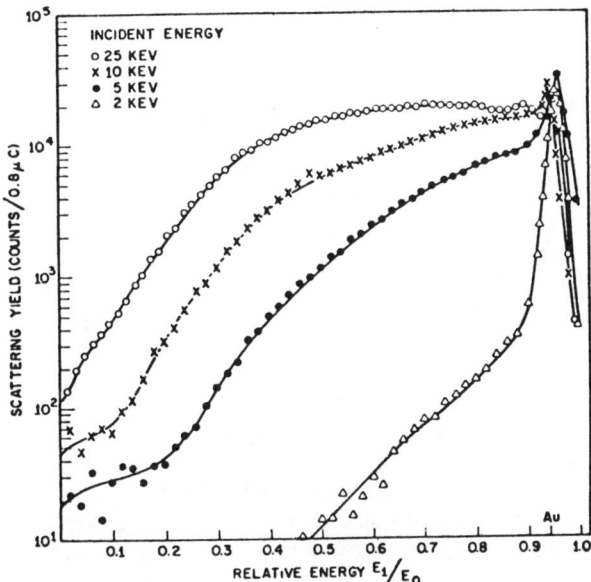

FIG. 10. Yield of He⁺ scattered at 120° by a Au surface for various beam energies. (From Ball et al.[49])

such information may not be complete because a special set of levels is involved (Section 4.1.4). Neutralization mechanisms considered to date have dealt with processes occuring as the ion leaves the surface. These include Auger and resonance transfer mechanisms,[52,53] both of which give rise to

$$P_i = e^{-v_i/v_1}, \tag{4.1.16}$$

where v_i is a characteristic velocity and v_1 is the component of the exit velocity normal to the surface. Estimates for P_i based on Eq. (4.1.16) are of the order 10^{-3} and may vary by an order of magnitude for different elements. These mechanisms are believed to be applicable to the cases where the yield varies monotonically with bombarding energy. However, oscillatory behavior in the ion yield has also been observed, namely, for He⁺ on Ga, Ge, In, Sn, Sb, Pb, and Bi[54] and other collisions.[55] The case of He⁺ on In is shown in Fig. 11. Erickson and Smith[54] have attributed the phenomenon

---

[52] A. Cobas and W. E. Lamb, *Phys. Rev.* **65**, 327 (1944).
[53] H. D. Hagstrum, *Phys. Rev.* **96**, (1954).
[54] R. L. Erickson and D. P. Smith, *Phys. Rev. Lett.* **34**, 297 (1975).
[55] T. W. Rusch and R. L. Erickson, in "Inelastic Ion–Surface Collions" (N. H. Tolk, J. C. Tully, W. Heiland, and C. W. White, eds.), p. 73. Academic Press, New York, 1977.

4. ION BEAM INTERACTIONS WITH SOLIDS

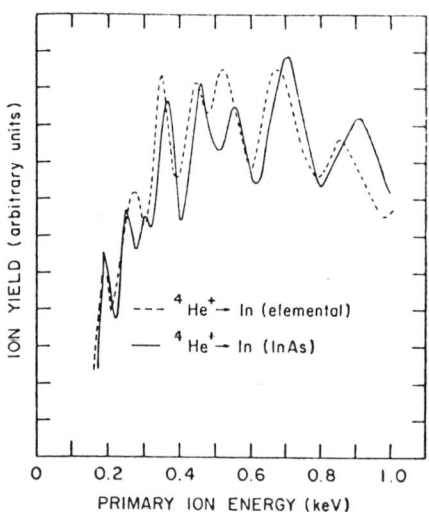

FIG. 11. Primary beam energy dependence of ^{4}He$^+$ scattering yield at 90° for In targets. The InAs curve has been doubled in magnitude. (From Erickson and Smith.[54])

to osciallations in neutralization probability similar to those studied in gaseous targets.[56] Such oscillations are viewed to arise from the phase interference of the wave functions associated with an electron transition near the crossing of levels between the ion and atom as they approach each other and form a quasi-molecule. Although many insights have been gained about the neutralization mechanisms, accurate predictions of the P_i are not yet possible.

Uncertainties in the neutralization probabilities and, to a lesser extent, in the scattering cross sections thus necessitate the use of calibration standards for quantitative analysis of surface composition. Chemical-bonding effects may be present, as can be seen, for example, in Fig. 11. Such effects are not too large, however, and the use of pure elements to calibrate elemental sensitivities may be adequate for most purposes.[40,41] Suggestions have also been made that the different patterns in Fig. 11 may be used as the basis for chemical identification and that the different patterns for He$^+$ scattering from Pb and Bi,[54] for example, may also be used to distinguish neighboring mass species that cannot be resolved in the elastic scattering spectrum.

Strictly speaking, some knowledge of the composition and lattice structure at the surface is necessary if the yield is to be interpreted unambiguously. This stems from the shadowing effect, whereby atoms protruding out of the surface may shield the scattering from the neighboring atoms, the

[56] H. S. W. Massey, E. H. S. Burhop, and H. B. Gilbody, "Electronic and Ionic Impact Phenomena," Vol. IV, 2nd ed. Oxford Univ. Press, (Clarendon), London and New York, 1974.

4.1. COMPOSITIONAL STUDIES

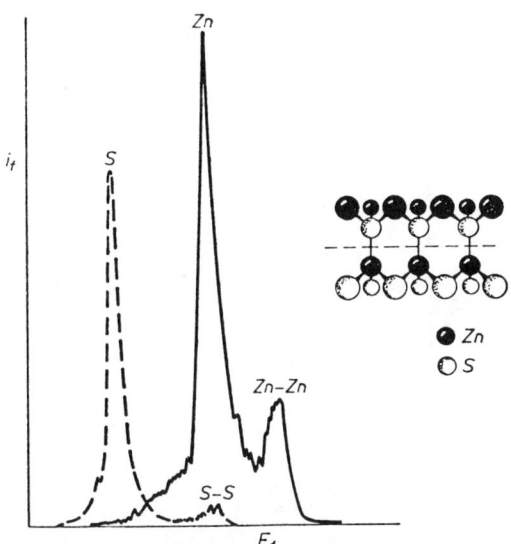

FIG. 12. Detected ion current as a function of the scattered Ne$^+$ energy for Ne$^+$ incident on two corresponding faces of a ZnS crystal. The (111) face consists entirely of Zn atoms while the ($\overline{1}\overline{1}\overline{1}$) face consists exclusively of S atoms. (From Brongersma et al.[43])

degree of which depends on the incidence and scattering angles. A dramatic example is the absence of backscattering signals for a surface highly contaminated with hydrogen. In such a case, the hydrogen overlayer completely shadows the substrate atoms, and the single binary collision of an ion with hydrogen cannot result in scattering angle of $\theta > 90°$. Another example is illustrated in Fig. 12 for 1-keV Ne$^+$ ions specularly scattered from a zinc sulphide polar crystal with $\theta = 45°$. Consistent with the theoretically expected structure,[43] only Zn peaks are seen from the (111) face, while the ($\overline{1}\overline{1}\overline{1}$) face yields only S peaks. For each face, the smaller peak corresponds to double scattering. The exact energy of such a peak depends on the path of the ion that is determined by the ion–atom scattering potential and the relative position of the closest atomic neighbor. Double and plural scattering has been successfully interpreted in terms of sequential binary collisions,[50] and the associated features contain information on the two-body potential and the geometric details of the surface, including lattice defects.

4.1.1.4. Applications of LEIS. The specific sensitivity to the outermost surface atoms makes the LEIS technique an ideal tool in studies where the composition of the outermost layer is important, such as in catalysis, adhesion, and cathodic emission. Most applications have taken advantage of the qualitative information obtainable from few simple measurements. An example, taken from Brongersma and co-workers,[41] is shown in Fig. 13,

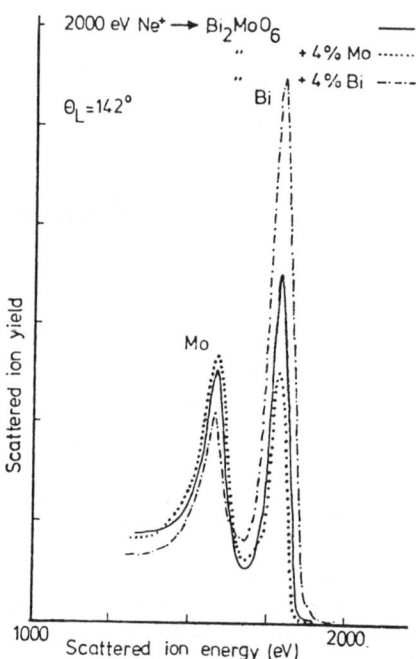

FIG. 13. Yield of 2000-eV Ne+ scattered from a bismuth molybdate (Bi_2MoO_6) catalysts having small differences in the bulk compositions. (From Brongersma et al.[41])

where Bi_2MoO_6 powder catalysts were examined with a 2000-eV Ne+ beam. Charge buildup at the insulator sample was alleviated by simultaneously flooding the sample with electrons. The addition of small amounts of MoO_3 (4 at. % Mo) and Bi_2O_3 (4 at. % Bi) to Bi_2MoO_6 is seen to produce pronounced change in the surface composition. The strong dependence of the catalytic activity on small changes in the bulk composition may thus be understood. It is known, for example, that the oxidation of propene to acrolein is enhanced by the addition of a small amount of MoO_3, but the opposite effect is observed for Bi_2O_3.

An application in which the scattering cross section and neutralization probability do not enter in the quantitative analysis is in the study of absorption or desorption of one monolayer or less of a foreign species on a clean substrate. As seen in Fig. 14 the time dependences of the elastic scattering signals have a character reminiscent of radioactive buildup and decay. In the absorption case, the buildup of the absorbate shadows the signals from the substrate, but in the impact desorption of the absorbate the intensity of the signals from the substrate increases. If the reasonable assumption is made that the probability of a scattered primary ion leaving as

4.1. COMPOSITIONAL STUDIES

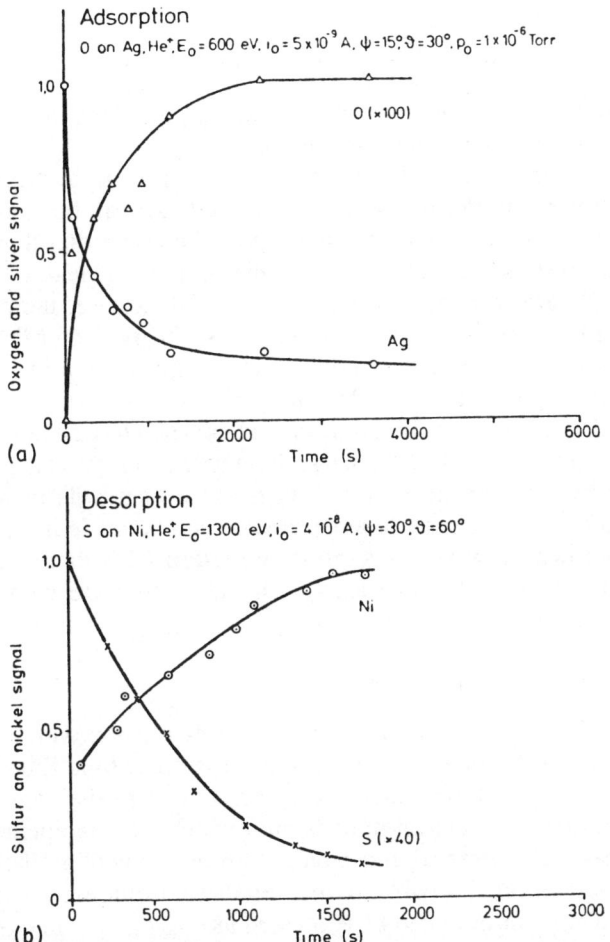

FIG. 14. Time dependence of the backscattering signals in (a) adsorption of O on Ag and (b) desorption due to He+ ion bombardment of S on Ni. (From Taglaver et al.[57])

an ion does not vary with the coverage and if the process is dominated by either absorption or desorption, the extraction of the corresponding cross section is straightforward. For example, in the ^{3}He$^+$ ion impact desorption of O absorbed on Ni (110) studied by Taglauer et al.[57] over the energy range of 700–1600 eV, the absorbate intensity $I_A(t)$ and substrate intensity $I_S(t)$ were shown to follow, respectively,

$$\ln[I_A(t)/I_A(0)] = -i_0\sigma_D t + \text{const} \qquad (4.1.17)$$

[57] E. Taglauer, G. Marin, and W. Heiland, *Appl. Phys.* **13**, 47 (1977).

and

$$\ln[1 - I_S(t)/I_S(\infty)] = -i_0 \sigma_D t + \text{const.} \tag{4.1.18}$$

For a known constant beam current density i_0, the slope in a semilog plot gives directly the desorption cross section σ_D.

With low-energy ion bombardment, the removal of atoms from the target by desorption or sputtering cannot be avoided, but can be controlled to a degree. For example, the use of a lighter ion like He$^+$ rather than a heavier one like Ar$^+$ results in a significantly smaller sputtering yield. Indeed, in the case of Ar$^+$ scattering from an oxidized Cu–Nb surface, the background from secondary ions is so large that the elastic Cu and Nb peaks cannot be observed in contrast to the scattering of He$^+$.[40] Such a condition would be favorable for analysis by secondary ion mass spectrometry, which is discussed in Section 4.1.4.2. In situations where the elastic scattering yields dominate over the sputtered ion yields, depth profiling can be achieved essentially by monitoring the time dependence of the elastic yields. The conversion of the time scale to a depth scale can be made if the sputtering yields are known. However, as is shown in Section 4.1.4, there are a number of problems associated with such a conversion that are common to all sputter etch techniques.

4.1.2. Nuclear Reaction Analysis

The great variety of nuclear reaction events offers a wide selection of highly specific techniques for the analysis of isotopic (and thus elemental) compositions in a sample. Such events are of course the consequence of close encounters of nuclear dimensions, which, for an incident ion of nuclear charge $Z_1 e$ directed at a target atom of $Z_2 e$ with a relative kinetic energy E, are more probable if the repulsive Coulomb barrier is surmounted. This condition can be expressed as

$$E \gtrsim 1.44 Z_1 Z_2 / r, \tag{4.1.19}$$

with r being the interaction radius in femtometers (10^{-15} m), and E in MeV. However, because of the finite probability for barrier penetration, nuclear reactions can proceed even if this condition is not satisfied; the reaction rates will generally be small except in instances where compound nuclear resonance conditions are satisfied or the target has a low-lying level that connects to the ground state with a large Coulomb-excitation amplitude.

With the more commonly available small accelerators producing ion beams of a few MeV, nuclear reaction analysis (NRA) is therefore usually practical only for the lighter elements. The relative insensitivity to heavy elements is a positive feature particularly in the analysis of light impurities in a heavy element substrate. Indeed, it is precisely this area which the other

4.1. COMPOSITIONAL STUDIES

"nondestructive" techniques do not adequately address. Thus NRA is complimentary to particle induced x-ray emission in impurity concentration studies and to RBS in depth-profiling analysis.

Nuclear reaction analysis may be broadly categorized under three headings for the purpose of discussion. These are (1) charged-particle activation analysis (CPAA), in which the ion beam is merely used to produce specific radioactive nuclei whose characteristic decays are followed after the bombardment is stopped; (2) prompt radiation analysis (PRA), in which the characteristic reaction products are detected during bombardment usually at a fixed energy; and (3) resonant reaction analysis (RRA) in which the energy shift and broadening of an otherwise isolated, narrow, compound nuclear resonance cross section measurement provides the depth-profile information of the corresponding isotope. There are of course many variations within these groupings. Since an excellent detailed review has been given by Wolicki,[58] the discussion that follows is brief and directed at the salient features. Literature on a variety of applications can be found in the proceedings of various conferences on ion beam analysis.[19-22]

4.1.2.1. Charged-Particle Activation Analysis. Since its discovery, radioactivity has been investigated for a number of purposes, and consequently many of the measurement techniques have become highly developed. For compositional study, neutron activation analysis (NAA) is perhaps the most familiar. The principles underlying CPAA differ from those of NAA only in the consequences of the use of an ion beam instead of neutrons to produce radionuclides. The range R of an ion beam with an incident laboratory energy of E_0 is finite, and the ion beam can produce radioactive nuclei only within this range. The total number Y of a particular species produced is then

$$Y = N \int_0^R n(z)\sigma(z) \, dz, \qquad (4.1.20)$$

where N is the number of incident beam particles and $n(z)$ and $\sigma(z)$ are, respectively, the pertinent number of target atoms per unit volume and the reaction cross section at a penetration depth of z. For materials with a uniform composition, n is independent of z, and one may write

$$Y = nN\bar{\sigma}R, \qquad (4.1.21)$$

where

$$\bar{\sigma} = \frac{1}{R} \int_0^R \sigma(z) \, dz = \frac{1}{R} \int_{E_0}^0 \frac{\sigma(E)}{-[dE/dz]} \, dE. \qquad (4.1.22)$$

---

[58] E. A. Wolicki, in "New uses of Ion Accelerators," (J. F. Ziegler, ed.), p. 159. Plenum, New York, 1975.

Although the evaluation of the thick target cross section $\bar{\sigma}$ requires a knowledge of $\sigma(E)$, it is nearly independent of the target material.[59] This is because at energies where $\sigma(E)$ is nonnegligible, the stopping power $[dE/dz]$ is dominated by electronic processes that, for a given ion in different materials, have a near universal dependence on the ion velocity with a scaling factor that is compensated for by the range parameter R. A catalog of thick target cross sections may thus be compiled, either by direct integration of Eq. (4.1.22) or by measurements using simple standards, with which the unknown amount of material in the sample to be analyzed may be calibrated by taking simple ratios.

Applications of CPAA have not been confined to any accelerator type. However, the CPAA of heavy elements using high-energy ion beams offers no general advantage over NAA. As already mentioned, the analysis of light elements using small accelerators, which is facilitated by a less interfering background, is particularly suited for surface layers. Moreover, since unambiguous analysis requires the monitoring of the characteristic decay curves, the easily implementable feature of pulsed beams with variable pulse durations may prove advantageous in following decays as short as a few nanoseconds. Most applications to date have generally focused on longer-lived radionuclide produced in p, d, t, and ^{3}He bombardments and on the detection of γ rays including those from the annihilations of positrons from β^+ emitters.

4.1.2.2. Prompt Reaction Analysis. Analysis by the detection of prompt reaction signals generally relies on the simplicity of the two-body final-state kinematics[11] in which the application of the energy and momentum conservation laws to the observed energy of one of the two final products is sufficient to identify the reaction unambiguously. Within this class of reactions the preferred ones are clearly those having a reasonably large cross section or those for which the measurement is simple to implement and interpret. Bombardment with simple ions and detection of γ rays or simple particles generally fit into this class. Typically, the energy spectrum will contain well-separated sharp peaks corresponding to the discrete low-lying states of the heavy final product. The simplicity alluded to lies in the simple detection scheme of one detector. Interferences from the more probable elastic scattering events are avoided by blocking out the charged particles in the case of γ or n measurements or, in the case of charged particle detection, by choosing a reaction with a positive Q value which places the reaction peak above the elastic peak.

In some instances several reactions can be used to analyze one particular species. The choice is clearly dictated by the available equipment, the nature

[59] E. Ricci and R. L. Hahn, *Anal. Chem.* **39**, 794 (1967).

4.1. COMPOSITIONAL STUDIES

of the material to be analyzed, and the degree of detailed information desired. All PRA can potentially yield depth-profile information because the energy of the detected particle directly reflects the energy-loss phenomena associated with the beam particles and, if charged, the detected particles also. The depth resolution attainable, however, differs for the different reactions.

At a fixed bombarding energy, considerations entering into the conversion of energy loss to the depth at which the reactions occur are analogous to those for RBS analysis discussed in Sec. 4.1.1.1, but due consideration must be given to the different nature of the particles detected. In the case of uncharged particles, such as in (p,γ) and (d,n) reactions, only the energy loss of the incident ion need be considered. In the charged-particle case, such as in (d,p) and $(^3He,\alpha)$ reactions, not only is the outgoing energy determined by a different kinematical relation, but the different charged species also entails a different set of stopping powers. The depth resolution is similarly determined by the overall detection resolution, which includes contributions from the incident energy, finite solid angle, detection system, and straggling effects.

The depth profile reflected by the peak shape in the energy spectrum is a convolution of the experimental factors and the reaction cross section. Extracting such a profile can be accomplished either by a deconvolution procedure,[60] by simulating the experimental data with a model-concentration profile, or by comparison with a standard in which the species of interest is uniformly distributed.[61] Whatever the procedure, errors can be minimized if the cross section varies smoothly with energy. Reactions with a sharp resonant cross section can be more profitably utilized in a slightly different approach described in the next section.

The majority of PRA applications has centered on the profiling of low-Z-elements.[58,62] In general, the use of photonuclear reactions is hampered by the relatively small cross section, whereas that of neutron-producing reactions suffers from the difficulties associated with detecting energetic neutrons with a reasonable energy resolution. These reactions also inherently yield a poorer depth resolution than reactions yielding charged particles at the same penetration depth because of the smaller overall energy loss associated with the reaction. However, the smaller energy loss can be used advantageously to probe deeper layers for which the increasing importance

---

[60] M. Hufschmidt, W. Möller, V. Heintze, and D. Kamke, in "Ion Beam Surface Layer Analysis" (O. Meyer, G. Linker, and F. Käppeler, eds.), Vol. 2, p. 831. Plenum, New York, 1976.
[61] D. W. Palmer, *Nucl. Instrum. Methods* **38**, 187 (1965).
[62] J. R. Bird, *Nucl. Instrum. Methods* **168**, 85 (1980).

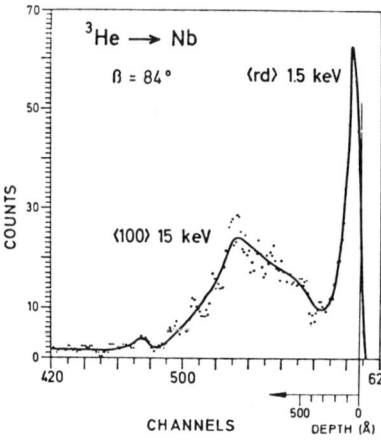

FIG. 15. Energy spectrum corresponding to the ^{3}He(d,α)^{1}H reaction for incident 1-MeV D$_2^+$. The target consists of a Nb single crystal implanted with 15-keV ^{3}He$^+$ in a $\langle 100 \rangle$ direction (dose = 1.5×10^{17} He cm^{-2}) and with 1.5-keV ^{3}He$^+$ in a random direction (dose = 6×10^{16} He cm^{-2}). (From Eckstein et al.[63])

of energy-straggling phenomena places a lesser demand on the detector resolution.

An example of depth profiling by the detection of charged products is shown in Fig. 15. The sample in this case is a niobium single crystal implanted with ^{3}He in a random direction at 1.5 keV and again in a $\langle 100 \rangle$ direction at 15 keV. A 1-MeV D$_2^+$ beam directed at normal incidence to the Nb crystal surface was used together with a 15-keV-resolution solid state detector positioned to observe α particles coming off at an angle $\beta = 84°$ relative to the surface normal. The spectrum, corresponding to the ^{3}He(d,α)^{1}H reaction from the two implanted layers, clearly shows the two separated profiles. There is virtually no interference from Nb. Eckstein et al.[63] quoted an achievable depth resolution of 10–20 Å for this case.

Finally, although the heavy reaction product left in an excited state may decay promptly—such as, for example, in (p,p'γ) and (p,$\alpha\gamma$)—the detection of such prompt decay at a fixed bombarding energy carries little depth information.

4.1.2.3. Resonant Reaction Analysis. An isolated compound nuclear resonance exhibited as a sudden enhancement in the cross section over a narrow interval of bombarding energy provides an alternative and sometimes superior method for depth profiling. The basic ideas underlying the technique can be seen by idealizing the resonance as a delta function $\delta(E - E_r)$ and neglecting straggling effects. The particular reaction yield will then result if the beam energy is exactly $E = E_r$ and with a strength proportional

---

[63] W. Eckstein, R. Behrisch, and J. Roth, in "Ion Beam Surface Layer Analysis" (O. Meyer, G. Linker, and F. Käppeler, eds.), Vol. 2, p. 821. Plenum, New York, 1976.

4.1. COMPOSITIONAL STUDIES

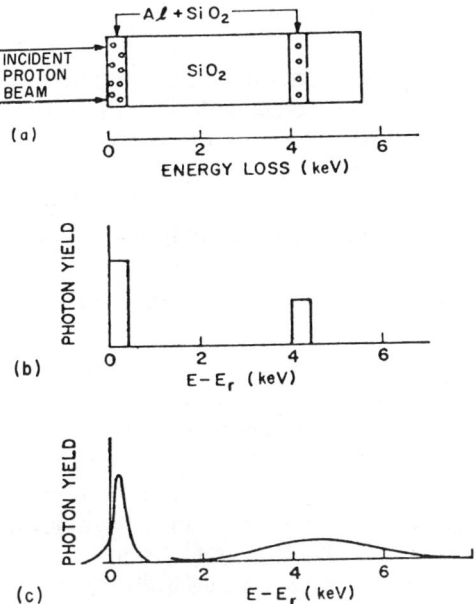

FIG. 16. Schematic diagrams of a SiO_2 sample with two thin layers of Al (a) and the corresponding idealized (b) and realistic (c) curves of photon yield from the $^{27}Al(p,\gamma)^{28}Si$ resonance reaction. (From Wolicki.[58])

to the areal density of the relevant target atoms, as will be the case if these atoms all lie on the surface. For atoms in a deeper layer z, the corresponding yield will result only if the bombarding energy is raised to $E = \Delta E + E_r$, where ΔE is the energy loss suffered by the beam particles penetrating into the depth z. The situation is illustrated schematically by Fig. 16b for a resonance in the $^{27}Al(p,\gamma)^{28}Si$ reaction for a target whose composition is indicated in Fig. 16a. Folding in the beam energy spread, finite resonance width, and straggling effects will broaden the profile as depicted in Fig. 16c. Unfolding techniques that determine the distribution profiles have been described by a number of authors.[58,64] In general the depth resolution is determined by the resonance width for the surface layer and by straggling effects for the deeper layers. The total probing depth for which the analysis remains simple is governed by the interval of bombarding energies in which the single resonance dominates the cross section.

[64] D. J. Land, D. G. Simons, J. G. Brennan, and M. D. Brown, in "Ion Beam Surface Layer Analysis" (O. Meyer, G. Linker, and F. Käppeler, eds.), Vol. 2, p. 851. Plenum, New York, 1976.

Depth profiling with the resonance technique was first reported by Amsel and Samuel[65] using the ^{27}Al(p,γ)^{28}Si resonance at 992 keV and ^{18}O(p,α)^{15}N resonance at 1167 keV in a study of anodic oxidation mechanisms. Since then, many other resonances including those in the elastic channel such as in ^{16}O(α,α)^{16}O at 3048 keV have been used[66] for a variety of applications. The technique is flexible in the sense that the signature may also be manifested in the secondary reaction products, such as in the ^{19}F(p,$\alpha\gamma$)^{16}O resonance at 1375 keV where the detection of the secondary γ rays of fixed energy is more convenient.

A notable application of RRA is in the elucidations of the role of hydrogen in various solid state phenomena.[67] There exist very few techniques for the microscopic profiling of hydrogen, and, within these limited choices, RRA can perhaps provide the most detailed information without destroying the sample. The high depth sensitivity of RRA in this case is achieved by reversing the usual procedure and bombarding the sample with a heavy ion beam to produce the known proton-induced resonances. The higher stopping power for heavy-ion beams magnifies the depth scale. A particular example is the ^{1}H(^{15}N,$\alpha\gamma$)^{12}C reaction with a corresponding resonance at 6385 keV. In a comparative study[68] of a number of ion beam techniques for hydrogen profiling in which identically prepared standards were used, this reaction yielded the highest depth resolution of ~ 40 Å for hydrogen at a depth of 4000 Å in Si. The standard consists of a high-purity Si wafer implanted with 10^{16} H cm^{-2} at 40 keV, and the 4.43-MeV γ-ray yield observed[69] as a function of ^{15}N bombarding energy is reproduced in Fig. 17. The raw data as they stand already reflect the theoretically expected implanted hydrogen profile. Hydrogen contamination on the surface, which is almost unavoidable, is seen to pose no problem because of the superior depth resolution.

4.1.3. Particle-Induced X-Ray Emission

In the collision of an ion with an atom, there is a significant probability for ejecting electrons from the inner shells of the colliding partners. The subsequent filling of such a vacancy can proceed by either x-ray emission or

---

[65] G. Amsel and D. Samuel, *J. Phys. Chem. Solids* **23,** 1707 (1962).

[66] G. Mezey, J. Gyulai, T. Nagy, and E. Kotai, *in* "Ion Beam Surface Layer Analysis" (O. Meyer, G. Linker, and F. Käppeler, eds.), Vol. 1, p. 303. Plenum, New York, 1976.

[67] W. A. Lanford, *Proc. 3rd Int. Conf. Ion. Beam Anal.*, p. 1. North-Holland Publ., Amsterdam, 1978.

[68] J. F. Ziegler *et al.*, *Proc. 3rd Int. Conf. Ion Beam Anal.*, p. 19. North-Holland Publ., Amsterdam, 1978.

[69] W. A. Lanford, H. P. Trautvetter, J. F. Ziegler, and J. Keller, *Appl. Phys. Lett.* **28,** 566 (1976).

4.1. COMPOSITIONAL STUDIES

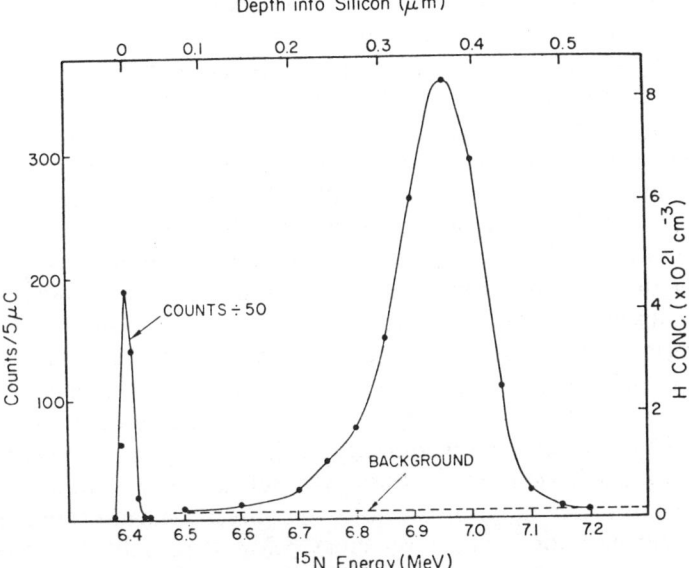

FIG. 17. The hydrogen profile of a silicon wafer implanted with 10^{16} H cm^{-2} at 40 keV deduced from the resonant ^{1}H(^{15}N,$\alpha\gamma$)^{12}C reaction. (From Lanford et al.[69])

radiationless Auger transition, both of which are characteristic of the emitting ion. The characteristic emission is of course the basis for the better-known techniques of x-ray analysis and Auger electron spectroscopy (AES) by electron and photon excitation. The impetus for the analysis by particle-induced x-ray emission (PIXE) can be traced to the expectation of lesser background signals, higher characteristic x-ray cross sections and the ready availability of energy-dispersive semiconductor x-ray detectors of sufficient resolution to resolve the K x rays of all but the lighter elements, as well as the L x rays of the heavier elements. The international conference at Lund[70] devoted solely to the subject of PIXE attests to its increasing acceptance as a sensitive analytical technique particularly for the detection of trace amount of impurities.

For analysis, the use of simple ion like H$^+$ and He$^+$ to excite K or L x rays is favored for a number of reasons. Perhaps the most important is the fact that the x-ray production mechanisms for such ions in solids are reasonably well understood. The creation of a vacancy in the inner shell of a target atom in this case is dominated by the Coulomb interaction between the bare nucleus of the projectile and the corresponding target electron, as evidenced

---

[70] *Proc. Int. Conf. Part.-Induced X-ray Emiss. Its Anal. Appl., Lund, 1976*, published in *Nucl. Instrum. Methods* **142**, (1977).

by the remarkable agreement over six orders of magnitude between the experimental K-vacancy cross sections and the predictions based on direct Coulomb ionization theories.[71] Not only is the perturbation due to the attached electron in the projectile small, but such projectiles are stripped bare of electrons when traversing solids with $E \gtrsim 100$ keV amu^{-1}. Additionally, the probability of creating multiple inner vacancies in a single collision is small and is adequately described within the context of multiple Coulomb ionization.[72] Since the fluorescence yield ω, which expresses the branching ratio for a particular x-ray emission in the vacancy filling, depends on the number of inner shell vacancies,[73] the simple vacancy produced by the simple ions means that the uncertainties in the ω are minimized. Chemical-bonding effects are also insignificant except perhaps for the lighter elements.

At a fixed incident velocity, the direct Coulomb ionization cross section for a given target atom is proportional to the square of the projectile's nuclear charge Z_1^2 and this would suggest higher analytical sensitivities with the use of heavier-ion beams. Moreover, substantial enhancement in the cross sections over those predicted by the direct ionization theories have been observed[71] for incident velocities below that of the corresponding target electron. Such enhancements have been attributed to the capture of target electrons into bound states of the projectile[74] and to ionization through electron promotion by means of the formation of transient molecular orbitals.[75] For compositonal analysis, the larger cross section must, however, be weighed against other factors. The enhancement is selective because it depends on the particular combination of colliding species and on their electronic structure at the time of collision, which, for heavy ions moving inside a solid, is only known in a statistical sense due to the many prior collisions.[76] This selectivity may be advantageous for the analysis of a particular element but introduces complications when the same beam is used for multielement analysis. Another complicating aspect can be seen in Fig. 18, where the Al K_α x-ray region has been scanned with a high-resolution curved crystal spectrometer during bombardment with e$^-$, H$_2^+$, and Ne$^+$ beams. One-electron transitions from an initial state having a single K and n L vacancies are denoted by KL^n. Thus KL^0 is the so-called diagram line, and

---

[71] D. H. Madison and E. Merzbacher, in "Atomic Inner Shell Processes" (B. Crasemann, ed.), Vol. 1, p. 1. Academic Press, New York, 1975.
[72] N. Cue, *Proc. 4th Conf. Appl. Small Accel., Denton, Tex.* (J. L. Duggan and I. L. Morgan, eds.), p. 299. IEEE Publ. No. 76CH 1175-9 NPS. IEEE, New York, 1976.
[73] C. P. Bhalla, *Proc. 4th Conf. Appl. Small Accel., Denton, Tex.* (J. L. Duggan and I. L. Morgan, eds.), p. 149. IEEE Publ. No. 76CH 1179-9 NPS. IEEE, New York, 1976.
[74] A. M. Halpern and J. Law, *Phys. Rev. Lett.* **31**, 4 (1973).
[75] M. Barat and W. Lichten, *Phys. Rev. A* **6**, 211 (1972).
[76] H. D. Betz, *Rev. Mod. Phys.* **44**, 465 (1972).

4.1. COMPOSITIONAL STUDIES

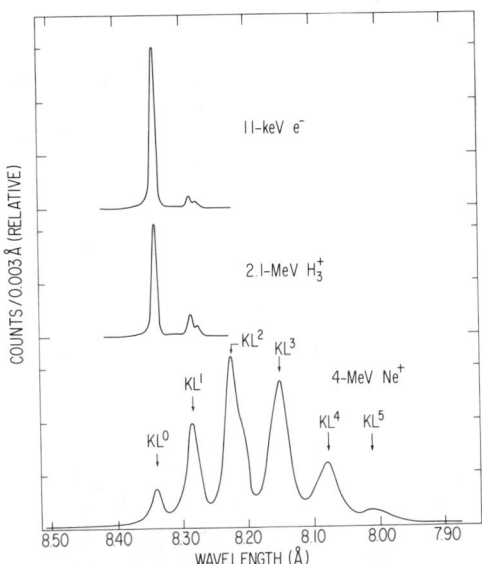

FIG. 18. High-resolution x-ray spectra obtained with a 4″-curved crystal spectrometer for e^-, H_3^+, and Ne^+ bombardments of an Al target.

its dominance in the cases of e^- and H_3^+ reflects the fact that the creation of only a single vacancy is likely. In the case of Ne^+, the shift of the intensity maximum away from the diagram line is a dramatic illustration of the comparatively large probability for multiple inner-shell ionization. The interpretation of the observed satellite structure must in addition take into account the variation in the fluorescence yields of the KL^n satellites and the vacancy-rearrangement process prior to the x-ray emission.[72] Observed chemical-bonding effects upon the satellite structure have been attributed[77,78] to the latter type of processes. With the poorer-resolution semiconductor detectors, the satellite structure will of course be merged into a single peak. Nevertheless, some sort of average of the effects described above will enter into the conversion of the x-ray intensity to the number of primary collision events. In most instances, the lack of basic information from which such averages can be obtained hampers the analysis with heavy-ion beams.

In PIXE as well as in most microanalysis techniques, the ultimate trace-element detection sensitivity in the case of no interference of characteristic lines, is governed by the background radiation, which arises mainly from the interactions of the probing beam with the most abundant elements in the

[77] F. Hopkins, A. Little, N. Cue, and V. Dutkiewicz, *Phys. Rev. Lett.* **37**, 1100 (1976).
[78] R. L. Watson, T. Chiao, and F. E. Jenson, *Phys. Rev. Lett.* **35**, 254 (1975).

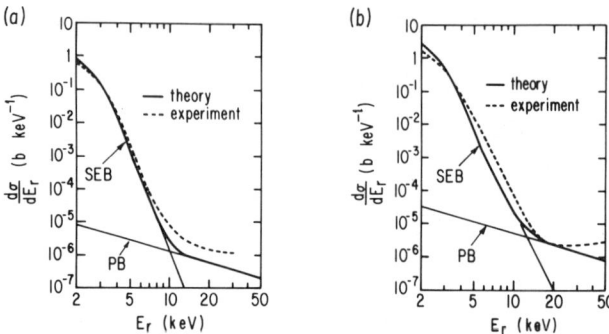

FIG. 19. Continuum background radiation at 90° in the 2-MeV proton bombardment of (a) carbon and (b) aluminum targets. (From Folkman et al.[80])

sample matrix. The background processes in PIXE have been identified[79] as (1) bremsstrahlung of secondary electrons (SEB), (2) projectile bremsstrahlung (PB), (3) Compton scattering of γ rays (CS), (4) radiative electron capture (REC), and (5) quasimolecular transition (MO).

Bremsstrahlung radiation arises from the deceleration of the charged particle in the field of the nucleus and is more probable for electrons than for other ions of the same velocity because the cross section varies inversely with the square of the particle mass. Since in the close encounters necessary to produce PB there is much larger probability for ionization of the target atoms, in which electrons are invariably ejected, SEB will dominate the low-energy continuum background radiation. This is shown in Fig. 19 for 2-MeV proton bombardment of a carbon and an aluminum matrix. The upper cutoff for SEB is due to the fact that secondary electrons with energy exceeding the value acquired by a free electron in a head-on collision with the incident projectile are increasingly difficult to produce because the ionization involves more tightly bound orbitals. Note that because the secondary electron yield is proportional to the ionization cross section, the larger characteristic x-ray yield with the heavier-ion beam will correspondingly be accompanied by a larger SEB background. A suggestion has been made[80] to choose a projectile such that in the PB region, the dipole radiation term, which is proportional to $(Z_1/A_1 - Z_2/A_2)^2$, cancels to zero. However, in practical situations, where a few MeV-per-amu ion beams are used, γ rays from nuclear reactions will be present and their Compton scattering inside the detector can produce a background level comparable to that for projectile bremsstrahlung.

[79] F. Folkmann, in "Ion Beam Surface Layer Analysis" (O. Meyer, G. Linker, and F. Käppeler, eds.), Vol. 2, p. 695. Plenum, New York, 1976.

[80] F. Folkmann, C. Gaarde, T. Huus, and K. Kemp, *Nucl. Instrum. Methods* **116**, 487 (1974).

The use of heavier ions also has the further disadvantage that the region corresponding to x-ray emission from the projectile will be masked. The projectile's characteristic lines are intense because the projectile can interact with all of the target constituents along its path, and additional structure can be present as a result of specific interactions. At high velocity, a vacancy in the projectile produced in a prior collision can be filled by radiative capture of a target electron. For the more probable capture of a free or loosely bound electron of velocity $v_e \ll v_1$, where v_1 is the projectile velocity, the radiated photon energy can be written as

$$h\nu \sim E_r + E_n + m_e \mathbf{v}_1 \cdot \mathbf{v}_e. \quad (4.1.23)$$

Here E_r is the energy of the target electron relative to the rest frame of the projectile that has a vacancy in a level n corresponding to an ionization energy E_n and m_e is the electron rest mass. If the vacancy is in the K shell, REC photons will lie above the K_β line of the projectile as can be seen in Fig. 20[81] for the bombardment of C and Al with S ions. According to Eq. (4.1.23) increasing the bombarding energy will shift the energy upward and broaden the REC distribution through the E_r and $\mathbf{v}_1 \cdot \mathbf{v}_e$ terms, respectively. This is

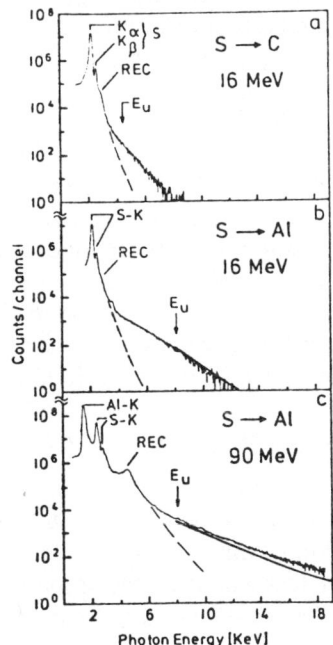

FIG. 20. X-ray spectra induced by sulfur ions bombarding 100-μg cm^{-2} C targets, corrected for absorption effects in the detector. The dashed and solid curves are calculated REC and MO tails, respectively. (From Betz et al.[81])

[81] H.-D. Betz, F. Bell, H. Panke, W. Stehling, E. Spindler, and M. Kleber, *Phys. Rev. Lett.* **34**, 1256 (1975).

seen in the two spectra for S on Al. The REC distribution should also reflect the momentum distribution of the electron being captured, but quantitative procedures for extracting such information remain to be worked out. Another notable feature in the spectra is the exponential tail above the REC distribution which has been attributed to quasi-molecular (MO) transitions. Within the frame-work of the electron-promotion theory[75] in which the levels of the separated atoms (SA) are correlated to the united atom (UA) levels as a function of internuclear separation, a K vacancy in the heavier partner of the SA pair is carried into the K shell of the UA. The filling of this vacancy with an electron from a higher orbital can occur throughout the range of the internuclear separation in which the K binding energy varies from the SA to UA limit and thus yields a band of MO radiations. That these MO transitions extend beyond the UA limit, indicated as Eu in the spectra, can be attributed[82,83] to the dynamical broadening caused by the finite interaction time. How such a K vacancy is created in the first place remains an active area of investigation. The richer features associated with the heavier ions, although of specific interest for the physics of collisions, will clearly complicate the analysis of impurity concentrations.

Most applications of PIXE have been made with light-ion beams and therefore much of the quantitative aspects of the analysis technique developed are based on such excitations. With protons, detection sensitivities of $10^{-7}-10^{-6}$ g in concentration or $10^{-16}-10^{-9}$ g in mass quantity have been quoted.[79] Since the energy dependence of the x-ray-production cross section is known and the stopping power is well characterized for light ions, elemental depth profiles can be obtained by observing the change in the x-ray yield as a function of the trajectory of the projectile.[84] The depth sensitivity, however, does not approach those achieved in RBS or NRA techniques, for example. Finally, because of the steep decrease in the fluorescence yield ω and the increased attenuation in the absorber for the x rays with decreasing Z, the detection of the Auger electrons rather than the x rays will be more sensitive in the analysis of light elements.

4.1.4. Secondary Particle Emission

When a solid is bombarded with an ion beam, there are emissions which cannot be described in terms of the immediate consequences of a single ion–atom collision event. When electrons are emitted such emission is called secondary electron emission (SEE); the emission of all heavier particles is encompassed by the term sputtering.

[82] W. Lichten, *Phys. Rev. A* **9**, 1458 (1974).
[83] J. H. Macek and J. S. Briggs, *J. Phys. B* **7**, 1312 (1974).
[84] L. C. Feldman and P. J. Silverman, *in* "Ion Beam Surface Layer Analysis" (O. Meyer, G. Linker, and F. Käppeler, eds.), Vol 2, p. 735. Plenum, New York, 1976.

4.1. COMPOSITIONAL STUDIES

Secondary electron emission has been extensively investigated at low bombarding energies ($\lesssim 1$ keV), where the features have been successfully interpreted in terms of the neutralization of the incoming ions by the electrons in the solid.[85] The dominant mechanism is the so-called Auger neutralization in which the incoming ionic vacancy is filled by an electron from the valence or conduction band of the solid with the simultaneous ejection of another electron from the same band during close approach of the ion to the solid surface. Since the intensity distribution as a function of the ejected-electron energy will in this case be governed by the local density of states in the surface region of the solid, the study of such distributions forms what is now called ion-neutralization spectroscopy (INS).[86] At higher bombarding energies, SEE can also arise from the direct interaction of the electrons in the solid with the incoming ion as a whole, but such processes are not well understood.[87] Clearly, SEE can provide important information on ion–solid interactions and the electronic structure of the solid, but the features are not sufficiently discriminating to form the basis for compositional analysis.

In sputtering, heavy particles in the form of atoms, ions, and clusters are ejected from the solid surface, and the relative intensities of these particles clearly reflect the composition of the surface layer under ion bombardment. In the following, the sputtering phenomenon will be reviewed briefly to set the stage for the discussion of the analysis techniques called secondary ion mass spectrometry (SIMS) and surface–composition analysis by neutral and ion-induced radiation (SCANIIR). The latter entails the detection of the characteristic optical emission from the sputtered particles.

4.1.4.1. *Sputtering.* Observed sputtered species range from atoms to molecules to clusters and these can be neutral or have positive or negative charge. Whatever the mass and charge distributions may be, the fundamental fact remains that the basic constituents of the solid are atoms, and an understanding of the sputtering mechanisms must first address the question of how such atoms are ejected without regard to their final configurations. The number of target atoms ejected per incoming ion is called the sputtering yield S. The sputtering yield of solids has been studied both experimentally and theoretically for a number of years and reviewed by a number of authors.[88-91]

[85] H. D. Hagstrum, *Phys. Rev.* **96,** 336 (1954).
[86] H. D. Hagstrum, *J. Vac. Sci. Technol.* **12,** 7 (1975).
[87] K. Dettman, *Proc. Conf. Interact. Energ. Charged Part. Solids, Istanbul* (A. N. Goland, ed.), p. 81. BNL No. 50336. Brookhaven Natl. Lab., Upton, New York, 1973.
[88] I. S. T. Tsong and D. J. Barber, *J. Mater. Sci.* **8,** 123 (1973).
[89] J. S. Colligon, *Vacuum* **24,** 373 (1974).
[90] G. M. McCracken, *Rep. Prog. Phys.* **38,** 241 (1975).
[91] P. Sigmund, *in* "Inelastic Ion–Surface Collisions" (N. H. Tolk, J. C. Tully, W. Heiland, and C. W. White, eds.), p. 121. Academic Press, New York, 1977.

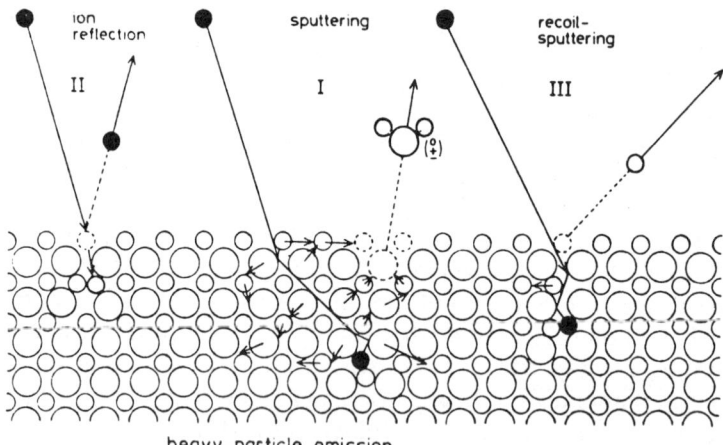

FIG. 21. Schematic of low-energy collisions leading to sputtering from a solid surface. (From Benninghoven.[92])

After the discovery that the sputtering yield depended on the incidence angle of the bombarding ions relative to the surface and that the average energy of the ejected atoms was much too large to be explained by evaporation, sputtering is now believed to be a consequence of atomic collision cascades. Referring to Figure 21,[92] a low-energy incident ion in its penetration of a solid can displace atoms from their lattice positions by collisions. The scattered ion and the displaced atoms, if sufficiently energetic, can in turn produce secondary displacements, which can produce further displacements and thus result in a collision cascade. Both the ion itself and the energetically displaced atoms have a finite probability of being scattered back toward the surface. If they leave the surface, they account for most of the sputtered energy, but constitute only a small fraction of the sputtered atoms. A majority of the displaced atoms have very low energies and small ranges and therefore can only get sputtered if they are located within a couple of atomic layers from the surface, but they constitute a major fraction of the sputtering yield because of their large number.

Collisions resulting in displacements of lattice atoms are most probable in the energy regime where the nuclear energy loss dominates. This fact and the relatively short range of incident ions with such an energy qualitatively explain[45] why the energy dependence of sputtering yield follows that of the nuclear stopping power for the same target material. At higher bombarding energies, S decreases because the penetrating ion must first lose energy by the dominant electronic excitation process, which results in little or no

[92] A. Benninghoven, *CRC Crit. Rev. Solid State Sci.* **6**, 291 (1976).

4.1. COMPOSITIONAL STUDIES

atomic displacement. When the ion is sufficiently slowed, the resulting collision cascade will lie at such a depth in which the energy of displaced atoms will be largely dissipated by the time they propagate back to the surface.

A detailed and comprehensive theory on sputtering of amorphous and polycrystalline monatomic solids was put forth by Sigmund.[93] Based on random collision processes in an infinite medium and the application of Boltzmann's equation and general transport theory, the yield calculation proceeded in four conceptual steps: (1) determining the amount of energy deposited by the energetic particles near the surface; (2) converting this energy into a number of low-energy displaced atoms; (3) determining how many of these atoms are transported to the surface, and (4) selecting those atoms that surmount the surface barrier. The theory has had considerable success in predicting many details of sputtering and absolute yields as a function of the energy, atomic number, and incidence angle of the bombarding ions.

Of particular interest in compositional analysis is the backsputtered yield, which, according to Sigmund, can be expressed for normal incidence as

$$S(E) = [3/(4\pi^2)]\alpha[4M_1M_2/(M_1 + M_2)^2](E/U_0) \quad (4.1.24)$$

for $E \lesssim 1$ keV, and

$$S(E) = 0.0420\alpha 4\pi Z_1 Z_2 e^2 a[M_1/(M_1 + M_2)]s_n(\epsilon)/U_0 \quad (4.1.25)$$

for $E > 1$ keV and heavy and intermediate-mass ions. The subscripts 1 and 2 refer, respectively, to the incident ion and target atom, α is a dimensionless factor dependent on the ratio M_2/M_1, U_0 is the surface binding energy, $a = 0.8853a_0 \ (Z_1^{2/3} + Z_2^{2/3})^{-1/2}$ with a_0 being the Bohr radius, $\epsilon = [M_2E/(M_1 + M_2)]/(Z_1Z_2e^2/a)$ is the reduced energy, and $s_n(\epsilon)$ is the reduced nuclear stopping cross section tabulated by Lindhard et al.[94] The degree of agreement between the predictions of Eqs. (4.1.24) and (4.1.25) and experimental results can be seen in Fig. 22 for inert gas ion bombardment of polycrystalline Cu. For not-too-oblique incidence, the dependence on incidence angle θ relative to the surface normal can be approximated by

$$S(E,\eta)/S(E,1) = (\eta)^{-f} \quad (4.1.26)$$

where $\eta = \cos\theta$ and f is a function of the mass ratio but has a near constant value of $\sim \frac{5}{3}$ for $M_2/M_1 \lesssim 3$. Figure 23 shows the prediction of the more accurate expression with the data for Ar+ bombardment of polycrystalline Cu. The agreement with data is excellent up to $\theta = 60°$. The experimentally

---

[93] P. Sigmund, *Phys. Rev.* **184**, 383 (1969).
[94] J. Lindhard, V. Nielsen, and M. Scharff, *Mat.-Fys. Medd.—K. Dan. Vidensk. Selsk.* **36**, No. 10 (1968).

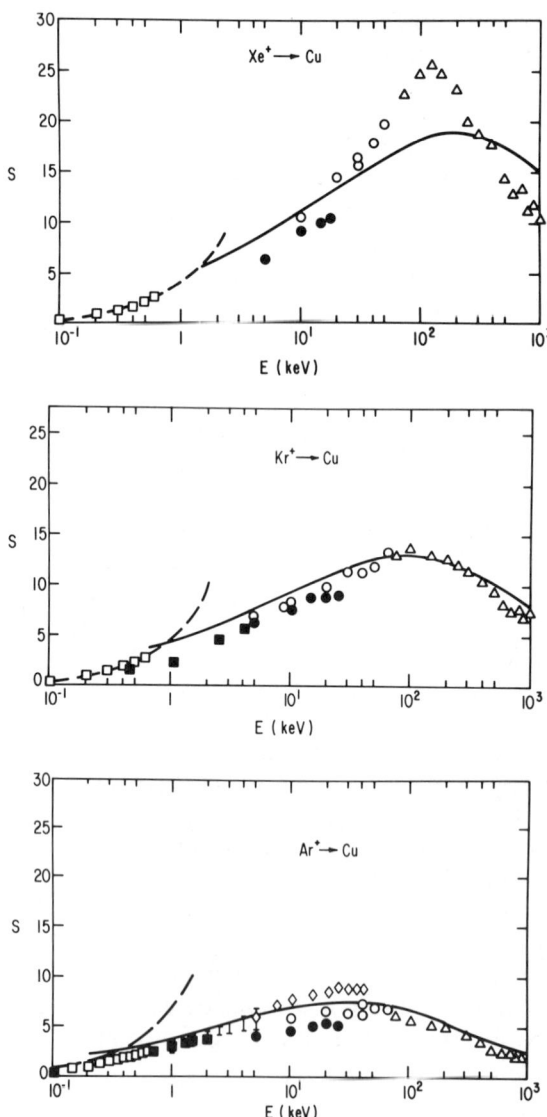

Fig. 22. Experimental and theoretical sputtering yields for Cu. The dashed and solid curves are based on Eqs. (4.1.24) and (4.1.25) respectively. (From Sigmund.[93])

observed maximum yield at 70°–80° and the decrease of the yield to zero at glancing incidence have been attributed to the repulsive action of surface atoms that prevents the ions from penetrating the solid and are not accounted for by the theory because of the assumption of an infinite medium.

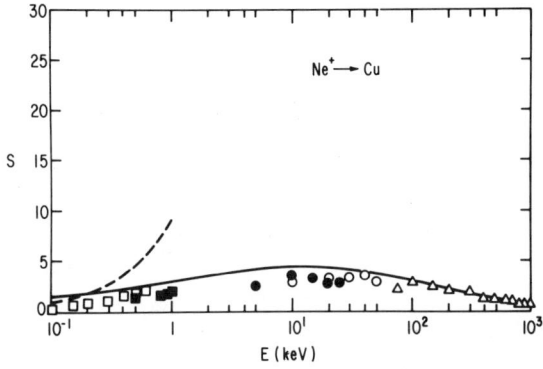

FIG. 22. (*continued*)

Studies of the energy and angular distribution of the sputtered atoms have provided additional insights on the sputtering mechanisms. In order to account for the observed energy distribution from polycrystalline targets, as shown for example in Figure 24,[95] a focused collision-chain mechanism was thought[96] to be necessary. However, the evidence to date suggests that this mechanism contributes only a small fraction to the sputtering yield.[90]

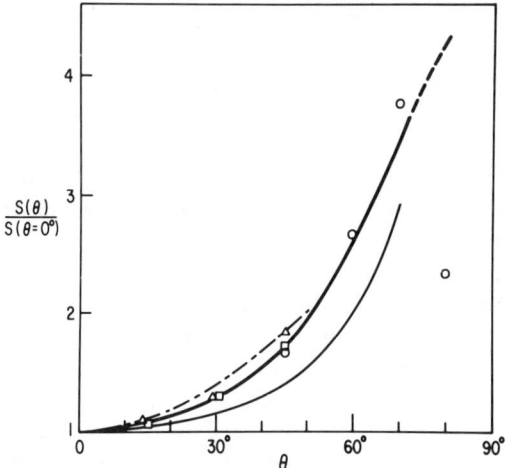

FIG. 23. Variation of the sputtering yield with angle of incidence for Ar^+ bombardment of polycrystalline copper. The solid curve is based on Eq. (4.1.26) while the thick solid curve is the more accurate prediction detailed in Sigmund.[93] The dash–dot curve together with the points are experimental data. (From Sigmund.[93])

[95] G. E. Chapman, B. W. Farmery, M. W. Thompson, and I. H. Wilson, *Radiat. Eff.* **13**, 121 (1972).
[96] M. W. Thompson, *Philos. Mag.* **18**, 377 (1968).

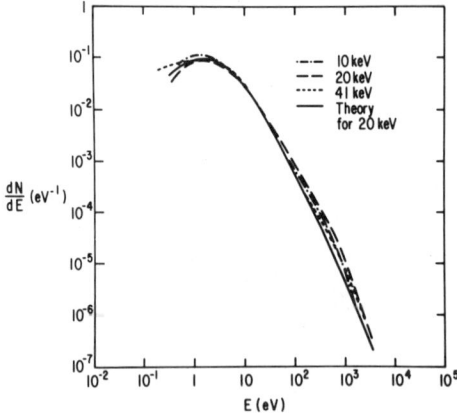

FIG. 24. Energy spectra for 10-keV, 20-keV and 41-keV argon on polycrystalline gold at room temperature compared with random cascade theory. (From Chapman et al.[95])

Further improvements on the random cascade model[97] would help to clarify the situation. The random cascade model predicts a cosine dependence for the angular distribution of the sputtered atoms,[93,96] and the observed deviation from this is likely to be largely due to the use of polycrystalline solids that may have preferential orientation of grains.

Bombardments of single crystals have shown significantly different orientation effects both with respect to the dependence on the incidence angle of the beam and the emission angle of sputtered atoms. In the former, an ion channeled between rows of atoms can penetrate deep into the crystal before undergoing a displacement collision. Thus the sputtering yield has deep minima in these directions. Elich et al.,[98] using the random cascade model and splitting the incident beam into a random and a channeled fraction, have obtained good agreement with experiment. In the case of angular distribution, sputtered atoms are ejected preferentially along close-packed rows and give rise to spots on a collector placed in front of the target. Although this may suggest the operation of a focused collision-chain mechanism, the surface-ejection model[99] based on the random collision cascade provides an equally acceptable explanation. In this case, the penultimate collision before the surface is viewed to be determining in the sense that the last collision results in an atom being sputtered if the energy transfer exceeds a certain threshold value. Ejections will thus occur most likely in directions where the binding energy is lowest or the energy transfer is the highest, and

---

[97] P. Sigmund, *Rev. Roum. Phys.* **17**, 1079 (1972).
[98] J. J. Elich, H. E. Roosendaal, and D. Onderdenlinden, *Radiat. Eff.* **14**, 93 (1972).
[99] C. Lehman and P. Sigmund, *Phys. Status Solidi* **16**, 507 (1966).

4.1. COMPOSITIONAL STUDIES

these correspond to situations where either the subsurface atom sees a clear path or has a direct collision with a nearest-neighbor surface atom. The model thus predicts that the average energy of the sputtered atoms varies inversely with the intensity, and this is consistent with the observation of Weijsenfeld[100] with 1-keV Ar$^+$ on a Ag single crystal. Additional support for the model can be found in Hofer's[101] study of HCP crystals in which it was shown that spot patterns exist in directions where there are no close-packed rows.

The random cascade model is seen to provide a basis for describing sputtering phenomena. Although the details have not been worked out for more complex solids, the applications of sputtering have, nevertheless, received significant development in a number of important areas such as in ion machining or etching, production of atomically clean surfaces and of thin films, and compositional analysis.

In compositional studies, the fact that a majority of the sputtered atoms comes from the surface layer means that the detection of sputtered species gives directly the surface composition. Moreover, as the ion beam continually erodes the surface, the attendant surface analysis will thus provide depth-profile information. Of course, depth profiling can equally be accomplished by other surface analysis methods such as AES, LEIS, and x-ray photoelectron spectroscopy (XPS), with sputtering merely serving as a microsectioning tool. In addition, the sputtered species can be detected by a number of methods depending on the experimental setup. Coburn and Kay[102] have made detailed comparisons of the various combinations of these methods. In line with our central theme here only SIMS and SCANIIR will be amplified since these use focused ion beams.

We conclude this section by pointing out that ion bombardment can alter the surface topography and structure. If the original surface is not flat, the variation of sputtering yield with angle of incidence will cause a faster erosion of the region having a large angle of incidence[103-106] and gives rise to the observed ridges, steps, cones, facets, and other irregularly shaped protrusions. The presence of surface impurities also results in cone-shaped protrusions[107] due to the different sputtering yields for the different species.

---

[100] C. H. Weijsenfeld, Thesis, Univ. of Utrecht, 1966.
[101] W. O. Hofer, *Radiat. Eff.* **19**, 263 (1973).
[102] J. W. Coburn and E. Kay, *CRC Crit. Rev. Solid State Sci.* **4**, 561 (1974).
[103] D. J. Barber, F. C. Frank, M. Moss, J. W. Steeds, and I. S. T. Tsong, *J. Mater. Sci.* **8**, 1030 (1973).
[104] G. Carter, *J. Mater. Sci.* **8**, 1473 (1973).
[105] P. Sigmund, *J. Mater. Sci.* **8**, 1545 (1973).
[106] I. H. Wilson, *Radiat. Eff.* **18**, 95 (1973).
[107] A. D. G. Stewart and M. W. Thompson, *J. Mater. Sci.* **4**, 56 (1969).

Another effect is the growth of extensive defects, such as dislocations during ion bombardment, which when exposed by the surface erosion cause variation in the sputtering yield across the surface leading to surface irregularities.[108] At higher doses, more dramatic blistering and exfoliation can occur.[109] There are also other more subtle effects, such as recoil implantation by which surface species are driven into deeper layers by the so-called knock-on process[110] and the bombardment-induced motion of impurities in insulators.[111] All these effects clearly affect the depth resolution of all sputter-based profiling techniques, and the degree to which they are present in a given situation must be assessed if quantitative composition information is to be extracted.

4.1.4.2. Secondary Ion Mass Spectrometry (SIMS). Reviews on this subject have been given by Coburn and Kay,[102] Benninghoven,[92] McHugh,[112] Blaise,[113] and Wittmaack.[114]

There are a number of reasons why the detection of sputtered ions is preferred in spite of the fact they constitute only a very small fraction of the total sputtered particles. An obvious one is the high sensitivity and the existing instrumental sophistication developed in connection with earlier mass spectroscopic studies with which low-energy ions can be mass analyzed. More important, however, is the excellent matching of the available detection techniques to the use of focused ion beams. Also, the measurement can be carried out in ultrahigh vacuum (UHV), thus avoiding surface contamination of the target, and the beam can be rastered to provide lateral-profile information. It should be noted that sputtered neutrals that have been postionized can be detected in same apparatus but that significant postionization can occur only when the neutrals interact with a relatively dense medium, a condition absent under a normal vacuum condition. Indeed, such a postionization technique for detecting sputtered neutrals has generally been made under a glow discharge condition in which the plasma serves both to sputter the target and to postionize the neutrals. Oechsner and Stumpe[115] have achieved excellent analysis results using such a technique to

---

[108] N. Hermanne, *Radiat. Eff.* **19**, 161 (1973).
[109] M. Kaminsky, *CRC Crit. Rev. Solid State Sci.* **6**, 433 (1976).
[110] F. Schulz, K. Wittmaack, and J. Maul, *Radiat. Eff.* **18**, 211 (1973).
[111] D. V. McCaughan, R. A. Kushner, and V. T. Murphy, *Phys. Rev. Lett.* **30**, 614 (1973).
[112] J. A. McHugh, *in* "Methods of Surface Analysis," (S. P. Wolsky and A. W. Czanderna, eds.), p. 223. Elsevier, Amsterdam, 1975.
[113] G. Blaise, *in* "Materials Characterization Using Ion Beams" (J. P. Thomas and A. Cachard, eds.), p. 143. Plenum, New York, 1978.
[114] K. Wittmaack, *in* "Inelastic Ion–Surface Collisions" (N. H. Tolk, J. C. Tully, W. Heiland, and C. W. White, eds.), p. 153. Academic Press, New York, 1977.
[115] H. Oechsner and E. Stumpe, *Appl. Phys.* **14**, 43 (1977).

4.1. COMPOSITIONAL STUDIES

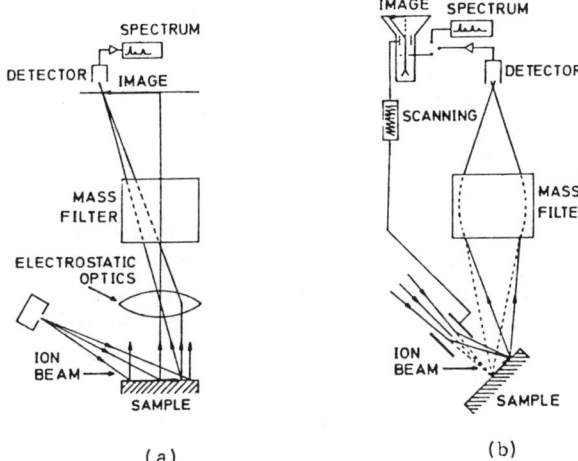

FIG. 25. Two general categories of instrumentation for SIMS analysis: (a) emission microscopy and (b) microprobe. (From Blaise.[113])

detect sputtered neutrals. Unlike SIMS, however, the capability of lateral profiling is not present.

Instrumentation for SIMS analysis can generally be grouped into the two categories[116] illustrated in Fig. 25. In emission microscopy a large area is exposed to the primary ion beam of uniform density and the secondary ion species of interest from a small section of this area is stigmatically imaged onto the focal plane and recorded. Typically, an image area of 250 μm in diameter is obtained, which translates into a spatial resolution of $\gtrsim 1$ μm on the target. In the microprobe mode, the finely focused primary beam (a few micrometers in diameter) is rastered across the sample by two pairs of deflection plates and the secondary ion species of interest is focused stigmatically into an exit slit and recorded. A point-by-point imaging of the surface can be accomplished, for example, through an xy-storage oscilloscope, with the axes corresponding to the amount of primary beam deflections. Typical spatial resolution of $\gtrsim 2$ μm is achieved in this mode. In profiling to a depth of 1 μm, a high primary ion beam intensity (~ 100 μA cm^{-2}) is used, which translates to a typical erosion rate of the order of one atomic layer per second. The analysis of secondary ions under high erosion rate is termed the "dynamic" procedure. Since the amount of erosion depends only on the total ion dose, the depth information is not lost by repeatedly stopping the intense beam bombardment and examining the surface under "static" procedure. This so-called static SIMS involves low

[116] H. Liebl, *J. Phys. E* **8**, 797 (1975).

ion-density bombardment ($\sim 10^{-2}\,\mu$A cm^{-2}) on a relatively large area (~ 0.1 cm^2) and typically erodes only a small fraction of a monolayer in a complete scanning of the mass spectrum. It can thus be used for surface processes and kinetic studies the details of which have been reviewed elsewhere.[93,113] Our discussion here is confined to the analysis of composition as a function of depth.

In the bombardment of multicomponent samples, the different elemental sputtering yields cause the elemental composition of the sputter-etched surface to be different from that of the bulk in general. This might seem to be a problem, but the surface composition adjusts so that, in steady-state conditions, the surface coverage θ_i times the sputtering yield S_i gives a sputtering flux $\theta_i S_i$ that is proportional to the atomic fraction of the element i in the bulk sample. This assumes of course that bulk-surface diffusion and thermal vaporization are absent. This capability of reflecting the composition of the source material directly in the sputtered flux in spite of the widely different elemental sputtering yields accounts for the popularity of SIMS for depth profiling.

In SIMS the central question to be addressed is how the recorded secondary ion intensities are related to the composition of the sample being eroded. In the absence of a comprehensive theory, a common approach has been based on the reasonable assumption that ion emission results from a sputtering process associated with an ionization process and that these two processes can be described separately. In Section 4.1.4.1 the sputtering yield of pure elemental solid was seen to be reasonably understood within the random cascade model. Although the theory can not account for the yield from a multielement target, the global composition of the sputtered products is, nevertheless, the same as that of the bulk solid when the steady-state conditions are obtained. Thus if the number of sputtered atoms of a particular species produced in the bombardment with N_p primary ions is N_i, the global sputtering yield can be defined as

$$S_T = \left(\sum_i N_i\right)/N_p \qquad (4.1.27)$$

and the atomic concentration of a particular species, say, M, is just

$$C(M)_{\text{sputtered}} = C(M)_{\text{solid}} = N_M/\left(\sum_i N_i\right). \qquad (4.1.28)$$

If the depth measured along the surface normal is denoted by x, the erosion rate can be written

$$dx/dt = S_T(x)N_p/n(x), \qquad (4.1.29)$$

where $n(x)$ is the atomic volume density at x and N_p is the incident flux of primary ions. Taking into account the ionization process, the intensity of an ionic species M^+ corresponding to a bulk concentration $C(M,x)$ is then

$$I(M^+,x) = P(M^+,x)C(M,x)S_T(x)I_p, \qquad (4.1.30)$$

with $P(M^+,x)$ being the ionization probability. Depth profiling is seen to be straightforward if $S_T(x)$ and $P(M^+,x)$ are constant during the erosion. Such a situation appears to be realized only[112] for elements in low concentration in single-phase matrices, where the major constituents are homogeneously distributed. In general, some sort of calibration procedures are required because the values for $P(M^+,x)$ and $S_T(x)$ are not known except in simple cases.

The ionization process is complex as evidenced by the strong dependence of the secondary ion yield on the surface condition and the primary beam used. Slodzian and Hennequin[117] first demonstrated that absorbed oxygen increased the secondary ion yield. Subsequently, Anderson and Hinthorne[118] have shown that O^+ primary ions enhance the yield of positive secondary, ions, whereas Cs^+ enhance that of the negative ones. Based on the premise that a majority of sputtered particles are ionized during the sputtering events, the above observations can be explained in terms of the neutralization of the ions at or near the surface. The factor that determines the neutralization efficiency is the "electron availability," which is suggested by the fact that oxygen is electronegative and cesium is electropositive. Another point of view is that the efficiency of sputtered-ion production is determined by the chemical bonding of the element of interest. For example, the ionically bonded alkali halides have large secondary ionization efficiencies. In systems in which chemical effects are not involved such as in noble gas bombardments of clean, metal surfaces, the secondary ion yield is much lower. The ionization under such conditions can proceed by means of autoionization of the highly excited sputtered neutrals leaving the surface.[119]

Proposed quantitative theories of secondary ionization have had varying degrees of success. Thermal models based on the thermodynamic equilibrium of hot plasma[118] and nonequilibrium surface ionization[120] require fitting parameters and have been criticized[113] on the grounds that ion emission really exhibits pronounced nonthermal features. Quantum models, although more realistic, have used approximations[119,121] whose general validity remains to be tested. The complexity of the surface process

[117] G. Slodzian and J. F. Hennequin, *C. R. Hebd. Seances Acad. Sci., Ser. B* **263**, 1246 (1966).
[118] C. A. Anderson and J. R. Hinthorne, *Anal. Chem.* **45**, 1421 (1973).
[119] G. Blaise and G. Slodzian, *J. Phys. (Orsay, Fr.)* **31**, 93 (1970).
[120] Z. Jurela, *Radiat. Eff.* **13**, 167 (1972).
[121] A. Blandin, A. Nourtier, and D. W. Hone, *J. Phys. (Orsay, Fr.)* **37**, 369 (1976).

and the lack of a comprehensive theory at the present time mean that, in general, the extraction of quantitative results from SIMS must still be based on the comparisons with standards or a universal scaling procedure.[121a]

Finally, there are the complications associated with the sputtering of multiply ionized and molecular ion species. Although these generally have lower intensities under clean surface conditions, the uncertainties introduced by the mass interference and the different ionization efficiencies impede the analysis. On the other hand, these complex ions carry chemical information that is of primary interest in surface kinetic studies.[92] The rich informational content of static SIMS can be seen in Fig. 26, where both the positive and negative secondary ion spectra of a contaminated vanadium surface are displayed.

4.1.4.3. Surface Composition Analysis by Neutral and Ion Impact Radiation (SCANIIR).

Reviews on the subject have been given by White et al.[122,123] and van der Weg.[124]

When a solid is bombarded with a low-energy ion beam, light emission is often visible to the naked eye even at a moderate beam intensity. It is therefore somewhat surprising that most of the quantitative studies have been made only in the past few years. This may be due to the extreme sensitivity of the light emission to the chemical nature and cleanliness of the surface, which made the data difficult to reproduce until the advent of clean vacuums and the development of target-cleaning techniques.

Most studies of light emission have been in the visible and near-uv region. The apparatus is relatively simple; a monochromator or a spectrograph is optically coupled to a chamber containing the target irradiated by a beam of particles. In depth-profiling measurements, the characteristic emission lines of the sputtered species are monitored as a function of the amount of surface erosion. The bombardment condition is usually chosen to have a relatively high erosion rate in order to complete the measurement in a reasonable length of time. At such a rate, residual gases even at moderate vacuum of 10^{-6} torr usually pose no serious problem.

The optical spectra recorded under ion bombardment of solid materials always consist of a series of sharp lines. In some cases, a broad structure is

---

[121a] P. Williams, W. Katz, and C. A. Evans, Jr., *Nucl. Instrum. Methods* **168**, 373 (1980).

[122] C. W. White, D. L. Simms, and N. H. Tolk, *in* "Characterization of Solid Surfaces" (P. F. Kane and G. B. Larrabee, eds.), p. 641. Plenum, New York, 1974.

[123] C. W. White, E. W. Thomas, W. F. Van der Weg, and N. H. Tolk, *in* "Inelastic Ion–Surface Collisions" (N. H. Tolk, J. C. Tully, W. Heiland, and C. W. White, eds.), p. 201. Academic Press, New York, 1977.

[124] W. F. van der Weg, *in* "Material Characterization Using Ion Beams" (J. P. Thomas and A. Cachard, eds.), p. 81. Plenum, New York, 1978.

4.1. COMPOSITIONAL STUDIES

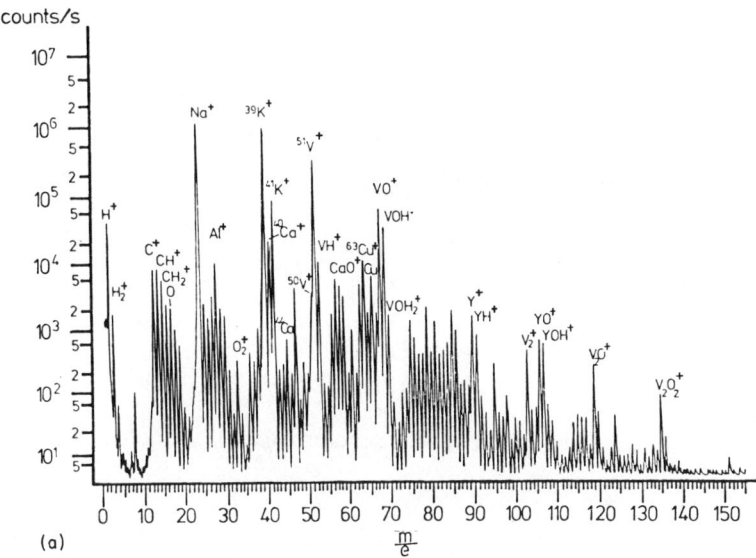

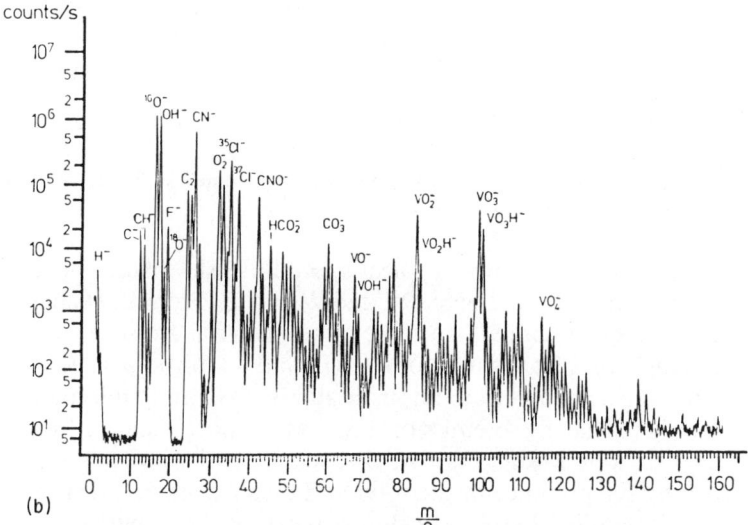

FIG. 26. Secondary ion spectra from a contaminated vanadium surface ($I_p = 5 \times 10^{-8}$ Å, $A = 0.1$ cm^2): (a) positive ions and (b) negative ions. (From Benninghoven.[92])

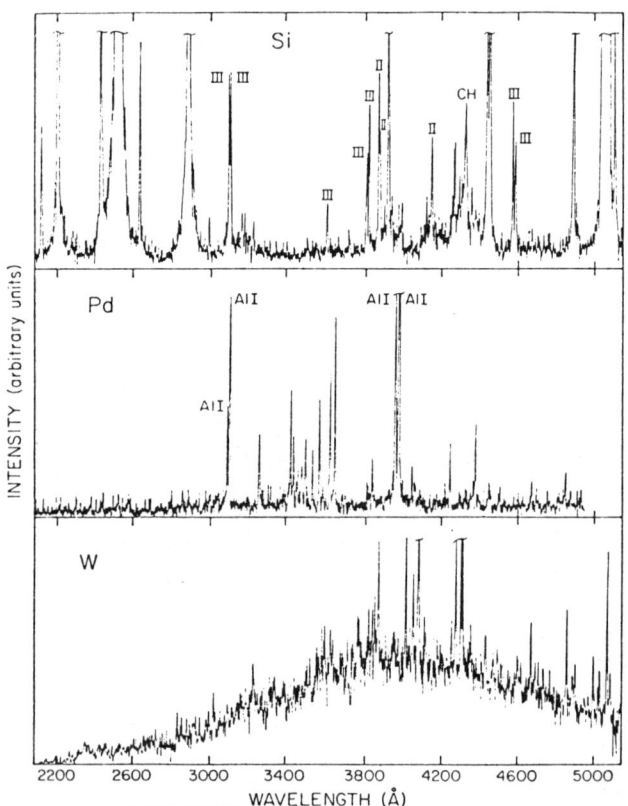

FIG. 27. Optical spectra obtained for 40-keV Ar+ bombardment of Si, Pd, and W targets. (From van der Weg and Lugujjo.[125])

superimposed on these lines. Typical spectra are shown in Fig. 27[125] for 40-keV Ar+ bombardment of Si, Pd, and W targets. Prominent characteristic atomic lines (unmarked) from the target material are seen to be abundant, although lines from singly ionized (II) and doubly ionized (III) species are also present as well as from impurity atomic (AlI) and molecular species (CH). The fact that these emission lines are characteristic of free atoms, molecules, or ions means that the de-excitation of these particles is not perturbed by the presence of the solid and thus occurs at a distance away from the solid surface. In the cases shown, the sputtering yields are large, and thus the spectra are dominated by emission from sputtered-target particles. The basis for compositional analysis is clearly seen to lie in the relative ease

[125] W. F. van der Weg and E. Lugujjo, in "Atomic Collisions in Solids" (S. Datz, B. R. Appleton, and C. D. Moak, eds.), p. 511. Plenum, New York, 1975.

4.1. COMPOSITIONAL STUDIES

with which the sputtered species can be unambiguously identified by the many characteristic lines associated with a particular species.

Line emission is not confined to the sputtered particles. In the case of light projectiles (H or He) on a heavier target where the sputtering yield is low, the line spectrum is dominated by the emission of the backscattered projectiles in either neutral or ionized states. Broad-band emission is also observed in heavy-ion bombardments and is particularly strong for transition-metal targets as can be seen in the W case in Fig. 27. Studies[126,127] of the spatial distribution of this type of broad-band emission have shown that the radiation originates from outside the solid surface. Additional studies by Rausch et al.[128] as a function of oxygen partial pressure in the target chamber strongly indicates that the broad band can be attributed to the sputtered metal oxides. Broad-band emission can also originate from the surface as observed in light-ion (or neutral) bombardment of solids, particularly insulators. In this case, the evidence points to the radiative recombination of electrons with the self-trapped hole created in the inelastic collision of the projectile with the electrons in the crystal.

Since the focus here is on compositional study, only the features of sputtered-particle emission are discussed further. Factors influencing the line-emission intensity are expected to be similar to those encountered in SIMS since the processes determining whether the particle finally leaving the surface is ionized, excited, or both are intimately related to each other. In SIMS, the question is whether the ionized particle is ejected in the sputtering process itself or is originally a neutral sputtered species subsequently ionized by surface processes. In SCANIIR, one is further concerned with the relative population of the excited electronic levels of the sputtered particle and the various de-excitation probabilities of these levels particularly in the surface region where the perturbation is strong. Although significant progress has been made toward the understanding of surface processes, the quantitative theories are more of a semiempirical nature.

The emission of light from sputtered atoms from clean metals is perhaps the best understood. Both the Doppler-broadened and shifted line shape[129] and the intensity variation as a function of bombarding energy[130] are generally accounted for by explicitly considering the competition of radiationless deexcitation processes that have been highlighted in Hagstrum's

---

[126] C. W. White, N. H. Tolk, J. Kraus, and W. F. van der Weg, *Nucl. Instrum. Methods* **132**, 419 (1976).
[127] C. B. Kerkdijk, K. H. Schartner, R. Kelly, and F. W. Saris, *Nucl. Instrum. Methods* **132**, 427 (1976).
[128] E. O. Rausch, A. I. Bazhin, and E. W. Thomas, *J. Chem. Phys.* **65**, 4447 (1976).
[129] W. F. van der Weg and D. J. Bierman, *Physica (Utrecht)* **44**, 206 (1969).
[130] N. H. Tolk, D. L. Simms, E. B. Foley, and C. W. White, *Radiat. Eff.* **18**, 221 (1973).

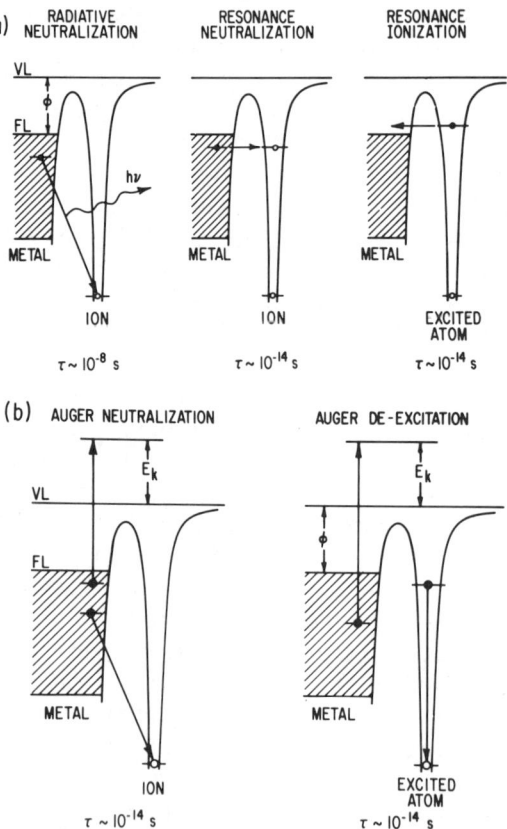

FIG. 28. Possible electronic transition processes for a slow-moving ion or excited atom near a solid surface: (a) one-electron processes and (b) two-electron processes. VL indicates the vacuum level; FL, the Fermi level; ϕ, the work function; E, the kinetic energy of the excited electron that may be ejected into vacuum; and τ, in each case, the initial-state lifetime of the process. (From Hagstrum.[86])

studies[131] of secondary electron emission. Two of the most likely ones are resonance ionization and Auger de-excitation depicted in Fig. 28 for an excited atom. The probability that a sputtered atom in an excited state escapes without undergoing radiationless transition then has the form

$$P(v_1) = e^{-A/(av_1)}, \qquad (4.1.31)$$

where v_1 is the velocity of the sputtered atom along the surface normal. Thus radiative transitions are more probable for the faster particles and explain the fact that the observed radiation is due mainly to the most energetically

[131] H. D. Hagstrum, *Phys. Rev.* **123**, 758 (1961).

4.1. COMPOSITIONAL STUDIES 273

sputtered particles as evidenced by the observed large Doppler shift.[129] It also accounts for the growth of the relative intensity of lines from the ionized particles over those from the neutral particles with increasing energy at high bombarding energies (~ 100 keV).[132] This follows from the fact that fast-recoil target particles are responsible for the light emission. At high primary energies, these particles are increasingly in the ionized states because of the violent collisions on the surface.

It should not come as a surprise that the probability for escaping neutralization in LEIS [Eq. (4.1.16)] and the probability for ionization in SIMS [Eq. (4.1.31)] have the same velocity dependence. In LEIS, the incident low-energy ion can be neutralized by the transfer of an electron from the target to the ion either resonantly or through the Auger mode as can be seen in Fig. 28. In SIMS, similar neutralization processes would occur when ions are sputtered. The details of the processes mentioned clearly depend on the electronic structure and population of states of both the solid and the moving particle. That these neutralization processes are efficient can be seen in the equal efficiency with which photons due to sputtering are produced in the bombardment of metals by neutrals and ions of the same species.[133] This follows from the fact that the low-velocity ions with sufficiently large ionization potentials striking a metal surface are neutralized several angstroms in front of the surface before the sputtering interaction occurs.

Strong chemical effects are observed in the light emission as well as in SIMS. For example, the line intensities of the metal species are very much enhanced compared to the clean-metal case when an oxidized metal is bombarded[134] as well as when oxygen is admitted into the sputtering chamber.[125] The enhancement can be explained as inhibition of the competing radiationless de-excitation processes due to the formation of the insulating oxide layer. The resulting band gap in the solid leaves few loose electron for Auger de-excitation and few levels for resonance electron transfer. As a consequence, slow sputtered particles which compose the majority of the sputtered particles will de-excite by radiative transitions. This explanation is also consistent with the observation that the excitation efficiency is much higher in insulators than in metals.[130] The observed narrower Si Doppler-line profile from SiO_2 over that from Si[134] may be similarly explained since the emission in the case of the insulator is dominated by the slow sputtered particles in contrast to the fast particles in Si.

In summary, the ease with which the light emission of the sputtered species can be detected and identified is appealing for surface composition

[132] M. Braun, B. Emmoth, and I. Martinson, *Phys. Scr.* **10**, 133 (1974).
[133] C. W. White, D. L. Simms, and N. H. Tolk, *Science* **177**, 481 (1972).
[134] C. W. White, D. L. Simms, N. H. Tolk, and D. V. McCaughan, *Surf. Sci.* **49**, 657 (1975).

analysis and also for depth profiling if the amount of target erosion by the beam is also monitored. Neutral-beam bombardment has been shown[130] to be particularly attractive in the analysis of insulators in avoiding the charge-build-up problem. As in the case of SIMS, however, the surface processes are incompletely understood, and the quantification generally requires the calibration of the experimental set up with known standards.

4.2. Channeling Studies of Lattice Defects

By L. M. Howe, M. L. Swanson, and J. A. Davies

Atomic Energy of Canada Limited Research Company,
Chalk River Nuclear Laboratories,
Chalk River, Ontario, Canada

4.2.1. Introduction

Several nuclear techniques have been developed for the study of lattice defects. Most of these methods, such as positron annihilation,[1] hyperfine interactions (Mössbauer effect[2] and perturbed angular correlation[3]), nuclear magnetic resonance,[4] and muon-spin precession,[5] as well as other methods such as electrical resistivity[6] and electron paramagnetic resonance,[7] are used to probe the electronic environment of the nuclei rather than the actual positions of the atoms in the lattice. However, an important characteristic of lattice defects is that they produce a localized spatial rearrangement of lattice atoms.[8] The technique of ion channeling is especially suited to studying these displacements of lattice atoms in the vicinity of lattice defects.[9-11] Other complementary methods of determining the properties of defects are low-energy electron diffraction (LEED),[12] neutron[13] or x-ray

---

[1] R. R. Hasiguti and K. Fujiwara, eds., *Proc. 5th Int. Conf. Positron Annihilation.* Jpn. Inst. Met., Sendai, 1979.
[2] G. Vogl, *Hyperfine Interact.* **2**, 151 (1976).
[3] *Hyperfine Interact.* **4** (1978).
[4] C. Minier, M. Minier, and R. Andreani, *Phys. Rev. B* **22**, 28 (1980).
[5] A. Seeger, in "Hydrogen in Metals" (G. Alefeld and J. Volkl, eds.), Topics in Applied Physics, Vol. 28, Vol. 1, p. 349. Springer-Verlag, Berlin and New York, 1978; also *Proc. 5th Int. Conf. Positron Annihilation,* p. 771. Jpn. Inst. Met., Sendai, 1979.
[6] A. Seeger, D. Schumacher, W. Schilling, and J. Diehl, eds., "Vacancies and Interstitials in Metals." North-Holland Publ., Amsterdam, 1970.
[7] J. W. Corbett, *Solid State Phys., Suppl.* **7** (1966).
[8] H. G. Van Bueren, "Imperfections in Crystals." North-Holland Publ., Amsterdam, 1960.
[9] D. S. Gemmell, *Rev. Mod. Phys.* **46**, 129 (1974).
[10] J. A. Davies, in "Channeling: Theory, Observation, and Applications" (D. V. Morgan, ed.), p. 391. Wiley (Interscience), New York, 1973.
[11] S. T. Picraux, in "New Uses of Ion Accelerators" (J. F. Ziegler, ed.), p. 229. Plenum, New York, 1975.
[12] P. J. Estrup and E. G. McRae, *Surf. Sci.* **25**, 1 (1971).
[13] G. E. Bacon, "Neutron Diffraction." Oxford Univ. Press, London and New York, 1962.

diffraction,[14] diffuse x-ray scattering,[15] extended x-ray absorption fine structure (EXAFS),[16] mechanical relaxation,[17] and electron or field ion microscopy.[18,19]

Ion channeling is the steering of a beam of energetic ions into the open spaces (channels) between close-packed rows or planes of atoms in a crystal (see Fig. 1 and Section 4.2.2).[9,20] The steering action occurs by a series of low-angle screened Coulomb collisions with the atoms bordering the channel. Because channeled ions do not penetrate closer than the Thomas–Fermi screening distance a (i.e., ~ 0.01 nm) to the atomic cores, close-encounter processes, such as large-angle Rutherford collisions, nuclear reactions, or inner-shell x-ray excitations are strongly attenuated when the ions are directed at a small angle ψ_{in} to a channeling direction. The normalized yield χ for such processes is defined as the ratio of the yield from a beam directed at an angle ψ_{in} to the yield for a randomly directed beam. Values of χ may be as low as 0.01 for perfect alignment ($\psi_{in} = 0$) along good axial channels; that is, only 1% of the ions interact with atoms on normal lattice sites. Channeling parameters are calculated by considering that the ions move in a continuum potential of the Thomas–Fermi form, created by the strings of atoms bordering the channel. The results of this classical, two-body collision theory agree well with experiment,[9] as is outlined in Section 4.2.2.

If host or solute atoms are displaced from substitutional lattice sites, they project into channels and interact with channeled ions, causing an increase in the yield χ. Thus channeling can be used to "see" lattice defects with a minimum of model fitting.[10,11] For example, if solute atoms are on substitutional lattice sites, the normalized yield χ_s from solute atoms varies with incident angle ψ_{in} in the same way as does χ_h, the normalized yield from host atoms (Fig. 1). However, if solute atoms are displaced into positions near the center of the channel where the ion flux is high, χ_s becomes much greater than χ_h, and a peak in yield may occur at the aligned direction. Such a peak is unambiguous evidence that the solute atoms are positioned close to the center of the channel.

A simple and quick method of determining positions of a solute atom in a

---

[14] C. S. Barrett and T. B. Massalski, "Structure of Metals." McGraw-Hill, New York, 1966.
[15] P. H. Dederichs, *J. Phys. F* **3**, 471 (1973).
[16] J. Stöhr, D. Denley, and P. Perfetti, *Phys. Rev. B* **18**, 4132 (1978).
[17] A. S. Nowick and B. S. Berry, "Anelastic Relaxation in Crystalline Solids." Academic Press, New York, 1972.
[18] S. Amelinckx, R. Gevers, G. Remaut, and J. Van Landuyt, eds., "Modern Diffraction and Imaging Techniques in Material Science." North-Holland Publ., Amsterdam, 1970.
[19] D. Seidman, *J. Phys. F* **3**, 393 (1973).
[20] J. Lindhard, *Mat.-Fys. Medd.—K. Dan. Vidensk. Selsk.* **34**, No. 14 (1965).

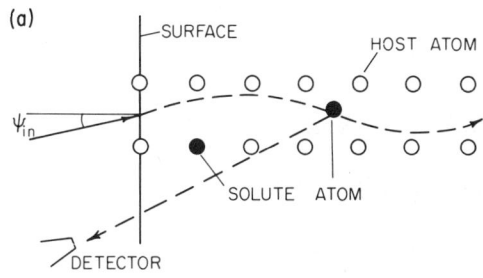

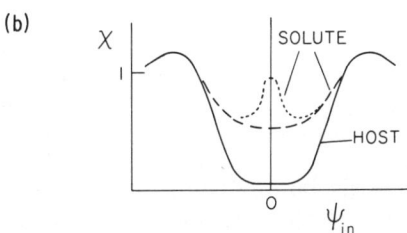

FIG. 1. (a) Schematic view of the channeling of ions directed at an angle ψ_{in} to a low-index direction in a crystal. (b) The normalized yield χ of ions that are backscattered from host atoms shows a strong dip at $\psi_{in} = 0$ (solid line). If 50% of the solute atoms are displaced into the channel, the normalized yield of ions backscattered from the solute atoms is approximately 0.5 at $\psi_{in} = 0$ (---). If the displaced solute atoms are located near the center of the channel, a narrow peak in yield may occur (· · ·).

crystal lattice is thus to measure backscattering yields from host and solute atoms in only two or three different channels. By a triangulation procedure, the position of the solute atoms in the lattice can be obtained. An illustration of this method is shown in Fig. 2 for a two-dimensional model of a simple cubic lattice.[10] Solute atoms in substitutional positions are completely shadowed in both $\langle 10 \rangle$- and $\langle 11 \rangle$-type channels. The × interstitial positions (halfway along the sides of the unit cells) are 50% shadowed in both the [10] and [01] channels, but lie in the center of the [11] and [1$\bar{1}$] channels. The □ interstitial positions (in the center of the square unit cell) are completely shadowed in both $\langle 11 \rangle$-type channels, but lie in the center of $\langle 10 \rangle$-type channels. It will be noted that all equivalent interstitial positions must be considered when observing along a given channel because of symmetry considerations.

In three dimensions, the interstitial position equivalent to □ of Fig. 2 is shown in Fig. 3 for a face-centered-cubic (fcc) lattice. This is the body-centered or octahedral position. Solute atoms in this position are exactly in the

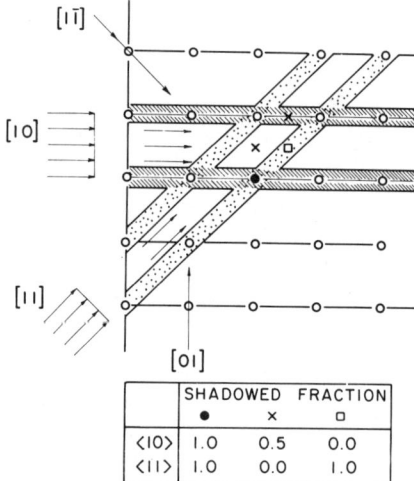

FIG. 2. A two-dimensional model illustrating how the channeling effect may be used to locate foreign atoms in a crystal. As shown by the table, three possible sites for a foreign atom (●, ×, and □) can be readily distinguished by comparing the channeling behavior along $\langle 10 \rangle$- and $\langle 11 \rangle$-type directions.

center of all $\langle 110 \rangle$ channels, but are completely shadowed in both $\langle 100 \rangle$ and $\langle 111 \rangle$ channels. Thus to identify solute atoms in body-centered positions for a fcc crystal, only two of these three channels need be used.

In more difficult cases, where a mixture of interstitial sites is present or small displacements from lattice sites are involved, it is necessary to do detailed angular scans through several axial and planar channels.[9-11] This procedure is discussed in more detail in Section 4.2.5. In principle, rather complicated positions or combinations of positions of solute atoms can be sorted out by comparing the angular dependence of yields near axial and planar channels with theoretical calculations obtained from flux profiles.

The basic apparatus required for channeling experiments[21] consists of a

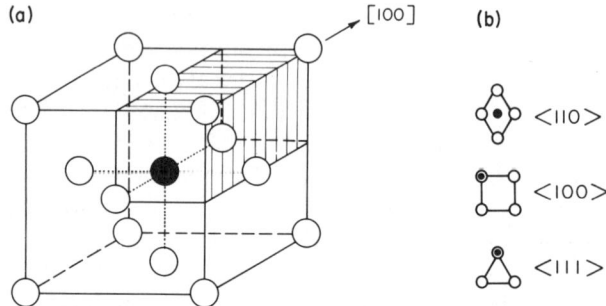

FIG. 3. (a) Octahedral interstitial site in a face-centered-cubic crystal and (b) the projections of this site into $\langle 110 \rangle$, $\langle 100 \rangle$, and $\langle 111 \rangle$ axial channels.

[21] W. K. Chu, J. W. Mayer, and M.-A. Nicolet, "Backscattering Spectrometry." Academic Press, New York, 1978.

4.2. CHANNELING STUDIES OF LATTICE DEFECTS

small accelerator, a target chamber containing a two- or three-axis goniometer, a particle detector, and electronic equipment for energy analysis of the detected particles. The ion beams normally used, 100–3000 keV, are easily obtained from a Cockroft–Walton or a Van de Graaff accelerator. Details of the target chamber, beam steering, sample-temperature control, current integration, and associated analytical equipment are given in Section 4.2.3.

As mentioned earlier, the use of ion channeling to study lattice defects in solids is based on the ability of channeled ions to "see" displacements of atoms from lattice sites. The various lattice defects are characterized by the type of displacement which they produce. Lattice defects can be classified according to their dimension in the following way.

(1) *Point defects:* vacant lattice sites (vacancies), self-interstitial atoms (extra host atoms squeezed between normal lattice sites), and solute atoms (Fig. 4). The solute atoms may be located on normal (substitutional) lattice sites, on interstitial sites, or at some intermediate point when they are associated with vacancies or self-interstitials. In metals, self-interstitials usually have a split configuration in which two host atoms share one lattice

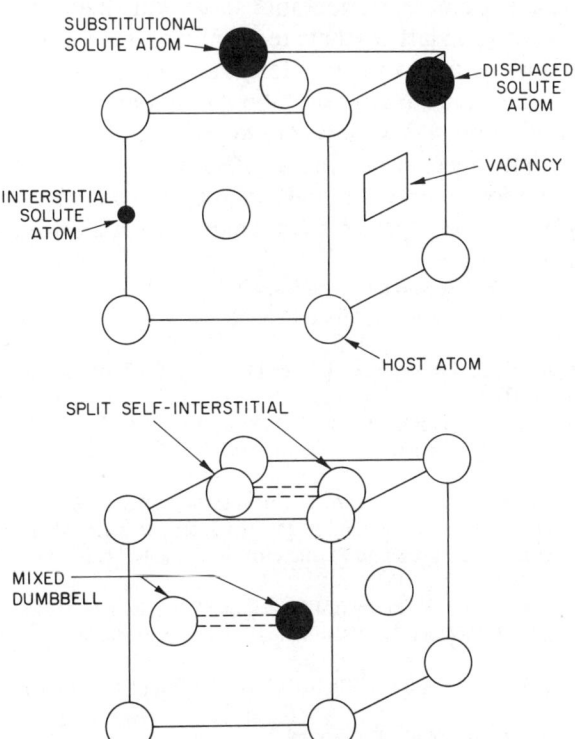

FIG. 4. Point defects in a face-centered-cubic crystal.

site (Fig. 4).[22] Often, small clusters of atomic-size defects are also considered to be point defects. The relaxation of lattice atoms adjacent to point defects can vary greatly. The calculated displacement of nearest neighbors toward vacant lattice sites in fcc metals[23] is less than 0.01 nm, but may be as large as 0.06 nm for diamond structures.[24] If clusters of vacancies are trapped by solute atoms, the solute atoms may relax into interstitial positions.[25] The relaxation of nearest neighbors to self-interstitial atoms[26] is about 0.03 nm for the case of $\langle 100 \rangle$ self-interstitials in Cu. A small solute atom may trap a self-interstitial atom to form a mixed dumbbell configuration, in which the solute atom and host atom straddle a lattice site; each atom in such a mixed dumbbell is displaced more than 0.1 nm from a normal lattice site[22,27,28] (Fig. 4).

(2) *Line defects:* dislocations,[29,30] which in the simplest case (edge dislocations) are the demarcation lines between extra planes of atoms and the perfect lattice. These defects are surrounded by regions of large lattice strain, which extend several interatomic distances away in the form of bending of atomic planes.

(3) *Planar defects*[8]: grain boundaries (the transition region between two crystals having different orientations), stacking faults, twin boundaries (the plane between two lattices that are the mirror images of each other), and surfaces. Because of the asymmetric forces at surfaces, the surface atoms are sometimes relaxed a small amount (about 0.003 nm) perpendicular to the surface.[31] In addition, the surface monolayer for specific crystallographic surface planes may reconstruct into a different structure.[32] This reconstruction can be retarded by only a partial monolayer of adsorbed foreign atoms.

(4) *Volume defects:* precipitates, voids[33] (large clusters of vacancies),

---

[22] W. Schilling, *J. Nucl. Mater.* **69/70**, 465 (1978).

[23] A. C. Damask and G. J. Dienes, "Point Defects in Metals." Gordon & Breach, New York, 1963.

[24] A. Seeger and M. L. Swanson, *in* "Lattice Defects in Semiconductors" (R. R. Hasiguti, ed.), p. 93. Univ. of Tokyo Press, Tokyo, 1968.

[25] M. L. Swanson, L. M. Howe, and A. F. Quenneville, *Phys. Rev. B* **22**, 2213 (1980).

[26] J. B. Gibson, A. N. Goland, M. Milgram, and G. H. Vineyard, *Phys. Rev.* **120**, 1229 (1960).

[27] M. L. Swanson, L. M. Howe, and A. F. Quenneville, *J. Nucl. Mater.* **69/70**, 372 (1978).

[28] N. Matsunami, M. L. Swanson, and L. M. Howe, *Can. J. Phys.* **56**, 1057 (1978).

[29] A. H. Cottrell, "Dislocations and Plastic Flow in Crystals." Oxford Univ. Press (Clarendon), London and New York, 1953.

[30] J. P. Hirth and J. Lothe, "Theory of Dislocations." McGraw-Hill, New York, 1968.

[31] J. A. Davies, D. P. Jackson, N. Matsunami, P. R. Norton, and J. U. Andersen, *Surf. Sci.* **78**, 274 (1978).

[32] P. R. Norton, J. A. Davies, D. K. Creber, C. W. Sitter, and T. E. Jackman, *Surf. Sci.* **108**, 205 (1981).

[33] D. I. R. Norris, *Radiat. Eff.* **14**, 1 (1972).

amorphous regions,[34] and defect-rich regions such as displacement cascades.[35] Precipitates are small crystals embedded in the host crystal; these may be incoherent, having a different structure from the host, or coherent, having the same structure as the host but a different lattice parameter.[36] Displacement cascades are created by heavy-ion bombardment; they consist of vacancy-rich regions surrounded by self-interstitial-rich regions.

When considering the use of channeling to study lattice defects, it is useful to distinguish between single scattering and multiple scattering events. Any displaced atoms associated with defects interact with channeled ions, giving rise to an increase in scattering events over a range of impact parameters; that is, both large-angle collisions and small-angle collisions occur. The effect of many small-angle collisions will be to increase the random component of the ion beam, i.e., the part of the beam that is deflected out of the channel, or dechanneled.[9,20] For example, the effect of thermal vibrations is to increase the dechanneling. At the same time, the number of ions backscattered from those vibrating atoms that project into the channel is increased (the size of the channel is effectively reduced). For purposes of quantitative evaluation of atomic displacement it is clear that analysis based on single collision properties is favored; an example was the location of solute atoms, as described earlier.

Specific examples of channeling studies of lattice defects are described under the headings of investigation of lattice disorder (Section 4.2.4), lattice site location of solute atoms (Section 4.2.5), and surface studies (Section 4.2.6). In the last two of these categories, often a large fraction of the atoms to be studied are displaced from normal lattice sites so that the positions of these atoms can be determined relatively easily. However, the study of host atoms that are displaced by radiation damage or plastic deformation is hampered by the fact that in general the saturation concentration of intrinsic defects is less than 1%. It is then difficult to determine the positions of the displaced atoms, even using double alignment (i.e., both the incident and outgoing trajectories aligned with major crystallograhic directions), because the observed change in yield represents a small perturbation on a large background. In this case, the rate of dechanneling (i.e., the rate at which χ increases with depth because of scattering effects) is a good measure of bulk damage in the crystal and is much more sensitive than the normalized yields near the surface. An indication of the predominant type of defect can be obtained from the energy dependence of the increase in dechanneling rate

[34] L. M. Howe and M. H. Rainville, *Proc. Int. Conf. Ion Beam Modif. Mater., Albany, N.Y., 1980*, published in *Nucl. Instrum. Methods* **182/183**, 143 (1981).
[35] A. Seeger, *in* "Radiation Damage in Solids," p. 101. IAEA, Vienna, 1962.
[36] A. Guinier, *Solid State Phys.* **9**, 293 (1959).

(e.g., E^{-1} for point defects and $E^{1/2}$ for dislocations).[37,38] However, microscopic details of displacements cannot be obtained at present.

If 10–100% of host atoms are displaced from lattice sites (e.g., during amorphization of semiconductors by heavy-ion bombardment), channeling provides a quick and reliable method of measuring the fraction of the host atoms which are in a noncrystalline or different crystalline form (precipitates).[39,40]

Channeling has been used to measure solute atom positons for solute concentrations as low as 10^{-4}–10^{-5} atomic fraction, by measuring yields of backscattered particles,[9-11,41] characteristic x rays[42] or nuclear reaction products[43] (Section 4.2.5). When an appreciable fraction of the solute atoms are displaced from lattice sites, for example, by association with radiation-induced defects, the position of the displaced atoms can be determined by channeling. In this way the following studies of solute atoms have been performed: solute solubility, equilibrium lattice positions (e.g., interstitial sites of gas atoms in metals[44]), segregation or preprecipitation, and defect–solute interactions. In this last category, several examples of the trapping of intrinsic defects (vacancies and self-interstitials) by solute atoms will be given.[25,27,45] Insight into the microscopic details of radiation damage has been obtained by such studies.

Channeling also provides a useful analytical technique for characterizing crystal surfaces[31,32,46-48]; when used in conjunction with LEED, Auger spectroscopy, and nuclear microanalysis, it permits an accurate picture of various surface structures to be determined. Any displacement of the surface host atoms relative to the bulk lattice is readily detectable, and, in many

[37] Y. Quéré, *J. Nucl. Mater.* **53**, 262 (1974).
[38] S. T. Picraux, *in* "Advanced Techniques for Characterizing Microstructures," p. 283. The Metallurgical Society of AIME, Warrendale, Pennsylvania.
[39] E. Bøgh, *Can. J. Phys.* **46**, 653 (1968).
[40] J. W. Mayer, E. Eriksson, and J. A. Davies, "Ion Implantation in Semiconductors." Academic Press, New York, 1970.
[41] J. W. Mayer and E. Rimini, eds., "Ion Beam Handbook for Material Analysis." Academic Press, New York, 1977.
[42] J. F. Chemin, I. V. Mitchell, and F. W. Saris, *J. Appl. Phys.* **45**, 532 (1974).
[43] J. U. Andersen, E. Laegsgard, and L. C. Feldman, *Radiat. Eff.* **12**, 219 (1972).
[44] H.-D. Carstanjen, *Phys. Status Solidi A* **59**, 11 (1980).
[45] M. L. Swanson, *in* "Advanced Techniques for Characterizing Microstructures," p. 305. The Metallurgical Society of AIME, Warrendale, Pennsylvania.
[46] E. Bøgh, *in* "Channeling: Theory, Observation, and Applications" (D. V. Morgan, ed.), p. 435, Wiley (Interscience), New York, 1973.
[47] F. W. Saris and J. F. Van der Veen, *Proc. 7th Int. Vac. Congr., 3rd Int. Conf. Solid Surf.*, Vienna (R. Dobrozemsky, F. Rüdenauer, F. P. Viehböck and A. Breth, eds.) p. 2503. Self-published, Vienna, 1977.
[48] L. C. Feldman, *CRC Crit. Rev. Solid State Mater. Sci.* **10**, 143 (1981).

cases, the magnitude and direction of the displacement can be established. Thus, for a clean surface, channeling may yield quantitative information on lateral reconstruction, surface relaxation (or contraction), and even the vibrational amplitude of the atoms in the surface plane. Furthermore, in the case of an adsorbate-covered surface, channeling may also be used to pinpoint the exact location of the adsorbed atoms on the crystal surface.

4.2.2. Basic Channeling Theory

4.2.2.1. The Continuum Potential and the Continuum Model of Channeling.
Presented here is a condensation of some of the basic theory that is required to use the channeling technique to study the displacement of atoms from their normal sites in crystal lattices. For more extensive treatments of channeling theory, we refer the reader to several books and articles.[9,20,49-53]

Inherent in the classical description of channeling is the concept that because of the ordered structure of the single-crystal target, a charged particle incident near a major crystal direction or plane will undergo a correlated series of small-angle collisions with the rows or planes of atoms in the crystal. In the continuum model[20,49-51] of channeling, the motion of channeled particles is determined by a continuum potential, i.e., a potential obtained by replacing the actual periodic potential of the row or plane by a potential averaged over a direction parallel to the row or plane. For axial channeling, Lindhard[20] has shown that the continuum approximation is valid if the angle of incidence of the particle on a row is less than

$$\psi_1 = (2Z_1 Z_2 e^2/dE)^{1/2} \quad \text{for} \quad E > E' \quad (4.2.1)$$

and

$$\psi_2 = (Ca\psi_1/\sqrt{2}d)^{1/2} \quad \text{for} \quad E < E', \quad (4.2.2)$$

where Z_1 and Z_2 are the atomic numbers of the channeled ions and crystal atoms, respectively, e the electronic charge, d the spacing between atoms in the row (see Fig. 5), E the incident energy of the ion beam, a a screening radius, C an adjustable parameter normally set equal to $\sqrt{3}$, and $E' = 2Z_1 Z_2 e^2 d/a^2$. For He ions incident along the $\langle 110 \rangle$ axes in Al, Cu, Ag, and Au, $E' = 70$, 215, 525, and 1190 keV, respectively. In this chapter most of

---

[49] C. Lehmann and G. Leibfried, *J. Appl. Phys.* **34**, 2821 (1963).
[50] J. Lindhard, *Phys. Lett.* **12**, 126 (1964).
[51] C. Erginsoy, *Phys. Rev. Lett.* **15**, 360 (1965).
[52] D. V. Morgan, ed., "Channeling: Theory, Observations, and Applications." Wiley (Interscience), New York, 1973.
[53] D. Van Vliet, *in* "Channeling: Theory, Observation, and Applications" (D. V. Morgan, ed.), p. 37, Wiley (Interscience), New York, 1973.

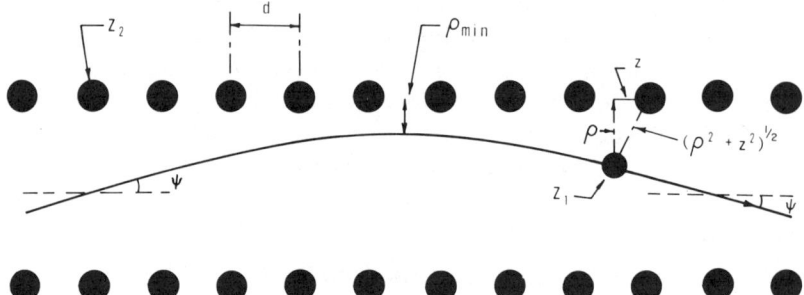

FIG. 5. Schematic diagram illustrating how a correlated sequence of collisions with an aligned row of crystal atoms (atomic number Z_2) can gently steer (channel) a particle (atomic number Z_1) moving at a relatively low angle ψ to a low-index direction. Channeling will occur for particles having trajectories such that the incident angle ψ_{in} at the midchannel plane is less than a critical angle $\psi_{1/2}$.

the cases being considered will fall in the high-energy regime described by Eq. (4.2.1), where ψ_1 values are typically $\sim 1-2°$.

Consider an ion moving at an angle ψ to a row (or string) of atoms of spacing d along the z axis and in a plane containing them (Fig. 5). Let $\rho(z)$ be its distance from the atomic string. The average interatomic potential (continuum potential) experienced by an ion at ρ is given by

$$U_1(\rho) = \frac{1}{d} \int_{-\infty}^{\infty} V(\rho^2 + z^2)^{1/2} \, dz. \qquad (4.2.3)$$

At small separations between the incident ion and a crystal atom, the interaction potential $V(r)$ is normally described by the Coulombic repulsion of the bare nuclei modified by some function of the separation that describes the electronic screening. Two of the interaction potentials that are most commonly used in channeling theories are

(1) the Lindhard standard potential[20]

$$V(r) = (Z_1 Z_2 e^2/r)[1 - r/(r^2 + C^2 a^2)^{1/2}] \qquad (4.2.4)$$

and
(2) the Molière potential[54]

$$V(r) = (Z_1 Z_2 e^2/r)(0.1 e^{-6r/a} + 0.35 e^{-0.3r/a} + 0.55 e^{-1.2r/a}). \qquad (4.2.5)$$

The screening radius a is actually an adjustable parameter but in practice it is usually determined by an analytical expression, the most common being:

[54] G. Molière, *Z. Naturforsch. A.*, **2**, 133 (1947).

4.2. CHANNELING STUDIES OF LATTICE DEFECTS

$$a = 0.8853a_0(Z_1^{2/3} + Z_2^{2/3})^{-1/2}, \quad (4.2.6)$$

where $a_0 = 0.0528$ nm is the Bohr radius.

Substituting the expression for the Lindhard standard potential Eq. (4.2.4) into Eq. (4.2.3) yields

$$U_1(\rho) = (2Z_1Z_2e^2/d)[\tfrac{1}{2}\ln\{(Ca/\rho)^2 + 1\}], \quad (4.2.7)$$

and this is often referred to as the standard continuum potential. The corresponding expression derived from the Molière potential Eq. (4.2.5) gives

$$U_1(\rho) = (2Z_1Z_2e^2/d)\{0.1K_0(6\rho/a) \\ + 0.55K_0(1.2\rho/a) + 0.35K_0(0.3\rho/a)\}, \quad (4.2.8)$$

where K_0 is a reduced Hankel function of order 0. A treatment analogous to that applied to atomic strings may be applied to the scattering of an ion from a plane of atoms. The continuum potential at a distance y from a dense plane of atoms is given by

$$Y_1(y) = (N_p) \int_0^\infty 2\pi r\, dr\, V(r^2 + y^2)^{1/2}, \quad (4.2.9)$$

where N_p is the number of atoms per unit area in the plane. Shown in Fig. 6[53] are continuum potentials for strings and planes calculated for the Lindhard

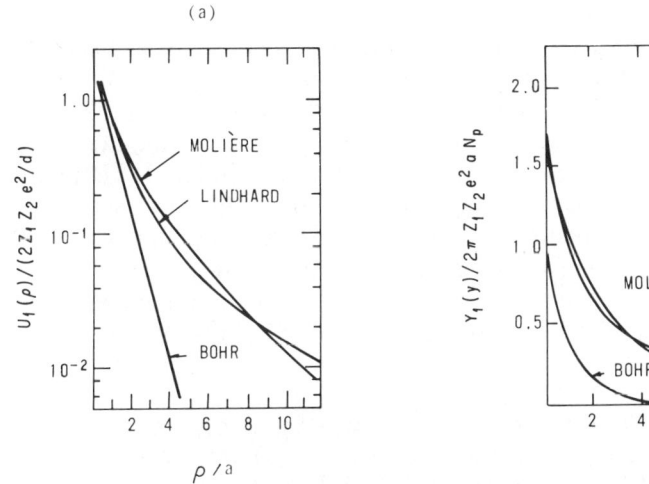

FIG. 6. (a) The continuum string potentials for the Lindhard standard, Molière, and Bohr potentials. (b) The planar continuum potentials as calculated by the Lindhard standard, Molière, and Bohr potentials. (Adapted from Van Vliet.[53])

standard and Molière potentials as well as for the Bohr[55] potential, which is given by $V(r) = Z_1 Z_2 e^2 e^{-r/a}/r$. The continuum potentials derived using the Lindhard and Molière potentials are in reasonably good agreement over most of the region of ion-crystal atom separations applicable to channeling behaviour except at relatively large separations, where the Lindhard potential is considered to be too strong. Both of these potentials are more suitable for use in channeling calculations than is the Bohr potential, which falls off too rapidly with increasing separation.

In a crystal lattice the continuum potential experienced by a channeled particle will be the sum of the potentials due to all of the rows (or planes) of atoms in the crystal. For a static lattice the total potential measured at a position ρ in the plane normal to the axial direction is given by

$$U(\rho) = \sum_i U_1(\rho_i) - U_{\min}, \qquad (4.2.10)$$

where the ρ_is are the positions of the atomic rows with respect to the position ρ (also measured in the normal plane) and $U_{\min}$ is the minimum potential energy at any point in the transverse plane. For a thermally vibrating lattice the continuum potential can be modified in various ways.[56-59] One of the simplest approaches[28] is to replace $U_1(\rho)$ by some maximum value of the thermally modified string potential $E_\perp^B$ when $U_1(\rho) > E_\perp^B$, where

$$E_\perp^B = \int_0^\infty \frac{d\rho^2}{u_2^2} \exp\left(\frac{-\rho^2}{u_2^2}\right) U_1(\rho) \qquad (4.2.11)$$

and u_2 is the root-mean-square thermal displacement in a plane. This procedure affects the continuum potential only very close to the atomic rows. An example of the application of the above procedure for the calculation of the continuum potential (based on a Molière potential) experienced by He$^+$ ions in a $\langle 110 \rangle$ axial channel of Al is shown in Fig. 7.[28] Because of symmetry, only one quarter of the channel is shown. Note that for the $\langle 110 \rangle$ channel in the fcc structure the potential minimum does not occur at the center of a channel, whereas for a channel such as $\langle 100 \rangle$ it does.

4.2.2.2. Particle Trajectories in the Crystal Lattice. For particles incident on a single crystal, three basic particle trajectories can be distinguished, as illustrated in Fig. 8.[60]

---

[55] N. Bohr, *Mat.-Fys. Medd.—K. Dan. Vidensk. Selsk.* **18**, No. 8 (1948).
[56] B. R. Appleton, C. Erginsoy, and W. M. Gibson, *Phys. Rev.* **161**, 330 (1967).
[57] D. V. Morgan and D. Van Vliet, *Radiat. Eff.* **5**, 157 (1970).
[58] J. H. Barrett, *Phys. Rev. B* **3**, 1527 (1971).
[59] Y. Quéré, *Phys. Rev. B* **11**, 1818 (1975).
[60] L. M. Howe and J. A. Davies, in "Site Characterization and Aggregation of Implanted Atoms in Materials" (A. Perez and R. Coussement, eds.), p. 241, Plenum, New York, 1980.

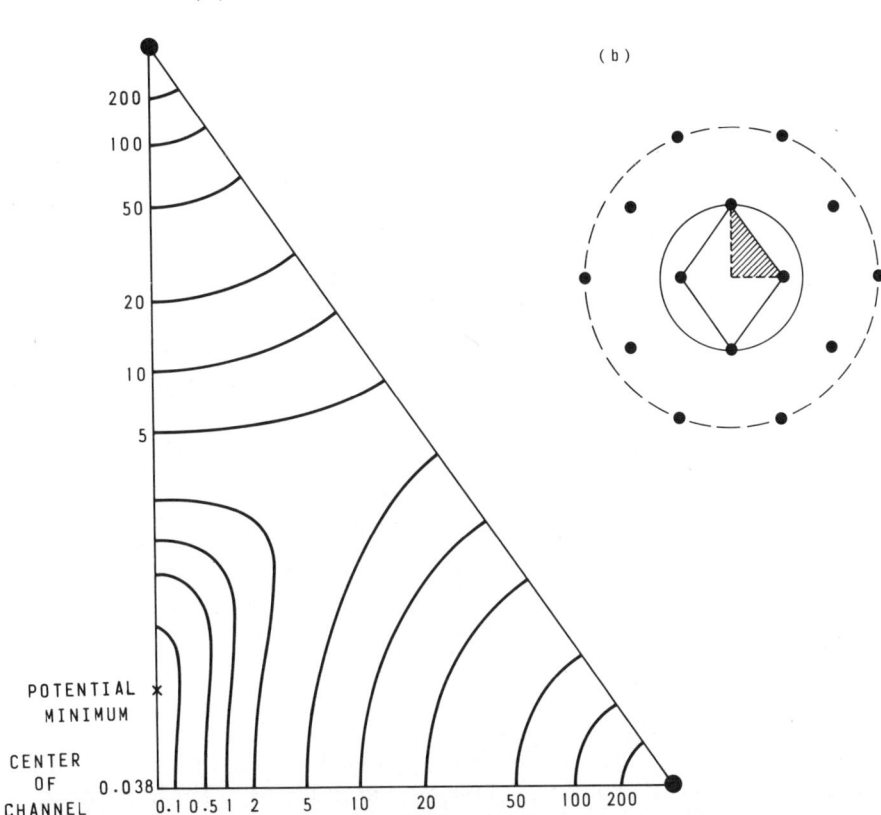

FIG. 7. (a) The equipotential contours (in units of eV) for ^{4}He$^+$ ions in Al for one quarter of the $\langle 110 \rangle$ channel [as shown by the shaded region in (b)]. The results were obtained for the 14 strings of atoms indicated by filled circles in (b). In (b) the first and second cells of strings are designated by solid and dashed lines, respectively. (From Matsunami et al.[28])

(1) An energetic charged particle entering a crystal lattice at an angle ψ_{in}, less than a predictable critical angle $\psi_{1/2}$, of the atomic row (or plane) is steered by successive gentle collisions (trajectory A) and is thereby prevented from entering a forbidden region around each lattice row. The radius ρ_{min} of this forbidden region may be equated roughly to the two-dimensional vibrational amplitude u_2 in the transverse plane. Within the crystal, the motion of a particle moving at a relatively low angle ψ to a low-index direction (as in Fig. 5) may be described by the Hamiltonian

$$H = U(\rho) + (p_x^2 + p_y^2 + p_z^2)/2m_1, \qquad (4.2.12)$$

where p_x, p_y, p_z are its components of momentum and m_1 is its mass. As a consequence of the continuum approximation, $U(\rho)$ is independent of z,

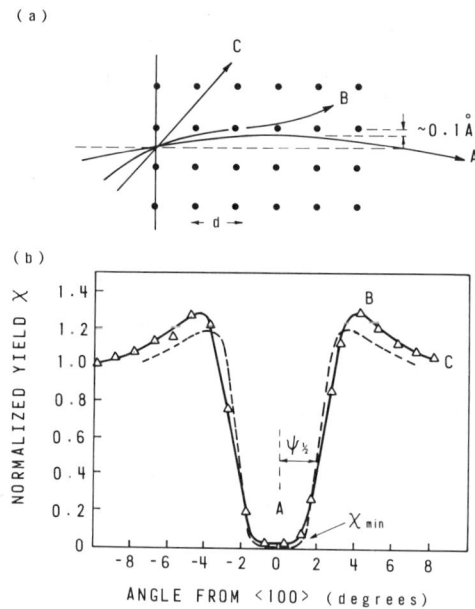

FIG. 8. (a) Schematic diagram of three typical charged-particle trajectories in a crystal. (b) Experimental (△) and calculated (---) angular dependence of the yield of a typical close-encounter process (in this instance, Rutherford scattering of 480-keV protons in $\langle 100 \rangle$ W). (From Howe and Davies.[60])

and hence the momentum component p_z parallel to the atomic row and the parallel velocity v_z are constants of the motion and $z = v_z t$. The trajectory of the particle can therefore be completely described by determining its transverse motion in the xy plane. In this transverse motion the ion moves with a velocity $v_\perp = v \sin \psi \approx v\psi$ in the potential field $U(\rho)$. The total energy of the transverse motion is

$$E_\perp = U(\rho) + (p_x^2 + p_y^2)/2m_1. \qquad (4.2.13)$$

For particles near the crystal surface where energy losses and multiple scattering can be neglected, $E_\perp$ is constant and is given by the following expression:

$$E_\perp = U(\rho_{\text{in}}) + E\psi_{\text{in}}^2,$$

where ρ_{in} is the incident position of the particles. Ions of transverse energy $E_\perp$ can only reach that portion of the transverse plane where $U(\rho_{\text{in}}) < E_\perp$, i.e., are contained within an accessible area $A(E_\perp)$.

(2) When the incident angle is much larger than $\psi_{1/2}$ the particle has no "feeling" for the existence of a regular atomic lattice, and so has a random trajectory (C).

4.2. CHANNELING STUDIES OF LATTICE DEFECTS

(3) If the incident angle ψ_{in} is only slightly larger than $\psi_{1/2}$, then the particle trajectory (B) actually has an enhanced probability of being close to the atomic rows, and hence of undergoing violent collisions.

A very important consequence of the above model is that all physical processes requiring smaller impact parameters than ρ_{min} (~ 0.01 nm) are completely prohibited for a channeled beam. Consequently, the yield of such a process is a quantitative measure of the nonchanneled fraction of the beam, and so provides a sensitive "detector" for studying the transition between channeled and nonchanneled trajectories. The experimentally observed and theoretically predicted orientation dependence of a typical close-encounter process is shown in Fig. 8b; the yields in regions A, B, and C arise from particles with the corresponding trajectories in Fig. 8a. The normalized yield χ_{min} at $\psi_{in} = 0$ does not fall quite to zero because there is still a small random fraction (~ 0.01), determined by the point of impact on the crystal surface.

4.2.2.3. Aligned Yields and Critical Angles. The normalized yield χ is defined as the ratio of the yield in a given direction to that in a random direction; it is therefore a direct measure of the unchanneled fraction of the beam. The minimum value χ_{min} occurs for perfect alignment and shallow depths. The observed critical angle $\psi_{1/2}$ and normalized yield χ both depend on the crystal temperature and the depth beneath the surface at which the measurements are made. Various calculations have been made of $\psi_{1/2}$ and χ_{min}. In one of the most detailed investigations, Barrett[58] performed Monte Carlo computer calculations from which he derived empirical formulas to fit the calculated results. For axial channeling, Barrett[58] obtained

$$\chi_{min} = Nd\pi[C(\Delta)u_2^2 + C'(\Delta)a^2], \qquad (4.2.14)$$

where N is the atomic density and Δ is the variance of the Gaussian distribution of the beam directions used in the simulations. For $\Delta = 0$, $C(0) = 3.0 \pm 0.2$ and $C'(0) = 0.2 \pm 0.1$. Equation (4.2.14) is valid for fairly shallow depths of penetration of the ion beam in the crystal (i.e., before dechanneling effects become important) and represents the contribution to χ exclusive of the "surface peak" contribution due to disordered surface region of the crystal (i.e., surface oxides, reconstructed or relaxed surface layers). Similarly, he obtained

$$\psi_{1/2} = k[U_1(mu_1)/E]^{1/2}, \qquad (4.2.15)$$

where U_1 is the static string potential [Eq. (4.2.3)], u_1 the root-mean-square thermal displacement in one direction, and k and m are adjustable fitting parameters (for protons, $k = 0.80$ and $m = 1.2$ give the best overall fit to the data). Eq. (4.2.15) may be rewritten

$$\psi_{1/2} = k\psi_1[f(mu_1/a)]^{1/2}, \qquad (4.2.16)$$

where ψ_1 is the Lindhard angle given in Eq. (4.2.1) and $f(mu_1/a)$ is a function of the interatomic potential chosen (see Table 1 in Barrett[58]). In Eqs. (4.2.14) and (4.2.16), for the screening radius a, Barrett[58] used $a = 0.8853a_0(Z_1^{1/2} + Z_2^{1/2})^{-2/3}$. Experimental values of χ_{min} and $\psi_{1/2}$ obtained from good-quality crystals are usually in quite good agreement with the values calculated from Eqs. (4.2.14) and (4.2.16). Values of $\psi_{1/2}$ vary from a few hundredths of a degree to a few degrees and values of χ_{min} are typically about 0.2–0.4 for planar channeling and about 0.01–0.05 for axial channeling. The angles ψ_1 and ψ_2, as determined by Lindhard[20] [Eqs. (4.2.1) and (4.2.2)], were intended to be characteristic of the angular extent of continuum string effects but not necessarily to have a precise experimental interpretation. However ψ_1 and ψ_2 are in qualitative agreement with experimentally measured half-widths and do predict quite accurately the dependence of $\psi_{1/2}$ upon Z_1, Z_2, E, and d, as can be seen from Eq. (4.2.16) for example. Note that even in an undamaged crystal $\chi_{min} \simeq 0.01$. (i.e, the background level is ~ 1%). Therefore, for measuring the total number of displaced atoms at least 1–2% of the lattice atoms have to be displaced in order to produce a reasonable signal-to-noise ratio.

The expression for χ_{min} given in Eq. (4.2.14) applies to a single-alignment situation, in which the incident beam is aligned with an axial direction in the crystal and the detector recording the yield of some close-encounter process is along a random crystal direction (or conversely). For a double-alignment situation in which the incident beam is aligned with an axial direction and the emergent beam is subject to blocking, by also having the detector aligned with an axial direction, the double-alignment minimum yield χ'_{min} is considerably reduced compared to the single-alignment minimum yield χ_{min}. In fact, it has been shown[58,61] that

$$\chi'_{min} = \nu(\alpha)\chi^2_{min}, \qquad (4.2.17)$$

where $\nu(\alpha)$ is a geometrical constant. Feldman and Appleton[61] found that $\nu(\alpha)$ was in the region 1.5 to 2.0 depending upon the angle α between the detector and beam directions. Barrett's computer calculations[58] yield $\nu(90°) = 1.1 \pm 0.1$ and $\nu(180°) = 1.2 \pm 0.1$. More recently, Kerkow et al.[62] measured backscattered yields of 300-keV protons from Si single crystals under channeling ($\chi_{ch}(z)$), blocking ($\chi_{bl}(z)$) and double-alignment ($\chi'_{da}(z)$) conditions for various depths of penetration of the incident beam z and found that the relation

$$\chi'_{da}(z) = \chi_{ch}(z)\chi_{bl}(z) \qquad (4.2.18)$$

---

[61] L. C. Feldman and B. R. Appleton, *Appl. Phys. Lett.* **15**, 305 (1969).
[62] H. Kerkow, H. Pietsch, and F. Täubner, *Radiat. Eff. Lett.* **50**, 169 (1980).

was obeyed within an experimental error of 10%. Hence with the double-alignment technique, an overall attenuation factor of 10^3-10^4 can be achieved in the background level from the undamaged crystal.

4.2.2.4. The Flux Distribution of Channeled Particles and the Yield from Atoms Displaced from Normal Lattice Sites. Within an axial or planar channel the flux of beam particles just below the crystal surface (e.g., $z \lesssim 100$ nm for 1-MeV He in $\langle 100 \rangle$ of Cu) is not uniform but oscillates with depth, as can be demonstrated by computer simulations.[63] Hence at these shallow depths the flux profiles should be obtained from Monte Carlo simulations of ion trajectories if one wishes to obtain information about atom displacements at discrete depths, i.e., rather than information averaged over larger depth intervals (i.e., $\Delta z \gtrsim 100$ nm). As the ion beam progresses more deeply into the crystal there is a trend towards statistical equilibrium, and once this is reached it has been shown by Lindhard[20] that there is an equal probability $P_0(E_\perp, \rho)$ of finding an ion anywhere within its accessible transverse space. Hence one can write

$$P_0(E_\perp, \rho) = \begin{cases} 1/A(E_\perp) & \text{for } U(\rho) \leq E_\perp, \\ 0 & \text{for } U(\rho) > E_\perp. \end{cases} \quad (4.2.19)$$

If the ions are incident uniformly on the crystal surface and the transverse energy at the crystal is conserved, then the flux distribution at a position ρ in the channel for an incident angle ψ_{in} (the tilt angle from the major axis) is given by

$$F(\rho, \psi_{in}) = \int_{A_0} \frac{dA(\rho_{in})}{A(E_\perp)}. \quad (4.2.20)$$

Here $A(E_\perp)$ is the accessible area where the transverse energy $E_\perp = U(\rho_{in}) + E\psi_{in}^2$ is greater than $U(\rho)$, and $dA(\rho_{in})$ is the small area around the incident position of ions ρ_{in}. The integration is performed over the area A_0 of the channel where $U(\rho) \leq E_\perp$ is satisfied. $U(\rho)$ is obtained by summing the string potentials (as in Fig. 7). Since the particle distribution in transverse energy $g_0(U(\rho_{in}))$ for incidence along a major axis is related to $dA(\rho_{in})$ by

$$dA(\rho_{in})/A_0 = g_0(U(\rho_{in}))dU(\rho_{in}), \quad (4.2.21)$$

Eq. (4.2.20) can be reduced to one-dimensional integration

$$F(\rho, \psi_{in}) \int_{U(\rho_{in}) \geq U(\rho) - E\psi_{in}^2} \frac{A_0 g_0(U(\rho_{in})) \, dU(\rho_{in})}{A(E_\perp)}. \quad (4.2.22)$$

In order to avoid the divergence of the integral due to the fact that $A(0) = 0$,

---

[63] D. V. Morgan and D. Van Vliet, *Radiat. Eff.* **12**, 203 (1972).

a term $\delta E_\perp$ can be added to $E_\perp$ which corresponds to the average increase of $E_\perp$ due to scattering of the ion by electrons as it penetrates a distance δz.[28,64] Also the angular spread $\Delta\psi$ of the incident beam may be incorporated into $g_0(E_\perp)$.[28]

Various calculations have been made for flux distributions of channeled ions in different crystals under a variety of experimental conditions. In Fig. 9, for example, equiflux contours are shown for 1-MeV He ions in a $\langle 110 \rangle$ axial channel of Al at 30 K.[28] Results are presented for analytical calculations [Eq. (4.2.22)] as well as for Monte Carlo simulations by Barrett[65] that were averaged over the depth interval 50–150 nm for consistency with the analytical calculations. The agreement between the results obtained by the two methods is good. Note that the flux near the center of the channel, or at

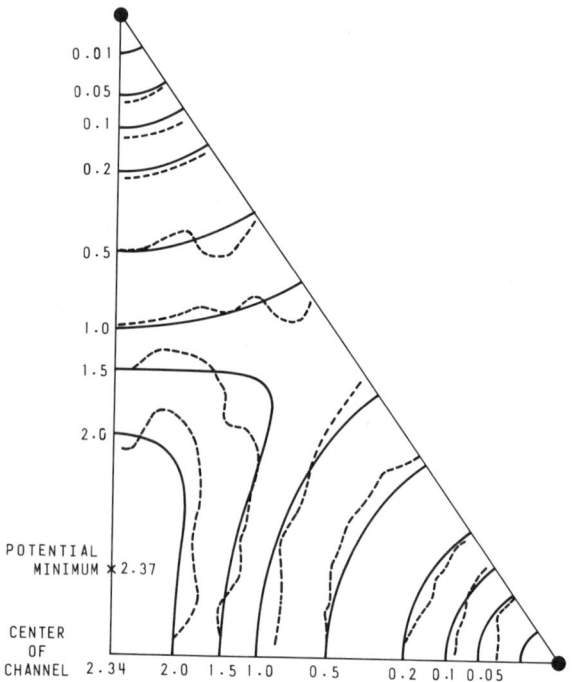

FIG. 9. Equiflux contours of one quarter of a $\langle 110 \rangle$ channel for 1-MeV ^{4}He$^+$ ions in Al at 30 K. Results of an analytical calculation are shown as solid curves and those of Barrett's Monte Carlo simulation are shown as dashed lines. The flux values have been normalized to the value for random incidence. (From Matsunami et al.[28])

[64] D. Van Vliet, *Radiat. Eff.* **10**, 137 (1971).
[65] J. H. Barrett, *Proc. 4th, Conf. Appl. Small Accel., Denton, Tex.* (J. L. Duggan and I. L. Morgan, eds.), p. 571. IEEE Publ. No. 76CH 1175-9 NIS. IEEE, New York, 1976.

the potential minimum, is greater than twice the value of the flux for random incidence of the ion beam and that the flux near the atomic rows is extremely small.

The yield from atoms displaced from their normal sites may be obtained by averaging the flux around the equilibrium positions of the displaced atoms; hence

$$Y(s) = \int F(\rho,\psi_{in}) \, P_s(\rho) \, d\rho, \qquad (4.2.23)$$

where s is the displacement of solute atoms from a lattice site along a particular $\langle lmn \rangle$ direction and $P_s(\rho)$ is the probability for the atoms being at ρ in the channel, as given by

$$P_s(\rho) = (d\rho/\pi u_2^2) \exp(-|\rho - s|^2/u_2^2). \qquad (4.2.24)$$

Shown in Fig. 10 are calculated yields [using Eqs. (4.2.23) and (4.2.24)] from displaced atoms for a $\langle 110 \rangle$ channel in Al at 30 K.[28] The calculated yields are given for various values of the displacements along $\langle 100 \rangle$ directions from the lattice point. For this $\langle 110 \rangle$ channel the yields are averaged since one third of the $\langle 100 \rangle$ displacements are along the long diagonal and two thirds are along the short diagonal of the $\langle 110 \rangle$ channel cross section. It can be seen that, for large values of s, the yield in the $\langle 110 \rangle$-aligned direction can be more than twice the random yield. On the other hand, in the $\langle 100 \rangle$- and $\langle 111 \rangle$- aligned directions, these $\langle 100 \rangle$ displacements produce a considerably smaller yield.[28]

4.2.2.5. *The Dechanneling of a Channeled Ion Beam.* As channeled ions penetrate a crystal, they can be deflected out of the channels (dechanneled) by the multiple scattering effect of electronic or nuclear collisions. Nuclear collisions that contribute to dechanneling may arise from thermal vibrations or from crystal-lattice defects. Dechanneling can be treated using the continuum approximation by solving the diffusion equation for the multiple electronic and nuclear scattering of ions to find the distribution $g(E_\perp, z)$ of the transverse ion energy $E_\perp$ as a function of depth.[66-70] An ion is assumed to be dechanneled when its transverse energy exceeds a critical value $E_\perp^c$. The

---

[66] C. Ellegaard and N. O. Lassen, *Mat.-Fys. Medd.—K. Dan. Vidensk. Selsk.* **35**, No. 16 (1967).

[67] K. Morita and N. Itoh, *J. Phys. Soc. Jpn.* **30**, 1430 (1971).

[68] E. Bonderup, H. Esbensen, J. U. Andersen, and H. E. Schiøtt, *Radiat. Eff.* **12**, 261 (1972).

[69] H. E. Schiøtt, E. Bonderup, J. U. Andersen, and H. Esbensen, *in* "Atomic Collisions in Solids" (S. Datz, B. R. Appleton, and C. D. Moak, eds.), Vol. 2, p. 843. Plenum, New York, 1975.

[70] N. Matsunami and L. M. Howe, *Radiat. Eff.* **51**, 111 (1980).

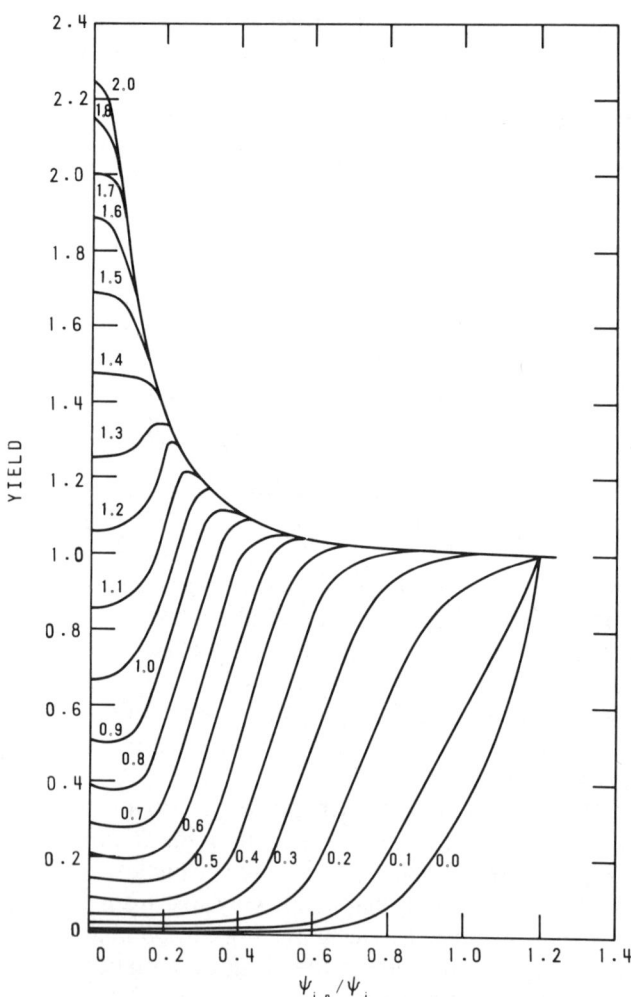

FIG. 10. Calculated yields in a ⟨110⟩ channel for atoms displaced along ⟨100⟩ directions as a function of the incident angle ψ_{in} normalized to the Lindhard characteristic angle ψ_1. The numbers on the curves designate values of the displacements in angstroms. The calculations are for 1-MeV ^{4}He$^+$ ions in a ⟨110⟩ channel of Al at 30 K at a depth of 100 nm and for $\Delta\psi = 0.048°$. (From Matsunami et al.[28])

dechanneled fraction of ions is

$$\chi(z) - \chi(0) = \int_{E_\perp^c}^{\infty} g(E_\perp, z) \, dE_\perp. \quad (4.2.25)$$

Calculations of dechanneling using the diffusion equation have been confined mainly to investigating the temperature and energy dependence of dechanneling in crystals that were relatively free of lattice defects.[66-70] Matsunami and Itoh[71,72] have shown, however, that permanently displaced atoms may alter $g(E_\perp, z)$ so as to enhance greatly the dechanneling caused by lattice vibrations and electronic collisions. This effect may modify considerably the determination from channeling data of the depth profile of lattice defects near the crystal surface and at relatively large depths. Also, Matsunami et al.[73] have performed the diffusion calculation for some general types of atomic displacements.

According to a phenomenological dechanneling model, the rate at which the channeled fraction $(1 - \chi)$ of ions becomes dechanneled by a concentration n_d of irradiation-produced defects is given by[39,74]

$$d(1 - \chi)/dz = -(1 - \chi)(n_d \sigma_d + \xi_{th}), \qquad (4.2.26)$$

where σ_d is an effective defect cross section for dechanneling and ξ_{th} is the effect of thermal vibrations. If n_d is independent of depth, then

$$\chi(z, n_d, T) = 1 - [1 - \chi(0)] \exp[-(n_d \sigma_d + \xi_{th})z], \qquad (4.2.27)$$

where $\chi(z, n_d, T)$ is now expressed as a function of depth z, defect concentration n_d, and temperature T and where $\chi(0)$ is the normalized yield at the surface. For small χ (<0.2) and if σ_d is independent of T, Eq. (4.2.27) implies that the contributions of point defects and thermal vibrations to dechanneling are independent. In some cases this has been found to hold, whereas in others it certainly does not.[75] An example of the latter is shown in Fig. 11 for a Cu–0.06 at. % Au crystal, where χ is plotted as a function of depth z for two different temperatures before and after irradiation. It is evident that the effect of the irradiation-produced defects on the dechanneling becomes larger as the crystal temperature is increased.

In general, in a crystal containing lattice defects the normalized yield χ_d will be due to a scattering contribution χ_R from the dechanneled fraction of the beam by all the atoms plus a contribution from the channeled part of the beam which may be directly backscattered (into the detector) by the defects. Then[38]

$$\chi_d(z) = \chi_R(z) + (1 - \chi_R(z))(fn_d/N), \qquad (4.2.28)$$

---

[71] N. Matsunami and N. Itoh, *Phys. Lett. A.* **43A**, 435 (1973).

[72] N. Matsunami and N. Itoh, in "Atomic Collisions in Solids" (S. Datz, B. R. Appleton, and C. D. Moak, eds.), Vol. 1, p. 175. Plenum, New York, 1975.

[73] N. Matsunami, T. Goto, and N. Itoh, *Radiat. Eff.* **33**, 209 (1977).

[74] K. L. Merkle, P. P. Pronko, D. S. Gemmell, R. C. Mikkelson, and J. R. Wrobel, *Phys. Rev. B* **8**, 1002 (1973).

[75] L. M. Howe, M. L. Swanson, and A. F. Quenneville, *Nucl. Instrum. Methods* **132**, 241 (1976).

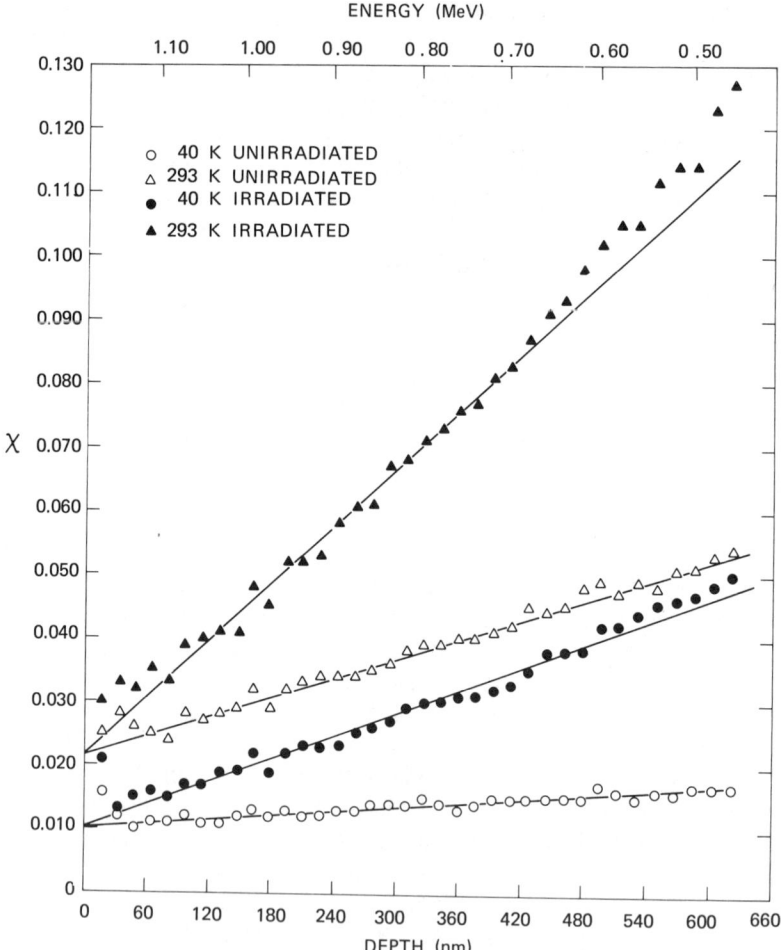

FIG. 11. Normalized backscattering yields χ of 1.5-MeV ^{4}He$^+$ ions at temperatures of 40 K and 293 K as a function of backscattered ion energy (hence also depth) for a Cu–0.06 at. % Au crystal. Values of χ are shown for the unirradiated crystal and for the same area of the crystal after a random irradiation of 8×10^{15} 1.5-MeV ^{4}He$^+$ ions cm^{-2} at 70 K followed by a 10-min anneal at 300 K. (From Howe et al.[75])

where f is the fractional atomic scattering per defect, which accounts for any difference between the number of defects and the effective number of scattering centers per defect. If n_d is a function of depth, then the scattering contribution from the dechanneled fraction of the beam may be expressed as[38]

$$\chi_R(z) = \chi_v(z) + (1 - \chi_v(z))[1 - \exp(-\int_0^z \sigma_d n_d(z')\, dz')], \quad (4.2.29)$$

where $\chi_v(z)$ is the contribution from the perfect crystal. This equation is applicable to cases where the contributions of point defects and thermal vibrations to dechanneling are essentially independent and where the probabilities for dechanneling in a disordered and a perfect crystal are additive.

4.2.3. Experimental Techniques

The essential features of a typical experimental channeling facility are illustrated schematically in Fig. 12. As noted in Section 4.2.1 three basic components are required[21]:

(i) a small accelerator—typically in the 0.5–3.0-MeV energy region and capable of producing well-collimated beams of H^+, D^+, or He^+ in the intensity range 10^{-10}–10^{-7} A;

(ii) a high-vacuum target chamber-equipped with a two-axis or three-axis precision goniometer, a suitable faraday cup for beam-current integration, and other auxiliary features such as target temperature control;

(iii) one or more nuclear-particle detectors plus the associated electronic equipment for energy analysis of the detected particles.

4.2.3.1. Accelerator System.
A 2–3-MV Van de Graaff or Dynamitron accelerator, such as was developed originally for low-energy nuclear physics, is an ideal type of machine for channeling studies. The beam divergence is small, and the available current is usually several orders of magnitude greater than required, thus making it a relatively easy task to achieve the desired beam collimation of 0.05° or better. Magnetic deflection (Fig. 12) is required in order to select the desired ion species and also to maintain a constant energy beam. In most channeling experiments, however, energy stability is not a particularly critical requirement, and the ~1-2keV energy control that can readily be achieved with a set of stabilizer slits plus a feedback amplifier is usually quite adequate.

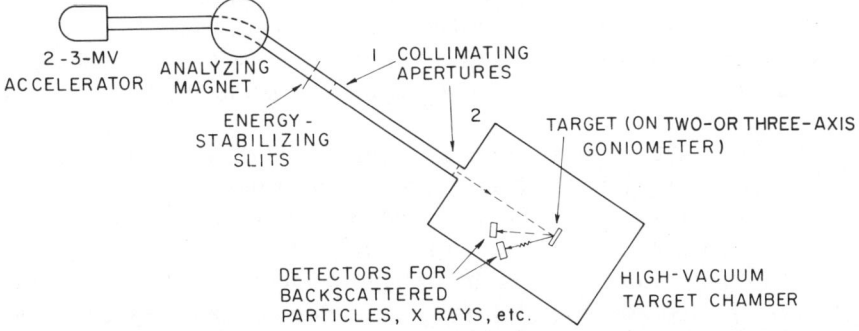

FIG. 12. Schematic outline of an experimental beam line for channeling experiments.

4.2.3.2. Target-Chamber–Goniometer System. In the target chamber, the most important feature obviously is the goniometer on which the single-crystal target is mounted. Since critical angles for channeling studies are usually of the order of $0.1–1°$, the angular precision of the goniometer should be $\lesssim 0.02°$. Two-axis and three-axis units with suitable precision are now available commercially, but many groups prefer the flexibility of one constructed to their own specifications. The degree of sophistication can vary all the way from simple manually operated units to fully automated computer-controlled goniometers with shaft-encoder readout of each axis. However, it is probably worth noting that most groups (including the authors of this chapter) are still using relatively simple manually operated two-axis goniometers (with tilt motion θ and azimuthal rotation ϕ controllable to $\sim 0.01°$) and find them completely adequate for almost all channeling studies.

Another important problem, and one which is often underestimated, is that of accurate current integration. Although a precision of 10–20% is fairly easy to achieve (e.g., by using a negatively charged grid in front of the target to suppress secondary electron emission), any greater precision requires that the target be mounted inside a properly designed faraday cup. One way to avoid the complication of mounting a faraday cup on the goniometer is to use an intermittent beam chopper—i.e, a rotating or vibrating paddle—to intercept the beam downstream of collimating aperture 2 (Fig. 12). By integrating the current falling on this paddle, one obtains a fairly reproducible *relative* measure of the integrated beam intensity during each run; this chopper can be calibrated from time to time by inserting a proper faraday cup into the beam path just ahead of the target. Since additional facilities for heating and cooling the target are often required, the use of an intermittent beam chopper for current measurement greatly simplifies the construction of a thermally isolated target stage on the goniometer.

There is a twofold reason for equipping the target stage with suitable cooling and heating facilities. First, this enables annealing studies to be performed in situ and monitored by means of channeling. Second, the detection sensitivity of the channeling technique itself can be enhanced considerably by cooling the target to liquid-nitrogen temperature or even lower.

One big disadvantage of most accelerators and their associated beam transport systems is that they operate under rather modest vacuum conditions, typically $\sim 1 - 4 \times 10^{-4}$ Pa, and hence for low-temperature experiments the target chamber requires some form of differential pumping. Even at room temperature, a target surface will rapidly become contaminated with an unacceptable (hydro)carbon deposit unless suitable steps are taken

4.2. CHANNELING STUDIES OF LATTICE DEFECTS

to reduce drastically the partial pressure of unwanted hydrocarbons in the chamber. One simple and particularly effective way of eliminating this surface contamination problem is to surround the target almost completely by a large cylinder cooled to ~ 20 K; this cryoshield maintains ultraclean conditions[76] at the target surface with respect to all condensable gases (N_2, O_2, CO, hydrocarbons, etc.) since at 20 K the only gases having vapor pressures greater than 10^{-8} Pa are H, He, and Ne. An additional advantage of this large shield is that if electrically isolated, it simultaneously provides an excellent faraday cup around the target for quantitative beam-current integration. Such a cryoshield system has been extensively used in low-temperature channeling studies. It has even been used in some of the first channeling applications[31] to surface studies (Section 4.2.6) since it provides a quick and easy way to achieve almost ultrahigh-vacuum (UHV) conditions within an ordinary vacuum system.

For more flexibility in surface studies, it is preferable to use a conventional UHV target chamber, containing the necessary channeling goniometer and detector systems, plus other conventional surface analysis probes such as low-energy electron diffraction (LEED), Auger spectroscopy, and an Ar^+ sputter gun. This UHV chamber must then be connected to the accelerator beam transport system by means of a differentially pumped section of beam line, as shown in Fig. 13.[77] Such systems, although rather

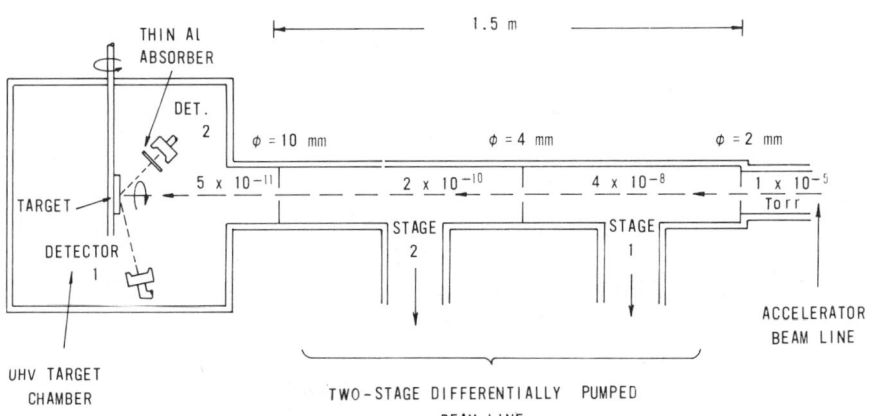

FIG. 13. Schematic diagram of a UHV target chamber, plus the differentially pumped beam line to a 2.5-MV Van de Graaff accelerator. Detector 1 is for RBS measurements, detector 2 (+ Al absorber) is for nuclear microanalysis. (Adapted from Davies and Norton.[77])

---

[76] J. Bøttiger, J. A. Davies, J. Lori, and J. L. Whitton, *Nucl. Instrum. Methods* **109**, 579 (1973).

[77] J. A. Davies and P. R. Norton, *Nucl. Instrum. Methods* **168**, 611 (1980).

complex, enable us to achieve the highly desirable goal of combining several different surface analytical techniques in a single investigation.

4.2.3.3. Nuclear Detector System. The third essential component in a channeling experiment is the detector system and its associated electronics. Particles that have undergone wide-angle ($\sim 160°$ in Fig. 12) Rutherford scattering may be detected by means of a solid state surface-barrier detector and their energy spectrum recorded on a multichannel analyzer. This backscattered energy spectrum contains detailed information[21,41] on the mass and depth distribution of the scattering centers in the crystal, as shown in Fig. 14. In this particular case, the solute atoms (Sb) are not uniformly distributed throughout the sample, but are implanted just beneath the Si surface. Since ^{4}He ions lose $\sim 41\%$ and $\sim 11\%$ of their energy in scattering from ^{28}Si and ^{122}Sb, respectively, we observe a sharp peak at $0.89E_0$ due to the Sb-implanted atoms and a continuous spectrum with a sharp threshold at $0.59E_0$ due to the thick Si substrate.

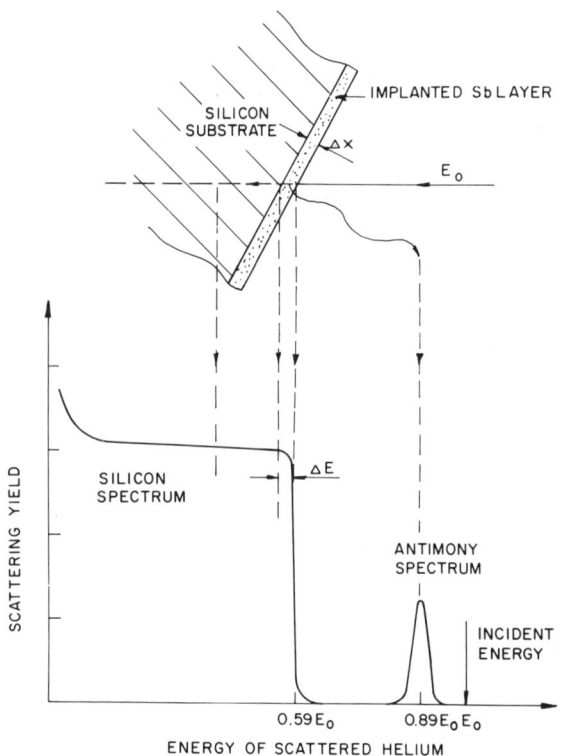

FIG. 14. Typical RBS energy spectrum for 1-MeV ^{4}He$^+$ in a randomly oriented Si substrate containing some heavy impurity atoms (Sb) in the near-surface region.

4.2. CHANNELING STUDIES OF LATTICE DEFECTS

Rutherford backscattering (RBS) is by far the most useful close-encounter process for investigating channeling behavior since it permits simultaneous observation of the depth dependence of lattice disorder (Section 4.2.4) and of the lattice location of heavier solute atoms (Section 4.2.5). Furthermore, the RBS cross sections are large, compared to nuclear reaction cross sections, and they vary in a smooth and accurately predictable manner[78] with atomic number and beam energy, thus making a wide range of systems easily accessible to quantitative analysis. Further details on RBS analysis are given in Section 4.1.1.

In some studies an alternative close-encounter process such as a nuclear reaction (Section 4.1.2) or inner-shell x-ray production (Section 4.1.3) is needed to detect the interaction of the channeled beam with specific foreign atoms in the lattice. This is particularly true whenever the foreign atoms are lighter than the host atoms. If the reaction produces energetic charged particles—as, for example, in (d,p) or (p,α) reactions—then the same type of surface barrier detector may be used as for backscattering; in other cases, it would be necessary to include a suitable γ- or x-ray detector in the target chamber. Note that these different close-encounter processes may be used interchangeably in studying channeling effects, as long as their impact parameters are all less than the mean vibrational amplitude (typically ~ 0.01 nm) of the lattice atoms.

The Ion Beam Handbook for Material Analysis[41] can serve as a very useful guide to the various close-encounter processes available; it also contains an extensive compilation of relevant cross-section data, tables, graphs, etc., which greatly facilitates the subsequent data handling.

Figure 15a illustrates the role of channeling in a typically thick target RBS experiment. The random spectrum was obtained by tilting the crystal away from perpendicular incidence and then rotating it continuously during the measurement. The aligned spectrum shows the large yield attenuation that occurs whenever a crystal axis, $\langle 111 \rangle$ in this case, is parallel to the incident beam direction. Note the small peak at the high-energy edge of the aligned spectrum; its area is an accurate measure of the number of unshadowed Si atoms in the surface region of the crystal and, as we discuss in Section 4.2.6, this sometimes provides very useful structural information about the crystal surface.

Detailed orientation scans from two different depth regions in the Si crystal are shown in Fig. 15b. These were obtained by recording the yield in the two narrow energy regions 1 and 2 (Fig. 15a) while rotating the crystal stepwise through the $\langle 111 \rangle$ direction. Thus, the two parts of Fig. 15 depict two alternative ways of presenting the same channeling data: in (a) we select an angle of incidence and then look in detail at the energy (i.e., depth)

---

[78] J. L'Ecuyer, J. A. Davies, and N. Matsunami, *Nucl. Instrum. Methods* **160**, 337 (1979).

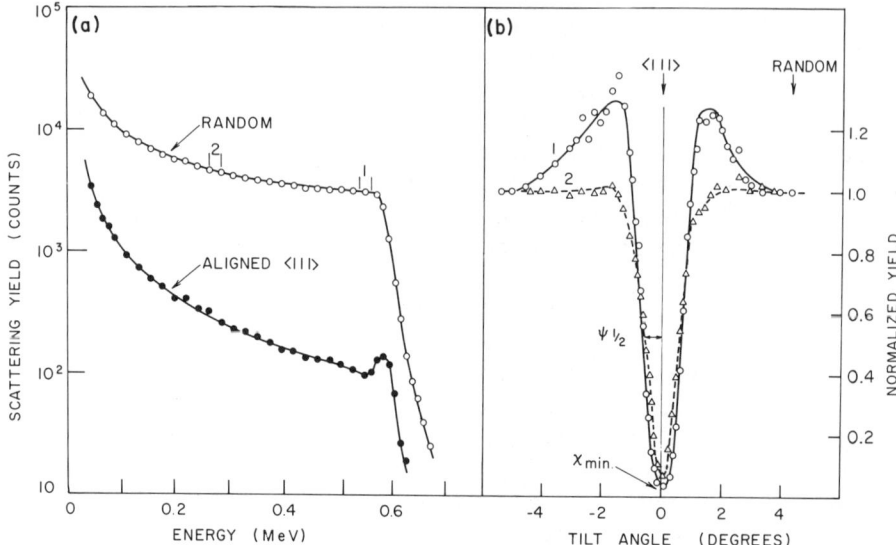

FIG. 15. RBS spectra for 1-MeV ^{4}He$^+$ ions in silicon. (a) Detailed RBS energy spectra for random and aligned $\langle 111 \rangle$ directions of incidence. The energy regions 1 and 2 correspond to mean depths of 90 nm and 600 nm, respectively. (b) Orientation dependence of the normalized (i.e., aligned/random) yield measured in the energy regions 1 and 2 in (a). Tilt angles marked $\langle 111 \rangle$ and RANDOM show the orientations at which the two energy spectra in (a) were taken.

dependence of the RBS yield, whereas in (b) we select one or more depth regions and look in detail at the angular dependence. The optimal choice in each case will obviously depend on the particular application, and many examples of both are found in subsequent sections.

4.2.3.4. *Target Preparation.* Since critical angles for channeling are typically $\sim 1°$ or less, it is evident that even a small amount of anomalous scattering—due to surface contamination, mosaic spread, mechanical damage, lattice defects, etc.—could significantly perturb the channeled flux distribution. For example, a 5-nm surface oxide layer (typical of many metals) would reduce the mean flux of channeled MeV ^{4}He$^+$ by several percent. Considerable care in target preparation is therefore required to minimize anomalous scattering effects.

Various techniques exist for cutting and polishing single-crystal specimens. The cutting procedures (diamond wheel, spark cutting, etc.) invariably produce a highly damaged surface, which must subsequently be removed by standard metallurgical polishing methods or, in some cases, by chemical or electrochemical polishing. Note that this damaged region often extends to depths of several micrometers, and hence it is usually best to start

4.2. CHANNELING STUDIES OF LATTICE DEFECTS

with a fairly coarse polishing procedure and end up using 0.5-μm Al_2O_3 powder in a vibratory polisher. Such polishing treatments may still leave a thin, highly damaged surface region that must finally be removed either by (electro)chemical etching or by Ar^+ sputtering followed by a high-temperature annealing in vacuum to remove the Ar-induced damage.

In the applications to surface studies described in Section 4.2.6, even a monolayer of oxide or other impurity is of course completely unacceptable. In such systems, facilities for Ar-sputtering and high-temperature annealing are part of the ultrahigh-vacuum target-chamber–goniometer system, so that the final cleaning may be performed in situ.

Each material has its own special properties, and it is not possible to give a universal recipe for target preparation. Electropolishing of metals, for example, often produces an unacceptably thick oxide layer on the polished surface, and this layer must then be removed by other methods. Similarly, the high-temperature annealing stage, although necessary, may cause certain bulk impurities to segregate preferentially at the surface; if so, then further Ar sputtering may be required. An excellent review of the target preparation problem for channeling studies has been given by Whitton.[79]

4.2.4. Investigation of Lattice Disorder

4.2.4.1. Amorphous Layers on Single Crystals.
Consider a thin amorphous layer (or a layer composed of randomly oriented polycrystallites) present on top of a perfect crystal. Such a layer could result from the formation of an oxide film, the evaporation of a foreign atom species onto the crystal surface, or ion implantation in the near-surface region of the crystal. A well-collimated beam of particles impinging along a low-index axis of this single crystal undergoes scattering by atoms in the overlying amorphous layer, and some of the beam particles are scattered outside of the critical angle $\psi_{1/2}$ for channeling in the underlying crystal. Consequently, the aligned yield from the near-surface region of the underlying crystal is increased. The angular distribution of the yield is also broadened (see Fig. 16[80]) because some particles are scattered into channeling directions even if the crystal is tilted by angles of more than the critical angle. The thickness of the amorphous layer can be found from the energy width ΔE taken from the full width at half-maximum (FWHM) of the signal from the amorphous layer. (For full details of backscattering analyses see Chu et al.[21] and Mayer et al.[81].)

---

[79] J. L. Whitton, *Proc. R. Soc. London, Ser. A* **311**, 63 (1969).
[80] E. Rimini, E. Lugujjo, and J. W. Mayer, *Phys. Rev. B* **6**, 718 (1972).
[81] J. W. Mayer, M.-A. Nicolet, and W.-K. Chu, in "Material Characterization Using Ion Beams" (J. P. Thomas and A. Cachard, eds.), p. 333. Plenum, New York, 1978.

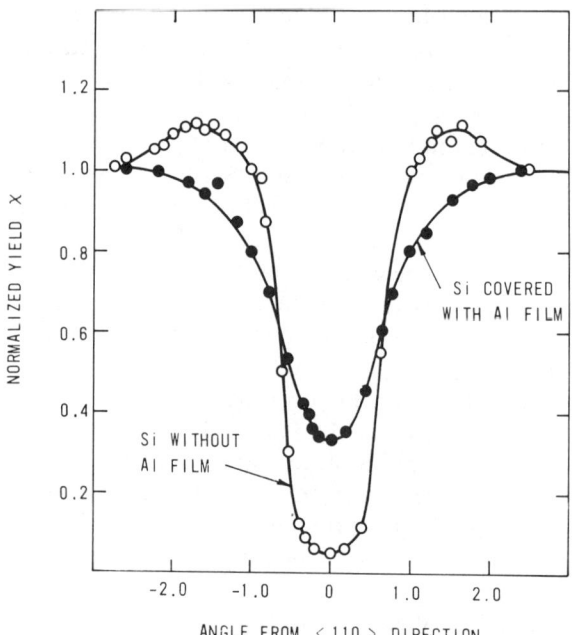

FIG. 16. The angular-yield profiles for 1.8-MeV ^{4}He backscattered from Si crystals that either had no Al film present or had a deposited Al film of 0.213 μm present. The yields were measured at about 0.1 μm below the Si surface. (Adapted from Rimini et al.[80])

In a relatively thin film, the probability that a particle is scattered through an angle more than $\psi_{1/2}$ in a single collision when traversing a film of Nt atoms cm^{-2} is given by

$$P(\psi_{1/2}) = \sigma_d(\psi_{1/2})Nt, \qquad (4.2.30)$$

where $\sigma_d(\psi_{1/2})$ is obtained from the Rutherford scattering law, i.e.,

$$\sigma_d(\psi_{1/2}) = \int_{\psi_{1/2}}^{\infty} \frac{d\sigma}{d\Omega}\, d\Omega = \frac{\pi Z_1^2 Z_2^2 e^4}{E^2 \psi_{1/2}^2}. \qquad (4.2.31)$$

This increases the minimum yield from the underlying single crystal to approximately $P(\psi_{1/2}) + \chi_v$, where χ_v is the minimum yield in the uncovered crystal. As the thickness of the amorphous film increases, the occurrence of multiple scattering events must be taken into account. The probability of scattering beyond the critical angle $\psi_{1/2}$ is given by

$$P(\tilde{\theta}_c, m) = \int_{\theta_c}^{\infty} 2\pi f_m(\tilde{\theta})\tilde{\theta}\, d\tilde{\theta}, \qquad (4.2.32)$$

where $f_m(\tilde{\theta})$ is the angular distribution of the particles, $m = \pi a^2 Nt$, $\tilde{\theta} = aE\theta/2Z_1Z_2e^2$, and $\tilde{\theta}_c = aE\psi_{1/2}/2Z_1Z_2e^2$. In experiments[80] with Al and Au thin films on Si, the observed minimum yield was in quite good agreement with the $P(\tilde{\theta}_c,m)$ values calculated from Eq. (4.2.32). The observed minimum yield has also been observed to decrease with an increase in the beam energy.[82] This energy dependence arises from two factors: the critical angle decreases as $E^{-1/2}$, and the beam angular distribution narrows more than the critical angle decreases.

4.2.4.2. Polycrystalline Layers on Single Crystals. In certain cases the crystalline nature of the polycrystalline film, present on an underlying single crystal, can be investigated by the channeling technique. If the polycrystallite size is larger than the beam-spot size and if the crystallites are highly oriented with respect to the substrate, then the aligned yield from the polycrystalline region has a value between that for a single crystal and that for a layer composed of random crystallites. In Fig. 17[83], for example, the film thickness dependence of χ_{min} is shown for Pt, Pd, and Ni atoms in PtSi, Pd$_2$Si, and NiSi$_2$ films grown on $\langle 111 \rangle$ Si. For all observed thicknesses of the Pd$_2$Si and NiSi$_2$ films and for thicknesses $\lesssim 30$ nm for the PtSi films the

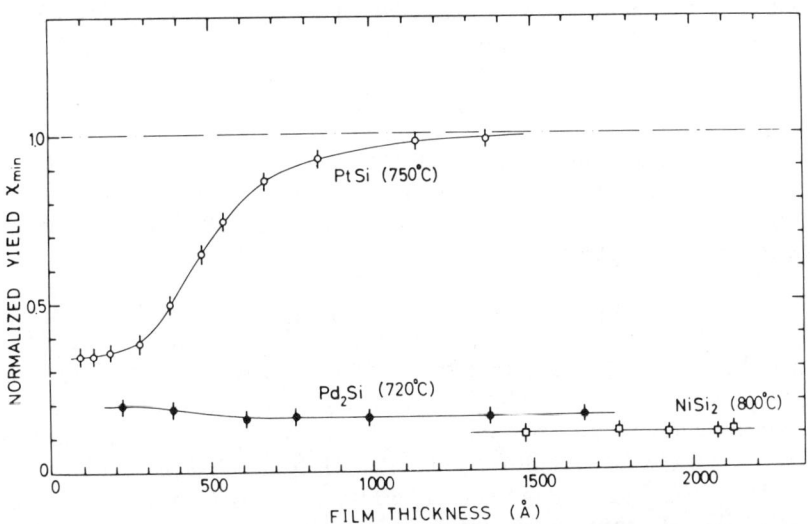

FIG. 17. Film-thickness dependences of the channeling minimum yields for PtSi, Pd$_2$Si, and NiSi$_2$ films grown on {111} Si. The films were annealed for 2 h at the indicated temperatures. The ordinate shows the normalized yield of 1.5-MeV ^{4}He$^+$ ions scattered from the metal atoms in the films. (From Ishiwara et al.[83])

[82] S. U. Campisano, G. Foti, F. Grasso, and E. Rimini, *Phys. Rev. B* **8**, 1811 (1973).
[83] H. Ishiwara, K. Hikosaka, M. Nagatomo, and S. Furukawa, *Surf. Sci.* **86**, 711 (1979).

results indicate that the crystallites in the silicide films grow epitaxially along the axes and planes of the underlying Si substrates with angular spreads smaller than the channeling critical angle. In the case of the PtSi films, χ_{min} increases rapidly for film thicknesses > 30 nm and approaches a random yield value at thicknesses > 100 nm. These results are quite consistent with the fact that the lattice parameter mismatch between orthorhombic PtSi and diamond-cubic Si crystals is ~ 12%, whereas the mismatch between Pd_2Si/Si and $NiSi_2/Si$ is ~ 2% and ~ 0.5%, respectively.

If one assumes a Gaussian distribution of the crystallite orientations and the channeling angular yield in single crystals, then the angular and energy dependences of the backscattering yield $\chi(\psi,E)$ in polycrystals can be expressed as[84]

$$\chi(\psi,E) = 1 - [(1 - \chi_0)/(1 + \sigma^2/\Delta_0^2)] \exp(-\psi^2/(\sigma^2 + \Delta_0^2)), \quad (4.2.33)$$

where $\chi(0,E) = \chi_{min}$, E is the incident energy, χ is the tilt angle of the beam from the most preferred axis in the polycrystals, σ is the standard deviation of the spread in crystallite orientation, and χ_0 is the minimum yield as $\sigma \to 0$ and may be considered as a measure of the density of scattering centers (grain boundaries and other lattice defects). Δ_0 is the standard deviation of the angular yield in single crystals, which is known to satisfy the relation[20]

$$\Delta_0(E) = C_1(Z_1/E)^{1/2}, \quad (4.2.34)$$

where C_1 is a constant. Figure 18 shows some results for the angular dependence of the yield of 1.0- and 2.0-MeV $^4He^+$ ions backscattered from a PtSi film.[83] From Eqs. (4.2.33) and (4.2.34), the energy dependence of $\psi_{1/2}$ may be expressed as[83,85]

$$\psi_{1/2}^2 = \ln 2(\Delta_0^2 + \sigma^2) = \sigma^2 \ln 2(1 + E_c/E), \quad (4.2.35)$$

where ln 2 is the conversion factor of the standard deviation to the half-width at half-maximum value and $E_c = C_1 Z_1/\sigma^2$. Using Eq. (4.2.35), the standard deviation σ can be obtained from the plot of $\psi_{1/2}$ versus Z_1/E, as shown in Fig. 18 for PtSi films on {111} Si. The estimated values of σ for the samples A and B (annealed for 2 h at 400°C and 750°C, respectively) were 0.87° and 0.59°, respectively. Information on χ_0 and σ can also be obtained from the energy dependence of χ_{min} as[83]

$$(1 - \chi_{min})^{-1} = (1 - \chi_0)^{-1} (1 + E/E_c). \quad (4.2.36)$$

---

[84] H. Ishiwara and S. Furukawa, *J. Appl. Phys.* **47**, 1686 (1976).
[85] H. Ishiwara, K. Hikosaka, and S. Furukawa, *J. Appl. Phys.* **50**, 5302 (1979).

4.2. CHANNELING STUDIES OF LATTICE DEFECTS

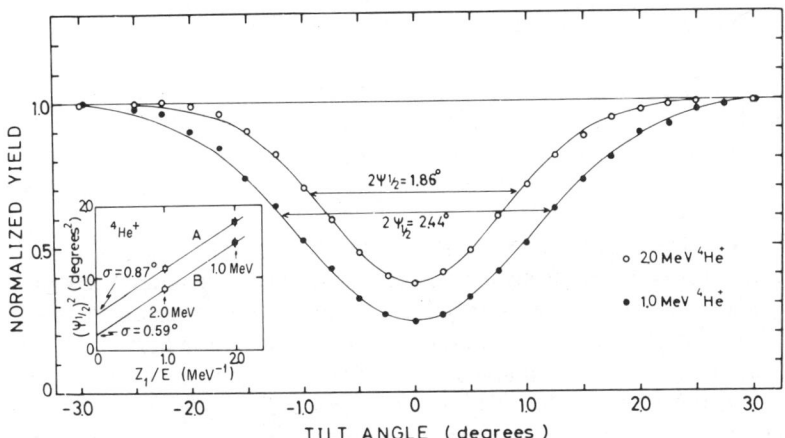

FIG. 18. Angular dependence of the backscattering yield of ⁴He⁺ ions around the $\langle 111 \rangle$ channeling axis in PtSi films 28 nm thick (annealed at 750°C). The solid lines represent a Gaussian fit to the experimental data points. Also shown (in the inset in the lower left-hand corner) is a plot of $(\psi_{1/2})^2$ versus Z_1/E. (From Ishiwara et al.[83])

Figure 19 shows the energy dependence of $(1 - \chi_{min})^{-1}$ for PtSi, Pd$_2$Si, and NiSi$_2$ films.[83] For the Pd$_2$Si and NiSi$_2$ films, $(1 - \chi_{min})^{-1}$ is essentially independent of incident beam energy, and hence σ is essentially zero. For the PtSi films, $(1 - \chi_{min})^{-1}$ has the energy dependence predicted by Eq. (4.2.36). The variation in the orientaton of the crystallites is always larger

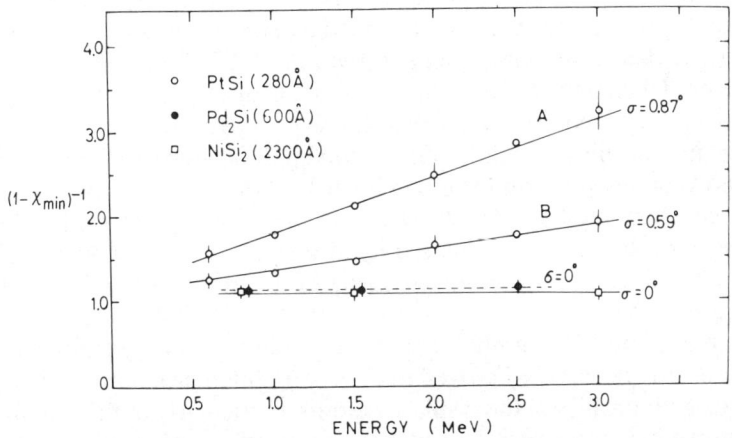

FIG. 19. The energy dependence of $(1 - \chi_{min})^{-1}$ for the backscattering of ⁴He⁺ ions from various deposited films. The Pd$_2$Si film was annealed at 720°C for 2 h, the NiSi$_2$ film at 800°C for 1 h, and the PtSi films A and B at 400°C and 750°C, respectively, for 2 h. (From Ishiwara et al.[83])

than 0.5° under any conditions of annealing time, temperature, and film thickness.[83]

Eq. (4.2.33) applies to situations where there is not an appreciable shoulder in the outer regions of the angular profiles. If a shoulder is present, then Eq. (4.2.33) will have to be modified somewhat, and it will contain a parameter representing the magnitude of the shoulder, as discussed by Ishiwara and Furukawa.[84]

4.2.4.3. Displacement of Host Crystal Atoms.

The application of the channeling–backscattering technique to provide information about host atoms displaced from their normal lattice sites will now be considered. This applicaton is of particular interest in the ion implantation field. As discussed in Section 4.2.2.5, in the analysis of disorder in a crystal one assumes that the beam particles can be divided into two parts: a channeled component that can interact with displaced atoms and a random component that is scattered by all the atoms of the crystal [see Eqs. (4.2.28) and (4.2.29)]. In the development of Eq. (4.2.29), the term $1 - \exp(-\int_0^z \sigma_d n_d(z') \, dz')$ represents the probability $P(n_d z)$ that a beam of intensity $1 - \chi_v(z)$, initially having a δ-function distribution, is deflected to angles larger than $\psi_{1/2}$ after traversing a number $n_d z$ of displaced atoms cm^{-2}. This is based upon a single scattering model.[39] In many of the applications of the method, it has been assumed that the disorder can be treated as randomly displaced atoms, essentially a dilute concentration of atoms in amorphous zones that are located within a perfect crystal that has a well-defined value of $\psi_{1/2}$. In this case the value of $P(n_d z)$ can be found from the procedure used to describe the amount of dechanneling in a perfect crystal overlaid by an amorphous layer (see Section 4.2.4.1). It is also usually assumed that for single randomly displaced atoms the defect scattering factor f [see Eq. (4.2.28)] is unity. In general, however, detailed knowledge about the spatial distribution of the displaced atoms would be required in order to carry out a rigorous analysis. Unfortunately, it is usually difficult to obtain data on the spatial distribution of host atoms (as opposed to solute atoms) because of the relatively low fraction of displaced host atoms usually present at any instant, particularly if one wishes to investigate the initial spatial distributions of individual displaced atoms before any appreciable clustering of the atoms into more complex defects occurs.

Despite some of the limitations of the randomly displaced atoms model, as presented above, it has proved to be very useful in providing information on the depth profiles of the displaced atoms, particularly in the ion implantation of semiconductors. In order to obtain the depth profile of the displaced atoms [i.e., $n_d(z)$ versus z] it is necessary to determine the random component $\chi_R(z)$ as a function of depth. The analysis for a disordered region situated inside the crystal (as in Fig. 20a[86]) usually involves using an iterative

4.2. CHANNELING STUDIES OF LATTICE DEFECTS

procedure starting at the surface and moving in succeeding increments of depth to determine the dechanneled fraction χ_R in the next layer, which then enables the direct channeling contribution and defect density $n_d(z)$ to be determined at that layer (see Chu et al.[21] and Rimini[87] for more details). If the damage is located near the surface and there is a well-defined peak in the aligned spectrum, $\chi_R(z)$ can be approximated by drawing a straight line from a point on the aligned curve for the virgin crystal, near the beginning of the damaged region, to a point on the spectrum for the aligned damaged crystal, just behind the damaged region. In the latter case the number N_d of displaced atoms cm^{-2} can be obtained from the area A_d of the disordered peak (in integrated counts) from the relationship

$$N_d = A_d \delta E_1 H^{-1}[K_{M_2}\epsilon(E) + \epsilon(K_{M_2}E)/\cos\theta]^{-1}, \quad (4.2.37)$$

where δE_1 is the energy width of a channel in the backscattered spectrum, H is the height of the random spectrum (counts per channel) near the crystal surface, K_{M_2} is the kinematic factor giving the ratio of the projectile energy after the elastic collision (with a crystal atom of mass M_2) to that before the collision, θ is the angle through which the incident projectile is scattered, and ϵ is the stopping cross section given by $\epsilon = (1/N)(dE/dX)$ where N is the atomic density of the crystal and dE/dX is the energy loss per unit path length experienced by the projectile as it penetrates the crystal.

Single scattering accounts for dechanneling in a reasonable manner for low amounts of disorder, but as the amount of disorder increases, plural or multiple scattering of the channeled ions becomes important. In Fig. 20b[86], for example, a comparison is shown between a disorder distribution obtained using the straight-line approximation, as referred to above, and one obtained using a plural scattering treatment developed by Keil et al.[88]

The defect concentration extracted in the above treatments represents an average of the defect distribution weighted by the channeled-ion flux. As discussed in Section 4.2.2.4, the channeled beam is not uniformly distributed across the channel between the channel rows. The flux distributions can be changed by tilting the crystal axis to an angle ϕ with respect to the beam direction (where $\phi < \psi_{1/2}$) and the variations in the measured yields as a function of ϕ may then be related to the distribution across the channel of the displaced atoms. This represents an extension of the flux-peaking effect

---

[86] F. H. Eisen, in "Channeling: Theory, Observation, and Applications" (D. V. Morgan, ed.), p. 415. Wiley (Interscience), New York, 1973.
[87] E. Rimini, in "Material Characterization Using Ion Beams" (J. P. Thomas and A. Cachard, eds.), p. 455. Plenum, New York, 1978.
[88] E. Keil, E. Zeitler, and W. Zinn, Z. Naturforsch., A 15, 1031 (1960).

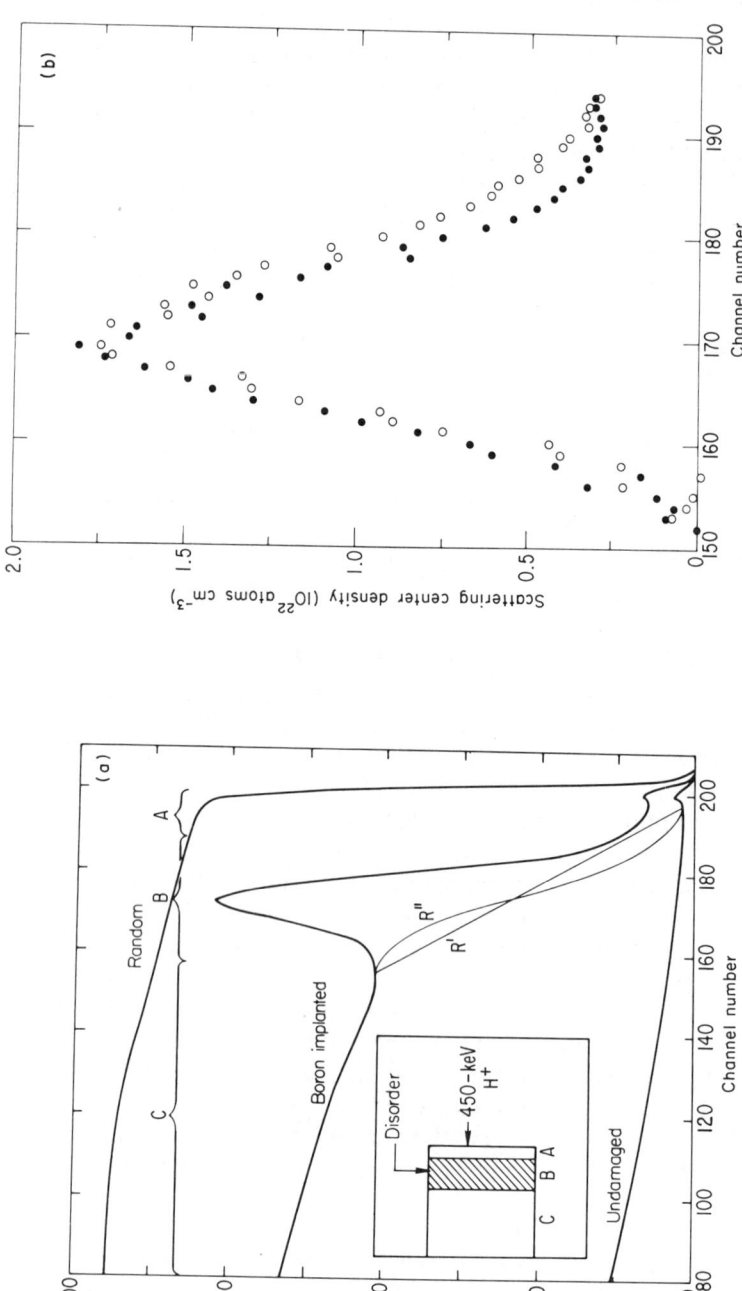

FIG. 20. (a) Aligned (⟨110⟩) and random backscattering spectra for a 450-keV proton beam incident on silicon. The spectrum showing a damage peak was obtained from a sample implanted with 8×10^{15} B ions cm^{-2}. The line labeled R' is the linear approximation to $\chi'(z)$ discussed in the text, and the curve labeled R" was calculated using the Keil plural scattering theory. The labeled regions of the spectrum for the B-implanted sample correspond to the regions with the same label in the schematic disorder distribution shown in the inset. (b) Comparison of disorder distributions calculated using the straight-line approximation (●) and the Keil plural scattering theory (○) to obtain $\chi'(z)$.[56,59] (From Eisen.[86])

used for impurity atom location (discussed in Section 4.2.5) to the location of displaced host atoms. Shown in Fig. 21, are defect-density profiles for the damage created in Si by 300-keV ^{14}N implants.[89] The profiles are angle dependent and indicate that the scattering centers are not uniformly distributed across the channel. Assuming that the channeled particles are distributed uniformly over the accessible areas A_i (which depend on the tilting angle), the distribution of defects across the channel can be obtained by the following procedure. For perfect alignment the $n_d(z,0)\Delta z$ defects visible to the channeled beam are distributed over the area A_0, whereas for the tilting angle ϕ_1, the $n_d(z,\phi_1)\Delta z$ defects visible to the beam are distributed over the area A_1. The difference $[n_d(z,\phi_1) - n_d(z,0)]\Delta z$ gives the number of defects in the depth Δz distributed over the area $(A_1 - A_0)$. Radial distributions of defects obtained from the data of Fig. 21 by applying the above procedure are shown in Fig. 22.[89] For the 3.2×10^{15} ions cm^{-2} fluence, the density of displaced Si atoms increases from the channel center towards the row at all depths. The 9.5×10^{15} ions cm^{-2} fluence results in a nonuniform distribution of defects over the depth range 0–200 nm but in an apparent random distribution of defects at depths greater than 400 nm. In the analysis, the ion fluxes were evaluated assuming statistical equilibrium and were considered to be depth independent. Scattering by defects will spread the channeled distribution of ions and as a consequence the ion fluxes will be smeared out.

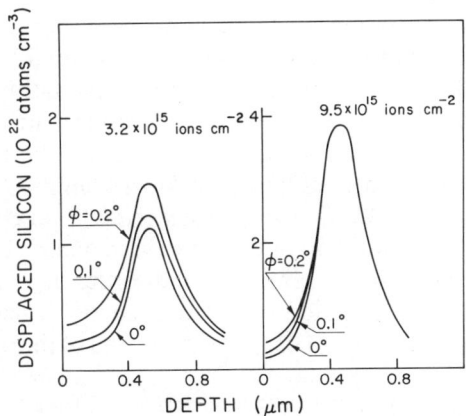

FIG. 21. Defect profiles in $\langle 111 \rangle$ Si implanted at room temperature with 3.2×10^{15} and 9.5×10^{15} ions cm^{-2} of 300-keV ^{14}N. The profiles are shown for different angles of tilt of the crystal axis with respect to the incident beam direction. (From Baeri et al.[89])

[89] P. Baeri, S. U. Campisano, G. Ciavola, and E. Rimini, *Nucl. Instrum. Methods* **132**, 237 (1976).

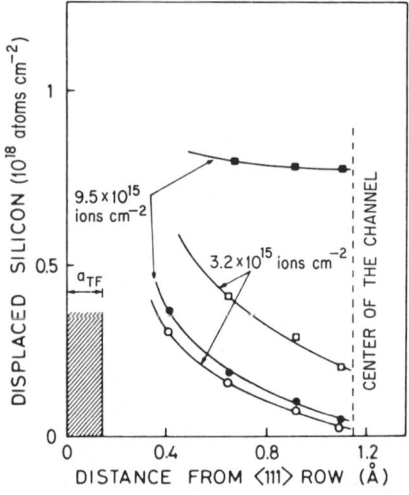

FIG. 22. Radial distributions of displaced atoms integrated over two different depth ranges: 0–2000 Å (circles) and 4000–6000 Å (squares), respectively, for the 3.2 and 9.5 × 10^{15} cm^{-2} ^{14}N ion implants shown in Fig. 21. (From Baeri et al.[89])

For example, after a 2-MeV ^{4}He ion beam has traversed a disordered region of 5 × 10^{17} displaced atoms cm^{-2} the ion flux distribution becomes nearly independent of tilting angle. Consequently, the apparent random distribution of displaced atoms observed at depths greater than 400 nm for the 9.5 × 10^{15} ions cm^{-2} implant can be due either to a true random distribution of the defects or to a smearing out of the ion flux profiles by small-angle scattering from defects.

Apart from actually extracting the defect-distribution profile, the channeling–backscattering technique has proven to be useful in following the buildup of disorder in a crystal during ion implantation as well as the removal of the disorder during subsequent annealing, particularly in the ion implantation of semiconductors (see, for example, the books edited by Namba[90] and by Chernow et al.[91] and the Proceedings of the International Conference on Ion Beam Modification of Materials[92]). In effect, it is providing a fast and simple evaluation of the crystalline quality of the sample. In Fig. 23, the removal of irradiation-induced disorder in a Si crystal by laser annealing is shown as a function of different energy densities of the laser beam.[93] For energy densities > 1 J cm^{-2} virtually all of the disorder has been

[90] S. Namba, ed., "Ion Implantation in Semiconductors—Science and Technology." Plenum, New York, 1975.

[91] F. Chernow, J. A. Borders, and D. K. Brice, eds., "Ion Implantation in Semiconductors 1976." Plenum, New York, 1977.

[92] Proc. Int. Conf. Ion Beam Modif. Mater., Budapest, 1978, published in Radiat. Eff. 47, Nos. 1–4 (1980).

[93] C. W. White, P. P. Pronko, B. R. Appleton, S. R. Wilson, and J. Narayan, Radiat. Eff. 47, Part 1, 31 (1980).

4.2. CHANNELING STUDIES OF LATTICE DEFECTS

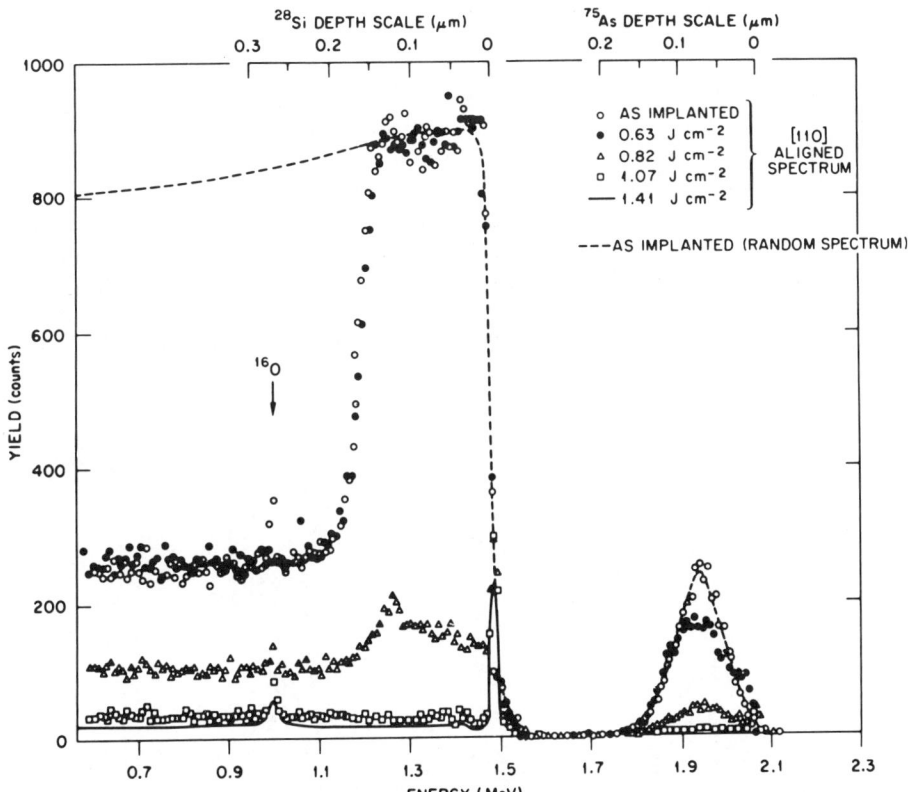

FIG. 23. Laser annealing of arsenic-implanted silicon using different energy densities. The Si was initially bombarded with 100-keV ^{75}As to a fluence of 1.4×10^{16} ions cm^{-2}. The analysis was along $\langle 110 \rangle$ using 2.5-MeV ^{4}He$^+$ ions. (From White et al.[93])

removed. The behavior of solute atoms during the implantation and annealing process can be followed at the same time, as for the As in Si in Fig. 23.

Although the concept of displaced atoms in random positions has proved to be very useful in evaluating the extent of disorder in crystals containing amorphous regions, the assumptions of randomly displaced atoms and unique critical angles do not hold generally. In crystals containing defects the displaced atoms may be present as dislocations, stacking faults, vacancy or interstitial clusters, and other complex configurations, as well as singly displaced atoms, and all of these defect configurations can give different contributions to the aligned spectra. If one wishes to obtain quantitative information using the disorder-analysis procedure outlined above, it is advisable to test the procedure to see if a consistent disorder distribution can be obtained using different incident ion beams and energies as well as using different axial and planar directions. Wherever possible, transmission electron microscopy should be used to provide detailed information on defects such as dislocations, stacking faults, annealing twins, and relatively large clusters of vacancies or interstitials. Even in the case of amorphous regions, it is worth noting that the channeling technique is essentially giving information on the average number of scattering centers, and it does not in general give information about the spatial distribution of the disordered regions. For example, in Fig. 24 the channeling–backscattering technique

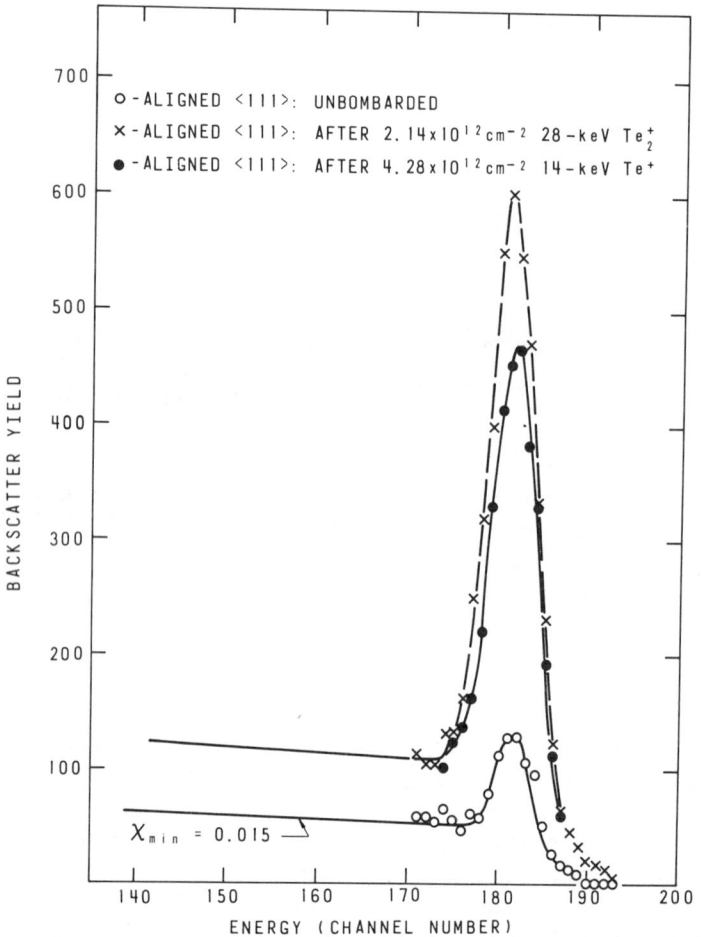

FIG. 24. Backscattering energy spectra for aligned ⟨111⟩ incidence from two Si crystals implanted at 50 K with the same atomic dose (4.28 × 10^{12} ^{128}Te atoms cm^{-2}) of 14-keV Te and 28-keV Te$_2$ ions, respectively. The analyzing beam was 1.0-MeV ^{4}He. The spectrum from an unimplanted crystal is included for comparison. (From Thompson et al.[94])

4.2. CHANNELING STUDIES OF LATTICE DEFECTS

has been used to investigate the role of varying the average deposited-energy density $\bar{\theta}_v$ in a collision cascade (thus the atomic–diatomic-ion comparison) on the number of displaced atoms formed in individual collision cascades in silicon crystals (hence the relatively low bombarding-ion fluences).[94] In collision cascades of fairly high $\bar{\theta}_v$ values (as produced by 14-keV Te and 28-keV Te$_2$ ions in Si) one expects most of the damage to be concentrated in a highly disrupted core region, i.e., disordered zone. The size and spatial distribution of these regions cannot be obtained from the channeling data, but these zones, which are typically $\sim 2-5$ nm in diameter, can be imaged in the electron microscope.[34,95] On the other hand, the displacements of single atoms (or very small clusters of atoms) in the peripheral regions of the highly damaged core will also be included in the disorder peak in the channeled case but will not be visible in the electron microscope. Hence if annealing studies are performed in situ following low temperature implants, the channeling investigations can provide information on the temperature regimes where point defects are annihilating as well as where the disordered zones disappear, particularly when the annealing of the disordered zones is also followed by electron microscopy. In Fig. 25, for example, it can be seen that there is quite a good correlation between electron microscope results and channeling data for the annealing out of "amorphous" zones in ion implanted Si.[95,96] In the electron microscopy case, a damage parameter was introduced that took into account both the change in the number density and the size of the defects during annealing, i.e., a parameter that enabled a meaningful comparison to be made with the change in the number of displaced atoms measured by the channeling technique.

For all of the specific cases discussed so far in this section, the defects present produced a relatively large contribution to direct scattering of the incident ion beam. There are many types of defects, however, that produce a relatively small contribution to direct scattering. Such is the case for dislocations, stacking faults, bubbles, and voids. It also is the case for point defects and defect clusters in metals. In Fig. 11, for example, the irradiation of a Cu–0.06 at. % Au crystal has resulted in a large increase in dechanneling without the creation of a disorder peak. The defects present in the region of the crystal being analyzed are believed to be mainly small point-defect complexes, i.e., clusters of several point defects. Referring to Eqs. (4.2.28) and 4.2.29), when f, the fractional atomic scattering per defect, is essentially zero, these equations reduce to

$$\chi_d(z) = \chi_R(z) \qquad (4.2.38)$$

[94] D. A. Thompson, R. S. Walker, and J. A. Davies, *Radiat. Eff.* **32**, 135 (1977).
[95] L. M. Howe, M. H. Rainville, H. K. Haugen, and D. A. Thompson, *Nucl. Instrum. Methods* **170**, 419 (1980).
[96] D. A. Thompson, A. Golanski, H. K. Haugen, L. M. Howe, and J. A. Davies, *Radiat. Eff. Lett.* **50**, 125 (1980).

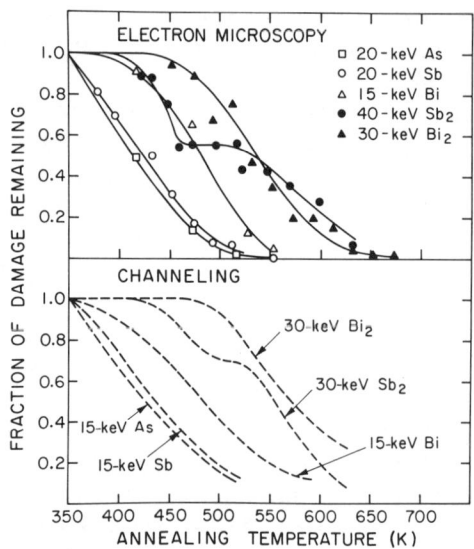

FIG. 25. A comparison of annealing results for ion-bombarded silicon from an electron microscopy investigation[95] (upper part) with those from a channeling study[96] (lower part). 10-min anneals in all cases. (From Howe et al.[95])

and

$$n_d(z) = \frac{1}{\sigma_d} \frac{\partial}{\partial z} \ln\left[\frac{1 - \chi_v(z)}{1 - \chi_d(z)}\right]. \qquad (4.2.39)$$

The total disorder, N_d, per unit area between the surface and depth $z = t$ is

$$N_d(z) = \frac{1}{\sigma_d} \ln\left[\frac{1 - \chi_v(t)}{1 - \chi_d(t)}\right]. \qquad (4.2.40)$$

For very small defect concentrations, one has as an approximation

$$n_d(z) = \frac{1}{\sigma_d} \frac{d\chi_d}{dz}, \qquad (4.2.41)$$

in which case the defect concentration is directly given by the slope of the normalized channeling spectrum divided by the dechanneling cross section for that defect.

In general, σ_d values are not known at present for most types of defects, although by combining dechanneling measurements with electron microscope investigations on systems where there is just one type of defect present it is possible to obtain estimates of σ_d as well as the dependence of σ_d on projectile energy and charge and the crystal temperature. For example, for dislocation lines σ_d increases as $E^{1/2}$; for stacking faults and voids σ_d is approximately independent of energy; and for randomly distributed interstitials or amorphous zones σ_d decreases as $\sim E^{-1}$. For a single displaced atom, Matsunami and Itoh[71] calculate that σ_d is on the order of 10^{-18} cm^2, for the point-defect complexes giving rise to the dechanneling in Fig. 11, $\sigma_d \sim 4 \times 10^{-18}$ cm^2 (based on an estimate of the defect concentration), whereas for relatively large vacancy clusters (2–10-nm diameter as determined by transmission electron microscopy) in irradiated gold,[97] $\sigma_d \sim 10^{-13}$–10^{-14} cm^2.

Some specific examples of dechanneling analyses performed in conjunction with transmission electron microscope (TEM) investigations are given below to illustrate the application of the technique for studying disorder in ion-implanted systems. In Fig. 26a the backscattering spectra are shown for a P-implanted Si crystal.[98] A well-defined region of disorder giving rise to an increased dechanneling rate without a direct scattering peak is observed in the channeled spectra for the depth region 200–500 nm. This effect is particularly pronounced for the $\{110\}$ planar channel. The TEM analysis showed that the disordered region contains primarily perfect dislocation loops of edge character lying on $\{111\}$ and $\{110\}$ planes. The mean dislocation-loop diameter was 27.5 nm, and the total projected length of dislocation lines per unit area was 2.4×10^5 cm cm^{-2}. For dislocation lines the planar dechanneling cross section per unit dislocation length is λ_p and is given approximately by[99]

$$\lambda_p = (Eb/\alpha Z_1 Z_2 e^2 N_p)^{1/2}, \qquad (4.2.42)$$

where b is the magnitude of the Burgers vector of the dislocation loops, α is a constant dependent on dislocation orientation and type, and N_p is the atomic density in the plane. Assuming negligible direct scattering and the additivity of dechanneling processes, the density of disorder as a function of z is given by

$$\int_0^z n_d(z)\, dz = \frac{-1}{\lambda_p} \ln\left[\frac{1-\chi_d}{1-\chi_v}\right], \qquad (4.2.43)$$

where the integral in this particular case pertains to the total projected length of dislocation lines at depth z. From Eq. (4.2.42) it can be seen that for dislocations λ_p should be proportional to $E^{1/2}$, and this is shown to be the case in Fig. 26b. Eq. (4.2.42) was actually derived for the case of straight nondissociated dislocations and is not expected to describe accurately the dechanneling by small dislocation loops. The dechanneling cross section per

---

[97] P. P. Pronko and K. L. Merkle, *in* "Applications of Ion Beams to Metals" (S. T. Picraux, E. P. EerNisse, and F. L. Vook, eds.), p. 481. Plenum, New York, 1974.

[98] S. T. Picraux, D. M. Follstaedt, P. Baeri, S. U. Campisano, G. Foti, and E. Rimini, *Radiat. Eff.* **49**, 75 (1980).

[99] J. Mory and Y. Quéré, *Radiat. Eff.* **13**, 57 (1972).

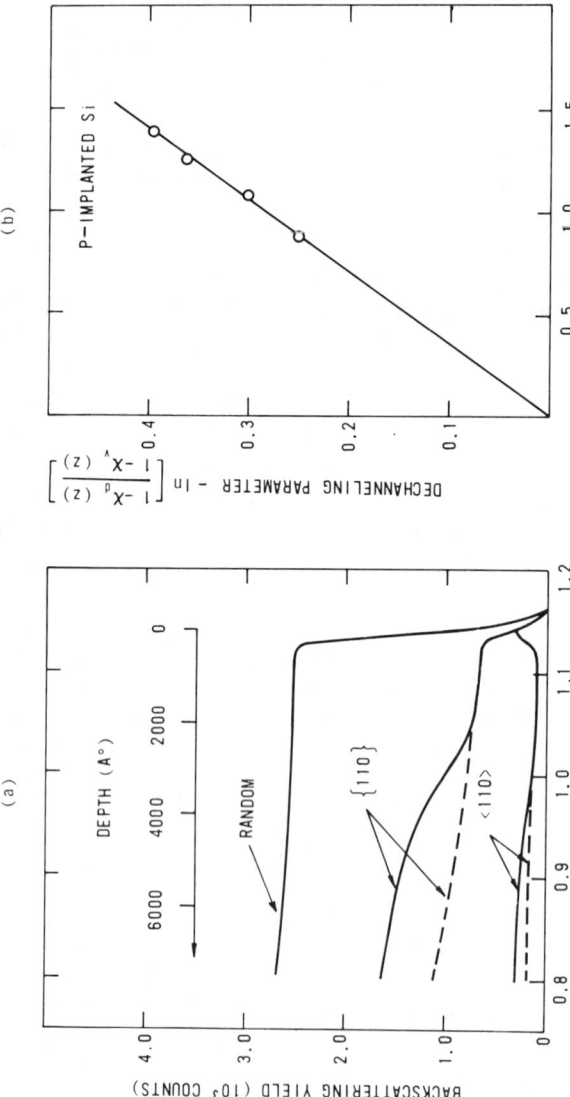

FIG. 26. (a) Spectra for 2-MeV ^{4}He backscattered from a Si crystal, which was implanted with 200-keV ^{31}P (while aligned along the $\langle 110 \rangle$ axis) and then annealed to 900°C, are shown for $\langle 110 \rangle$ axial and $\{110\}$ planar channels as well as for a random orientation (solid curves). Spectra for the unimplanted crystal are shown as dashed lines. (b) The dechanneling parameter $-\ln\{[1 - \chi_d(z)]/[1 - \chi_v(z)]\}$ versus $E^{1/2}$ at $z = 500$ nm for a $\{110\}$ planar channel. (Adapted from Picraux et al.[98])

4.2. CHANNELING STUDIES OF LATTICE DEFECTS

unit dislocation length λ should be smaller for dislocation loops than for straight dislocation lines, and this reduction has been observed experimentally.[74,100] Actually, λ should be a function of the diameter D of the dislocation loops, and Eq. (4.2.42) can be considered as an asymptotic value λ_∞ when D tends to infinity. Quéré suggests, as a semiempirical approximation, the following expression for λ for dislocation loops[101]

$$\lambda = \lambda_\infty(D/(D + D_0)), \qquad (4.2.44)$$

where D_0 is a constant. For perfect loops the dechanneling cross section will still be expected to vary as $E^{1/2}$, whereas for imperfect loops containing a stacking fault the dechanneling cross section is expected to have an energy dependence slower than $E^{1/2}$.

In ion-implanted materials, it is of considerable interest to be able to determine the depth distribution of the induced disorder, as well as to characterize the nature of the disorder as completely as possible. This helps

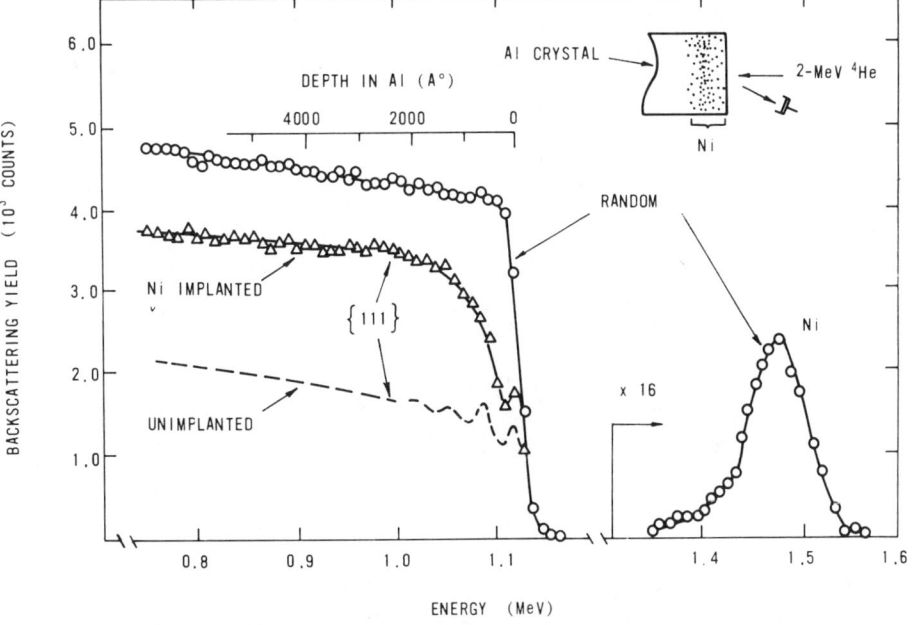

FIG. 27. Spectra for 2-MeV ^{4}He backscattered from an Al crystal before and after implantation with 7×10^{15} ^{58}Ni ions cm^{-2}. (Adapted from Picraux.[38])

[100] W. F. Tseng, J. Gyulai, S. S. Lau, J. Roth, T. Koju, and J. W. Mayer, *Nucl. Instrum. Methods* **149**, 615 (1978).
[101] Y. Quéré, *Radiat. Eff.* **38**, 131 (1978).
[102] D. K. Brice, "Ion Implantation Range and Energy Deposition Distributions." Vol. 1. IFI Plenum, New York, 1975.

to clarify the origin of induced material property changes as well as to provide information on the degree to which the disorder remains localized in the region of primary energy deposition. In Fig. 27, the energy spectra are shown[38,98] for a {111} planar channel of an Al crystal before and after implanting with 7×10^{15} ^{58}Ni atoms cm^{-2}. After implantation an appreciable increase in dechanneling has occurred over the depth region 0–200 nm. TEM investigations revealed the presence of tangled dislocation networks of mixed type (i.e., of both edge and screw character), which resulted from the formation, growing, and interaction of dislocation loops

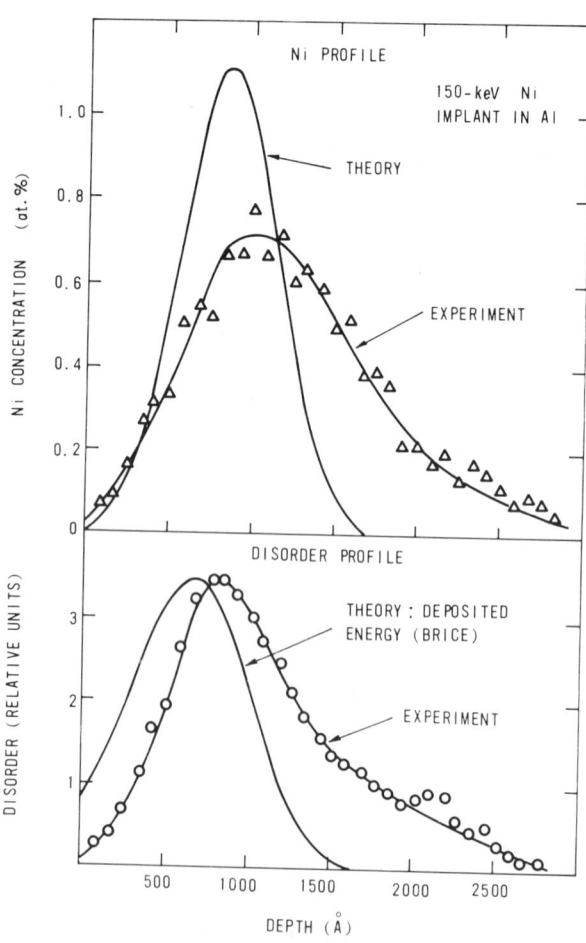

FIG. 28. Experimental depth profiles of Ni atom concentration and lattice disorder as obtained from the analysis of the backscattering spectra in Fig. 27. Also shown are the corresponding theoretical depth profiles, as determined from calculations by Brice.[102] (Adapted from Picraux.[38])

4.2. CHANNELING STUDIES OF LATTICE DEFECTS

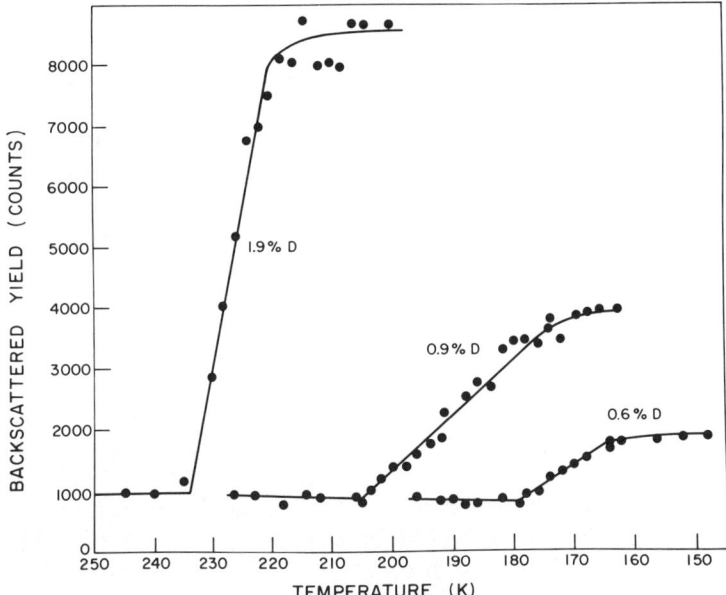

FIG. 29. Yield versus temperature plots for 1-MeV ^{4}He backscattered from three $\langle 100 \rangle$-aligned Nb crystals containing the deuterium concentrations shown. The changes of transition temperature, slope, final disorder level, and precipitate size can be directly related to the deuterium concentration. (Adapted from Whitton et al.[103] Copyright 1976, Pergamon Press, Ltd.)

during the implantation. Consistent with the TEM observations was the fact that the dechanneling cross section in the implanted crystal varied as $E^{1/2}$. The depth profile of the disorder was extracted from the {111} planar channeling data of Fig. 27 by using Eq. (4.2.39) and (4.2.42) and is shown in Fig. 28. Also shown in Fig. 28 is the experimental Ni atom concentration depth profile as well as theoretical depth profiles of deposited energy (damage) and Ni atom concentration. In this particular case the disorder introduced by the Ni ion implantation extends over the same depth region as the Ni atoms. The Ni atom profile is spread to somewhat greater depths than predicted by random stopping theory,[102] presumably because of some channeling of the implanted Ni. The corresponding spreading is observed in the disorder profile. There is also a reduction in the experimentaly observed disorder in the first ~ 50–100 nm below the crystal surface, as compared with the theoretical deposited energy profile. This is possibly due to the surface acting as a competing sink for migrating point defects.

The channeling technique can also be used to monitor the disorder accompanying phase changes or precipitation phenomena, especially if appreciable strain is produced in the lattice by these transformations. Shown in Fig. 29 are channeling–backscattering measurements of the increase in lattice disorder accompanying precipitation of Nb–D during the $\alpha \rightarrow \alpha + \beta$ phase change.[103] This increase in lattice disorder can be related to a combination of mismatch of the coherent Nb–D β-phase precipitates in the Nb lattice, the formation of incoherent precipitates and the generation of prismatic dislocations in the vicinity of the incoherent precipitates. From the data the terminal solid solubility of H and D in Nb was measured over the concentration range of 0.24–3.45 at. % and temperature range 161–249 K. The role of stress fields on the temperature of precipitation was also shown to be significant.

4.2.5. Lattice Site Location of Solute Atoms

4.2.5.1. Introduction.
Solute atoms are considered to be point defects in a crystal. They may occur in substitutional or interstitial lattice sites, or in association with other defects such as vacancies, self-interstitials, dislocations, grain boundaries, or surfaces. A knowledge of the position of solute atoms in the lattice is essential for an understanding of defect properties. Channeling is one of the most direct and unambiguous methods of locating solute atoms in a crystal, as outlined in Section 4.2.1.

If solute atoms have almost the same size and valence as the host atoms, the solubility is usually high; substitutional solid solutions result, in which solute atoms replace host atoms on normal lattice sites (Hume-Rothery rules).[104] Solute atoms having diameters more than 15% different from those of the host atoms are relatively insoluble, leading to precipitation. In general, solute atoms either attract one another, resulting in segregation or precipitation, or attract solvent atoms, resulting in ordered solid solutions.[105]

If very small solutes, such as H, C, N, or O, fit into the interstices of the crystal, there is often a considerable interstitial solubility.[106] Such interstitial solid solutions as C in Fe or O in Ti are relevant for structural materials.

Solute atoms can also trap vacancies and self-interstitials. In general, large solute atoms attract vacant lattice sites, and small solute atoms attract self-interstitial atoms,[22,27] although electronic effects may modify this rule. This interaction between solute atoms and intrinsic defects is of great importance for technology, especially in semiconductor devices and in radiation environments.

[103] J. L. Whitton, J. B. Mitchell, T. Schober, and H. Wenzl, *Acta Met.* **24**, 483 (1976).
[104] W. Hume-Rothery and G. V. Raynor, "The Structure of Metals and Alloys." p. 97. Inst. Met., London, 1962.
[105] L. Guttman, *Solid State Phys.* **3**, 145 (1956).
[106] H. J. Goldschmidt, "Interstitial Alloys." Butterworth, London, 1967.

4.2. CHANNELING STUDIES OF LATTICE DEFECTS

We shall illustrate the channeling technique for the lattice location of solute atoms by considering several possible positions of solute atoms:

(1) substitutional sites,
(2) small displacements from lattice sites,
(3) large displacements from lattice sites (low-symmetry interstitial sites),
(4) large displacements into interstitial sites of high symmetry, and
(5) random sites.

Representative examples of such sites are illustrated in Fig. 4 for a face-centered-cubic lattice. A random site is one in which the positions of solute atoms are distributed randomly in the lattice. For example, the solute atoms could be located in small precipitates having a different crystal structure from that of the matrix (noncoherent precipitates).

The angular scans of backscattering yields show characteristic profiles when solute atoms project various distances into the channels.[9-11,63,64] Such profiles are shown for a low-index axial channel in Fig. 30.

(a) For solute atoms in substitutional lattice sites, the yields from solute and host atoms are identical for all channels (assuming the same vibrational amplitude).

(b) For solute atoms displaced a few hundredths of nanometers into a channel, the dip in yield from solute atoms is narrower than that from host atoms.

(c) For large displacements (~ 0.1 nm), small double peaks or a small central peak may be observed.

(d) For displacements into the center of a channel, a large, narrow peak in yield from solute atoms is seen.

(e) For randomly located solute atoms, a flat solute profile is observed.

Combinations of substitutional plus small displacements or substitutional plus interstitial sites are also shown in Fig. 30. These curves are schematic only. (A specific example of yield profiles calculated directly from the calculated flux contours was shown in Fig. 10.) The lattice positions (1)–(3) and (5) correspond qualitatively to profiles (a)–(c) and (e), respectively. For high-symmetry interstitial positions (4), the effect of the crystal symmetry is to give profiles of types (a), (c), and (d) for different channels: i.e., a high-symmetry interstitial position often lies in the center of one set of channels, off-center in another set, and completely shadowed in another set. (Lower-symmetry positions project into all channels.) The projections of tetrahedral and octahedral interstitial sites into major channels for bcc, fcc, diamond-cubic, and hcp structures are shown[41] in Fig. 31. For example, the octahedral interstitial site in a hcp structure would show profiles of type (a)

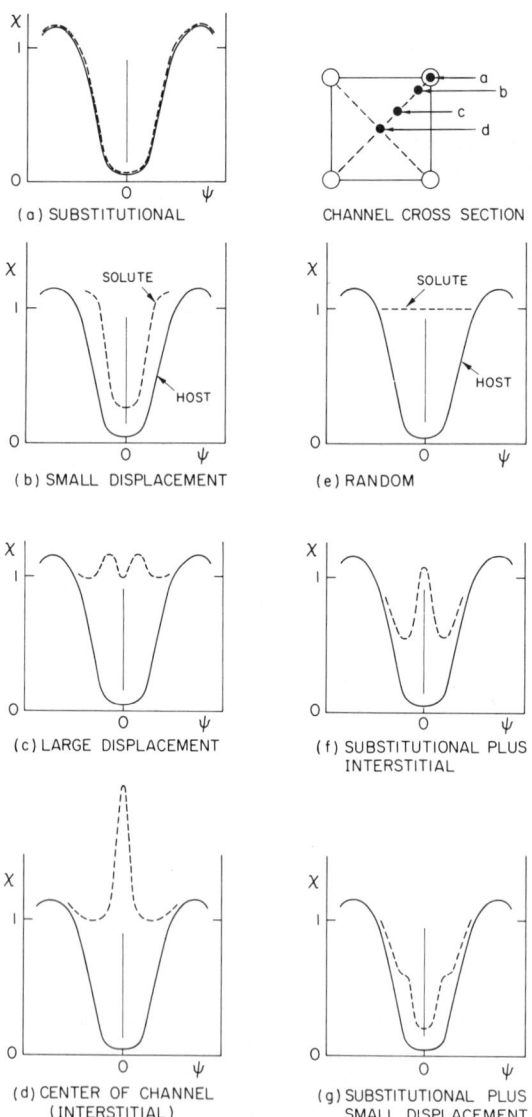

FIG. 30. Angular-scan profiles for different projections of a solute atom into an axial channel.

for $\{11\bar{2}0\}$ planar channels, type (c) for $\langle 10\bar{1}0\rangle$ axial channels, and type (d) for $\langle 0001\rangle$, $\langle 11\bar{2}0\rangle$, $\{0001\}$, and $\{10\bar{1}0\}$ channels.

The general procedure for locating solute atoms is thus to determine the type of displacement that predominates, using selected angular-scan data of the type shown in Fig. 30. Symmetry considerations dictate how much data

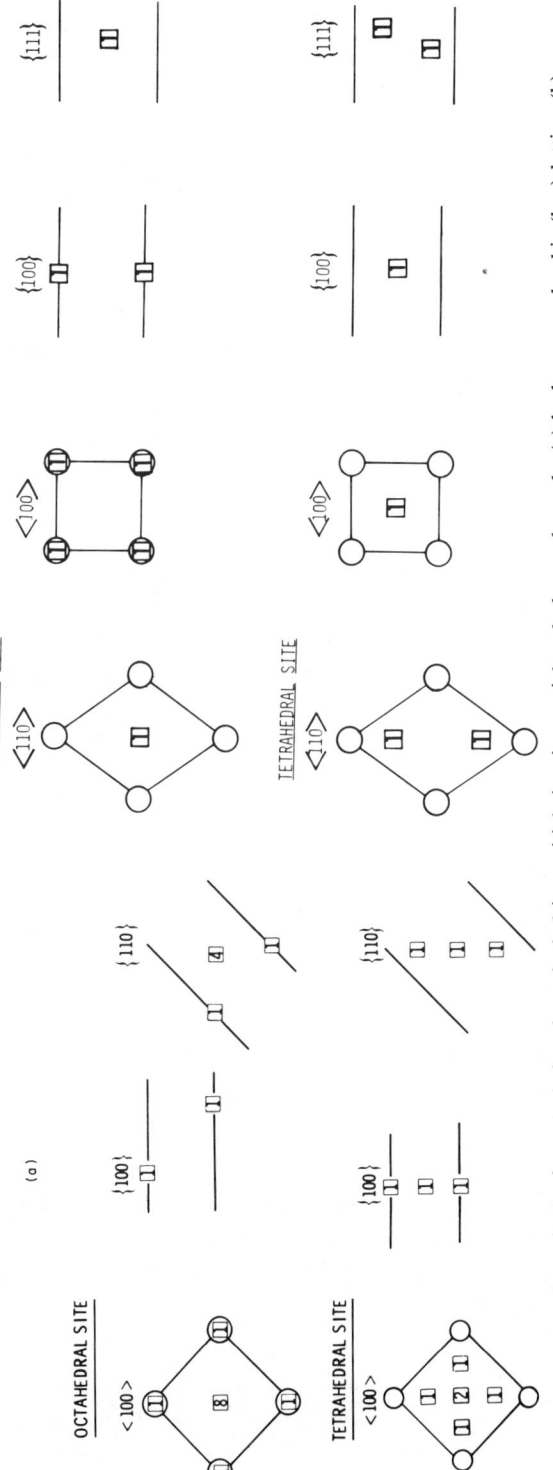

FIG. 31. Projections of tetrahedral and octahedral interstitial sites into axial and planar channels: (a) body-centered-cubic (bcc) lattice, (b) face-centered-cubic (fcc) lattice, (c) diamond-cubic lattice (tetrahedral site only, Δ), (d) hexagonal close-packed (hcp) lattice. In (a) and (b) the numbers within the squares refer to the relative number of sites for that projected position. (From Mayer and Rimini.[41])

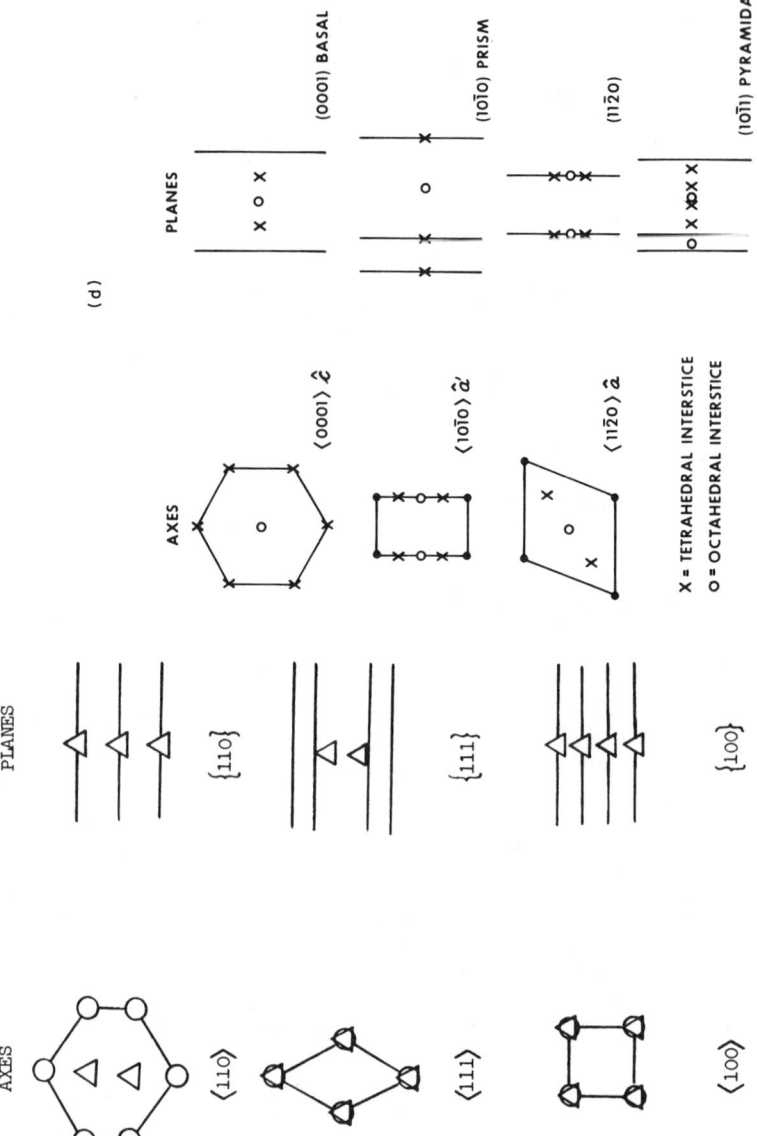

Fig. 31 (continued)

4.2. CHANNELING STUDIES OF LATTICE DEFECTS

of this nature are required. A considerable number of scans may be necessary when a low-symmetry position or several different solute atom positions occur. However, if the solute atoms are in a high-symmetry position, channeling data along only two or three axes usually are sufficient to determine that position (as outlined in Section 4.2.1). In the fcc lattice, for example, when peaking in the yield from solute atoms is observed for a $\langle 110 \rangle$ axis, as shown in Fig. 30d, and complete shadowing is observed for a $\langle 100 \rangle$ axis (Fig. 30a), it can be inferred that the solute atoms are in the body-centered interstitial position shown in Fig. 3a.

To determine exact lattice sites of solute atoms, it is necessary to compare angular scans of the type shown in Fig. 30 with those calculated from flux contours, as outlined in Section 4.2.2. When unique, high-symmetry positions are involved, it is possible in this way to specify the position of the solute atom to less than 0.01 nm. If a variety of sites occurs, this procedure inevitably leads to a variety of possible combinations of lattice sites, and considerable model fitting may be required. In such cases, the physical insight gained by performing a systematic series of experiments is helpful. An example is discussed in Section 4.2.5.5

Both qualitative and quantitative information on lattice location of solute atoms is obtained from measurements of the normalized yields $\chi_h^{\langle lmn \rangle}$ from host atoms and $\chi_s^{\langle lmn \rangle}$ from solute atoms for perfect alignment along various $\langle lmn \rangle$ channels. A useful empirical quantity from such measurements is the apparent fraction of solute atoms $f_{ds}^{\langle lmn \rangle}$ that are displaced from lattice sites into the given channel:[107]

$$f_{ds}^{\langle lmn \rangle} = (\chi_s^{\langle lmn \rangle} - \chi_h^{\langle lmn \rangle})/(1 - \chi_h^{\langle lmn \rangle}). \quad (4.2.45)$$

This quantity is equal to the sum of the actual displaced fractions f_i of solute atoms in the various lattice sites, designated as i, each being weighted by the effective ion flux $g_i F_i$ at those sites,

$$f_{ds}^{\langle lmn \rangle} = \sum_i f_i g_i^{\langle lmn \rangle} F_i^{\langle lmn \rangle}. \quad (4.2.46)$$

Here $g_i^{\langle lmn \rangle}$ is a geometric factor giving the fraction of displacements for the site which project into the given $\langle lmn \rangle$ channel, and $F_i^{\langle lmn \rangle}$ is the calculated normalized ion flux for that site. The normalization is with respect to only the average channeled flux.

The derivation of Eq. (4.2.46) follows. Consider an ion beam of intensity I to be divided into a channeled component I_c and a nonchanneled component $I - I_c$. Since only non-channeled ions can interact with host atoms,

$$\chi_h^{\langle lmn \rangle} = (I - I_c)/I. \quad (4.2.47)$$

[107] M. L. Swanson, L. M. Howe, and A. F. Quenneville, *Phys. Status Solidi A* **31**, 675 (1975).

Here it is assumed that the displaced fraction of host atoms can be neglected. If solute atoms are in perfect lattice sites, and their vibrational amplitude is the same as that of the host atoms, then $\chi_s^{\langle lmn \rangle} = \chi_h^{\langle lmn \rangle}$. However when a fraction $f_i g_i^{\langle lmn \rangle}$ of solute atoms are displaced to site i in the channel, the channeled ions can interact with these atoms, and the nonchanneled ions can interact with all solute atoms. Thus

$$\chi_s^{\langle lmn \rangle} = \frac{I - I_c}{I} + \sum_i f_i g_i^{\langle lmn \rangle} \frac{I_{ci}}{I}, \qquad (4.2.48)$$

where the summation is over all solute atom sites i. Here I_{ci} is the channeled beam intensity at site i. But

$$I_{ci}/I = (I_{ci}/I_c)(I_c/I) = F_i^{\langle lmn \rangle} (1 - \chi_h^{\langle lmn \rangle}) \qquad (4.2.49)$$

from Eq. (4.2.47) and the definition of $F_i^{\langle lmn \rangle}$. Thus Eq. (4.2.48) can be written

$$\chi_s^{\langle lmn \rangle} = \chi_h^{\langle lmn \rangle} + \sum_i f_i g_i^{\langle lmn \rangle} F_i^{\langle lmn \rangle} (1 - \chi_h^{\langle lmn \rangle}), \qquad (4.2.50)$$

and from Eq. (4.2.45)

$$f_{ds}^{\langle lmn \rangle} = \sum_i f_i g_i^{\langle lmn \rangle} F_i^{\langle lmn \rangle}. \qquad (4.2.51)$$

In using this equation, it is necessary to recall that the normalized flux

$$F_i^{\langle lmn \rangle} = I_{ci}/I_c$$

is defined as the ratio of the channeled beam intensity at the displaced solute atom site to the total *channeled* beam intensity. This is the quantity normally calculated by analytical or Monte Carlo methods for an ideal lattice,[28,63,64] that is, when dechanneling is not considered.

When Eq. (4.2.51) is used for determination of solute atom positions it must be realized that values of $F_i^{\langle lmn \rangle}$ are extremely sensitive to factors such as beam divergence, depth of analysis, lattice vibrations, other crystal imperfections, and surface conditions of the crystal, especially when considering flux peaking. Thus complete angular scans give more detailed information.

4.2. CHANNELING STUDIES OF LATTICE DEFECTS

continuous solid solubility, with short- and long-range ordering. For small concentrations ($\sim 1\%$) of Au in Cu, it is expected that the Au atoms are on random substitutional lattice sites. Evidence for this is obtained from x-ray data as well as from many other physical properties. Channeling data[109] also indicate for a Cu–2 at. % Au alloy that the Au atoms occupy perfect substitutional lattice sites, as shown in Fig. 32. Here the $\langle 110 \rangle$ angular dependences of χ_{Cu} and χ_{Au} are identical, as expected for this alloy.

For larger solute concentrations in substitutional solid solutions, ordered structures often occur (e.g., Cu_3Au, CuAu). In an ordered lattice, a given channel may be bordered by strings of only one element or by mixed strings

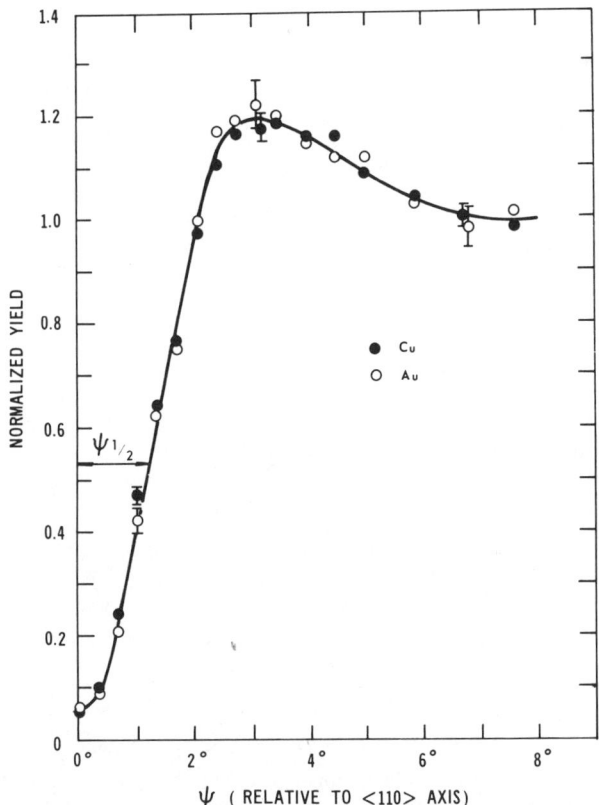

FIG. 32. Angular dependence of the normalized backscattering yields of 1.2-MeV $^4He^+$ ions from Cu and Au atoms in a single crystal of Cu containing ~ 2 at. % Au. The yields were measured for the depth interval 40–90 nm below the surface. (From Alexander and Poate.[109])

[109] R. B. Alexander and J. M. Poate, *Radiat. Eff.* **12**, 211 (1972).

of both elements. In the former case, a difference in angular half-width $\psi_{1/2}$ occurs for the different elements because of the relation [see Section (4.2.2.3)] $\psi_{1/2} \propto (2Z_1 Z_2 e^2/Ed)^{1/2}$ where Z_1 and Z_2 are the projectile and target atomic numbers, E is the projectile energy, and d is the atomic spacing along the strings of atoms bordering the channel. (This difference has been observed[9,110,111] also in many compounds containing elements of very different Z_2 values such as UO^2 and GaP.) Such an effect has been observed in ordered Cu_3Au, as shown in Fig. 33. Thus perfectly substitutional solute atoms may show different angular widths from those of the host. The effect

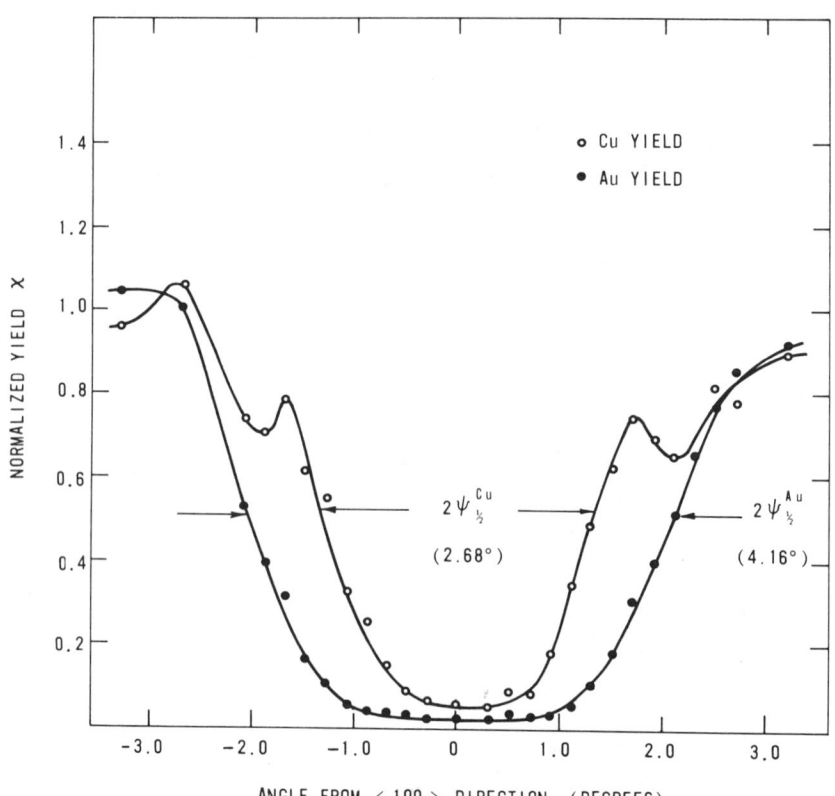

FIG. 33. Angular dependence of the normalized yield χ of backscattered 1-MeV ^{4}He$^+$ ions from Cu and Au atoms for a scan through the $\langle 100 \rangle$ axial channel in a Cu–28 at. % Au crystal. The backscattering measurements were taken at 35 K.

[110] L. Eriksson and J. A. Davies, *Ark. Fys.* **39**, 439 (1969).
[111] S. T. Picraux, J. A. Davies, L. Eriksson, N. G. E. Johansson, and J. W. Mayer, *Phys. Rev.* **180**, 873 (1969).

4.2. CHANNELING STUDIES OF LATTICE DEFECTS

of short-range ordering, which may also occur for more dilute solid solutions, has yet to be investigated.

By ion implantation, the equilibrium solubility of solute atoms may be exceeded by a large amount, creating metastable solid solutions. The solute atom positions in such alloys can be studied by channeling.[112]

The segregation of solute atoms to grain boundaries or to precipitates may or may not produce a decrease in the substitutional component of the solute atoms. An interesting study of segregation was performed in a low-alloy steel.[113] An Fe crystal was implanted with both Ti and Sb and then annealed 1 h at 873 K. This treatment produced fcc TiC precipitates by the gettering of C impurity atoms. One $\langle 100 \rangle$ axis of each precipitate was aligned with a $\langle 100 \rangle$ axis of the host bcc Fe crystal, but the other two $\langle 100 \rangle$ TiC axes were aligned with $\langle 110 \rangle$ Fe axes. Thus in the channeling experiment, attenuation of Ti yields occurred only along $\langle 100 \rangle$ Fe axes. However, attenuation of Sb yields was observed for $\langle 100 \rangle$, $\langle 110 \rangle$, and $\langle 111 \rangle$ Fe channels, thus showing that an appreciable fraction of the Sb atoms occupied substitutional Fe lattice sites adjacent to TiC precipitates.

4.2.5.3. Small Displacements. When narrowing of yield profiles from solute atoms is observed (Fig. 30b) for small solute concentrations ($<1\%$), a displacement of solute atoms of $\lesssim 0.05$ nm from lattice sites is indicated. Such a displacement is expected when solute atoms are associated with lattice defects, such as vacancies or self-interstitials. Narrowing could also be due to an enhanced thermal vibrational amplitude of the solute atoms or to the appearance of small coherent precipitates. No direct evidence for this latter effect has been obtained yet.

Small displacements of solute atoms are indicated for Bi implanted in Si (Fig. 34). In this case, the narrowing of the dip in Bi yield occurred mainly near the aligned position rather than over the complete dip.[114] This behavior is a strong indication that only a portion of the Bi atoms were displaced from lattice sites (see Fig. 30g). The data are in good agreement with 50% of the Bi atoms lying on perfect lattice sites and 50% of the Bi atoms displaced ~ 0.045 nm in $\langle 110 \rangle$ directions.

Double-alignment data (Section 4.2.2.3) in this case provided additional information concerning the lattice sites of Bi. Because the yield from Bi atoms in $\langle 110 \rangle$ double alignment was reduced by a factor of 3 as compared with the yield in single alignment, it appears that at least 95% of the Bi atoms were located within 0.06 nm of $\langle 110 \rangle$ lattice rows.

Further information is required to determine the reason for this small

---

[112] J. M. Poate, *Radiat. Eff.* **49**, 81 (1980).

[113] J. A. Knapp and D. M. Follstaedt, *Proc. Int. Conf. Ion Beam. Modif. Mater., Albany, N.Y. 1980*, published in *Nucl. Instrum. Methods* **182/183**, 1017 (1981).

[114] S. T. Picraux, W. L. Brown, and W. M. Gibson, *Phys. Rev. B* **6**, 1382 (1972).

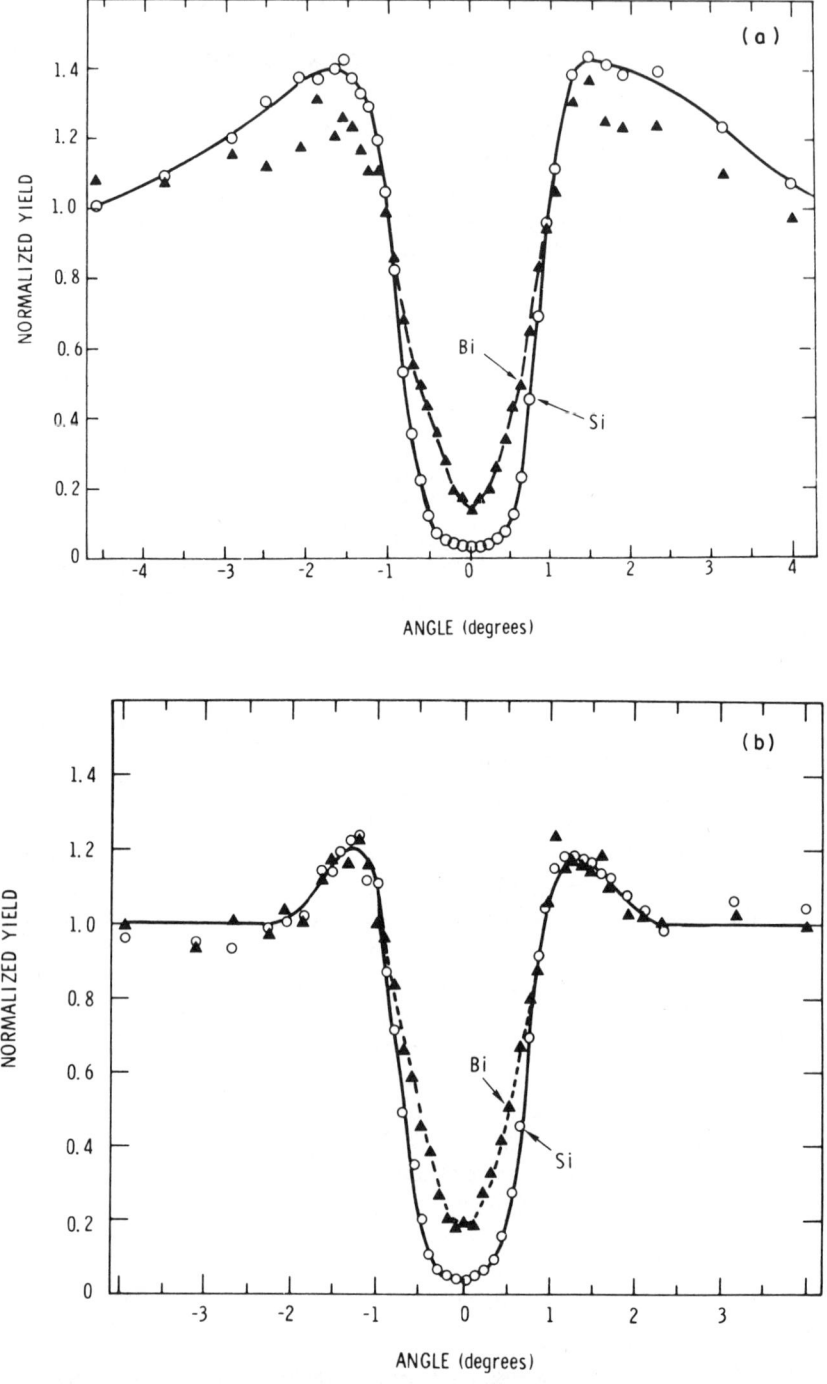

FIG. 34. Angular dependence of the normalized yields of backscattered 1-MeV ^{4}He$^+$ ions from Si and Bi atoms in a Bi-implanted Si crystal at 80 K. The crystal was implanted at 293 K with 150-keV ^{209}Bi to a fluence of 1.8×10^{14} cm^{-2} and then annealed for 1800 s at 923 K. (a) $\langle 110 \rangle$ angular scan, (b) $\langle 111 \rangle$ angular scan. (From Picraux et al.[114])

displacement of Bi atoms from lattice sites. For example, an annealing experiment that showed that the displacement occurred at a temperature where a specific defect became mobile (e.g., the divacancy) would indicate that defect-solute interactions were responsible for the observed displacement. Alternatively, since the Bi concentration in the implanted region was ~ 0.1 at. % (100 times greater than the solubility limit), the narrowing in Bi yield profiles could be due to solute ordering or preprecipitation.

Narrowing of channeling dips for solute atoms has been observed for other solute atoms in Si, when the solute concentrations exceed the solubility limit by a large factor,[115] as, for example, in laser-annealed samples of Si. In these cases, it is possible that the narrowing is due to solute ordering, solute preprecipitation or interaction of solute atoms with other defects.

In this context it is worth noting that the formation of Guinier-Preston (G-P) zones can be studied by channeling. These coherent precipitates have the same structure as the host lattice, but the concentration of solute atoms is higher than the average concentration in the alloy. Thus some differences in angular width of channeling dips are expected because of the Z_2 variation in Eq. (4.2.16). In this way one can, in principle, determine whether the G-P zone platelets in quenched and aged Al-2 at. % Cu alloys occur as monolayers of pure Cu or as several Cu-rich monolayers.

Narrowing of solute atom yields was also observed[116] for B in Si, as shown in Fig. 35. The narrowing occurred for $\langle 110 \rangle$ axial channels but not for $\langle 100 \rangle$ channels. In this case, the B was implanted at an energy of 25 keV to a fluence of 10^{15} cm^{-2}, and the sample was ruby laser annealed (1.6 J cm^{-2}). The crystal was then irradiated with 0.7-MeV ^{1}H$^+$ at 35 K and isochronally annealed up to 293 K. The major displacement of B atoms occurred during annealing near 240 K. It had been determined previously by EPR measurements that this annealing stage corresponded to migration of a defect caused by the interaction of interstitial Si atoms with B atoms.[117] This defect was either a Si-B split interstitial or a B interstitial. Thus in this case the displacement of B atoms from lattice sites was unambiguously due to defect-solute interactions.

The analysis of B in Si is an example of the use of nuclear reactions to determine lattice positions of light solute atoms (see Section 4.2.3). The yield of α particles from the ^{11}B(p,α)^{8}Be nuclear reaction was compared with the yield of backscattered 0.7-MeV ^{1}H$^+$ ions from the Si atoms. Because of

---

[115] C. W. White, S. R. Wilson, B. R. Appleton, and F. W. Young, Jr., *J. Appl. Phys.* **51**, 738 (1980).

[116] M. L. Swanson, L. M. Howe, F. W. Saris, and A. F. Quenneville, in "Defects in Semiconductors" (J. Narayan and T. Y. Tan, eds.), p. 71. North-Holland Publ., Amsterdam, 1981.

[117] G. D. Watkins, *Phys. Rev. B* **12**, 5824 (1975).

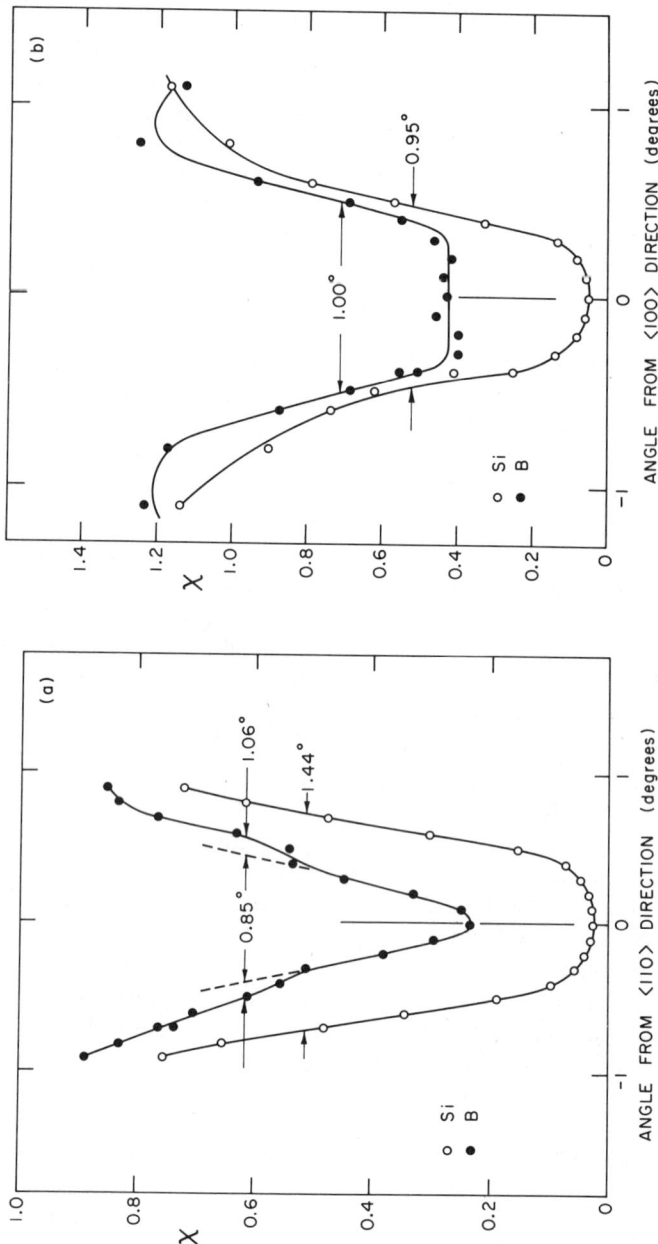

FIG. 35. Angular scans for a laser annealed Si–0.1 at. % B crystal after irradiation at 35 K with 0.7-MeV ^{1}H$^+$ ions to a fluence of 5×10^{16} cm^{-2} followed by a 20-h anneal at 293 K. The normalized yields χ_{Si} of ^{1}H$^+$ ions (incident energy 0.7 MeV) backscattered from Si atoms and the normalized yield χ_B of α particles from the ^{11}B(p,α)^{8}Be nuclear reaction are plotted as a function of the angle from the indicated axes. The measurements were taken at 35 K. (a) $\langle 110 \rangle$ channel, (b) $\langle 100 \rangle$ channel. (From Swanson et al.[116])

the complex nature of this particular nuclear reaction, it is difficult to extract depth information from an energy analysis of the emitted α particles. However, the laser-annealing treatment distributes the B atoms almost uniformly up to a depth of only 200–300 nm. When the yields from B atoms and Si atoms are compared for this shallow depth interval, the problem of increased yields at greater depths because of dechanneling is eliminated.

4.2.5.4. Large Displacements: Low-Symmetry Interstitial Sites. When solute atoms are displaced ~ 0.1 nm from lattice sites, the angular-scan profiles may have double peaks as in Fig. 30c, or they may be almost flat as in Fig. 30e. For displacements somewhat larger than 0.1 nm, a small central peak may also be seen. As mentioned previously, such large displacements into low-symmetry positions are the most difficult to analyze by channeling, or by any other technique. However, detailed channeling results provide the best method for analysis.

Large displacements of solute atoms into low-symmetry positions are often due to the trapping of intrinsic point defects, especially if the solute atoms are injected by ion implantation, which simultaneously introduces many point defects. Solute atoms may trap either vacancies or self-interstitials. Because small solute atoms contract the lattice, but self-interstitial atoms expand the lattice, it is expected that these defects will be strongly bound to each other. Such strong trapping has been found by channeling[27,45,118,119] and by other methods, in particular the Mössbauer effect.[120] Conversely, large solute atoms attract vacancies strongly (Section 4.2.5.5).

As an example, we shall consider the displacement of small solute atoms (Cu, Mn, Zn, or Ag) in Al, which is caused by the trapping of self-interstitial Al atoms. When an alloy of Al containing about 0.1 at. % of a small solute is irradiated at a temperature below that of self-interstitial migration (~ 45 K), relatively few solute atoms are displaced from lattice sites. However, during annealing through 45 K, the self-interstitials become mobile and are trapped by solute atoms (Fig. 36).[45,120,120a] The predominant trapping configuration is found by channeling to be the $\langle 100 \rangle$ mixed dumbbell (Fig. 4), in which a solute atom and a host atom straddle a lattice site.[22,27,28,45] Other trapping configurations have been inferred from internal friction data.[120b]

Because the displacement of solute atoms in such mixed dumbbells is large (0.1–0.14 nm), the solute atoms project well into $\langle 110 \rangle$ channels,

[118] M. L. Swanson, L. M. Howe, and A. F. Quenneville, *Can. J. Phys.* **55**, 1871 (1977).
[119] L. M. Howe, M. L. Swanson, and A. F. Quenneville, *Radiat. Eff.* **35**, 227 (1978).
[120] W. Mansel, H. Meyer, and G. Vogl, *Radiat. Eff.* **35**, 69 (1978).
[120a] K. Herschbach and J. J. Jackson, *Phys. Rev.* **153**, 694 (1967).
[120b] L. E. Rehn, K.-H. Robrock, and H. Jacques, *J. Phys. F* **8**, 1835 (1978).

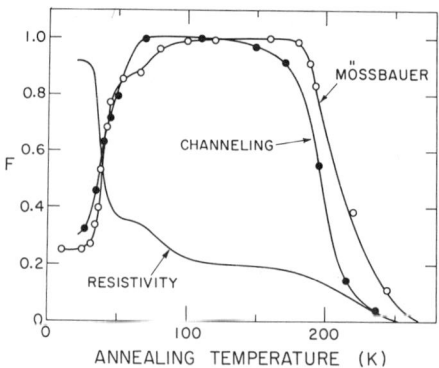

FIG. 36. A comparison of annealing data in irradiated Al and dilute Al alloys[45] (600-s pulse anneals). For the different experiments, the ordinate F is defined as follows. (a) Fraction of the electrical resistivity increment remaining after annealing 20-MeV deuteron-irradiated[120a] pure Al. (b) Fraction of ^{57}Co solute atoms that had trapped self-interstitials in a neutron-irradiated Al–0.0016 at. % Co alloy (fluence 0.7×10^{18} cm^{-2} at 4.6 K), as measured by the area of the Mössbauer defect line.[120] The results were normalized against the fraction 0.38 at 160 K. (c) Fraction of Ag atoms which had formed mixed dumbbells in 1-MeV ^{4}He$^+$ irradiated Al–0.082 at. % Ag as measured by channeling.[45] The results were normalized against the fraction 0.39 at 110 K, and the data below 70 K were taken from a separate experiment. The increase in dechanneling caused by the irradiation also annealed out completely near 200 K, indicating that most of the irradiation-induced point defects had been annihilated.

giving a large increase in backscattering yield from the solute atoms in these channels. An angular scan through a $\langle 110 \rangle$ channel for Al–0.13 at. % Cu, in which ~ 40% of the Cu atoms have formed Al–Cu mixed dumbbells (and the other 60% are still on lattice sites), shows a well-defined peak in yield from Cu atoms for $\langle 110 \rangle$ alignment (Fig. 37). A comparison of $\langle 110 \rangle$, $\langle 100 \rangle$, and $\langle 111 \rangle$ yields with calculated ion fluxes $F_i^{\langle lmn \rangle}$, using Eqs. (4.2.45) and (4.2.46) and assuming a unique position of displaced solute atoms, indicated that the Cu atoms in Al–Cu mixed dumbbells were displaced 0.148 ± 0.005 nm from lattice sites[28] in $\langle 100 \rangle$ directions. The calculated yields for these displacements, using the analytical method outlined in Section 4.2.2 are also shown in Fig. 37 and agree well with the experimental data.[28] If the displacement of solute atoms were less than 0.12 nm, this peaking effect would vanish (see Fig. 10). Thus a rather accurate measure of displacements is provided by angular-scan data.

Channeling measurements can be used to study the evolution and annihilation of the mixed dumbbell trapping configuration as a function of irradiation and annealing conditions. In this way, detailed information can be obtained about defect production rates, defect mobility, trapping effi-

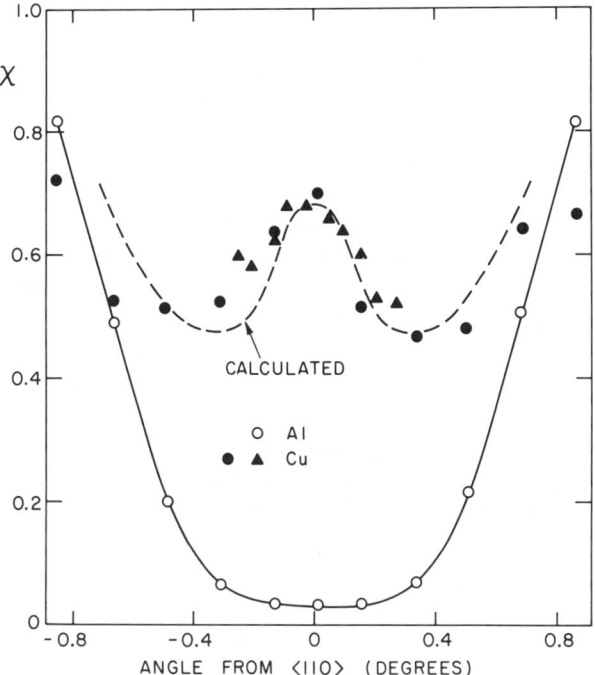

FIG. 37. Experimental and calculated yields for an angular scan through a $\langle 110 \rangle$ channel in an Al–0.13 at. % Cu crystal, using 1.5-MeV ^{4}He$^+$ at 30°K. The crystal had been irradiated at 70 K with 1.5-MeV ^{4}He$^+$ ions to a fluence of 9.0×10^{15} cm^{-2}. The calculated curve was obtained by assuming that 40% of the Cu atoms had been displaced 0.148 nm along $\langle 100 \rangle$ directions and that 60% of the Cu atoms remained on lattice sites.[28] (Compare Fig. 30f.) (Adapted from Matsunami et al.[28])

ciencies, defect stability, and annihilation mechanisms.[27,45,107,121] For example, it can be seen from Fig. 36 that the mixed dumbbells are stable from ~ 50 K to ~ 200 K, when they are annihilated by combination with vacancies.

During annealing, the concentration of mixed dumbbells (e.g., Al–Ag) can be increased by the release of self-interstitials from weaker traps (e.g., Mg atoms, which are large solutes in Al). In this way the binding energy of self-interstitials to other solute atoms can be found.[122]

It appears that very small solute atoms can also be trapped in low-symmetry positions in crystals. Deuterium atoms implanted into body-centered-cubic metals often occupy unique sites that are large distances from

[121] M. L. Swanson and L. M. Howe, *Radiat. Eff.* **41,** 129 (1979).
[122] M. L. Swanson, L. M. Howe, and A. F. Quenneville, *J. Phys. F* **6,** 1629 (1976).

normal lattice sites, but are close to interstitial sites. One example is D implanted into Fe at 90 K. Since D is very mobile in Fe at 90 K, it can be retained only in a trapped state. The nature of the trapping was studied by a series of anneals, which showed that the deuterium was released from the sample in two pronounced stages, near 260 K and 400 K (Fig. 38).[123] The position of the trapped D at 200 K was found to be ~0.04 nm from octahedral sites, as shown in Fig. 39. This displaced octahedral site fits the channeling data of Fig. 40 best; in particular, the peak in D yield observed for a {112} plane was consistent only with this site. The results indicated that D was trapped by single vacancies. A change in configuration occurred above 260 K, perhaps reflecting stronger trapping at vacancy clusters.

Many experiments with ion-implanted Si have shown that the implanted solutes can occupy a variety of lattice sites, but it is often not clear whether these are equilibrium or defect-associated sites. One example is Yb implanted into Si; the solute atoms lie near the tetrahedral interstitial sites, producing strong peaking in Yb yields for $\langle 110 \rangle$ alignment and small dips in Yb yields for $\langle 100 \rangle$ and $\langle 111 \rangle$ channels.[124]

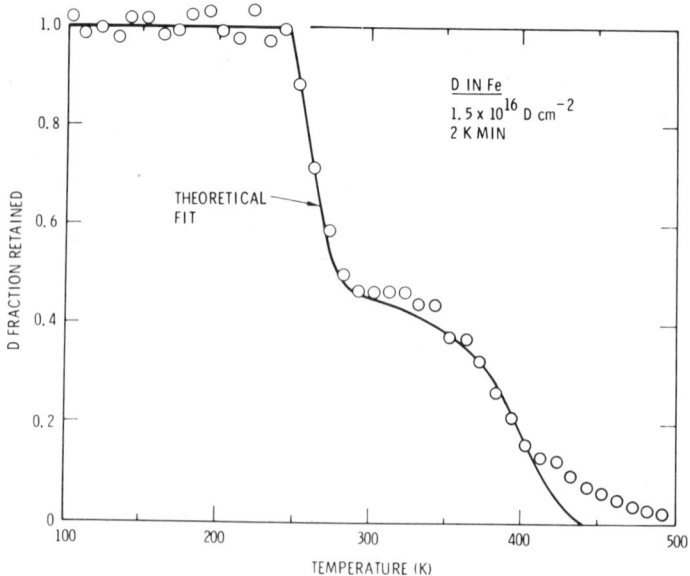

FIG. 38. Release of implanted D from the implanted region in Fe upon ramp heating. The theoretical calculation corresponds to two trap energies of 0.48 and 0.8 eV. (From Picraux.[123])

[123] S. T. Picraux, *Proc. Int. Conf. Ion Beam Modif. Mater., Albany, N.Y., 1980*, published in *Nucl. Instrum. Methods* **182/183**, 413 (1981).

[124] J. U. Andersen, O. Andreasen, J. A. Davies, and E. Uggerhøj, *Radiat. Eff.* **7**, 25 (1971).

4.2. CHANNELING STUDIES OF LATTICE DEFECTS

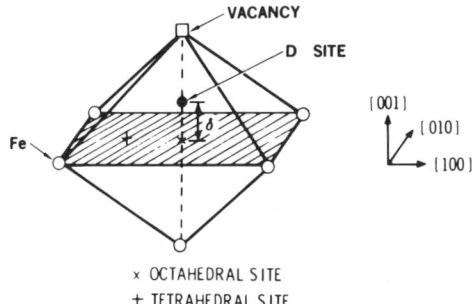

FIG. 39. Position of D in Fe corresponding to the distorted O site in the bcc lattice. (From Picraux.[123])

Substitutional solutes in Si are displaced from lattice sites by irradiation.[125] Strong evidence has been obtained that group V solutes trap vacancies, and group III solutes trap self-interstitials. This trapping is thought to be the result of Coulomb interaction since vacancies and group III solutes are negatively charged, whereas self-interstitials and group V solutes are positively charged. The trapping configurations have not yet been identified by channeling; one example, B in Si, was given in the previous section. In these cases, the channeling results can be compared with the definitive EPR data, which have identified the electronic states of many trapping configurations,[126] but which do not specify atom locations.

4.2.5.5. High-Symmetry Interstitial Sites. The interstitial sites to be considered here have the same or almost the same symmetry as that of the host crystal. For example, the body-centered site in a fcc lattice (Figs. 3 and 31b) has cubic symmetry. The significance for channeling studies is that only two or three channels need be used for a definitive lattice location study since the interstitial site often is exactly in the center of at least one channel and completely shadowed in other channels.

4.2.5.5.1. GAS ATOMS. Small gas atoms often occupy octahedral or tetrahedral interstitial sites in metals.[44,106] One example[127] that illustrates the use of the channeling method for determining high-symmetry interstitial positions is the lattice location of oxygen in Ti. It has been established by x-ray methods that O occupies the octahedral interstitial site.[128] This position lies in the center of [0001] and $\langle 11\bar{2}0\rangle$ axial channels and (0001) planar channels (Fig. 31d). Axial scans through these three channels (Fig. 41)

[125] L. W. Wiggers and F. W. Saris, *Radiat. Eff.* **41**, 149 (1979).
[126] G. D. Watkins, *in* "Radiation Effects in Semiconductors" (F. L. Vook, ed.), p. 67. Plenum, New York, 1968.
[127] R. B. Alexander and R. J. Petty, *Phys. Rev. B* **18**, 981 (1978).
[128] B. Holmberg, *Acta Chem. Scand.* **16**, 1245 (1962).

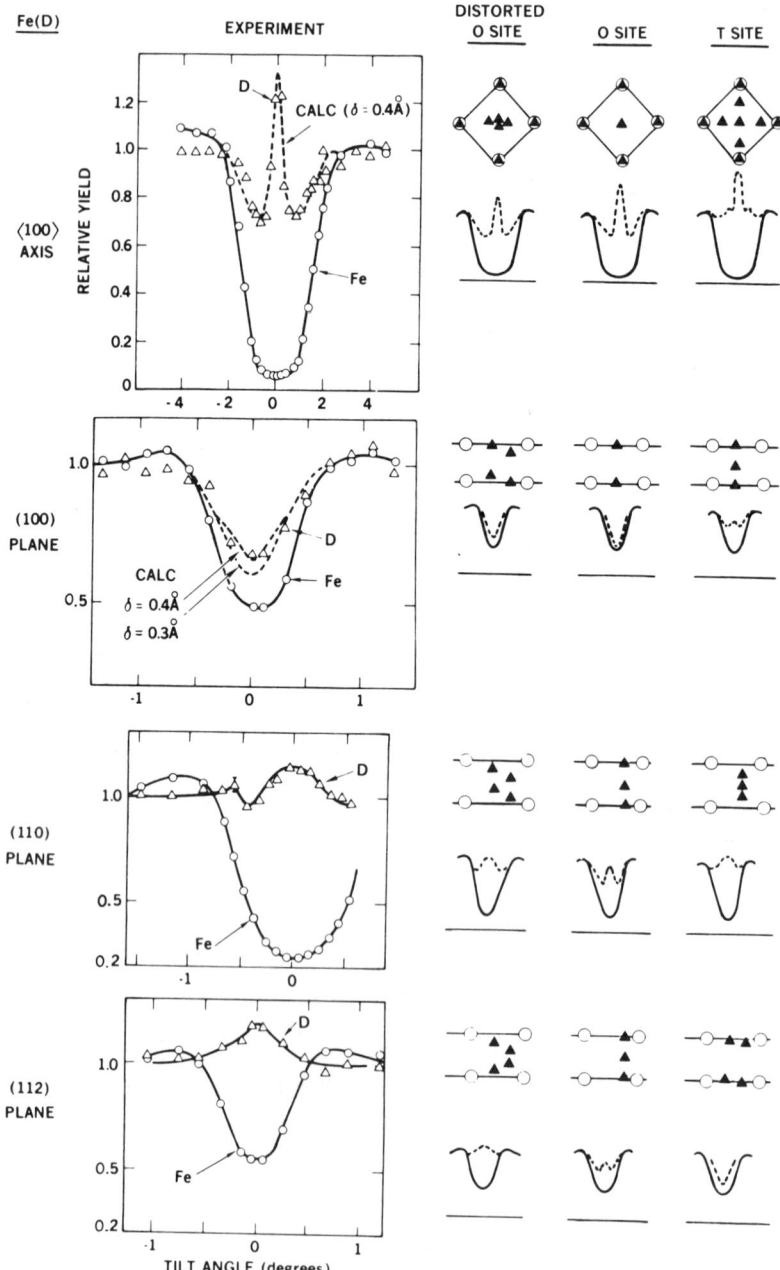

Fig. 40. Angular scans for D in Fe showing distorted O-site occupancy after annealing to 200 K. (From Picraux.[123])

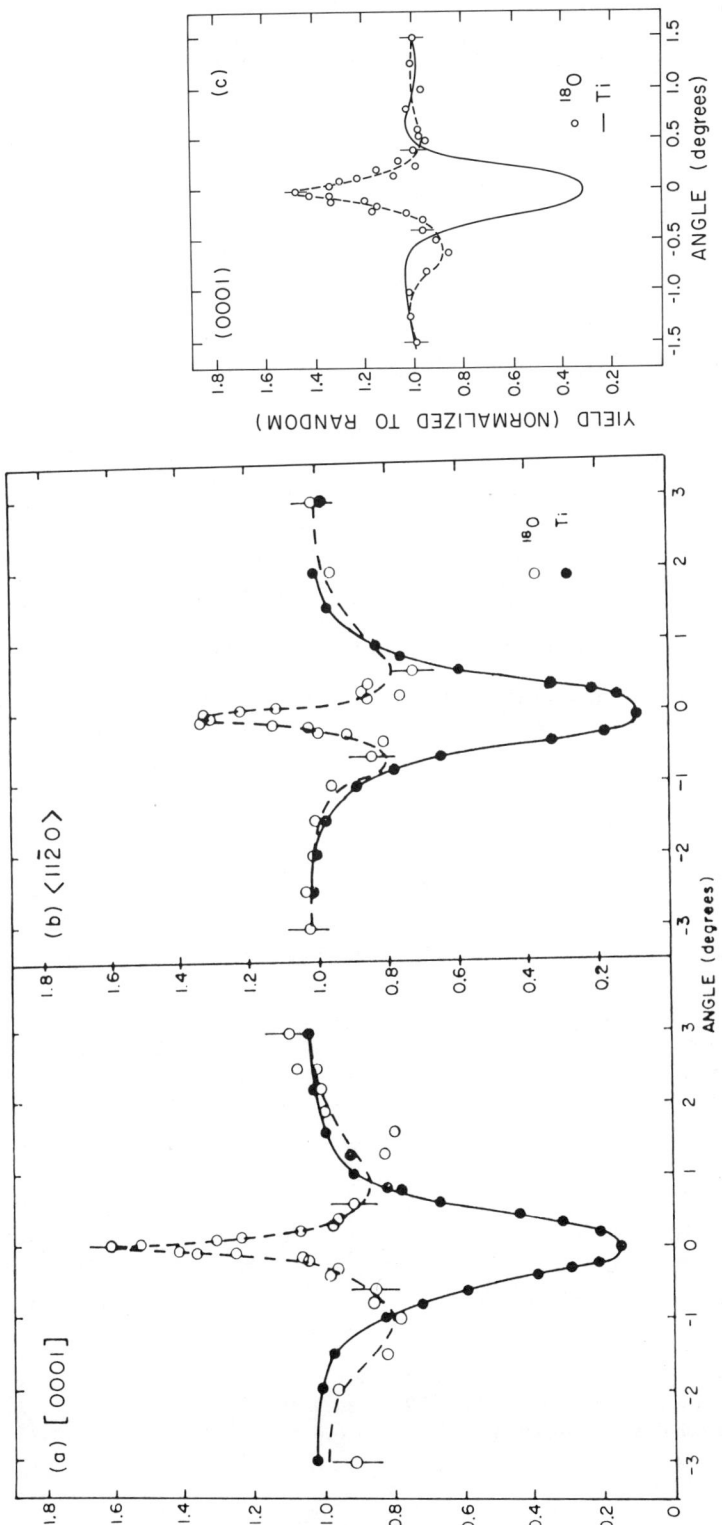

FIG. 41. Angular scans through (a) [0001], (b) $\langle 11\bar{2}0 \rangle$ axes, and (c) a (0001) plane, using 730-keV protons in ^{18}O-implanted Ti crystals. The unfilled circles correspond to the α-particle nuclear-reaction yield from ^{18}O, and the filled circles to the proton backscattering yield from Ti. Some of the data points have been omitted for clarity. (From Alexander and Petty.[127])

indeed showed strong peaking in yields from O atoms, for a crystal implanted in the surface region to an average concentration of ~ 0.14 at. % ^{18}O. The nuclear reaction ^{18}O(p,α)^{15}N was used to detect the ^{18}O.

In this case a slight discrepancy between calculated and experimental flux peaks was observed: the peak was greater for (0001) planar channels than for $\langle 11\bar{2}0 \rangle$ axial channels. There are several possible explanations for this difference.

(a) The calculations were performed for a Cu $\langle 110 \rangle$ axis and {111} plane, which although similar to a Ti $\langle 11\bar{2}0 \rangle$ axis and a (0001) plane, could give quantitatively different results. (Compare the calculations of Vianden et al.[129] for Be.)

(b) The effect of the radiation damage introduced by the angular scans could perturb the positions of the O atoms (see Kaim and Palmer[130]).

(c) When yields from implanted solute atoms are measured, the depth dependence of the ion flux, and its variation with different channels, must be considered.

(d) Anisotropic thermal vibrations of the O atoms could be significant.

In regard to the last point, Carstanjen[44] has studied the effect of the measuring temperature on the position of deuterium in Pd. In this case, D was diffused into Pd to a concentration of 0.7 at. % and analyzed by the D(^{3}He,p)^{4}He reaction. The presence of a large peak in yield from D for a $\langle 110 \rangle$ axial channel and {111} planar channel, together with complete shadowing for $\langle 100 \rangle$ axial and {100} planar channels showed that the D was in octahedral interstitial sites. The vibrational amplitude of D was estimated by measurement of the decrease in half-width of the $\langle 100 \rangle$ channeling dip as a function of temperature in the range 295–446 K.

The positions of small solute atoms, mostly gases, in metals as obtained from channeling studies have been summarized by Picraux[123] and Carstanjen.[44] In most cases, these positions agree with data from other sources. In some cases, it appears that radiation-induced defects (vacancies) stabilize a higher-energy interstitial position; for example, implanted D in Al occupies the tetrahedral site,[131] whereas the octahedral one is considered to be the normally stable site.

4.2.5.5.2. DEFECT-ASSOCIATED SITES. Large solute atoms may also occupy high-symmetry interstitial sites if a suitable number of vacancies are trapped by the solute atom. For example, it has been observed that the irradiation of an Al–0.02 at. % In crystal causes a large fraction of the In

---

[129] R. Vianden, E. N. Kaufmann, and J. W. Rodgers, *Phys. Rev. B* **22**, 63 (1980).
[130] R. E. Kaim and D. W. Palmer, *Philos. Mag., Part A* **40**, 279 (1979).
[131] J. P. Bugeat, A. C. Chami, and E. Ligeon, *Phys. Lett.* **58A**, 127 (1976).

4.2. CHANNELING STUDIES OF LATTICE DEFECTS

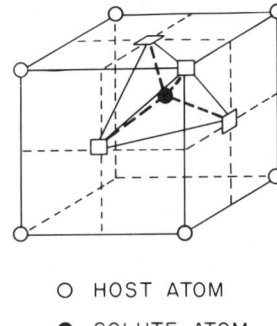

○ HOST ATOM
● SOLUTE ATOM
□ VACANCY

FIG. 42. Proposed tetravacancy-solute atom complex in a face-centered-cubic crystal.

atoms to be displaced into tetrahedral interstitial sites. In this case, symmetry considerations indicate that four vacancies surround each of the displaced In atoms, to form the configuration shown in Fig. 42.

In order to show that the trapping of vacancies by the In atoms produces this defect configuration, a homogeneous alloy of Al–0.02 at. % In was irradiated with 1-MeV α particles at 35 K, and then annealed in isochronal steps, as shown in Fig. 43. The In atoms became displaced from lattice sites

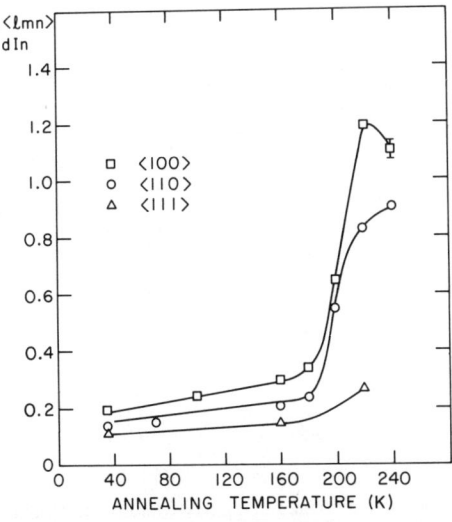

FIG. 43. The effect of isochronal annealing (600-s pulses) on the apparent displaced fraction $f_{dIn}^{\langle lmn \rangle}$ of In atoms into $\langle 100 \rangle$, $\langle 110 \rangle$, and $\langle 111 \rangle$ axial channels of an Al–0.02 at. % In crystal. The backscattering measurements were made with 1-MeV ^{4}He$^+$ at 35 K. Before these anneals, the crystal was irradiated at 35 K with 1-MeV ^{4}He$^+$ in a random direction to a fluence of 6 × 10^{15} cm^{-2}.

during annealing near 200 K. This temperature corresponds to the vacancy-annealing stage, as shown by the annihilation of Al–Ag mixed dumbbells at the same temperature (Fig. 36) and also by other radiation damage experiments. Furthermore, the displaced fraction of In atoms was strongly reduced by subsequent irradiation at 70 K, where the trapped vacancies were annihilated by self-interstitial Al atoms.

The growth of the peak in $\langle 100 \rangle$ yield from In atoms as a function of irradiation fluence is shown in Fig. 44. This peak is strong evidence that the In atoms occupy the tetrahedral interstitial site, which lies in the center of $\langle 100 \rangle$ channels (Fig. 31). For the highest fluence, the peak-to-valley ratio was 2.3, which is close to that calculated for tetrahedral sites.[28] Thus a large

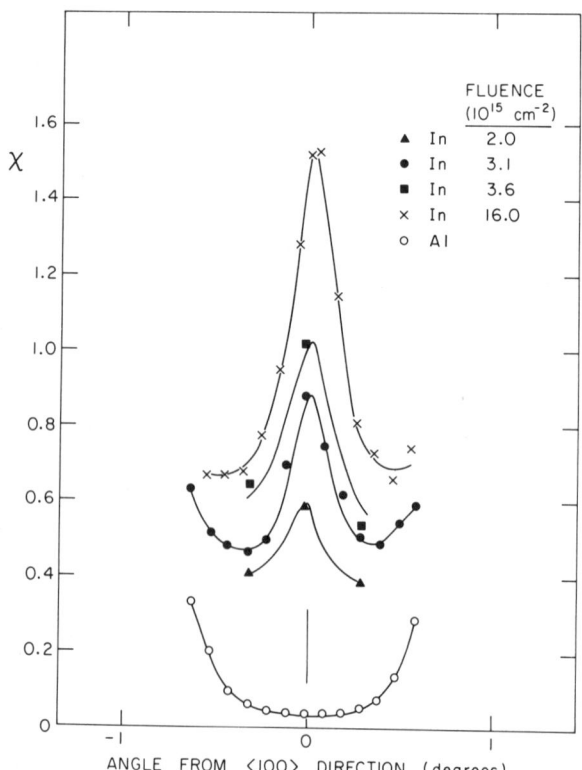

FIG. 44. Angular dependence of the normalized yields of backscattered 1-MeV ^{4}He$^+$ ions from Al and In atoms near a $\langle 100 \rangle$ axis of an Al–0.02 at. % In crystal. The sample was irradiated in a random direction to the indicated fluence of 1-MeV ^{4}He$^+$ ions at 35 K. Each irradiation was followed by a 600-s anneal at 220 K before the measurements. The yields were measured at 35 K for a depth interval of 50–280 nm from the surface. Since the Al yields were almost the same for each irradiation, only one set of data is shown.

fraction (~ 0.60) of the In atoms were in tetrahedral interstitial sites. Most of the remaining In atoms occupied substitutional or random sites.

In the case of Sn solute atoms in Al, similar vacancy clustering occurred,[25] with the presence also of interstitial Sn atoms surrounded by groups of three and six vacancies. The evolution of such vacancy clusters as a function of irradiation fluence was monitored by measurements of $\langle 100 \rangle$, $\langle 110 \rangle$, and $\langle 111 \rangle$ channeling yields.[132]

4.2.5.6. Random Sites. When solute atoms are in random sites, the apparent displaced fractions $f_{ds}^{\langle lmn \rangle}$ for all channeling alignments are equal, and no structure is seen in angular scans (Fig. 30). In this case, the solutes may be present in precipitates of random alignment or in association with large defect clusters. The solutes may also have diffused to the surface. If the solute atoms are incorporated in precipitates with some preferred orientation with respect to the host crystal, attenuation of yields occurs in some channels. The case of solute atoms in coherent precipitates has been discussed earlier.

4.2.6. Surface Studies

4.2.6.1. Introduction. Previous sections of this chapter dealt with the use of channeling to investigate the location of solute atoms (4.2.5) and also of displaced host atoms (4.2.4) within the bulk of a single crystal. In this section, we consider a simple extension of these channeling techniques that enables similar information to be obtained for displaced *surface* atoms. Of particular interest is the use of single- and double-alignment geometry to measure changes in surface structure, vibrational amplitude, or interplanar spacing (relaxation) relative to the underlying lattice. Such information is basic to our understanding of the nature of surface adsorption on various crystal surfaces and hence has obvious application in the fields of catalysis and chemisorption.

The idea of using MeV ion channeling for surface studies was first suggested by Bøgh[133] in 1967. He pioneered the early development of the technique and subsequently published an excellent review thereon.[46] Nevertheless, for many years, widespread application was severely hampered by the technical difficulties of performing channeling experiments under sufficiently high quality (UHV) vacuum conditions (i.e., $< 10^{-8}$ Pa) for meaningful comparison with other surface studies. As noted above (in Section 4.2.3), these technical problems have been overcome, and several groups now have excellent UHV channeling facilities for surface studies.

---

[132] M. L. Swanson, L. M. Howe, J. A. Moore, and A. F. Quenneville, *J. Phys. F. Met. Phys.* **11**, L185 (1981).
[133] E. Bøgh, *Phys. Rev. Lett.* **19**, 61 (1967).

4.2.6.2. Basic Principles. The principle involved in applying channeling for surface analyses may be illustrated simply by considering the flux distribution at the first two atoms in an aligned atomic row (Fig. 45). The first atom in each row is completely unshadowed, and hence, for all angles of incidence, it contributes normally to the observed backscattered yield. The flux distribution at all subsequent atoms, however, is strongly suppressed whenever the beam is aligned with a crystal axis, due to the correlated steering effect of large impact parameter collisions with the preceding atom(s) along the row. In the idealized *static* lattice (shown in Fig. 45), the flux at the second atom falls to zero within a well-defined distance of closest approach, R. For Coulomb scattering,

$$R_c = 2(Z_1 Z_2 e^2 d/E)^{1/2}, \qquad (4.2.52)$$

where d is the atomic spacing, E the incident energy, and Z_1 and Z_2 are the atomic numbers of beam and target atoms, respectively. For 1.0-MeV He$^+$ incident along the $\langle 100 \rangle$ direction of Pt, the value of R_c (0.0187 nm) is considerably larger than the bulk value of u_2, the two-dimensional thermal vibrational amplitude, which for Pt at room temperature is 0.0089 nm. When one uses a more realistic screened Coulomb potential, such as the Molière potential, [Eq. (4.2.5)], the calculated distance-of-closest-approach value, $R_M = 0.0154$ nm, is decreased somewhat; nevertheless, it is still much larger than u_2, and hence, even when thermal vibrations are included, shadowing at the second and all subsequent atoms in the aligned row should be a dominant effect. Consequently, a clearly resolvable surface peak is

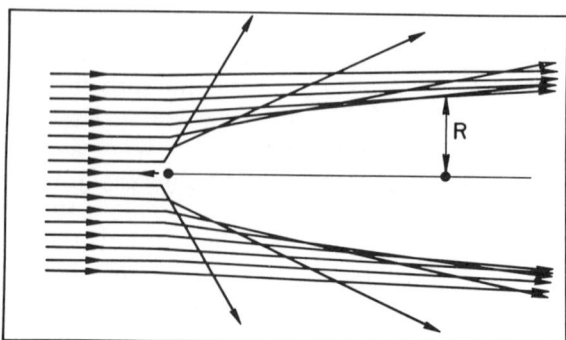

FIG. 45. Formation of a shadow cone at the second atom in an aligned row. The shadow cone radius at this second atom is denoted R. (From Feldman.[48])

4.2. CHANNELING STUDIES OF LATTICE DEFECTS

observed in the RBS energy spectrum, as shown in Fig. 46. A similar peak, but with much poorer depth resolution, was also present in Fig. 15.

Since RBS cross sections are accurately known, this surface-peak area provides a quantitative measure of the number of unshadowed lattice atoms per square centimeter; this number in turn contains useful information about structural changes in the surface region, such as the outward relaxation shown in Fig. 47. For perpendicular incidence ($\langle 111 \rangle$ in Fig. 47), the second atom in each row is shadowed, and hence the observed surface-peak area will correspond to 1 atom per row (4.5×10^{15} Pt atoms cm^{-2}) at all energies, provided the shadow cone radius R at the second atom is much larger than u_2. Note that a surface relaxation Δd, or an enhanced vibrational amplitude perpendicular to the surface, has no effect on this $\langle 111 \rangle$ shadowing, but any lateral displacement within the surface plane would be clearly detected.

For nonperpendicular axes, such as the $\langle 110 \rangle$ in Fig. 47, an outward (or inward) displacement Δd shifts the shadow cone relative to the underlying row of atoms, thus causing the surface peak to increase towards a value of 2 atoms per row. Since R varies approximately as $E^{-1/2}$ [Eq. (4.2.52)], this $\langle 110 \rangle$ surface peak (in a static lattice) would increase from a value of 1 atom

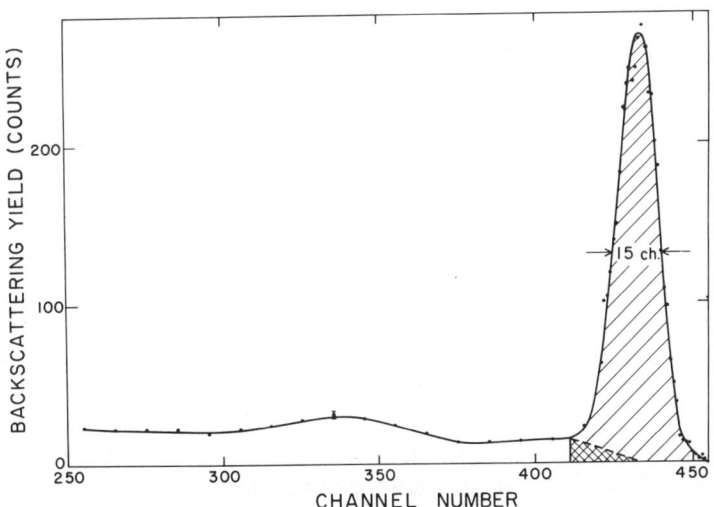

FIG. 46. Backscattered energy spectrum for 2.0-MeV ^{4}He$^+$ incident on a Pt(111) crystal along the [110] direction at 60 K. The energy scale has been expanded, with a large zero-offset. The width (FWHM) of the surface peak is equivalent to a depth resolution of 3.5 nm; its area is equivalent to 1.39 Pt atoms per $\langle 110 \rangle$ row. (From Davies et al.[31])

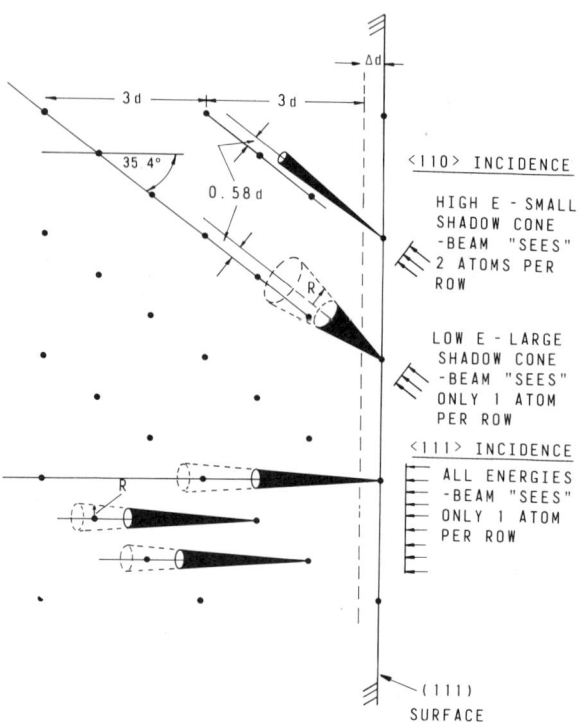

FIG. 47. Atomic configuration near the surface of a Pt(111) crystal, illustrating the use of channeling to investigate the surface relaxation Δd.

per row at low E to approximately 2 atoms per row at high E. Thus the magnitude of the surface relaxation Δd (and also of any lateral displacements) can be obtained by measuring the surface-peak area for various low-index directions as a function of energy.

In practice, a more sensitive experimental procedure is to perform detailed angular scans through each of the chosen close-packed axes, as shown in Section 4.2.6.3.

In most cases of practical interest, thermal vibration effects are not negligible, and the observed surface peak may include significant contributions from several underlying atoms per row. Quantitative analytical estimates are then no longer possible, except in the two-atom limit of Fig. 45, where Feldman et al.[134] have shown that the surface-peak area N in atoms per row is given by

$$N = 1 + (1 + R^2/2u_2^2)e^{-R^2/2u_2^2}. \qquad (4.2.53)$$

[134] L. C. Feldman, R. L. Kauffman, P. J. Silverman, R. A. Zuhr, and J. H. Barrett, *Phys. Rev. Lett.* **39**, 38 (1977).

4.2. CHANNELING STUDIES OF LATTICE DEFECTS

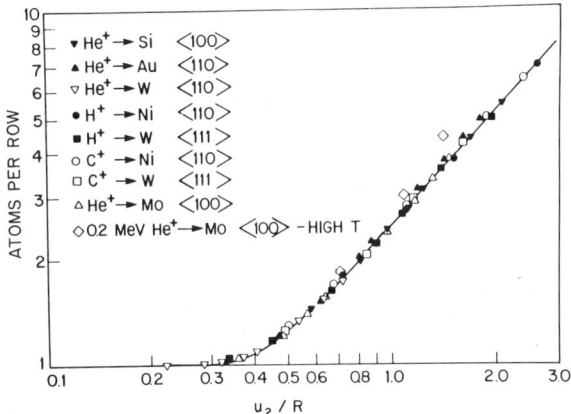

FIG. 48. Calculated surface-peak area in atoms per row versus u_2/R. The high-temperature points correspond to u_2 values up to four times the room-temperature value; all other points are for room-temperature conditions. (Adapted from Feldman.[48])

The first term in Eq. (4.2.53) represents the unit contribution from the first atom in the row and the second term the temperature-dependent contribution from the second atom. Although this two-atom approximation is often not adequate for detailed comparison with experiment, it has provided a very useful scaling parameter (u_2/R) for correlating various surface peak intensity measurements, as seen in Fig. 48.

For quantitative interpretation of surface scattering studies, one normally relies on Monte Carlo computer simulation in which a large number of incident ion trajectories are computed and the contribution to the surface-peak yield is calculated for each successive layer,[58] as shown in Fig. 49. An additional advantage of the computer simulation technique is that specific

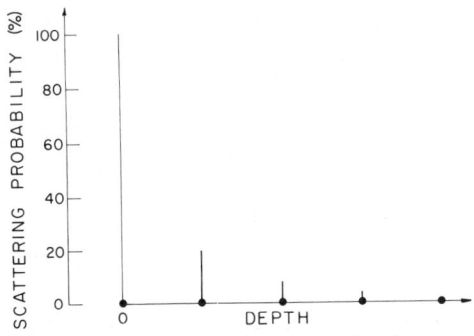

FIG. 49. Calculated normalized backscattering probability from successive planes in an Au(011) crystal using 0.5-MeV ^{4}He$^+$ ions at 300 K. (From Bøgh.[46])

surface structure models, including enhanced vibrational amplitude, are easily incorporated for comparison with experiment.

A further complication of any quantitative comparison between simulation and experiment is the existence of significant correlations in the thermal vibration of adjacent atoms in the lattice. The effect of such correlations is to decrease slightly the relative displacement between neighboring atoms, thus increasing the shadowing and reducing the magnitude of the surface peak. This correlation effect has recently been included by Jackson and Barrett[135] in their Monte Carlo simulations.

So far, we have been discussing how the shadow cone along the *incident* ion trajectory may be used to obtain quantitative information about the structure of a crystal surface relative to the bulk lattice. An equally important and closely related process is to utilize the shadow cone along the emerging (backscattered) trajectory—i.e., the so-called "blocking" process.[136] The concept, which is illustrated in Fig. 50, is essentially a double-alignment experiment in which channeling of the incident beam is used to obtain a well-resolved surface peak (Fig. 50b) and, by means of a movable, well-collimated detector, this is combined with blocking of the emitted beam along some other direction.

Note that, along the incident beam direction in Fig. 50a, only 50% of the atomic rows originate in the surface plane; the other 50% originate in the second plane. Consequently, at least 50% of the backscattered ions contributing to the surface peak can be blocked along certain emission directions by the surface plane of atoms, thus producing a blocking dip in the surface peak yield. Any relaxation Δd of the surface plane will rotate the position of this blocking dip through an angle $\Delta\theta$ relative to the underlying bulk blocking direction, as shown in Fig. 50c. Hence the magnitude and sign of Δd may be obtained from purely geometrical considerations. Another big advantage of this blocking technique is that it can be readily adapted to provide information on the location of adsorbate atoms since these too may produce observable blocking dips in the backscattered yield from the underlying crystal. One major disadvantage is that the acceptance angle to the detector must be very small, i.e., less than the blocking angle R/d. Hence rather large analyzing beam fluences are used and considerable care is required in order to minimize radiation damage and beam-induced desorption effects.

It should perhaps be noted that another very important application of MeV ion beams in surface adsorbate studies is the availability of various nuclear reactions, such as $^{12}C(d,p)$, $^{16}O(d,p)$, $^{14}N(d,\alpha)$, and $D(^3He,p)$, to

---

[135] D. P. Jackson and J. H. Barrett, *Phys. Lett.* **71A**, 359 (1979).
[136] W. C. Turkenburg, W. Soszka, F. W. Saris, H. H. Kersten, and B. G. Colenbrander, *Nucl. Instrum. Methods* **132**, 587 (1976).

4.2. CHANNELING STUDIES OF LATTICE DEFECTS

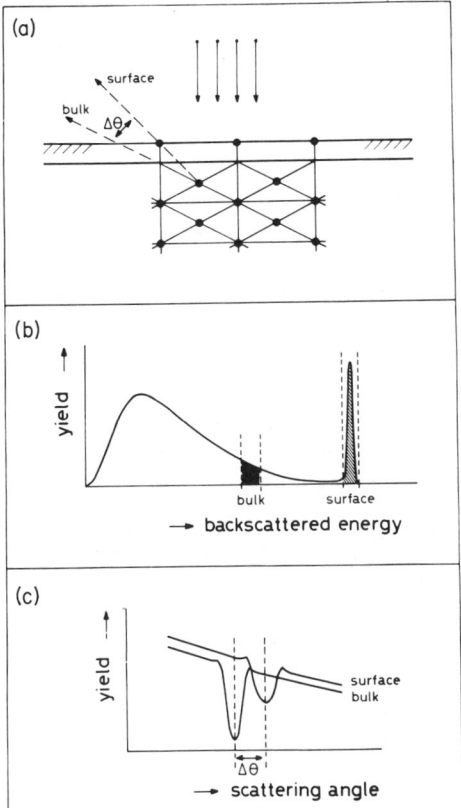

FIG. 50. (a) Schematic diagram of a combined channeling–blocking experiment. (b) Channeled energy spectrum showing a clearly resolvable surface peak. (c) Detector scan showing the blocking dips for the two shaded energy regions in (b). (From Turkenburg et al.[136])

determine the absolute coverage of many common adsorbates with an accuracy[77] of 5–10%. Although such nuclear microanalyses do not involve channeling directly, they nevertheless play an extremely important complementary role in several of the surface scattering studies that are being discussed here.

In the following sections, examples of applications of channeling in studying surface structure are presented in order to illustrate better the basic concepts already discussed.

4.2.6.3. Surface Relaxation of Pt. MeV ion scattering is ideally suited to measure extremely small changes Δd in interplanar spacing at the surface. Whenever Δd is comparable in magnitude to the two-atom shadow cone radius R, then the surface-peak intensity for all channeling directions, except normal incidence, will be increased considerably—as was shown in Fig. 47. Hence lattice expansion (or contraction) of even a few percent is easily detected. It turns out that the most sensitive and accurate technique is to measure the surface-peak intensity as a function of angle for small angular deflections about a given bulk channeling direction. In such an angular scan along a nonperpendicular axis (for example, the $\langle 110 \rangle$ in Fig. 47), any outward or inward displacement of the surface plane produces a significant asymmetry in the observed surface-peak minimum relative to the underlying bulk channeling axis. A small outward displacement or relaxation shifts the minimum value slightly from the bulk ($\langle 110 \rangle$) direction *towards* the surface normal. Similarly, an inward displacement or contraction would shift the minimum *away* from the surface normal. At the same time, any enhancement in surface vibrational amplitude increases the absolute magnitude of the surface peak and also increases the width of its angular scan, but does not displace the minimum value from the bulk ($\langle 110 \rangle$) direction. Hence detailed angular scans contain quantitative information not only on Δd but also on the surface Debye temperature θ_s.

An example[31] of this type of measurement, for the Pt(111) surface in the presence of adsorbed H_2, is shown in Fig. 51. The experimental data indicate a small but readily detectable anisotropy, with the minimum being shifted several tenths of a degree from the $\langle 110 \rangle$ *towards* the surface normal. Comparison with the set of Monte Carlo curves for various Δd values suggests that an outward relaxation of approximately 0.003 nm is involved. Also, the observed agreement between the data and the simulation curve

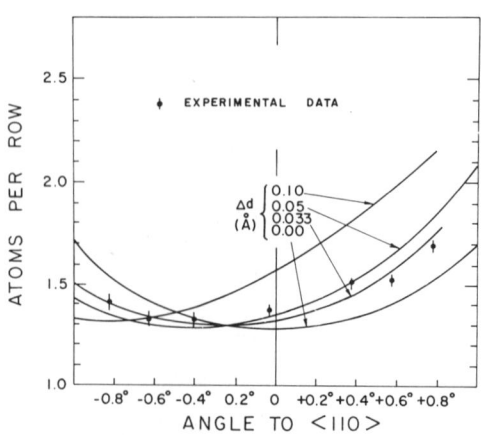

FIG. 51. Angular scans of the surface peak about the $\langle 110 \rangle$ channeling direction in Pt(111) at 78 K, using a 2.0-MeV ^{4}He beam. Solid curves were obtained by Monte Carlo simulation, using various values of Δd and assuming isotropic surface vibrations characterized by the bulk Debye temperature $\theta_D = 239$ K. (From Davies et al.[31])

4.2. CHANNELING STUDIES OF LATTICE DEFECTS

($\Delta d = 0.0033$ nm) indicates that in this case the surface vibrational amplitude is quite accurately characterized by the bulk Debye temperature.

A rather different example,[137] the Pt(100) surface plus adsorbed H_2, is shown in Fig. 52. Here, the shift in surface-peak minimum is barely detectable, indicating a much smaller outward relaxation (i.e., $\Delta d \sim 0.001$ nm). On the other hand, the experimental data points are $\sim 20\%$ higher than a Monte Carlo simulation based on the bulk Debye temperature value (curve A), thus suggesting an enhanced vibrational amplitude in the surface and near-surface region. Included in Fig. 52 are several simulation curves obtained for various combinations of surface (θ_1) and subsurface (θ_2) Debye temperatures, chosen to agree with the magnitude of the observed surface

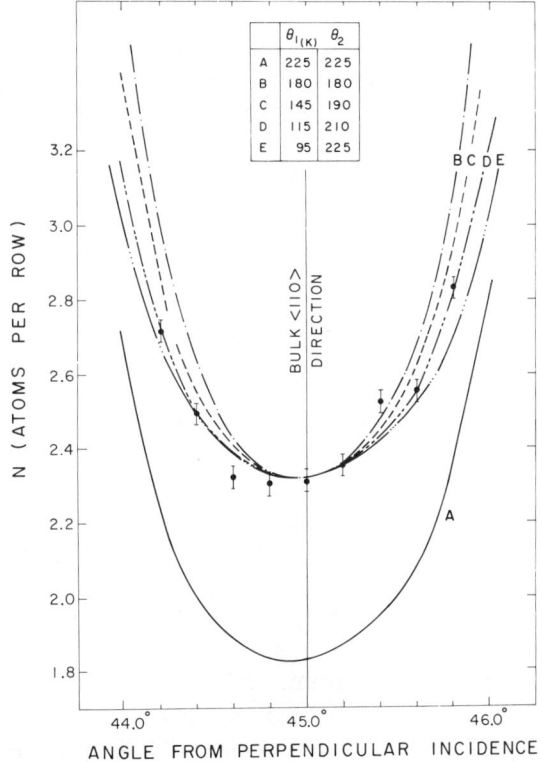

FIG. 52. Angular scans of the surface peak about the $\langle 110 \rangle$ channeling direction in H_2-covered Pt (001) at 150 K, using 2.0-MeV ^{4}He$^+$ ions. The curves were obtained by Monte Carlo simulation, using $\Delta d = 0.001$ nm and various combinations of surface (θ_1) and subsurface (θ_2) Debye temperatures. (From Davies et al.[137])

[137] J. A. Davies, T. E. Jackman, D. P. Jackson, and P. R. Norton, Surf. Sci. 109, 20 (1981).

peak at $\langle 110\rangle$ incidence. It is evident that the predicted angular width is largest (curve E) when the enhanced vibration is restricted to the surface plane and is narrowest (curve B) when the enhancement is distributed uniformly (i.e., $\theta_1 = \theta_2$) over all planes contributing to the surface peak. Clearly, curve D gives a quite reasonable fit to the experimental angular-scan data, suggesting that the enhanced vibration of this hydrogen-covered Pt(100) surface is best characterized by strong enhancement ($\theta_1 = 115$ K) in the surface plane coupled with a considerably weaker enhancement ($\theta_2 = 210$ K) in the underlying ones. Hence, in favorable cases, angular-scan data may even distinguish between surface and subsurface contributions to enhanced vibration.

It is interesting to note the complementary role that LEED plays in all these MeV ion scattering experiments. In the surface relaxation measurements discussed above, for example, it is always necessary to establish whether or not the crystal surface has undergone significant lateral reconstruction and this is most easily and reliably monitored by viewing the LEED pattern. On the other hand, LEED measurements themselves are considerably more difficult to calibrate on an absolute scale and are almost an order of magnitude less sensitive in their ability to detect small relaxations or changes in the Debye temperature of the crystal surface.

4.2.6.4. Lateral Reconstruction of W(001). LEED studies have shown that many crystal surfaces, when clean, undergo lateral reconstruction to produce a surface plane of atoms with a completely different structure to that of the bulk lattice. Again, MeV ion scattering is an ideal technique and can be used to measure both the number of atoms participating in this reconstruction and the magnitude of their displacement. In the case of W(001) for example, the clean, reconstructed surface has a $c(2 \times 2)$ LEED pattern, but returns to the bulk (1×1) structure as hydrogen is adsorbed on the surface. Feldman et al.[138] have recently carried out an extensive study of this system, using MeV ion scattering in conjunction with LEED and nuclear microanalysis. The observed change in surface peak as a function of hydrogen exposure is shown in Fig. 53. The sharp drop at an exposure of $0.3-0.4 \times 10^{-6}$ Torr s correlates well with the change in LEED pattern. The hydrogen-saturated value (~ 1.4 atoms per row) agrees well with the computer simulation value of 1.3 atoms per row, based on the bulk vibration amplitude without lateral reconstruction, thus confirming that the hydrogen-covered surface is nonreconstructed. It should be pointed out that at

---

[138] L. C. Feldman, P. J. Silverman, and I. Stensgård, *Surf. Sci.* **87**, 410 (1979); I. Stensgård, L. C. Feldman, and P. J. Silverman, *Phys. Rev. Lett.* **42**, 247 (1979).

4.2. CHANNELING STUDIES OF LATTICE DEFECTS

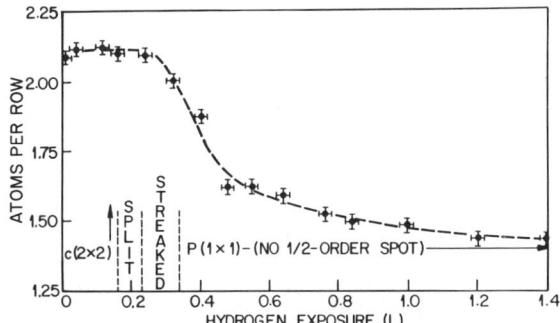

FIG. 53. Intensity of the $\langle 111 \rangle$ surface peak in W(001) as a function of hydrogen exposure, (L), using 1-MeV ^{4}He$^+$ for analysis. The symmetry of the LEED pattern and the nature of the $\frac{1}{2}$-order spots is also indicated. The arrow indicates the coverage where maximum intensity of the $\frac{1}{2}$-order spots is observed. Note that 1 L is equivalent to 1 × 10^{-6} torr s. (From Stensgård et al.[138])

MeV energies any shadowing effect of a low-Z adsorbate such as hydrogen is extremely small and can be safely neglected in the computer simulation.

The total change in scattering yield with H$_2$-coverage is ~ 0.67 atoms per row for $\langle 111 \rangle$ incidence and ~ 0.34 atoms per row for perpendicular ($\langle 100 \rangle$) incidence. Simple crystallographic considerations show that this 2 : 1 ratio is to be expected for displacements that are not aligned with either direction of incidence and that the reconstruction therefore involves the lateral displacement of roughly two thirds of a monolayer of W atoms. Since R_c varies as $E^{-1/2}$, Feldman et al.[138] also extended their scattering measurements over a wide energy range (0.1 – 2.0 MeV) in order to obtain information on the magnitude of the actual displacements. Their results are given in Fig. 54, together with computer simulation curves for various assumed displacements. Appropriate intercomparison shows that the clean, reconstructed surface contains 0.5 – 0.6 monolayers of atoms laterally displaced by 0.023 ± 0.002 nm.

By adsorbing deuterium instead of hydrogen, and using the nuclear reaction D(^{3}He,p)^{4}He, which has a known cross section, Feldman et al.[138] were also able to determine accurately the deuterium coverage as a function of exposure. They found that the saturation coverage is 2 × 10^{15} cm^{-2}, which is consistent with the expected number of bridge sites in the surface. They also showed that a sharp decrease in surface peak associated with reordering occurs at a deuterium coverage of only ~ 3 × 10^{14} cm^{-2}, which seemingly is too small a value for chemical bonding to be the driving force for the reconstruction.

Other similar examples of surface reconstruction in which MeV ion

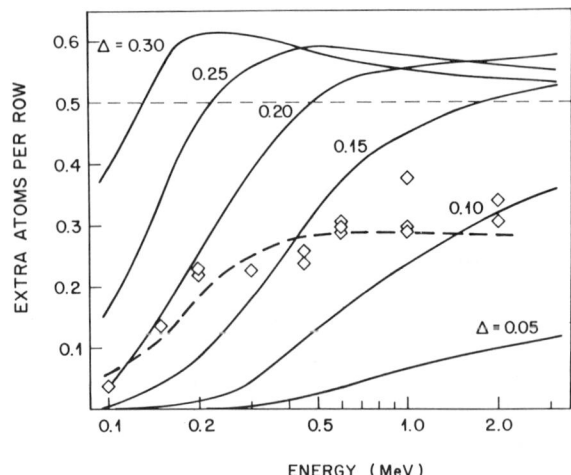

FIG. 54. Energy dependence of the difference between clean (reconstructed) and hydrogen-saturated (nonreconstructed) surface peak intensities, using ^{4}He$^+$ ions incident along a $\langle 001 \rangle$ direction in W (001). Solid curves show the calculated dependence, assuming a full monolayer is displaced a distance Δ (angstroms). The dashed curve is a similar calculation, with 0.5 monolayers displaced 0.23 Å. Note that, for a $\langle 001 \rangle$ direction in W, one monolayer is equivalent to 0.5 atoms per row. [From Feldman et al.[138]]

scattering has been extensively used are the Pt(001), Pt(110), and Au(001) surfaces.[139-141]

4.2.6.5. Location of Sulfur on Ni (110). As noted earlier, one special feature of the combined shadowing and blocking technique developed by Saris[136] and co-workers is the ability to locate the position of impurity atoms adsorbed on a crystal surface. An excellent example of this is the study of sulfur on Ni(110) by Van der Veen et al.,[142] shown in Fig. 55. The basic scattering geometry was illustrated schematically in Fig. 50; note, however, that in Fig. 55 the incident beam direction ([101]) is now tilted 60° with respect to the normal ([110]) direction. The shadowing effect causes the backscattering to occur mainly from the first Ni atom in each row (and also from the adsorbed S atoms, but these give a peak at much lower backscat-

---

[139] P. R. Norton, J. A. Davies, D. P. Jackson, and N. Matsunami, *Surf. Sci.* **85**, 269 (1979).
[140] E. Bøgh, and I. Stensgård, *Proc. 7th Int. Vac. Congr., 3rd Int. Conf. Solid Surf., Vienna* (R. Dobrozemsky, F. Rüdenauer, F. P. Viehböck, and A. Breth, eds.), p. 1. Self-published, Vienna, 1977.
[141] D. M. Zehner, B. R. Appleton, T. S. Noggle, J. W. Miller, C. H. Jenkins, and O. E. Schow, *J. Vac. Sci. Technol.* **12**, 454 (1975).
[142] J. F. Van der Veen, R. M. Tromp, R. G. Smeenk, and F. W. Saris, *Surf. Sci.* **82**, 468 (1979).

4.2. CHANNELING STUDIES OF LATTICE DEFECTS

tered energy). For perfect shadowing, the Ni surface peak intensity would be exactly 1 atom per row; in Fig. 55, it is considerably higher (i.e., ~ 1.5 atoms per row), indicating that significant scattering contributions also come from the second and even the third monolayers. This extra yield arises partly because at room temperature the vibrational amplitude u_2 is not negligible compared to R and partly because of surface relaxation effects. Note that the clean surface exhibits a prominent blocking dip at 59° — i.e., displaced ~ 1° from the bulk [011] blocking direction — thus indicating a relaxation of −4% (a small contraction) for the surface plane.

Comparison of the clean and sulfur-covered spectra in Fig. 55 reveals two significant differences. First, there is a small dip in yield around 52° in the sulfur-covered case; this dip is attributed to those particles backscattered from Ni surface atoms and subsequently blocked along their outgoing path by the sulfur overlayer. From the angular position of the dip (52°), the authors deduced that the sulfur atoms lie 0.087 ± 0.003 nm above the Ni surface, in good agreement with the LEED analysis value of 0.093 ± 0.01 nm.

Second, the sulfur-covered surface exhibits a much broader dip around the [011] blocking direction, with the centroid being displaced to angles slightly *larger* than 60°. Detailed analysis in this case is very complex, due to the overlap of blocking contributions from both the surface plane of Ni and

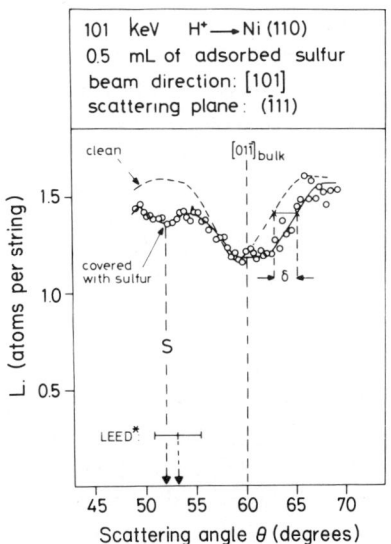

FIG. 55. Blocking angular scan for 101-keV $^1H^+$ scattered from clean and sulfur-covered Ni(011) surfaces. (From Van der Veen et al.[142])

the sulfur overlayer. However, the shift to larger angles suggests that the Ni surface undergoes an ~6% expansion in the sulfur-covered case, compared to the 4% contraction of the clean surface.

Other examples of the application of MeV ion channeling in surface and interface studies are given in Feldman's review article.[48]

4.2.7. Conclusions

We have shown how channeling measurements can be used to investigate lattice defects. Three main categories of defects were considered: (1) damaged regions in a crystal, including precipitates, dislocations, and zones amorphized by ion bombardment; (2) solute atoms, especially their interaction with vacancies and interstitials; and (3) surfaces.

The channeling method is especially useful for the determination of the atomistic nature of lattice defects, in particular defect–solute complexes and surfaces, because of its ability to see atomic displacements by direct geometric considerations. Thus the analysis of data is not very sensitive to unknown parameters such as the exact form of interatomic potentials. Atomic positions can be determined to an accuracy that often exceeds that of other methods. Channeling analysis of solute atom positions in the bulk and on the surface of crystals can be used for a wide range of elements. The ability to detect different elements at low concentration depends somewhat on the method used (backscattering, x rays, or nuclear reactions) and may be either higher or lower than that of other methods. In general, solute atom positions can be determined for concentrations as low as 10^{-4} atomic fraction. In some cases, ion beam techniques are the only method for studying certain elements, such as hydrogen.

Although the channeling technique is "nondestructive," the analyzing beam does introduce radiation damage, which must be considered carefully in any studies of lattice defects. The damage caused by analysis can be reduced by taking measurements near good channeling directions or at elevated temperatures, where irradiation-induced defects anneal out.

In channeling studies, the depth resolution along the beam direction is rather good (typically ~ 10 nm). However, in the transverse plane, the resolution is limited by the beam diameter, typically ~ 1 mm; an appreciable reduction of this dimension is restricted by the concomitant increase in radiation damage caused by the analyzing beam.

Channeling measurements require a relatively perfect single crystal. The sensitivity of the method is proportional to the minimum yield obtained for a given channel, which in turn varies as the perfection of the crystal. The crystal surface must be relatively free of gross deformation, although oxide layers a few atomic layers deep do not affect the bulk channeling properties

4.2. CHANNELING STUDIES OF LATTICE DEFECTS

greatly. In practice, chemically polished or electropolished crystals that exhibit sharp back-reflection x-ray diffraction spots are suitable for channeling studies.

When several types of lattice defect are present in a crystal, the analysis by channeling may be difficult. For example, the effects of dislocations, damaged regions, and mosaic blocks on channeling cannot easily be distinguished. However, the study of the interaction of specific solute atoms with surface layers or point defects is relatively independent of the presence of such a mixture of bulk lattice defects.

The channeling technique can be fruitfully applied to the study of lattice defects in many areas, some of which are the following:

(1) the characterization of near-surface regions of implanted semiconductors after various annealing treatments;

(2) the determination of the positions of solute atoms at surfaces, dislocations, voids, and precipitates;

(3) the study of the growth of small clusters of point defects that have been trapped by solute atoms;

(4) the determination of structural changes in crystal surfaces such as relaxation, reconstruction, and enhanced thermal vibration;

(5) the study of epitaxial growth in the initial stages and its relation to the atomic structure of surfaces;

(6) the determination of the atomistic nature of layered structures;

(7) the study of the relation of defect–solute atom interactions to near-surface properties, such as wear and corrosion; and

(8) the study of metastable alloys created by ion implantation or other means.

Many of these areas of application of ion channeling are reviewed in the recent monograph on materials analysis by Feldman et al.[143]

---

[143] L. Feldman, J. W. Mayer and S. T. Picraux, "Materials Analysis by Ion Channeling." Academic Press, New York, 1982.

5. MAGNETIC RESONANCE METHODS FOR STUDYING DEFECT STRUCTURE IN SOLIDS

By David C. Ailion and William D. Ohlsen

Department of Physics
The University of Utah
Salt Lake City, Utah

5.1. Introduction

5.1.1. Basic Phenomenon of Magnetic Resonance

Magnetic resonance has been a powerful tool in the study of defects in solids. Its power comes in no small part from the fact that it probes the microscopic structure of defects. The information content in a magnetic resonance spectrum is sufficiently rich to allow the drawing of important conclusions regarding defect structure. Both nuclear and electron resonance can easily distinguish different elements since the resonance signals for different elements occur at different magnetic fields for a specific irradiation frequency. Also, electron resonance and, to a lesser extent, nuclear resonance can often distinguish spins having different locations in the crystal. In contrast to other techniques (e.g., electrical conductivity), which measure average bulk properties, the contributions from different defects in the same sample are usually easily separated in magnetic resonance experiments. Much of this information arises from determination of the positions, shapes, widths, and intensities of the magnetic resonance lines. Information about the motions of atoms can be obtained from nuclear relaxation time measurements, which are discussed in Part 6 of this volume.

The essential features of magnetic resonance arise from the fact that nuclei of many elements and electrons are characterized by intrinsic magnetic moments.* The magnitude of this magnetic moment is different for

---

[1] C. P. Slichter, "Principles of Magnetic Resonance," 2nd ed. Springer-Verlag, Berlin and New York, 1978.

---

* An excellent introductory treatment to magnetic resonance theory is given in Slichter.[1] A comprehensive advanced treatment of nuclear magnetic resonance is given in Abragam.[2] A thorough treatment of electron resonance is given in Abragam and Bleaney.[3]

different nuclei and can be used to identify the nucleus. Furthermore, the magnetic moment μ can be related to the spin angular momentum operator **I** by the following relation:

$$\boldsymbol{\mu} = \gamma \hbar \mathbf{I}, \tag{5.1.1}$$

where γ is called the gyromagnetic ratio. If we apply a magnetic field $\mathbf{H}_0$ in the z direction, we obtain a set of evenly spaced levels, whose separation is given by

$$\hbar \omega_0 = \gamma \hbar H_0$$

or, equivalently,

$$\omega_0 = \gamma H_0. \tag{5.1.2}$$

This equation, commonly called the resonance condition, results from determining eigenvalues of the Hamiltonian operator

$$\mathcal{H} = -\boldsymbol{\mu} \cdot \mathbf{H}_0 = -\gamma \hbar H_0 I_z. \tag{5.1.3}$$

Equation (5.1.2) depends on the fact that the adjacent eigenvalues of I_z differ by unity and magnetic dipole transitions are allowed between adjacent levels.

Since $\hbar$ on both sides of the Eq. (5.1.2) can be canceled, we can apply a semiclassical description to this phenomenon. The application of a magnetic field to a *free* bar magnet will cause it to precess about the field at the Larmor frequency $\omega_0 = \gamma H_0$. However, this precession occurs at a fixed orientation θ of the magnetic moment relative to the field direction [see Fig. 1]. For an ensemble of equal magnetic moments having the same θ but differing in phase, there will be a net magnetization and corresponding susceptibility along $\mathbf{H}_0$ but no observable effects at the Larmor frequency. It would be extremely difficult to distinguish this magnetization from other contributions to the susceptibility by normal nonresonance techniques. In order to observe the Larmor precession, we need to modulate the energy of the precessing spins somehow.

Now, a clue for accomplishing this task can be gained by recognizing that the effect of the magnetic field is to cause the spins to precess. As a result, if we consider a new set of coordinate axes (the $x'y'z'$ axes of Fig. 2) whose z component is along $\mathbf{H}_0$ but whose x and y components are rotating at the Larmor frequency, the magnetic moment is fixed in this frame. Thus, in this

---

[2] A. Abragam, "The Principles of Nuclear Magnetism." Oxford Univ. Press (Clarendon), London and New York, 1961.

[3] A. Abragam and B. Bleaney, "Electron Paramagnetic Resonance of Transition Ions." Oxford Univ. Press (Clarendon), London and New York, 1970.

5.1. INTRODUCTION

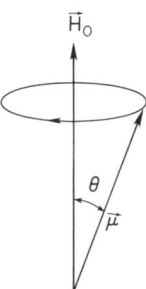

FIG.1. Precession of classical magnet μ about magnetic field H_0.

rotating frame, it is as though the magnetic field were zero. (We describe this phenomenon by saying that the rotating-frame transformation effectively "transforms away the magnetic field.") Now, suppose we also apply a smaller magnetic field H_1, rotating at a frequency ω_0. Such a field would be static in the Larmor rotating frame. Furthermore, suppose H_1 were along the x' direction. Its effect in the rotating frame on the magnetic moment would be exactly the same as the effect of H_0 in the lab frame; the magnetic moment would precess about it. As viewed in the original lab frame, the magnetic moment would have a complicated motion consisting of a rapid rotation about H_0 superimposed on a slow precession about H_1. This second precession results in alternating gain and loss of energy since θ is changing. This energy change can now be detected easily by electronic methods that are described in detail in Sections 5.2 and 5.4. (We should note that the rate at which energy alternates is governed by $\omega_1 = \gamma H_1$.) This energy modulation resulting from the application of a large static field and small rotating field is called *magnetic resonance absorption.*

Consider what would happen if ω, the frequency of rotation of H_1, were slightly different from the Larmor frequency ω_0. Then, in the Larmor rotating frame, H_1 would no longer be fixed, and its effectiveness in causing energy modulation would be sharply reduced. We can still observe magnetic

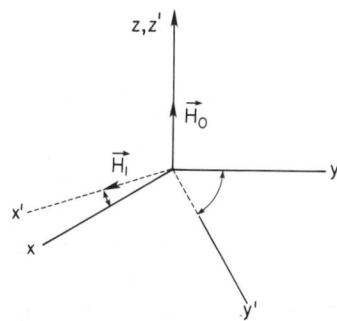

FIG. 2. Rotating coordinate axes.

resonance absorption provided the amount by which ω is off resonance, $\omega - \omega_0$, is smaller than the linewidth.

It should be noted that a quantum-mechanical treatment will yield exactly the same results as we obtained here. This result is due to the fact that the quantum-mechanical expectation value for the magnetic moment can be shown[1] to obey the same equation of motion as that of a classical magnetic moment.

5.1.2. Relaxation Times T_1 and T_2

We have been introduced to the basic phenomenon of magnetic resonance. However, our treatment is incomplete in two important respects. No discussions have been included of sources of line broadening or of interactions between our spin system and other systems. Much of the information about both the static and dynamic properties of defects can be obtained from measurements of linewidths and relaxation times (i.e., the times which characterize the flow of energy between the spin system and other systems).

5.1.2.1. Linewidth and Spin–Spin Relaxation Time T_2. Typically the linewidth of a magnetic resonance line is due to inhomogeneities in the magnetic field, either external or internal. An important exception can occur for a spin system that is tightly coupled to an electrostatic system, for example, a nucleus with an electric quadrupole moment in a noncubic environment or an electron spin with nonnegligible spin–orbit coupling. In these cases the electrostatic broadening may dominate all other sources of line broadening. In electron resonance, lifetime broadening[1] may be the dominant effect.

Consider a collection of noninteracting spins, having half-integral spin ($I = \frac{1}{2}$), in an inhomogeneous applied field. Suppose this field varies spatially by an amount ΔH, such that the actual field varies from $H_0 - \Delta H/2$ to $H_0 + \Delta H/2$ over the sample. In that case the resonance frequencies of spins in different parts of our sample would likewise vary from $\omega_0 - \Delta\omega/2$ to $\omega_0 + \Delta\omega/2$, where

$$\Delta\omega = \gamma \, \Delta H. \tag{5.1.4}$$

This spread in Larmor frequencies results in a width of the resonance line that is characteristic of the inhomogeneity in the applied field and is thus referred to as *inhomogeneous* broadening.

We can view this phenomenon in another way. Suppose we apply a radio-frequency (rf) field $\mathbf{H}_1$ for a time just long enough to tilt the magnetization into the x–y plane and then turn it off. This $\mathbf{H}_1$ pulse is called a 90° (or $\pi/2$) pulse. The magnetization will then precess about the z direction at the Larmor frequency. However, there is a spread in Larmor frequencies $\Delta\omega$

5.1. INTRODUCTION

over the sample. As a result, the spins gradually get out of phase and the resultant magnetization goes to zero. This decreasing magnetization, termed a *free-induction decay* (FID), is shown in Fig. 3a. The time for the magnetization to go to $1/e$ of its initial value is called T_2, the spin–spin relaxation time. It is clear that the linewidth and T_2 are inversely related since $\Delta\omega T_2 \sim \pi$. This result can be seen by noting that the free-induction decay is the Fourier transform[2] of the normal resonance line (Fig. 3b). Thus a narrow linewidth means that T_2 is long and vice versa.

It is a reasonable approximation to treat the nuclear spins in a monatomic liquid as being noninteracting, with the result that the NMR linewidth will be due to inhomogeneous broadening; values of T_2 for liquids in typical magnetic fields range from 1 ms to 10 s or longer, depending on field homogeneity. (For inhomogeneously broadened lines, the inverse of the linewidth is called T_2^*.) However, in a solid or in a polyatomic liquid, one cannot neglect the dipole–dipole interaction between neighboring spins. In this case, there will be a spread in dipolar local fields over the spin system and a corresponding broadening of the resonance line. We can estimate the size of this broadening since the field due to a dipole μ at a distance r is given by $H_{loc} \sim \mu/r^3$, and is of the order of 1–100 G for typical atomic distances in solids. The corresponding range for T_2 in solids is 1–100 μs. Accordingly, this broadening (called *homogeneous* broadening) may easily dominate inhomogeneous broadening in a solid.

We can also describe homogeneous broadening using the language of quantum mechanics. Treating the dipole–dipole interaction as a perturbation on the N-atom Zeeman Hamiltonian for isolated atoms has the effect of replacing each of the previously degenerate Zeeman levels with a spread of energy levels. The linewidth results from this spread.

A further source of line broadening arises if an atom that has a quadrupole

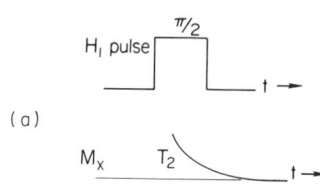

(a)

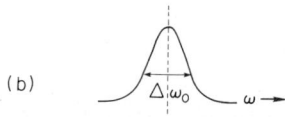

(b)

FIG. 3. Magnetic resonance signals. (a) Upper picture shows the envelope of a $\pi/2$ pulse and the lower picture shows the associated free-induction decay (FID). (b) Normal resonance absorption line (Fourier transform of FID).

moment is placed in an electic field gradient (such as that which results from a noncubic crystal structure). For a spin-$\frac{3}{2}$ nucleus, this quadrupole interaction will result to first order in the original resonance line being split into three lines.[4] This splitting can be highly anisotropic; accordingly, for a powder or polycrystalline sample we can get a large number of overlapping lines that appear as a considerably broadened line. Similar anisotropic interactions also occur in electron paramagnetic resonance (EPR) systems. Furthermore, if the quadrupole interaction is sufficiently strong, even the central line can be broadened. This is called second-order quadrupole broadening[2,4] and may give rise to a resonance linewidth of 50–100 G or more for the central line.

For EPR, the dipole–dipole interactions between an electron and many nuclei can determine the breadth of the resonance line. In EPR this broadening is also called inhomogeneous broadening since electrons in different nuclear-spin environments are isolated in that they cannot engage in energy-conserving mutual spin flips. This inability to exchange energy is also the essential feature of inhomogeneously broadened NMR lines.

5.1.2.2. Spin–Lattice Relaxation Time T_1. The process by which a spin system exchanges energy with its surroundings is called spin–lattice relaxation. Consider a thermal equilibrium magnetization M_0 along the H_0 direction. Now suppose we tilt this magnetization into the xy plane as a result of applying an rf pulse. It is clear that the rf pulse has given energy to the spin system, thereby disturbing the system from thermal equilibrium. Even the process of spin–spin relaxation, which involves no change in energy for the spin system, will not restore the original thermal equilibrium. This equilibrium will be restored only when the original magnetization reappears along H_0. This process requires the existence of another system (*the lattice*) to absorb the energy given up as the magnetization is reestablished along H_0. The time constant characterizing this reestablishment is called T_1. More precisely, if there is no rf field present, we define T_1 by

$$\frac{dM_z}{dt} = \frac{M_0 - M_z}{T_1}. \quad (5.1.5)$$

We can view rf excitation, spin–lattice relaxation, and spin–spin relaxation as thermal processes involving transitions between energy levels. Consider a spin $\frac{1}{2}$, which, in a static magnetic field, will have two energy levels. If it is in thermal equilibrium at a temperature T, the probability of occupancy of the two levels will be governed by a Boltzmann factor with the result that the lower level will be more heavily occupied than the upper. If the temperature is reduced, the population difference between the two levels

---

[4] T. P. Das and E. L. Hahn, *Solid State Phys., Suppl.* **1** (1958).

will be increased. Now, the application of a weak rf field H_1 may be viewed as a perturbation that induces transitions between the two levels. Since the transition probabilities for up and down spin transitions are the same and since there is an excess population in the lower state, the effect of this perturbation is a net excess of transitions that are upward in energy. Thus the effect of the rf field is to decrease the population difference. Since there is a net flow of energy into the spin system, the rf field may be regarded to be "heating" the spin system. (This point of view is not appropriate for rf fields whose magnitude is stronger than internal fields.[5])

Spin–lattice relaxation, on the other hand, characterizes the flow of energy between the spin system and its surroundings so as to result in changing the temperature of the spin system to match that of the surroundings. If the spin system is hotter initially, say, as the result of the rf field, then the process of spin–lattice relaxation will result in net transitions of the spin system that are downward in energy and a corresponding flow of this energy into the surroundings. Thus the process of spin–lattice relaxation normally "cools" the spin system and reestablishes a population difference corresponding to the lower temperature of the lattice.

The process of spin–spin relaxation does not involve a net flow of energy between the spin system and some other system but rather an exchange of energy between different parts of the spin system. If initially we had a spin system whose energy-level populations were not characterized by a Boltzmann distribution (i.e., an exponential dependence of the populations on energy), we could still achieve a Boltzmann distribution by simultaneous transitions that are both upward *and* downward in energy such that the total spin energy is conserved. We may regard the process of spin–spin relaxation then as establishing a Boltzmann distribution and corresponding temperature.

5.1.3. Bloch Equations

If we combine the effects of the rf field H_1 with spin–spin and spin–lattice relaxation, we obtain the Bloch equations:

$$\frac{dM_x}{dt} = \gamma(\mathbf{M} \times \mathbf{H})_x - \frac{M_x}{T_2},$$

$$\frac{dM_y}{dt} = \gamma(\mathbf{M} \times \mathbf{H})_y - \frac{M_y}{T_2}, \qquad (5.1.6)$$

and

$$\frac{dM_z}{dt} = \gamma(\mathbf{M} \times \mathbf{H})_z + \frac{(\mathbf{M}_0 - M_z)}{T_1}.$$

[5] A. G. Redfield, *Phys. Rev.* **98**, 1787 (1955).

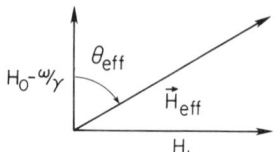

FIG. 4. Effective magnetic field $\mathbf{H}_{\text{eff}}$ and its components in the rotating frame.

The terms involving relaxation times simply express the facts that the transverse magnetization M_x, M_y will decay to zero in a time characterized by T_2 and the longitudinal magnetization M_z will approach its thermal equilibrium value M_0 in a time T_1. The first term expresses the fact that an applied field $\mathbf{H}$ causes the magnetization to precess about it.

Suppose $\mathbf{H}$ arises from a static field $\mathbf{H}_0$ along the z direction and an rf field $2H_1 \cos \omega t$ along the x direction

$$\mathbf{H} = 2H_1 \mathbf{i} \cos \omega t + H_0 \mathbf{k}. \tag{5.1.7}$$

Because of the time dependence of the rf field, it is most convenient to transform the Bloch equations into a frame rotating at the rf frequency ω. In this frame $\mathbf{H}_1$ in static and we choose the new x direction to be along $\mathbf{H}_1$.* If we are off resonance (i.e., if $\omega \neq \gamma H_0$), the z component $h = H_0 - \omega/\gamma$. The total effective field $\mathbf{H}_{\text{eff}}$ in the rotating frame (Fig. 4) is thus given by

$$\mathbf{H}_{\text{eff}} = H_1 \mathbf{i} + h \mathbf{k}. \tag{5.1.8}$$

We make the further simplifying assumption that the relaxation times T_1 and T_2 are unchanged in the rotating frame. (This assumption is valid for small values of H_1, but has been shown[5] to be in serious error for H_1 large enough to cause saturation.) In this case, the transformed Bloch equations can be solved easily for small H_1 (i.e., $\gamma^2 H_1^2 T_1 T_2 \ll 1$) to give

$$M_x = M_0(\gamma T_2) \frac{(\omega_0 - \omega)T_2}{1 + (\omega - \omega_0)^2 T_2^2} H_1 \tag{5.1.9a}$$

and

$$M_y = M_0(\gamma T_2) \frac{1}{1 + (\omega - \omega_0)^2 T_2^2}. \tag{5.1.9b}$$

Figure 5 shows plots of M_x and M_y versus ω.

---

* Actually, the lab-frame rf field, $2H_1 \cos(\omega t) \mathbf{i} = H_1(\mathbf{i} \cos \omega t + \mathbf{j} \sin \omega t) + H_1(\mathbf{i} \cos \omega t - \mathbf{j} \sin \omega t)$, is the superposition of two counterrotating fields, one at the frequency ω and the other at the frequency $-\omega$. It is only the field at the frequency ω that will be static in the rotating frame. The other component field is still time dependent, but is so far off resonance (2ω) that it can be neglected.

5.1. INTRODUCTION

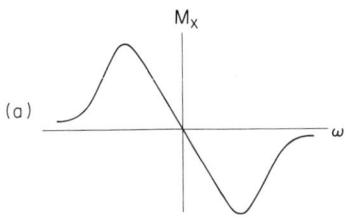
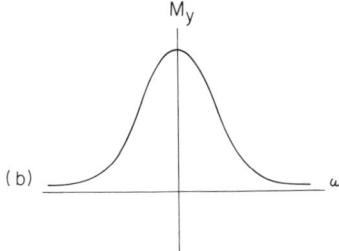

FIG. 5. Magnetization versus frequency. (a) M_x [from Eq. (5.1.9a)], (b) M_y [from Eq. (5.1.9b)].

If we define a complex susceptibility χ by $\chi = \chi' - i\chi''$, we can relate χ' and χ'' to M_x and M_y, respectively:

$$\chi' = \frac{M_x}{2H_1} = \frac{\chi_0}{2} \omega_0 T_2 \frac{(\omega_0 - \omega)T_2}{1 + (\omega - \omega_0)^2 T_2^2}, \qquad (5.1.10a)$$

and

$$\chi'' = \frac{\chi_0}{2} \omega_0 T_2 \frac{1}{1 + (\omega - \omega_0)^2 T_2^2}, \qquad (5.1.10b)$$

where $\chi_0 = M_0/H_0$. Normally χ' and χ'' are interrelated by the Kramers–Kronig relations,[1] provided H_1 is sufficiently small that the dependence of M_x on H_1 is linear, as in Eqs. (5.1.9a) and (5.1.9b). Hence it is sufficient in most resonance experiments to measure just one of these susceptibilities. For sufficiently large H_1 such that $\gamma^2 H_1^2 T_1 T_2 > 1$, χ' and χ'' will no longer be independent of H_1, and the linearity assumed in the Kramers–Kronig relations will not hold. The term *saturation* is applied to this case as it corresponds to the rf field being so large that its main effect is to equalize population differences.[2]

These changes in the real and imaginary parts of the complex susceptibility will affect the inductance of a pickup coil wound around the material. If we insert material of susceptibility $\chi(\omega)$ into a coil of inductance L_0, the inductance will change to

$$L = L_0[1 + 4\pi\chi(\omega)]. \qquad (5.1.11)$$

If the coil has resistance r_0, the complex impedance Z will be

$$Z = r_0 + i\omega L = r_0 + i\omega L_0[1 + 4\pi\chi'(\omega) - 4\pi i\chi''(\omega)]$$
$$= r' + i\omega L', \qquad (5.1.12)$$

where the resistance has changed to r' and the inductance to L', given by

$$r' = r_0 + 4\pi\omega L_0\chi''(\omega) \qquad (5.1.13a)$$

and

$$L' = L_0[1 + 4\pi\chi'(\omega)]. \qquad (5.1.13b)$$

We shall now derive an expression relating the electromotive force (emf) generated in the coil to the voltage across a tuned circuit containing the coil. If the coil is connected in parallel with a capacitor C, and this combination is driven by an ac current generator of amplitude I_0, the amplitude of the voltage across the coil is

$$V = I_0 \left(\frac{1}{i\omega L_0 + r_0} + i\omega C\right)^{-1}. \qquad (5.1.14)$$

For the case $H_0 \neq \omega/\gamma$ (so that $\chi = 0$) and $\omega = \omega_r (= 1/\sqrt{L_0 C})$ with $r_0 \ll \omega L_0$, this equation reduces to

$$V = V_0 = I_0(\omega_r L_0 Q), \qquad (5.1.15)$$

where the quality factor $Q = \omega L/r_0$.

When the magnetic field is tuned to nuclear resonance so that χ is nonzero, Eq. (5.1.14) becomes

$$V = I_0 \left(\frac{1}{i\omega L_0(1 + 4\pi\chi) + r_0} + i\omega C\right)^{-1}. \qquad (5.1.16)$$

If $4\pi\chi$ is assumed to be small compared to Q^{-1}, Eq. (5.1.16) at electrical resonance, $\omega = \omega_r$, becomes approximately

$$V = I_0\left[\omega_r L_0 Q\left(1 + \frac{4\pi\chi}{i} Q\right)\right] \qquad (5.1.17a)$$

$$= V_0 + V_0 \frac{4\pi\chi}{i} Q. \qquad (5.1.17b)$$

In other words, the magnetic resonance causes a change in the voltage across the electrically resonant LC circuit of magnitude

$$-V_0 Q\, 4\pi(\chi'' + i\chi'). \qquad (5.1.18)$$

Since V_0 and H_1 are both proportional to I_0, we see that the signal voltage, in the absence of saturation of χ' and χ'', is proportional to the product of H_1

and χ. Another way to look at this signal is to say that $H_1\chi$ is a magnetization, and this precessing magnetization induces the signal voltage in the coil.

Since Q is typically of order 100 for NMR or 5000 for EPR, the placing of the sample coil in a tuned circuit results in a considerable increase in sensitivity. We discuss these and other spectrometer design questions further in Sections 5.2.2 and 5.4.1.

5.1.4. The Nature of Magnetic Resonance Information

In the previous sections, we have described the basic phenomenon of magnetic resonance. Now, we briefly discuss the measurable parameters in a magnetic resonance experiment that are capable of yielding the most information about defect properties in solids. The principal sources of information in NMR experiments are (a) the position (i.e., the frequency) of the resonance line, (b) the strength of the line, (c) the linewidth, and (d) the various relaxation times. In EPR, nuclear quadrupole resonance (NQR), and chemical-shift studies, information is also obtained from the number of lines and their relative amplitudes as well as the variations of line position as the direction of H_0 relative to the crystal axes is changed. These quantities are not necessarily independent of one another. For instance, the area under a resonance line is proportional to the number of nuclei [through χ_0 in Eqs. (5.1.10)]; hence the line's intensity and width will be inversely related. Also, the spin–spin relaxation time T_2 and the linewidth are inversely related. In this part we discuss primarily the first three of these, whereas Part 6 of this volume concentrates more on relaxation times as a source of motional information.

The resonance condition can be achieved either by varying the frequency ω of the rf field H_1 or by varying the magnitude of H_0. In many experiments it is easy to obtain fixed rf and microwave sources of very high frequency stability. Furthermore, magnetic fields of acceptable stability that can be varied over wide ranges are readily available; accordingly, it is more common to meet the resonance condition by varying H_0 while holding ω fixed. (We shall assume ω to be fixed and H_0 varied in the following treatment.) Thus the resonance line's position will be determined by the total magnetic field seen by the spin. In addition to the external field H_0, there will be internal sources of magnetic field. This internal field usually arises from neighboring spins and other nearby sources of magnetic field; thus it is particularly sensitive to the presence of nearby defects. The interactions most responsible for determining the positions of NMR lines in solids are electric quadrupole interactions due to electric field gradients at the sites of the spins, chemical shifts due to orbital motion of electrons on neighboring atoms, Knight shifts in metals due to first-order hyperfine interactions with free-electron spins, and line shifts in insulators due to indirect nuclear–nuclear coupling by means of second-order hyperfine interactions. Electron resonance line positions, on the other hand, are determined primarily by spin-orbit interactions, electrostatic interactions with neighboring atoms (crystalline fields), spin-spin interactions, and hyperfine interactions with neighboring nuclei. These interactions and procedures for their measurement are discussed in more detail in Chapters 5.3 and 5.4.

In cases where a particular interaction results in a resonance line being split into several lines, each of the lines corresponds to a group of spins of which all are equivalent in that they are subject to the *same* value of the interaction. Thus the relative intensities of the different lines will be proportional to the number of spins in a particular environment or site. By comparing the experimentally measured value of the line shift with a calculated value, one can in principle determine the structure and symmetry of the site. However, in many cases, it is difficult to measure the absolute value of the line shift with great accuracy since the unshifted line's position is not always precisely known. Fortunately, in many of these cases the line shift is anisotropic, and the anisotropy is determined by the local structure. Thus a measurement of the dependence of the line position on the angles that describe the orientation of H_0 relative to the crystal axes may provide a definitive determination of the magnetic resonance coupling mechanism as well as the local environment.

The NMR linewidth in a solid is usually determined by magnetic dipole-dipole interactions, particularly for spin-$\frac{1}{2}$ nuclei. However, for $I > \frac{1}{2}$, the linewidth is often determined by electric quadrupole interactions, particularly in powdered samples of substances in which the nuclear site does not have cubic symmetry. The line shape may be changed even in cases where the second moment is unchanged. It can be shown[1] that the Ruderman-Kittel interaction,[6,7] which describes the second-order indirect internuclear coupling by means of the hyperfine interaction with electron spins, can give rise either to a line broadening in heteronuclear systems (i.e., containing two or more nuclear species) or to a narrowing of the central portion of the line in homonuclear systems (i.e., containing only one nuclear species). Since the linewidth in a solid is anisotropic, defect information can often be obtained from orientation measurements in a single crystal.

In systems in which there is appreciable translational diffusion or molecular reorientation, parts of the linewidth may be reduced[8] (*motional narrowing*), thereby allowing other interactions to be studied. Such studies can also be used to obtain dynamical information about atomic motions and

---

[6] M. A. Ruderman and C. Kittel, *Phys. Rev.* **96**, 99 (1954).
[7] N. Bloembergen and T. J. Rowland, *Phys. Rev.* **97**, 1679 (1955).
[8] N. Bloembergen, E. M. Purcell, and R. V. Pound, *Phys. Rev.* **73**, 673 (1948).

parameters (like barrier heights) that affect these motions. These techniques are discussed in detail in Part 6. A similar effect can be used to narrow the line of a stationary atom artificially by applying an alternating set of rf pulses, which simulates the fluctuating local field experienced by a diffusing nucleus.[9] Similar effects can be achieved by actually spinning a sample rapidly in the magnetic field.[10] Rotating frame studies at the "magic angle"[11] can also be used to narrow the line since many of the interactions that contribute to the linewidth are proportional to $1 - 3\cos^2\theta_{\text{eff}}$, where θ_{eff} is the angle between the effective field $\mathbf{H}_{\text{eff}}$ in the rotating frame and $\mathbf{H}_0$ (see Fig. 4). (The magic angle is just the value of θ_{eff} for which $1 - 3\cos^2\theta_{\text{eff}} = 0$.) Many NMR experiments utilize spin-decoupling techniques in which dipolar and other interactions between different spins are removed by the application of appropriate pulse sequences. (See Haeberlen,[12] pp. 82–89.)

5.2. Nuclear Magnetic Resonance Spectrometer Design

This chapter and the following chapter focus on NMR techniques typically requiring apparatus* operating in the radio frequency region (1 – 100 MHz). Chapter 5.4 describes electron resonance methods that require microwave apparatus since the electron's Larmor frequency is several thousand times those of nuclei in the same magnetic field. Techniques appropriate for studying static properties are described in this part, whereas methods for studying dynamic and motional properties are emphasized in Part 6.

An investigator who wishes to obtain apparatus for NMR measurements should know what nucleus he will study and what magnetic field and rf frequency ranges he will require.† He should have some approximate idea what his linewidth will be; in particular, he should know whether his nucleus is spin $\frac{1}{2}$ or whether there is likely to be substantial quadrupolar broadening. He should, if possible, know whether the spin–lattice relaxation time T_1 will

[9] J. S. Waugh, L. M. Huber, and U. Haeberlen, *Phys. Rev.* **20**, 180 (1968).
[10] E. R. Andrew, *Prog. Nucl. Magn. Reson. Spectrosc.* **8**, 1 (1971).
[11] M. Lee and W. I. Goldburg, *Phys. Rev.* **140**, A261 (1965).
[12] U. Haeberlen, *Adv. Magn. Reson., Suppl.* **1** (1976).
[13] E. Fukushima and S. B. W. Roeder, "Experimental Pulse NMR: A Nuts and Bolts Approach." Addison-Wesley, Reading, Massachusetts, 1981.

---

* An excellent, simple-to-read treatment of pulse NMR apparatus and techniques is found in Fukushima and Roeder.[13]

† Very convenient charts which show basic NMR features are readily available from NMR companies like Varian (Instrumentation Division, 611 Hansen Way, Palo Alto, CA 94393) and Bruker Instruments (Manning Park, Billerica, MA 01821). These charts contain useful information for each paramagnetic nucleus (e.g., Larmor frequencies, relative sensitivity, spin, natural abundance, quadrupole moment).

be extremely long or not. Finally, he should have some idea whether the nuclear abundance is low or the relative sensitivity is small so that he can decide whether he will need to use the signal-enhancement techniques like those that are described in Chapter 5.6.

5.2.1. Pulse versus Continuous-Wave Spectrometers

NMR spectrometers can be classified into two general types: (1) steady-state (cw) spectrometers, which measure directly the frequency dependence of the susceptibilities of Eqs. (5.1.10a) and (5.1.10b) by sweeping either the rf frequency ω or the static field H_0 through the resonance line, and (2) pulse spectrometers, which measure directly the time evolution of the magnetization following the turnoff of a pulse of rf power (H_1). In the early days of NMR, cw spectrometers were used primarily for investigations of static properties, whereas pulse techniques were applied primarily to atomic motions. Such a distinction is no longer valid because pulse methods have become quite widespread and have largely replaced cw methods, thanks largely to the development of Fourier transform techniques.* Today, the most common use of cw spectrometers as the primary spectrometer is for those systems in which the nuclei have long T_1 (e.g., hours in length). Since such studies are usually made on systems of very high purity, which may be quite rare, the large majority of NMR experiments currently are best performed with pulse spectrometers.

There are three principal advantages of pulsed over cw-swept excitation. In the pulse experiment, the signal-to-noise (S/N) ratio is often superior in that a major source of noise (oscillator noise) is absent since the NMR signal is observed after the H_1 pulse is turned off. This situation contrasts with that of the cw case in which H_1 is on all the time, including the time during which the signal is observed. A second, possibly more important, advantage is that pulse experiments are more efficient than swept experiments with regard to the gaining of information per unit time since all nuclei within the frequency bandwidth of the pulse are excited simultaneously and can be so studied. In contrast, the swept experiment is usually much slower since different resonances are excited sequentially. This feature can also result in an improved S/N ratio since a higher speed can result in more efficient signal averaging. Also, because of the simultaneity of the pulsed excitation, it is possible to use a multichannel spectrometer with more than one transmitter to excite different NMR lines simultaneously. In such an experiment the improvement in sensitivity[15] will be $(2n)^{1/2}$, where n is the number of channels. This

---

[14] D. Shaw, "Fourier Transform N.M.R. Spectroscopy." Elsevier, Amsterdam, 1976.
[15] P. Fellgett, *J. Phys. Radium* **19**, 187 (1958).

* A comprehensive treatment of Fourier transform techniques is found in Shaw.[14]

5.2. NUCLEAR MAGNETIC RESONANCE SPECTROMETER DESIGN

feature is of particular importance when studying chemically shifted species using Fourier transform methods. The third advantage of pulse NMR, of particular importance when measuring relaxation times, is that there is a definite starting point. The resulting phase coherence of the signal following the pulse makes relaxation time measurements straightforward.

5.2.2. Pulse NMR Spectrometer Design

Magnetic resonance spectrometers have three basic parts: (a) a large magnet, which is responsible for the static magnetic field H_0; (b) a transmitter, which provides the alternating field H_1; and (c) a receiver, which amplifies and detects the magnetic resonance signal. In addition to these basic components there will usually be auxiliary apparatus for improving the S/N ratio, cryostats and other apparatus for varying and controlling the temperature, and equipment for displaying and analyzing the data. Some of these functions can be handled by a digital computer, as we shall see. The essential features of a pulse NMR single resonance spectrometer are shown in Fig. 6.

The transmitter consists essentially of four parts: a cw rf source (oscillator), an rf gate, a pulse generator (which may be computer controlled), and a power amplifier. The oscillator provides a signal (at the Larmor frequency of the nucleus to be studied), which is fed into an rf gate. This gate is normally biased off so as to allow no rf output except during a pulse, which opens the gate, thereby providing a short burst of rf voltage. This voltage is then fed into a power amplifier, which supplies the rf current in the sample coil (in the magnet) that is responsible for H_1.

The receiver also consists of three principal parts. First, there is a high-gain (30–40 dB gain) low-noise rf amplifier that can amplify the small rf voltage induced by the magnetization arising from the precessing spins.

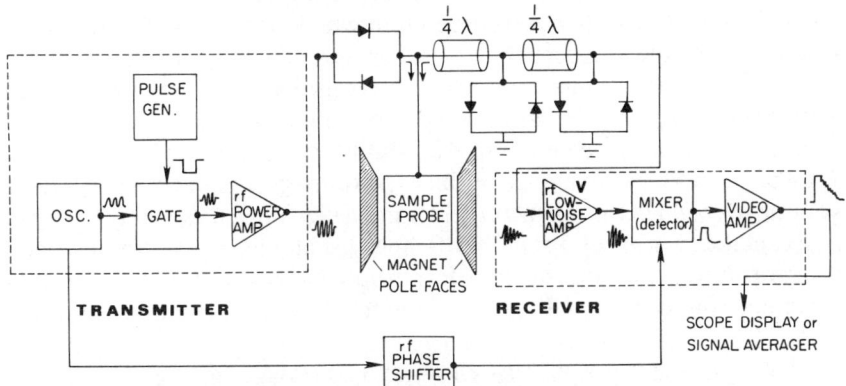

FIG. 6. Block diagram of typical pulse NMR spectrometer.

After several stages of amplification, this voltage is then mixed in a phase-sensitive detector[16] with an appropriately phase-shifted rf signal (the *reference*) obtained from the original oscillator. Thus the detected output is the envelope of both the free-induction decay and the original rf pulse. This envelope is then amplified and displayed, using either a scope or an X-Y recorder. In addition, as indicated in Fig. 6, there are quarter-wavelength cables and crossed diodes that protect the receiver from the large rf pulse.

In the following sections, we discuss these components in detail along with considerations to be used in their selection and/or construction.

5.2.2.1. The Magnet. The heart of almost every magnetic resonance spectrometer is a large magnet, which is the source for H_0. With the exception of pure quadrupole resonance (in which electric field gradients are responsible for quantization of the energy levels), nuclear resonance in ferromagnetic materials (in which the ferromagnetic electron is the source of H_0), and electron resonance in the earth's magnetic field, all magnetic resonance experiments in solids require an external magnet. These magnets are of three types: permanent magnets, electromagnets (usually iron core), and superconducting magnets. We now discuss the relative advantages and disadvantages of each.

The magnet that is simplest and lowest in cost (both to purchase and to operate) is clearly the permanent magnet. Its principal disadvantage is its lack of field variability, which necessitates the use of variable-frequency spectrometers. Also, the maximum field and the sample gap of these magnets tends to be limited, unless one goes to much larger magnets. In addition to low cost and simplicity, permanent magnets have the further advantage of high stability with regards to both absolute field and field homogeneity. They are rarely used in solid state research and should be purchased only when routine, repetitive measurements of abundant, large gyromagnetic-ratio nuclei are contemplated.

By far the most commonly used magnet in solid state magnetic resonance research is the iron-core electromagnet, in which power is supplied in coils wound around the iron core and excess heat is removed by circulating water. This type of magnet is more expensive and considerably more complex than a permanent magnet, but has the very important advantage of variable field control, which allows considerable flexibility in the choice of experiments. Another important advantage of this type of magnet is the relatively large volume over which the magnetic field homogeneity in a reasonably large gap is acceptable. Since solid state NMR lines usually have a width of several gauss or more, one can usually detect at full signal strength all nuclei that experience the same H_0 within approximately 1 G. A field of this homogene-

---

[16] J. J. Spokas and C. P. Slichter, *Phys. Rev.* **113**, 1462 (1959).

5.2. NUCLEAR MAGNETIC RESONANCE SPECTROMETER DESIGN

ity is experienced by all nuclei in a circular region, concentric with the pole faces, of about 3-cm diameter both in 38-cm (15 in.) pole-face-diameter magnets having a 10-cm gap and in 30-cm (12 in.) diameter magnets having an 8-cm gap. In choosing a magnet, there is usually a trade-off involving maximum field strength (which requires tapered narrow-gap pole faces), maximum homogeneity (which requires untapered narrow-gap pole faces), and ability to accommodate large samples as well as cryostats and other auxiliary apparatus (which require a large gap). Normally, for solid state work where ultrahigh homogeneity is less important, these magnets have gaps in the range 5–10 cm and corresponding maximum field strengths of order 1–2.5 T (1 T = 10^4 G).

Superconducting magnets are usually solenoidal in shape and have the advantages of being able to achieve very high fields (~ 9 T) with good stability. Since the coil consists of superconducting wire, which is usually maintained at liquid-helium temperatures, the operating costs of these magnets can be high, due to the cost of liquid helium, unless a closed-cycle helium-recovery system is used. Also, these magnets tend to be more expensive to manufacture than electromagnets. Since the bore of the solenoid is cylindrical, access is somewhat more awkward than with electromagnets. However, the ability to operate at high fields is very important in that the signal-to-noise ratio* is, in principle, proportional to $H_0^{3/2}$. Another advantage of high fields is that chemical shifts increase with H_0, as we shall see. For these reasons, superconducting magnets are used in chemical and biological applications as well as in solid state observations of nuclei that are hard to see (either because they are in low concentration or because they have small γ).

In biological applications involving animals and humans, air-gap magnets[17] without pole faces are sometimes used since they can accommodate much larger specimens, albeit at a much lower field. Nonsuperconducting magnets that are large enough to accommodate human beings are usually solenoidal, with a 70-cm-diameter bore, and can provide fields up to a maximum of about 0.1 T. Some companies (e.g., Oxford Instruments, 9130-H Red Branch Rd., Columbia, MD 21045) now build whole-body superconducting magnets for NMR imaging that can provide much higher magnetic fields (1.5 T).

In addition to the basic magnet and power supply, auxiliary apparatus is often required to (1) improve the field homogeneity and (2) monitor and improve the field stability.

In electromagnets to be used in solid state research, acceptable homogene-

---

[17] P. C. Lauterbur, *Nature* (*London*) **242**, 190 (1973).

* See p. 167 of Shaw.[14]

ity over a wide field range is achieved by the use of simple shim coils,[18] which are usually supplied as part of the magnet system at the time of purchase. In high-resolution applications, higher-order shim coils can provide better homogeneity but at a fixed field setting. The resulting loss of field flexibility is acceptible only if all observations will be on the same nucleus. Further improvements in homogeneity can be achieved by spinning a sample[19] in order to average the magnetic field inhomogeneities over a plane perpendicular to the axis of rotation.

Good field stability in an electromagnet clearly requires a stable current supply. Most commercial magnets include a field stabilizer consisting of a flux-sensing device (such as a Hall probe or a flip coil), which senses the field and sends an error signal into the current supply to compensate for drift in the field. Except for samples having very long T_1, the stability achieved with the above stabilizers (one part in 10^5 long term and one part in 10^6 short term) is usually adequate for most solid state experiments. In some cases (e.g., measurements of small chemical shifts), improved stability is required. Excellent long-term stability can be achieved by utilizing either an external or internal field/frequency lock. The external lock involves a second NMR experiment whose frequency is adjusted to be on exact resonance at the field at which the primary nucleus is at resonance. By setting the reference detector in the dispersion mode [i.e., to measure χ' of Eq. (5.1.10a)], the output voltage will be a linear function of H_0 and will be zero at resonance. The off-resonance voltage will be positive or negative depending on the sign of the magnetic field drift. Accordingly, the field stability can be improved by feeding the error voltage into the flux stabilizer of the magnet. Even higher stability can be achieved by use of an internal lock which involves excitation of another resonance line in the same sample.

Improved short-term stability against field jitter can be achieved by using a many-turn flux-sensing coil placed in the magnet gap since a time variation in H_0 will induce an emf proportional to dH_0/dt. The induced emf can be integrated in a low-drift operational amplifier, whose output can be fed to field correction coils wound on the magnet pole faces. By using negative feedback, dH_0/dt can be minimized.

5.2.2.2. The Sample Probe. The rf probe that contains the sample in the magnet consists of either a single coil or a pair of coils (the crossed-coil configuration). In the crossed-coil configuration, one coil (the transmitter coil) is connected (through an impedance-matching network) to the output of the power amplifier and its current is the source of H_1. The NMR signal is induced in a second coil (the receiver coil), which is connected to the input

---

[18] W. A. Anderson, *Rev. Sci. Instrum.* **32**, 241 (1961).
[19] F. Bloch, *Phys. Rev.* **94**, 496 (1954).

5.2. NUCLEAR MAGNETIC RESONANCE SPECTROMETER DESIGN

of a low-noise high-gain rf amplifier. In a single-coil probe, both of these functions are combined in a single coil.

The rf probe is probably the most critical component in determining the performance of the spectrometer. The ultimate S/N ratio is determined by the sensitivity of the receiver coil. On the other hand, the transmitter coil determines the size of H_1. In a pulse experiment it is normally desirable to rotate the magnetization into the xy plane in the shortest possible time, certainly in a time short compared to the dephasing time T_2^*. Since a $\pi/2$ pulse requires that

$$\gamma H_1 t_w = \pi/2, \qquad (5.2.1)$$

a short value of the pulse width t_w requires a large H_1. Since the rf pulse, which also gets coupled to the receiver, has a nonzero fall time, a large rf voltage from the transmitter will exist in the receiver at the same time as the nuclear free-induction decay and may prevent observations of the initial parts of the decaying magnetization. For solids having very short T_2 this leakage may actually prevent the direct observation of the NMR signal, even for strongly magnetic nuclei. Furthermore, steady-state leakage may result in saturation of the NMR signal if T_1 is long. However, a well-designed rf gate should prevent rf transmission except during the actual rf pulse and its fall time (See Section 5.2.2.3).

In order to minimize flux leakage between the two coils in a crossed-coil probe, the transmitter coil is normally oriented so that its axis is perpendicular to that of the receiver coil. Because of the need for a return lead for the receiver coil, perfect orthogonality cannot be achieved. However, flux leakage can be minimized by rotating the receiver coil about its axis for minimum coupling and inserting flux paddles,[20] which induce leakage of opposite sign. The rf recovery time can be reduced further by lowering the Q of the transmitter coil, provided sufficient H_1 is still available (since H_1 is proportional to the Q of the coil). Since the NMR S/N ratio is directly proportional to the Q of the receiver coil, a major advantage of the crossed-coil scheme is that the transmitter and receiver coils can be optimized separately. A major disadvantage of the crossed-coil arrangement is that the transmitter coil must be appreciably larger than the receiver coil to minimize coupling and obtain good H_1 homogeneity. This feature results in a loss of power efficiency since a considerably larger power amplifier will be required to provide a given H_1 over a large volume than over a small volume.

The single-coil arrangement is simpler and has many advantages and some disadvantages with respect to the crossed-coil probe. Probably the main advantage of the single-coil probe is its much higher power efficiency

---

[20] F. Bloch, W. W. Hansen, and M. G. Packard, *Phys. Rev.* **70**, 474 (1946).

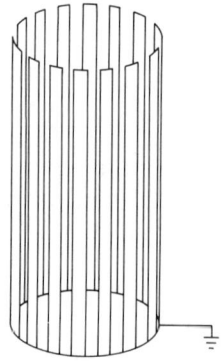

FIG. 7. Electrostatic shield for sample probe.

since the coil can be tightly wound around the sample, in contrast to the crossed-coil probe in which the transmitter coil must be considerably larger and adjusted for minimum coupling to the receiver coil.

However, it is not uncommon to insert an electrostatic shield (Fig. 7) between the sample container and the coil. This shield should be slotted in a direction perpendicular to the direction of the rf-coil windings in order to prevent eddy currents and thus allow rf penetration. This electrostatic shield prevents detuning of the coil by different samples.

Since the coil is common to both the transmitter and the receiver, special precautions must be taken to protect the first stage of the low-noise amplifier from damage by the large rf pulse without degrading the NMR signal. This protection can be achieved by the combination of crossed diodes and transmission lines[21] shown in Fig. 6. The crossed diodes appear to the high-voltage rf pulse as short circuits. The rf power across the diodes is strongly attenuated since the quarter-wavelength transmission line converts the short circuit into an open circuit, thereby causing most of the rf power to go into the sample probe. Further attenuation of the rf level at the receiver input is accomplished by the second quarter-wavelength line and crossed diodes. During the NMR signal detection, the rf levels are so small that the diodes appear as open circuits. Thus the series crossed-diode combination at the output of the power amplifier appears as a high impedance to the NMR signal, thereby forcing it to go entirely into the receiver. However, these diodes improve the system's S/N ratio since they remove low-level noise and other transients originating in the transmitter.

Since both the power amplifier and the receiver preamplifier are normally placed some distance from the sample coil in the magnet gap, the tuning is more normally performed very close to the coil in the magnet. In order to

[21] I. J. Lowe and C. E. Tarr, *J. Phys. E* **1**, 320 (1968).

5.2. NUCLEAR MAGNETIC RESONANCE SPECTROMETER DESIGN

minimize degradation of the circuit's L/C ratio and the corresponding Q (due to the large cable capacitance), the tuned coil is matched to have a resistive input/output equal to R (typically 50 Ω). The tuning–matching circuit for a typical sample probe is shown in Fig. 8. By assuming that the effective parallel resistance of the coil is much larger than R, the tuning and matching conditions can be satisfied easily by

$$\omega(C_1 + C_2) = 1/\omega L \qquad (5.2.2)$$

and

$$C_2 = 1/(RQ\omega^3 L)^{1/2}, \qquad (5.2.3)$$

where R is the equivalent resistance (usually 50 Ω) and Q is the quality factor of the coil. In order for the coil to have the largest possible L/C ratio consistent with Eq. (5.2.2), C_2 is chosen to be as small as possible. L is then determined from Eq. (5.2.3), and C_1 is then determined from Eq. (5.2.2).

In order to optimize the S/N ratio as well as the power transfer efficiency, the coil's Q normally is made as large as possible. However, since the ringing time is proportional to Q, a large Q may result in a long system-recovery time following the turnoff of the rf pulse, thereby making difficult the observation of signals having a broad linewidth and a correspondingly short T_2. As a result, for strong signals whose T_2 is very short, it may be advantageous to reduce the Q (by placing a resistor in parallel to the coil) and pay the price of a lower S/N ratio.

5.2.2.3. The Transmitter. 5.2.2.3.1. FREQUENCY SOURCE (OSCILLATOR). The oscillator shown in Fig. 6 provides a continuous rf voltage that is the source of H_1. In the usual pulse NMR application, it is desirable to have high frequency stability during the experiment. A crystal oscillator is probably the least expensive frequency source of high stability and is widely used in pulse NMR. It has the obvious disadvantage of lack of frequency versatility. In cases where more than one nucleus is studied or where it is desired to observe more than one resonance of the same species, a frequency synthesizer may be used. A synthesizer is also useful when it is necessary to sweep

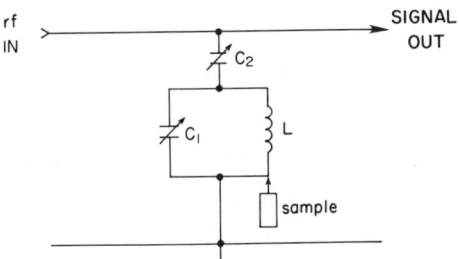

FIG. 8. Tuning–matching circuit for sample probe.

the frequency to a precisely determined value.[22] An LC oscillator also has the advantage of frequency variability, but has much poorer stability than crystals or synthesizers.

In many cases[23,24] it is desirable to obtain two rf signals from the same frequency source but phase shifted relative to each other. This can be accomplished in two ways. One is to feed the output of the oscillator into a power splitter, which produces two or three in-phase outputs with some decoupling of the signal paths. One of the outputs would then be followed by a phase shifter (which is often simply a delay line) to achieve the desired phase relation. [Examples of satisfactory power splitters are the Anzac THV-50 (Anzac Division, Adams-Russell, 80 Cambridge St., Burlington, MA 01803), a two-way divider with bandwidth 2–200 MHz, and the Anzac DS-308, a three-way divider with bandwidth 1–300 MHz]. In cases where the phase shift is to be 90°, an alternative method is to feed the oscillator output into a quadrature hybrid (e.g., Merrimac OH-1-30, Merrimac, 41 Fairfield Pl., West Caldwell, NJ 07006), which is a three-port device having one input and two outputs that are 90° out of phase with each other.

5.2.2.3.2. GATE. In order to form a pulsed rf signal, it is necessary to feed the oscillator output into an rf gate, which is controlled by pulses received from the pulse generator. The gate should have a very high on–off attenuation ratio since rf leakage during the off time can result in saturation of signals having a long T_1. At the very least, rf leakage can result in increased noise. An excellent rf gate for use in pulse NMR is the Watkins–Johnson S1, (Watkins-Johnson, 3333 Hillview Ave. Palo Alto, CA 94304), which has a 90-dB on–off attenuation ratio from 0.5 to 50 MHz and an 80-dB ratio up to 100 MHz.

5.2.2.3.3. PULSE GENERATOR. In order to provide the variety of rf pulses that are required in a multipulse experiment or even in a relaxation time measurement, it is necessary to have a versatile pulse generator to control the rf gate described in the previous paragraph. The pulse generators used in NMR have usually been hardware instruments[25] consisting of multivibrators whose timing is controlled internally by a stable oscillator (the clock). Pulse durations and pulse delays are then determined by either multivibrator RC time constants or digitally controlled time intervals whose timing is obtained from the clock. Changes in pulse durations and positions are then made manually by setting switches. Such pulse generators are generally inexpensive and easy to construct but limited in flexibility. Increasingly, however, the function of pulse generation has been performed by digital

[22] H. T. Stokes and D. C. Ailion, *Phys. Rev. B* **18**, 141 (1978).
[23] S. R. Hartmann and E. L. Hahn, *Phys. Rev.* **128**, 2042 (1962).
[24] J. Jeener and P. Broekaert, *Phys. Rev.* **157**, 232 (1967).
[25] H. T. Stokes and T. A. Case, *J. Magn. Reson.* **35**, 439 (1979).

computers. Enormous flexibility in pulse sequence can then be obtained by means of software control of pulse duration and interval. A minor complication in the use of computers as pulse generators is the need for an appropriate interface between the computer and the gate. A second complication is that the time intervals required for pulse duration (a few microseconds) may be orders of magnitude shorter than those required for pulse interval (from milliseconds to seconds). In order to minimize the wasteful use of large registers, which would be required if the pulse interval is determined by the (10 MHz) clock used for pulsewidth, a second, slower clock is usually used for pulse interval determination when using computer control.

5.2.2.3.4. POWER AMPLIFIER. In order to generate sufficient H_1 to rotate the magnetization in a time short compared to the spin-dephasing time T_2, a power amplifier is connected between the gate output and the same coil. The amount of power required should be such that the resultant H_1 is large compared to the NMR linewidth. Typically, an H_1 of order 10–100 G (corresponding to 10–1000 W depending on coil size) is required for studies in solids. (In a liquid or motionally narrowed solid, the resonance lines are narrow; thus much less power is required.) A second, related problem is that the FID signal from a broad resonance line will decay to zero very quickly following the turnoff of the rf pulse. (Typically, $T_2 \sim 1-100$ μsec.) It is, therefore, crucial to minimize the rf-pulse fall time; otherwise, the FID may be considerably reduced or even completely gone before the H_1 pulse has dropped to a value that allows observation of the magnetization. We thus have conflicting requirements for observations in solids — we need large H_1 pulses but also very short rise and fall times.

Probably the most satisfactory (but not the cheapest) way to meet the requirements for solid state NMR is to use commercial linear power amplifiers, such as those built by ENI, Inc. (3000 Winton Rd. South Rochester, NY 14623) or by Amplifier Research, Inc. (1600 School House Rd., Souderton, PA 18964). These amplifiers are highly linear and operate over a broad range of frequencies (up to 200 MHz, typically). Thus they offer considerable flexibility for studying more than one nuclear species and have short rise and fall times, but tend to be quite expensive. Since these amplifiers are operated Class A and are on all the time, they tend to generate spurious noise spikes during the FID. (These noise spikes can be reduced by using a filter following the power amplifier and/or gating off the amplifier immediately after the rf pulse. The resulting low-level noise may be reduced further by the series combination of crossed diodes shown in Fig. 6. These diodes also tend to reduce the rf-pulse fall time since they block out the smaller voltages of the decaying tail of the rf pulse because of their high resistance to small voltages.)

A much less expensive, though less flexible, approach is to use a Class C power amplifier tuned to a particular rf frequency. Class C amplifiers have high power-delivering efficiency and are capable of delivering appreciable peak power. Because of the low duty cycle resulting from the typical short* pulses used in pulse NMR, the power amplifier need not be capable of large *average* power delivery provided it can deliver large *peak* power. The tuning eliminates harmonics arising from Class C operation. This approach is economical and works well provided all experiments are to be done at the same frequency since a change in frequency would require retuning. A second disadvantage of this approach is that the high Q required for maximum H_1 results in a longer coil ringdown and thus a longer recovery time than with the broad-band Class A amplifiers described earlier.

5.2.2.4. The Receiver. The principal purpose of the receiver is to amplify the emf induced by the nuclear magnetization from its initial level (of order microvolts) to the level required for data handling and display (of order volts). In the process, the nuclear signal must be demodulated (i.e., the rf frequency removed) by an appropriate detector (normally a phase-sensitive detector). Receiver design plays a crucial role in determining the ultimate S/N ratio.

5.2.2.4.1. RADIO-FREQUENCY PREAMPLIFIER. The first part of the receiver consists of a low-noise rf amplifier with about 30–40 dB gain. Not only must this amplifier have a low noise figure and broad frequency response, but it must be able to recover quickly from the overload due to the rf pulse so that the decaying FID can be amplified as soon as possible after the pulse.

The first amplifier stage is most important in determining ultimate S/N. Accordingly, it should have a very low noise figure. The Avantek UA-142 (Avantek, 3175 Bowers Ave., Santa Clara, CA 95051) is an excellent choice for many applications. It has 16-dB gain over a 20–100 MHz bandwidth and a typical midband noise figure of 1.5 dB. Alternatively, the Anzac AM 107 amplifier has 10-dB gain over the range 1–100 MHz and a noise figure of 1.5 dB also. The later amplifier stages of the preamplifier become successively less critical in determining S/N, but must be able to operate at successively higher voltage levels without overload.

The recovery time of the rf amplifiers can be shortened by inserting the smallest practical coupling capacitors between stages and by adding shunt back-to-back diodes to ground at each output (Fig. 9). The diodes tend to prevent overload, whereas the small coupling capacitors allow a rapid recovery from a partial overload. Further improvement in the recovery time

---

* An exception occurs in $T_{1\rho}$ and $T_{1D\rho}$ experiments in which long pulses are required (see Part 6).

5.2. NUCLEAR MAGNETIC RESONANCE SPECTROMETER DESIGN

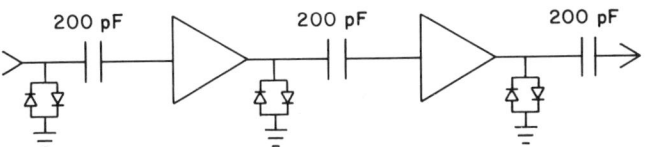

FIG. 9. Receiver section with recovery time reduced by insertion of small coupling capacitors and back-to-back shunt diodes between amplifier stages.*

(at the expense of low frequency response) can be achieved by replacing the two coupling capacitors on the circuit board inside the Anzac amplifier. Amplifiers designed with no coupling capacitors show even better recovery times. By the above techniques, overall system recovery times of 2–5 μs can be achieved at frequencies of around 20 MHz.

5.2.2.4.2. DETECTOR. The principal purpose of the detector is to remove the rf component from the nuclear signal (FID) with minimum noise insertion and distortion. Since detection is absolutely necessary if signal averaging is employed, a good detector has become an essential part of every pulse NMR spectrometer.

Probably the simplest detector is the diode detector, which is simply a diode–capacitor combination. It has two principal advantages.[26] (1) It requires no reference voltage, thus minimizing the problem of rf leakage during the time between pulses. (2) Signal averaging can be done without requiring high magnetic field stability. However, these advantages are more than offset by three serious disadvantages. (1) A diode detector, by its very nature, has a wide bandwidth and thus a poor S/N ratio. (2) A diode detector is nonlinear for small signals (because of the nonlinearity of the diode characteristics); thus, without calibration, it cannot reliably be used to measure the relative sizes of small signals (as in relaxation time measurements). (3) Since the diode detector is insensitive to phase, it cannot be used in Fourier transform and other experiments that require knowledge of the phase of the NMR signal. Because of these disadvantages, diode detection is rarely used.

Most state-of-the-art spectrometers employ phase-sensitive detection,[16] typically using double-balanced mixers. In phase-sensitive detection, a separate rf frequency f (the "reference") is mixed with the NMR signal (at f_0), thereby creating voltages at frequencies $(f_0 + f)$ and $(f_0 - f)$. By filtering, only the signal at $(f_0 - f)$ is kept. A principal advantage of this scheme is that the reference signal is usually sufficiently large that the diodes are biased into their linear region for all values of the NMR signal. To achieve this

---

[26] T. C. Farrar and E. D. Becker, "Pulse and Fourier Transform NMR," Chap. 3. Academic Press, New York, 1971.

advantage requires typically that the reference voltage be above approximately 0.5 V rms and that the NMR signal be less than approximately 0.2 V. It is usually convenient to use a reference signal obtained from the same oscillator as that which supplies the voltage. In that case, $(f_0 - f)$ will be zero when the nuclei are at exact resonance. The detected signal will still depend on the phase difference ϕ between the NMR signal and the reference at the detector and will be a maximum when $\phi = 0$. In order to ensure this condition, a variable phase shifter (typically consisting simply of a number of lengths of coaxial cable) is usually inserted in the reference channel between the oscillator and detector (see Fig. 6). The phase is then adjusted for maximum on-resonance FID. If the nuclei are off resonance, the FID will exhibit beats whose frequency equals $(f_0 - f)$ and whose position relative to the turnoff of the rf gate can be shifted by changing the reference phase ϕ. Since the reference frequency is the same as that of the rf pulse, the beat-frequency pattern in the FID will be stable for a given setting of ϕ, but will depend strongly on H_0 and will be very sensitive to fluctuations in H_0.

Phase-sensitive detection using a single mixer has a disadvantage in that there will be a dc component proportional to the amplitude of the reference voltage. Thus fluctuations in the reference amplitude will be a source of noise. This problem is minimized by the use of a double-balanced mixer since the output with no input signal will be at 0 V dc, independent of the reference level. A second possible source of noise in phase-sensitive detection is cw rf leakage from the oscillator, which is picked up in the sample probe and amplified in the preamplifier. Fluctuations in this rf level will then be a source of noise, thus partially negating a major advantage of pulse NMR — its insensitivity to oscillator noise. The most straightforward way to minimize this problem is to shield carefully all parts of the receiver (especially the sample probe) from the oscillator. One must be very careful to have good grounds (but avoid ground loops) and to filter all power supply inputs in the receiver. An alternative solution is the superheterodyne[27] receiver, i.e., the use of a relatively low-gain preamp that drives the local oscillator port of the double-balanced mixer with a frequency $(f + f_i)$, where f is the NMR oscillator frequency and f_i is a typical intermediate frequency (i.f.) (e.g., 30 MHz). The mixer could then be followed by a conventional, commercial narrow-banded i.f. strip.

An excellent choice for the double-balanced mixer is the Anzac MD-109. It has a bandwidth of 200 MHz and good signal-to-noise characteristics and requires a reference source of about 0.7-V peak amplitude. Transmission of 60-Hz pickup on the baseline can be minimized by floating the output of the

---

[27] J. D. Ellett, Jr., M. G. Gibby, U. Haeberlen, L. M. Huber, M. Mehring, A. Pines, and J. S. Waugh, *Adv. Magn. Reson.* **5**, 117 (1971).

5.2. NUCLEAR MAGNETIC RESONANCE SPECTROMETER DESIGN

mixer with respect to ground and by fitting the two output connectors directly onto the inputs of a differential amplifier (e.g. the Tektronix AM-502 amplifier, Tektronix, Beaverton, OR 97977).

5.2.2.5. Data Handling, Display, and Signal-Averaging Techniques. The previous sections describe the essential parts of a basic pulse NMR spectrometer; this section deals with overall questions of data handling and display, including signal-averaging techniques.

5.2.2.5.1. SAMPLE AND HOLD CIRCUITS. The most immediate requirement is to preserve the magnetic resonance information either for immediate or for later analysis. Probably the simplest way is to present the FID on an oscilloscope and either photograph it or plot it on a chart recorder. An inexpensive storage device useful for subsequent signal averaging is the gated integrator (or "boxcar integrator") in which a voltage, corresponding to the height of the FID at a convenient time after the turnoff of the rf pulse, is fed onto a storage device such as a capacitor[28] or a counter.[29]

If it is desired to transfer the information into a computer, eventually it is necessary first to digitize the FID voltage using an instrument like the fast-transient Biomation recorders of Gould, Inc. (4600 Old Ironside Dr., Santa Clara, CA 95050), which can be used with FIDs having short T_2 (~ 10 μs). The fast-transient recorder consists of a number (e.g., 256) of memory channels that are addressed sequentially; it therefore provides both sample-and-hold features and analog-to-digital conversion at relatively high speed. Since the fast-transient recorder cannot perform arithmetic operations, it is necessary to perform integration or signal averaging, if desired, subsequently with a boxcar, multichannel signal averager or computer. Essentially, all the functions of a fast-transient recorder plus signal averaging can be performed by a multichannel signal averager such as those of Nicolet Instrument Corp. (5225 Verona Rd., Madison, WI 53711); however, these latter instruments are much more expensive and may be slower than fast-transient recorders, although they have more channels and a higher bit resolution within each channel.

An alternative approach is to use CAMAC modules to interface the computer to the NMR system.[13] The costs associated with CAMAC interfaces are intermediate between those of homemade and commercial interfaces. Software writing is easier since the command structure for the commercially available CAMAC modules has become standardized. Both timing modules (used for generating pulse sequences) and digitizing modules (used for signal averaging) are commercially available from companies such as LeCroy Research Systems (700 S. Main St., Spring Valley, NY 10977).

[28] R. J. Blume, *Rev. Sci. Instrum.* **32**, 1016 (1961).
[29] G. Samuelson and D. C. Ailion, *Rev. Sci. Instrum.* **40**, 681 (1969).

5.2.2.5.2. INTEGRATION AND SIGNAL-TO-NOISE IMPROVEMENT. It is often important in solid state physics to look at nuclei for which the signal-to-noise ratio is small—because the nuclei have either low abundance, small γ, or broad linewidth. Of course, one should first optimize the S/N ratio of the basic spectrometer by careful choice of receiver and transmitter components, as has been described in the preceding sections. All noise, drift, rf pickup, acoustic ringing, and spurious oscillations should be eliminated.[13] Further improvement in the S/N ratio can be achieved by signal averaging. The basic principle of signal averaging is that if a noise signal is observed repeatedly and added coherently to the accumulated total of previous observations, the cumulative signal will be proportional to N, the number of observations, whereas the noise will increase as $N^{1/2}$. Thus the S/N ratio will be proportional to $N^{1/2}$. So, to get a factor-of-10 improvement in the S/N ratio will require 100 observations. A further factor-of-10 improvement requires 10^4 observations, and so on. Normally, the rapidity with which one can repeat an NMR observation is determined by T_1. Thus the achievement of substantial improvement in the S/N ratio by signal averaging can greatly lengthen the duration of the NMR experiment. As a result, increased requirements are placed on the stability of the spectrometer (as well as the patience of the experimenter). However, for systems having short T_1 (like metals[5] and some superionic[30] conductors), signal averaging can achieve substantial S/N ratio improvement without significant lengthening of the total time required.

The signal-to-noise ratio can also be improved by utilizing the fact that the NMR signal, after detection, has a limited spectrum of frequency components, whereas noise generally has a broad frequency spectrum. Accordingly, appreciable S/N ratio improvement can usually be achieved by placing a sharp-rolloff filter* after the detector in the receiver circuit.

In contrast to signal averaging and spectrometer design, other methods for S/N ratio improvement may be limited to special situations. In many cases it is possible to perform the experiment at a much lower temperature. Since thermal noise decreases at lower temperatures and since the equilibrium magnetization obeys Curie's law,

$$\mathbf{M}_0 = C\mathbf{H}_0/T, \qquad (5.2.4)$$

going to lower temperatures will increase the signal and also decrease the noise.

Furthermore, operating at a higher field (and correspondingly higher

---

[30] C. E. Hayes and D. C. Ailion, *Solid State Ionics* **5**, 233 (1981).
[31] J. Schaefer and E. O. Stejskal, *J. Magn. Reson.* **15**, 173 (1974).

* Baseline artifacts arising from filters are discussed in Schaefer and Stejskal.[31]

5.2. NUCLEAR MAGNETIC RESONANCE SPECTROMETER DESIGN

frequency) generally increases the S/N ratio (which can be shown* to be proportional to $H_0^{3/2}$). Additional S/N ratio improvement results from the fact that the coil ringdown time (and corresponding system-recovery time) decreases at higher frequencies. These advantages may be partially offset by a decrease in the sample coil circuit's L/C ratio arising from the fact that stray capacitance effectively limits the maximum value of L for tuning at a particular frequency.

In cases where the lindwidth is very broad, due either to dipolar or to quadrupolar interactions, the FID decay time may be so short that the signal is reduced considerably by the end of the recovery period following the rf pulse. The S/N ratio can then be enhanced by either decreasing the NMR linewidth (lengthening T_2^*) or by decreasing the spectrometer recovery time. The former can be achieved by raising the temperature to the region of motional narrowing[8] or by artificially narrowing the line.[9-12] For quadrupolar-broadened lines, the use of a single crystal can result in appreciable line narrowing. Of course all available means should be employed for minimizing the spectrometer recovery time (as we have discussed earlier). Beyond these techniques, information from the dead-time region can be obtained with more sophisticated techniques like the solid-echo[32] technique and the zero-time resolution (ZTR) method[33] (which are both discussed in Fukushima and Roeder[13]). In cases where the linewidth is so broad that the dead time is comparable to T_2, it may be necessary to use a cw spectrometer.

Another source of difficulty can arise from instrumental distortion of the NMR signal, which can cause significant errors in Fourier transform spectroscopy and in line-shape studies. Distortions in the FID can come from inhomogeneities in H_1 or in H_0 over the sample volume; additional coherent signals such as spurious ringing and rf pickup; and distortions in the shape of the rf pulse arising from such sources as probe arcing, transmitter power-supply sag, and detuning of the sample coil. If NMR distortion is encountered, its source should be determined and eliminated. Inhomogeneities in H_1 or H_0 should be checked with smaller samples, ringing and pickup can be identified by changing appreciably the static magnetic field, and rf pulse-shape deterioration can be seen easily by monitoring the rf pulse[13] in the transmitter coil (by using a capacitative voltage divider).†

5.2.2.5.3. FOURIER TRANSFORM (FT) NMR. As we have stated earlier, much structural information can be obtained from a knowledge of widths and central frequencies of NMR lines. Thus what we want is the NMR frequency spectrum (i.e., magnetization versus *frequency* for fixed field);

---

[32] J. G. Powles and P. Mansfield, *Phys. Lett.* **2**, 58 (1962).
[33] K. W. Vollmers, I. J. Lowe, and M. Punkkinen, *J. Magn. Reson.* **30**, 33 (1978).

* An expression for the S/N ratio is presented on p. 166 of Shaw.[14]
† See Fig. 13 of Part 6.

however, in a pulse experiment, what we measure is the magnetization versus *time* following an rf pulse. It can be easily shown[2] that the frequency spectrum is simply the Fourier transform of the pulse response. Thus an efficient way to perform cw NMR is to execute a pulse experiment and take the Fourier transform of the FID.

There are some important differences between FT NMR and cw NMR. FT NMR is intrinsically more efficient in that a pulse having a broad range of frequency components can excite *simultaneously*[34] all the lines in the spectrum, whereas a swept cw spectrometer excites them *sequentially*. Since the pulse can be repeated more frequently than can the line scan, there is an obvious S/N ratio advantage using signal averaging with pulse FT NMR. On the other hand, cw spectrometers must still be used for extremely broad lines whose T_2 is less than the spectrometer dead time (~ 5 μs, typically).

There are some potential experimental problems[13,14] in FT NMR that have no analog in cw NMR. For instance, the procedure of truncating the FID can result in additional fictitious lines in the Fourier transform. These artifacts can be reduced or eliminated by apodization. [Apodization is a procedure for cutting off the FID gently by multiplying it by a more slowly varying function (like a decreasing exponential).]

Another problem arises from the fact that in normal phase-sensitive detection it is impossible to tell on which side of resonance an off-resonance line lies. One solution is to set the rf frequency slightly off resonance for all the lines of interest; thus all the lines will be on the same side of resonance and can be distinguished easily by virtue of having different precession frequencies in the rotating frame of the rf field. This scheme has a disadvantage in that noise from the nonresonant region on the other side of the carrier is folded in, thereby increasing the noise power; also, transmitter power, which is centered at the rf frequency, is wasted in that it irradiates the nonresonant and resonant regions equally. These problems can be solved by using quadrature detection[35] (QD), a scheme utilizing two reference signals 90° out of phase with each other and two phase-sensitive detectors along with a complex (not real) fast Fourier transform program. However, in QD care must be taken to make the two phase-sensitive detectors identical and to make the two reference signals *exactly* 90° out of phase.

The process of taking the Fourier transform can be done in a variety of ways. The most obvious is to interface the receiver output to a digital computer and utilize a standard Fourier transform[36] routine. This process can be performed much more rapidly using an array processor, like those of Analogics, Inc. (Audubon Road, Wakefield, MA 01880). A method appro-

[34] R. R. Ernst and W. A. Anderson, *Rev. Sci. Instrum.* **37**, 93 (1966).
[35] E. O. Stejskal and J. Schaefer, *J. Magn. Reson.* **14**, 160 (1974).
[36] J. W. Cooley and J. W. Tukey, *Math. Comput.* **19**, 297 (1965).

priate for studying the shape of very broad lines is the method of Clark,[37] which consists of periodically feeding the receiver output to a boxcar integrator whose output (which is proportional to the detected amplitude of some portion of the FID) is plotted while the field (or frequency) is slowly swept through the resonance line.

5.3. Applications of Nuclear Magnetic Resonance

If a researcher wishes to study defects in solids by nuclear magnetic resonance, he is faced with two problems. First, magnetic resonance is a bulk technique whose sensitivity is proportional to the number of identical nuclei being studied. Defects on the other hand typically occur in low concentration. Accordingly, it is usually difficult to obtain sufficient sensitivity for direct observation of most defects. As a result, nuclear resonance often studies defects indirectly, by observing the effect of the defect on the resonance of other atoms. A second problem that occurs in all nuclear resonance studies in solids is that the linewidth is usually quite large ($\sim 10\text{--}100$ G) in contrast to the narrow linewidths characteristic of liquids; as a result, NMR signals may be difficult to see, even for abundant nuclei. This large linewidth in solids arises from magnetic dipole and electric quadrupole interactions. In this chapter we discuss methods for handling both of these problems in nuclear resonance studies of defect structure in solids. In Part 6 we discuss pulse NMR methods for studying atomic diffusion in solids.

In the following sections we discuss a number of examples that illustrate methods for using NMR to obtain information about static properties of solids. These include examples of linewidth investigations, quadrupole splittings, and Knight shifts, as well as discussions of chemical shifts, artificial line narrowing, and NMR imaging. We do not give a comprehensive survey of the literature in these areas, but rather, present examples that illustrate the types of information obtainable from the different types of NMR experiments.

5.3.1. Line-Shape Studies

The line shape can provide much information about the environment of a nuclear species. As we have seen in Chapter 5.1, the linewidth in a solid may be due to dipolar interactions (for $I = \frac{1}{2}$ nuclei) or to quadrupolar interactions (for $I > \frac{1}{2}$ nuclei in a noncubic environment). Since dipolar interactions depend on the distance between spins, a linewidth measurement can

---

[37] W. G. Clark, *Rev. Sci. Instrum.* **35**, 316 (1964).

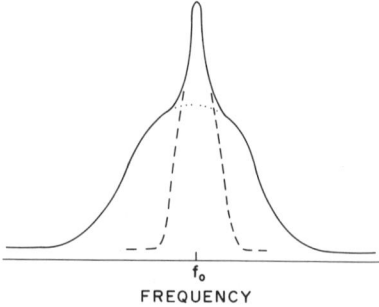

FIG. 10. Typical two-line spectrum consisting of a narrow line superimposed on a broad line.

determine the average spin spacing in a dipolar solid and can determine the size of electric field gradients in a quadrupolar solid. Furthermore, if atomic motions are present, as with translational diffusion[38] or molecular reorientations,[39] a measurement of the temperature dependence of the motionally averaged linewidth can determine the activation energy and mean jump time τ. In this section we present two examples: (a) the determination of proton microstructure in amorphous-silicon–hydrogen films and (b) the determination of the environment of sodium atoms in Na β''-alumina exposed to moisture.

5.3.1.1. Hydrogen NMR in a-Si:H Films. Figure 10 shows a typical narrow-line spectrum superimposed on a broad line. Such a spectrum suggests that the nuclei of the species under study are in two distinctly different environments characterized by very different local fields. In order to determine whether the different nuclei are in good thermal contact with each other, a "hole burning" experiment can be performed in which an rf field large enough to saturate the broad line (see Section 5.1.3) is applied at a frequency outside of the narrow line. Since the rf field heats the broad-line nuclei (i.e., equalizes the population of their energy levels), the broad-line signal will disappear. If the narrow line also disappears, then one concludes that there is good thermal contact between the two species and that the heating effect is transferred from the broad line to the narrow line by spin diffusion.* If, on the other hand, the narrow line persists upon saturation of the broad line, it is clear that the two species are thermally isolated (i.e., with regard to rf heating) from one another and must be in different particles or at least in very different magnetic environments.

[38] D. F. Holcomb and R. E. Norberg, *Phys. Rev.* **98**, 1074 (1955).
[39] H. S. Gutowsky and G. E. Pake, *J. Chem. Phys.* **16**, 1164 (1948).

* See Chapter 5 of Abragam.[2]

5.3. APPLICATIONS OF NUCLEAR MAGNETIC RESONANCE

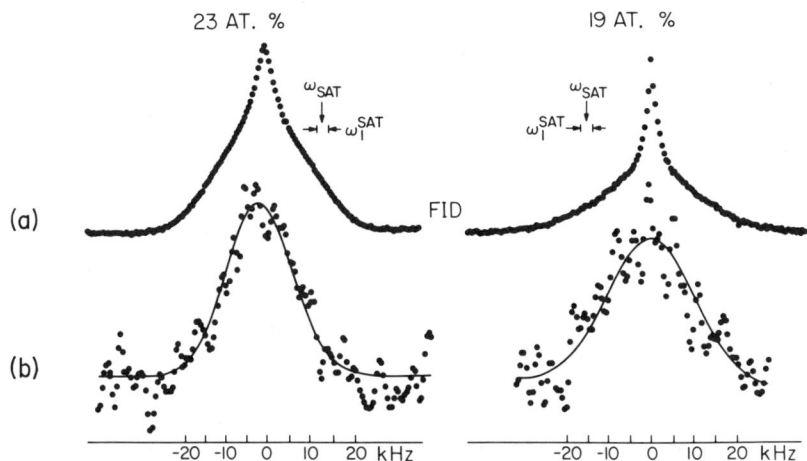

FIG. 11. Results of "hole burning" experiment in two different samples of amorphous Si:H films. (a) The spectrum after irradiation with a $\pi/2$ pulse. (b) The difference between the curves of (a) and curves obtained after a saturation pulse is applied at ω_{SAT}. (From Reimer et al.[40])

Figure 11 shows the results obtained by Reimer et al.[40] from hole burning in two samples of amorphous Si:H films each containing different amounts of hydrogen. Figure 11a shows two lines obtained for a 90° pulse at the resonance frequency ω_0 without any saturation. Figure 11b shows the *difference* between each spectrum of Fig. 11a and the spectrum obtained after a large saturation pulse is applied to the broad line region at ω_{SAT} prior to the 90° pulse at ω_0. Since no vestige of the narrow line is present in the difference spectrum (i.e., it was unaffected by the saturation pulse), the conclusion is drawn that the two components are spatially isolated. The areas under the two curves are proportional to the relative numbers of nuclei in the two phases. By determining the distance between spins in each phase from the measured linewidth and by comparing observed vibrational spectra with those of silicon monohydride, the authors were able to conclude that the broad component corresponds to structural inhomogenities [e.g., small voids (trivacancies), hydrogenated surfaces, $(SiH_2)_n$ regions, etc.] whereas the narrow component arises from interstitial monohydride groups distributed randomly throughout the film but separated from the regions containing the broad-linewidth protons.

5.3.1.2. Sodium in Na β''-Alumina Exposed to Moisture. A second illustration of the use of linewidths involves a nucleus whose linewidth is broadened by large quadrupolar rather than dipolar interactions. As discussed in Section 5.1.2.1, the linewidth of a quadrupolar nucleus in a

[40] J. A. Reimer, R. W. Vaughan, and J. C. Knights, *Phys. Rev. B* **24**, 3360 (1981).

394   5. MAGNETIC RESONANCE METHODS FOR DEFECT STRUCTURE

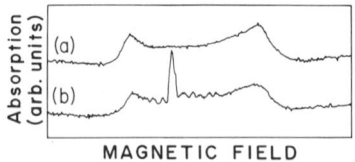

FIG. 12. Sodium line shapes in (a) dry and (b) wet powdered Na β''-alumina. (From Hayes and Ailion.[30])

noncubic environment in a powder sample may be broadened substantially. Figure 12a shows a typical second-order quadrupolar powder pattern; in this case, it represents the central (i.e., $m = \tfrac{1}{2} \leftrightarrow m = -\tfrac{1}{2}$) transition sodium resonance in a dry-powder sample of Na β''-alumina and has a linewidth of approximately 50 G. Figure 12b shows the sodium resonance in the same powder after 66-h exposure to 100% humidity. Sodium β'' alumina is technically important because of its use in high current-density batteries. The high sodium ionic conductivity, which takes place in certain crystallographic planes, is degraded by exposure to moisture. Hence it is important to study the interaction between sodium ions and water. There are two observable effects of exposure to moisture shown in Fig. 12b. First, there is some loss of structure and decrease in the overall width of the broad-line powder pattern of Fig. 12a. This effect is due to water that has penetrated the sodium conduction planes in the interior of the sample and changed the electric field gradients at the sodium sites. Second, a narrow sodium line appears in the middle of the broad-line spectrum. This narrow line arises from sodium ions associated with water complexes in a liquidlike region near the surface.[30] This example shows how the observation of linewidth changes associated with exposure to a chemical agent (in this case water) gives insight into the nature of the contact between the agent and the original substance.

5.3.2. Quadrupole Resonance

We have seen examples in the previous section of the kind of information available from line-shape studies for dipolar and quadrupolar interactions. For a nucleus possessing a quadrupole moment in a noncubic environment and hence experiencing an electric field gradient, considerable microscopic information is obtainable from the positions of the resonance lines.

5.3.2.1. Basic Theory.
The Hamiltonian for a nucleus experiencing both a Zeeman and a quadrupole interaction can be written

$$\mathcal{H} = \mathcal{H}_Z + \mathcal{H}_Q. \tag{5.3.1}$$

$\mathcal{H}_Z$ has been described in Chapter 5.1, and $\mathcal{H}_Q$ is given by

$$\mathcal{H}_Q = \mathbf{Q} \cdot \nabla \mathbf{E}, \tag{5.3.2}$$

where $\mathbf{Q}$ is the electric-quadrupole-moment tensor and $\nabla \mathbf{E}$ is the electric-field-gradient tensor.

5.3. APPLICATIONS OF NUCLEAR MAGNETIC RESONANCE

Let us choose the x,y,z axes to be principal axes of the field-gradient tensor. Furthermore, for an axially symmetric* field tensor (i.e., $\partial E_x/\partial x = \partial E_y/\partial y$), Eq. (5.3.2) can be written[1]

$$\mathcal{H}_Q = \frac{e^2qQ}{4I(2I-1)}(3I_z^2 - I^2), \qquad (5.3.3)$$

where Q is called the *quadrupole moment* of the nucleus and is defined in terms of the charge distribution of nucleons in the nucleus. The quantity q is sometimes called the *field gradient* and is defined by

$$eq = \frac{\partial E_z}{\partial z}. \qquad (5.3.4)$$

We should note that the z axis defined above is a principal axis of the electric field gradient and has nothing to do with the direction of the Zeeman field.

There are two extreme cases which we can consider: $\mathcal{H}_Q \gg \mathcal{H}_Z$ and $\mathcal{H}_Q \ll \mathcal{H}_Z$. In the first case the energy levels are determined essentially by the eigenstates of $\mathcal{H}_Q$ and are given by

$$E_m = \frac{e^2qQ}{4I(2I-1)}[3m^2 - I(I+1)] \qquad (5.3.5)$$

and are shown for a spin-$\frac{3}{2}$ nucleus in Fig. 13a. (The special case, $\mathcal{H}_Z = 0$, is commonly referred to as *pure quadrupole resonance*.) If we apply an rf magnetic field, we could observe resonance absorption (Fig. 13b) at the frequency $\omega_Q/\hbar$. We should note that the $+$ and $-$ levels are *degenerate* (in contrast to NMR) since a head-to-tail flip of a nucleus leaves quadrupole interactions unchanged. Pure quadrupole resonance in single crystals can be used to assign the orientations[4] of the principal axes of the field-gradient tensor. Furthermore, it can detect two or more *chemically inequivalent* sites (i.e., having different field-gradient components), which will appear as more than one closely spaced resonance. Pure quadrupole resonance spectrometers have a practical advantage in that a large static magnet is not used. However, there is often a practical problem in finding and identifying pure quadrupole resonances in that their frequencies are not known a priori. Also, pure quadrupole frequency measurements cannot be used alone to distinguish *physically inequivalent* sites (i.e., those having the same magnitude but different directions for the principal components of the principal axes). However, these sites can be distinguished in single crystals by applying a small Zeeman field (treated as a perturbation). The presence of magnetic

---

* The more general case in which the field tensor is not axially symmetric can be described in terms of a nonzero asymmetry parameter η (i.e., $\eta \neq 0$) and is discussed in Abragam[2] and in Das and Hahn.[4]

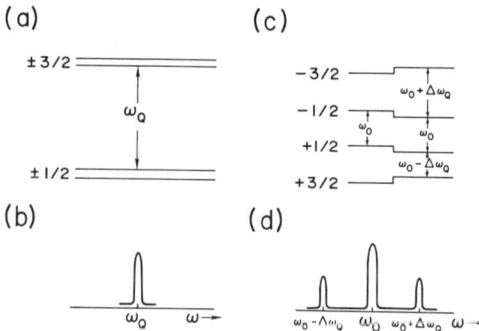

FIG. 13. Energy levels and NMR absorption lines for spin $\frac{3}{2}$ quadrupolar nucleus in noncubic environment. The (a) and (b) figures are, respectively, energy levels and absorption lines with no magnetic field present, whereas (c) and (d) refer to the same quantities in a large magnetic field.

dipolar interactions as well as dislocations and strains can result in significant broadening of the pure quadrupole lines.

In the high-field case ($\mathcal{H}_Z \gg \mathcal{H}_Q$), the quadrupole interaction can be treated as a perturbation on the well-known Zeeman levels (described in Chapter 5.1). If we calculate matrix elements of the quadrupole Hamiltonian in the Zeeman representation, we find that the levels are shifted to first order by an amount proportional to $[3m^2 - I(I + 1)]$, as in Eq. (5.3.5). Thus level shifts are the same for $+m$ levels and $-m$ levels, but are different for levels having different values of $|m|$ (Fig. 13c). The transitions m to $m - 1$ are no longer independent of m and may be at higher or lower frequency than the Larmor frequency. For a spin-$\frac{3}{2}$ nucleus, as shown in Fig. 13d, the central transition ($m = +\frac{1}{2}$ to $-\frac{1}{2}$) is unshifted, but extra lines (called *satellites*) result from the $m = +\frac{3}{2}$ to $+\frac{1}{2}$ and $m = -\frac{1}{2}$ to $-\frac{3}{2}$ transitions. The position of the satellites may be far removed from the Larmor frequency and thus not seen in an experiment that irradiates only the central transition. Furthermore, the frequencies of the satellites depend on crystal orientation relative to the applied magnetic field;[1,4] therefore, in a powder sample the individual particle satellites will be spread over an appreciable frequency range and will be difficult to observe. For sufficiently large quadrupole interactions, one must treat the quadrupole interaction to *second* order in perturbation theory, in which case even the central transition is affected. (The satellite transitions will still be determined essentially by first-order interactions.) The central line will now experience a frequency shift, which likewise depends on crystal orientation. For a powder sample, this second-order central transition shift will then result in a broadened line having the characteristic shape shown in Fig. 12a.

By comparing the positions of the satellites and the central shift in

5.3. APPLICATIONS OF NUCLEAR MAGNETIC RESONANCE

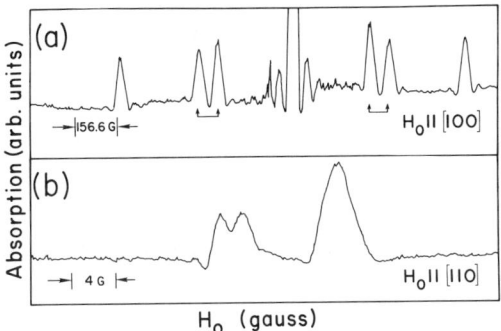

FIG. 14. Sodium resonance in a single crystal of NaCN. (a) First-order satellite spectrum with $H_0 \parallel [100]$ cubic direction. (b) Second-order central transition for $H_0 \parallel [110]$ cubic direction. (From Tzalmona and Ailion.[41])

single-crystal samples with field gradients calculated assuming a specific model, considerable microscopic detail about the charge distribution responsible for the field gradients at the nuclear sites can be obtained. The size of the satellite shifts provides a measure of the strength of the quadrupole coupling; a multiplicity of satellites may indicate the presence of more than one site for the nuclei.

5.3.2.2. Application to Structure Determination. As an illustration of the application of high-field quadrupole resonance to structure determination, we consider observations performed by Tzalmona and Ailion[41] of the ^{23}Na resonance in the orthorhombic phase of NaCN. NaCN undergoes an elastic-ordering phase transition at 288°K from a cubic phase at high temperatures to an orthorhombic phase at lower temperatures. In the orthorhombic phase, NaCN exhibits elastic order consisting of alignment of the CN groups. Since the orthorhombic unit cell lengths a, b, and c are inequivalent, alignment of the CN$^-$ ions along these directions are also inequivalent. Thus this crystal can form elastic domains in each of which all the CN$^-$ ions have a specific orientation which is different in the different domains. An important question concerns the number of different domains and the orientation of the CN$^-$ ions within each domain.

Quadrupole resonance is particularly useful for this kind of problem since the electric field gradient at a Na site depends sensitively on CN$^-$ ion orientation. Accordingly, both the first-order satellite spectrum and the second-order central line shifts should result in a multiplicity of lines due to the different domains. Figure 14a shows the ^{23}Na satellite spectrum taken in a single crystal of NaCN oriented with H_0 parallel to the cubic phase [100]

[41] A. Tzalmona and D. C. Ailion, *Phys. Rev. Lett.* **44**, 460 (1980).

direction. This spectrum shows three sets of satellites (corresponding to four domains each). Figure 14b shows, on a different scale, the central transition for $H_0 \parallel [110]$. Here, we see a three-line pattern corresponding to two domains, two domains, and eight domains, respectively. By studying these spectra at different crystal orientations and calculating the spectra for specific domain models, using first- and second-order perturbation theory, the authors[41] were able to show that NaCN is characterized by 12 domains (as opposed to six or eight) and were able to determine the CN^- orientation within each domain.

5.3.3. Electron–Nuclear Magnetic Interactions

In the last section we discussed electric quadrupole interactions as an example of electric interactions affecting nuclei, whereas in Section 5.3.1 we discussed magnetic interactions between different nuclei. In this section we consider magnetic interactions between nuclei and electrons. More particularly, we discuss here the effects of these interactions on nuclear resonance line positions and relaxation times. In Chapters 5.4 and 5.6 we discuss the effects of electron–nuclear interactions on the electron resonance, both in single and in double resonance (ENDOR). In Chapter 5.6 we also discuss how electron–nuclear interactions can be used to enhance the sensitivity of nuclear resonance in dynamic nuclear polarization experiments.

There are essentially two types of magnetic electron–nuclear interactions that affect the nuclear resonance: (1) interactions between electron spins and nuclear spins (hyperfine interactions) and (2) interactions between electron oribital magnetic moments and nuclear spins. The first of these gives rise to the so-called Knight shifts and the second to chemical shifts. Both of these interactions are discussed with great clarity in Slichter's book.[1] We merely paraphrase some of his arguments here.

5.3.3.1. Knight Shifts. It is observed[42] that the resonance frequency in a particular magnetic field H_0 of a nucleus in a metal is higher than the frequency in the same field H_0 of the same nucleus in a diamagnetic substance. For instance, the resonant frequency of ^{63}Cu in metallic copper is 0.23% higher than in CuCl. The observed fractional shift, called the Knight shift, is independent of H_0 and of the temperature, but is positive and increases with increasing nuclear charge Z. Since Knight shifts of most elements are catalogued,[42] a careful measurement of the line position and intensity can be used to indicate the presence of metallic regions in an otherwise insulating material since metals typically have unpaired electrons (in contrast to diamagnetic materials).

The Knight shift arises from interactions between s electrons and nuclei.

[42] W. D. Knight, *Solid State Phys.* **2**, 93 (1956).

5.3. APPLICATIONS OF NUCLEAR MAGNETIC RESONANCE

For an s electron, the orbital angular momentum is zero, which means that a classical electron orbit penetrates the nucleus. For zero separation r, the dipole–dipole approximation breaks down (since it diverges like r^{-3}). In this case the correct s-state Hamiltonian* can be derived[1] from relativistic quantum theory and shown to be

$$\mathcal{H} = \tfrac{8}{3}\pi\, \gamma_e \gamma_n \hbar^2\, \mathbf{I} \cdot \mathbf{S}\, \delta(\mathbf{r}), \qquad (5.3.6)$$

where $\mathbf{I}$ and $\mathbf{S}$ refer to the nuclear and electron spin operators, respectively, and γ_n and γ_e are the respective gyromagnetic ratios. Using first-order perturbation theory with this Hamiltonian for an unpaired conduction electron, the Knight shift can be calculated and is equivalent to having present an extra field ΔH, given by

$$\Delta H / H_0 = \tfrac{8}{3}\pi < |u_\mathbf{k}(0)|^2 >_{E_F} \chi_e^s. \qquad (5.3.7)$$

Here χ_e^s represents the electron susceptibility, $u_\mathbf{k}(0)$ the wave function of an electron of energy E_F and wave vector $\mathbf{k}$ at $\mathbf{r} = 0$. This hyperfine interaction gives rise to an analogous shift in the electron resonance, which, being proportional to the nuclear susceptibility χ_n^s, is much smaller than the Knight shift.

The hyperfine interaction also affects the nuclear spin–lattice relaxation time T_1, which, for a gas of noninteracting electrons, can be shown[43] to be related to the Knight shift by the formula

$$T_1 \left(\frac{\Delta H}{H_0}\right)^2 = \frac{\hbar}{4\pi k_B T}\, \frac{\gamma_e^2}{\gamma_n^2}, \qquad (5.3.8)$$

where T is the absolute temperature and k_B is Boltzmann's constant. (A more accurate expression[1] for the right-hand side of this equation depends on the electron susceptibility and the density of states at the Fermi energy.)

Thus the presence of unpaired electron spins, such as the conduction electrons of a metal, is manifested by a nonzero Knight shift and a spin–lattice relaxation time T_1 proportional to the reciprocal of the temperature.

For particularly large hyperfine interactions it is necessary to go to higher order in perturbation theory. In second-order theory, the electron–nuclear interaction is manifested by an apparent coupling between different nuclei, the so-called indirect coupling.[1] This mechanism can result in a splitting of resonance lines that is independent of temperature and field and can be shown[1] to arise, for s states, from an effective Hamiltonian of the form

$$\mathcal{H} = A_{12} \mathbf{I}_1 \cdot \mathbf{I}_2. \qquad (5.3.9)$$

[43] J. Korringa, *Physica (Utrecht)* **16**, 601 (1950).

---

* For electrons having higher orbital angular momentum (e.g., p, d, c, f), one can use the normal dipole–dipole interaction since the electron does not penetrate the nucleus.

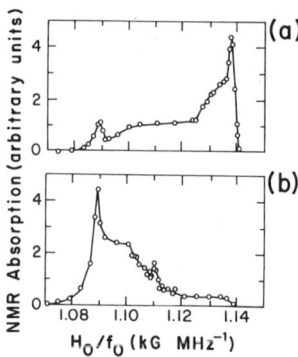

FIG. 15. ^{195}Pt resonance of small platinum particles. (a) A large Knight shift and small surface resonance for the 15% dispersion sample. (b) A smaller Knight shift and larger surface resonance for the 46% dispersion sample. (From Stokes et al.[44])

For non-s states the coupling is of the form

$$\mathcal{H} = B_{12}[\mathbf{I}_1 \cdot \mathbf{I}_2 - 3(\mathbf{I}_1 \cdot \mathbf{R}_{12})(\mathbf{I}_2 \cdot \mathbf{R}_{12})/R_{12}^2]. \quad (5.3.10)$$

The coupling constants A_{12} and B_{12} are independent of field and temperature. Expressions (5.3.9) and (5.3.10) are called *pseudoexchange* and *pseudodipolar* couplings, respectively. Since the couplings are proportional to the fourth power of the electron wave function of the nucleus,[1] these interactions are important only for heavier elements.

Figure 15 shows an exciting application, obtained by Rhodes et al.,[44] of Knight shifts to surface physics. In particular it shows the ^{195}Pt resonance of small platinum metal particles supported on alumina. The dispersion (fraction of Pt atoms on the particles' surfaces) was measured by hydrogen chemisorption to be 15% and 46% for two different samples. Nonmetallic Pt compounds typically exhibit ^{195}Pt resonances at H_0/f_0 between approximately 1.09 and 1.10 kG MHz^{-1}, whereas in bulk Pt metal, the large Knight shift results in the resonance being at $H_0/f_0 = 1.138$ kG MHz^{-1}. Figure 15a shows both these resonances in the 15% dispersion sample. The authors[44] assigned the nonmetallic resonance to surface Pt atoms and checked this idea by varying the particle size. Figure 15b shows their observation of a greatly enhanced nonmetallic resonance and a greatly reduced Knight-shifted metal resonance in 46% dispersion samples. By integrating the area under the resonance lines, they obtained dispersion values of 12% and 46% for the two samples, in excellent agreement with the chemisorption measurements.

5.3.3.2. Chemical Shifts. In addition to hyperfine interactions, nuclear spins experience an interaction with the magnetic moment associated with

---

[44] H. E. Rhodes, P.-K. Wang, H. T. Stokes, C. P. Slichter, and J. H. Sinfelt, *Phys. Rev. B* **26**, 559 (1982); H. T. Stokes, H. E. Rhodes, P.-K. Wang, C. P. Slichter, and J. H. Sinfelt, in "Nuclear and Electron Resonance Spectroscopies Applied to Materials Science" (E. N. Kaufmann and G. K. Shenoy, eds.), p. 253. Elsevier/North-Holland, New York, 1981.

5.3. APPLICATIONS OF NUCLEAR MAGNETIC RESONANCE

the orbital motions of the electrons. For isolated atoms this interaction is very large (equivalent to magnetic fields of hundreds of thousands of gauss). However, in liquids and solids this interaction is tiny (~ 0.1 G) since electron orbital angular momentum is largely quenched[1] (i.e., the expectation value of the angular momentum operator is close to zero) due to the presence of nearby electric charges. In the absence of an applied magnetic field, the orbital angular momentum is strictly quenched; however, the presence of a magnetic field of order 10^4 G results in a small unquenching due to mixing different states. We thus see a small shift in the position of the resonant field. This shift, called the *chemical shift*, is proportional to the applied field H_0 and is given by

$$\Delta H = -\sigma H_0. \qquad (5.3.11)$$

The chemical-shift coupling constant σ is of order 10^{-6}–10^{-5} for protons, but may be an order of magnitude larger for heavier nuclei, which have more electrons. The chemical shift depends sensitively on the electron wave function at the nucleus, which in turn is very sensitive to the molecular environment. Since σ is so small, the majority of reported studies have been performed in liquids where motional narrowing of the dipolar linewidth allows chemical shifts to be resolved. The accurate measurement of these small shifts requires highly homogeneous magnets (called high-resolution magnets), which are widely used by chemists to determine the structure of molecules in liquids. Sample spinning and multipulse[12,45] methods for narrowing resonance lines artificially have been developed for studying chemical shifts in solids.

An important feature of the chemical-shift interaction is that it usually has a paramagnetic ($\sigma < 0$) and a diamagnetic ($\sigma > 0$) part. Slichter shows[1] for a particularly simple arrangement of electric charges q (Fig. 16) that the paramagnetic and diamagnetic terms can be written, respectively,

$$\sigma_P = -\frac{2}{3} \frac{\hbar^2}{m} \frac{q^2}{mc^2} \frac{1}{\Delta} \left\langle \frac{1}{r^3} \right\rangle \qquad (5.3.12a)$$

and

$$\sigma_D = \frac{q^2}{3mc^2} \left\langle \frac{1}{r} \right\rangle, \qquad (5.3.12b)$$

where q and m are the charges and masses of the particles (presumably electrons) located a distance r from the nucleus in question and $\Delta/2$ represents the mean energy-level splitting for this geometry. [For a more complicated arrangement of charges, we can simply add their contributions[2] in Eqs. (5.3.12a) and (5.3.12b).] We note that atoms having more localized

[45] M. Mehring, "High Resolution NMR Spectroscopy in Solids." 2nd ed. Springer-Verlag, Berlin and New York, 1983.

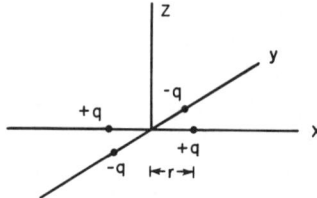

FIG. 16. Charge distribution for chemical-shift calculation. Charges $\pm q$ are all a distance r from the origin.

electron wave functions (smaller values of $\langle 1/r^3 \rangle$) are more likely to have paramagnetic chemical shifts.

The chemical-shift coupling is not a scalar quantity, as might be inferred from Eq. (5.3.11), but is a second-rank tensor and is defined more generally[45] by the shielding Hamiltonian

$$\mathcal{H} = \gamma \mathbf{H_0} \cdot \boldsymbol{\sigma} \cdot \mathbf{I}, \tag{5.3.13}$$

where the chemical-shift shielding tensor is given by

$$\sigma = \begin{vmatrix} \sigma_{xx} & \sigma_{xy} & \sigma_{xz} \\ \sigma_{yx} & \sigma_{yy} & \sigma_{yz} \\ \sigma_{zx} & \sigma_{zy} & \sigma_{zz} \end{vmatrix} \tag{5.3.14}$$

The tensor is usually anisotropic due to anisotropies in electron orbits.

Since the chemical shift depends on local environment, the observation of a spectrum of chemically shifted lines can provide a great deal of microscopic information about the molecular environment. The area under each resonance line arises from equivalent spins in the same environment; thus a comparison of the areas under different resolved lines determines the relative number of atoms in each environment. Determination of the principal axes of the chemical-shift shielding tensor can be used to determine bond directions[12] and other features of molecular electronic structure.

5.3.3.3. *Artificial Line Narrowing.* In order to measure Knight shifts or chemical shifts accurately, it is important that the linewidth be less than the shift to be measured. For dipolar broadened lines, there is usually no difficulty for Knight shifts or for chemical shifts of heavier elements, particularly in liquids. However, the chemical shifts of lighter elements of solids are often not resolvable by direct observations. For this reason a number of ingenious multipulse techniques[9,46,47] have been developed to narrow the lines artificially.

As discussed in Section 5.1.4, the earliest line-narrowing techniques involve magic-angle experiments and sample spinning. The multipulse

[46] P. Mansfield, *J. Phys. C* **4**, 1444 (1971).
[47] W. K. Rhim, D. D. Elleman, and R. W. Vaughan, *J. Chem. Phys.* **59**, 3740 (1973); also W. K. Rhim, D. D. Elleman, L. B. Schreiber, and R. W. Vaughan, *J. Chem. Phys.* **60**, 4595 (1974).

5.3. APPLICATIONS OF NUCLEAR MAGNETIC RESONANCE

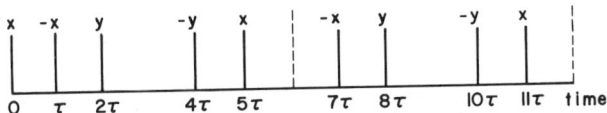

FIG. 17. WAHUHA multipulse sequence. This is a repeating four-pulse sequence consisting of $\pi/2$ pulses sequentially in the $-x, y, -y,$ and x directions with alternate spacings of τ and 2τ.

techniques all involve the application of pulse sequences that simulate motional narrowing. They are discussed in detail in the books by Haeberlen,[12] Mehring,[45] and Slichter.[1] A typical multipulse sequences sequentially flips the magnetization into different directions by a series of $\pi/2$ and $-\pi/2$ pulses about the x and y axes and then returns the magnetization to the original direction. In the different intermediate orientations the spins experience different dipolar interactions as with motional narrowing in liquids.

Figure 17 shows one of the earliest multipulse sequences (the WAHUHA sequence[9]) used for artificial line narrowing. All the pulses are $\pi/2$ pulses. The basic sequence is a repeated four-pulse sequence consisting of pulses sequentially in the $-x, y, -y,$ and x directions, with alternate spacings of τ and 2τ. Initially there is a preparation pulse, consisting of a $\pi/2$ pulse in the x direction at time $t = 0$, which tilts the magnetization from the z into the y direction. The remaining pulses of the sequence cause the magnetization to sample orientations both along the z direction (on-axis) and along x and y directions (side positions). An important feature that makes the dipolar energy vanish[1] is that spins spend twice as long in side positions as in the on-axis positions. This feature is common to this and to all other multipulse cycles.

Double resonance experiments, involving cross polarization techniques, have been developed for studying the high-resolution (chemical-shift) spectra of rare (i.e., of low abundance) spins. These, as with all double resonance techniques, require the presence of an abundant spin species (called I spins) and a dilute spin species (called S spins). The S spins' magnetization is enhanced by transferring polarization from the I spins to the S spins. The most famous technique for combining multipulse with double resonance for studying high-resolution spectra of weak spins, is the proton-enhanced nuclear-induction spectroscopy experiment of Pines et al.,[48] discussed in detail in Chapter 4 of Mehring's book.[45]

5.3.4. NMR Imaging

Spin imaging is an important new use of NMR, which, up to now, has seen primary application in medicine. This field has developed because of the need in medicine to determine *the spatial distribution* of a particular atomic or molecular species (e.g., water molecules in a living organism). In

[48] A. Pines, M. G. Gibby, and J. S. Waugh, *J. Chem. Phys.* **56**, 1776 (1972); **59**, 569 (1973).

contrast to x-ray techniques, NMR has the advantage of being nonhazardous and noninvasive and can result in a two-dimensional picture (similar to a medical x ray) of the water distribution. A further advantage over x-radiology is that the NMR image is unaffected by the presence of nonresonant nuclei; in contrast, an x-ray picture may be obscured or blocked by the presence of uninteresting bony structures. Imaging undoubtedly has many potential applications to materials science and solid state physics problems in which it is important to determine the macroscopic structure of a sample without destroying it. For instance, the geometric interface between oil and rock in oil shale could be studied by imaging.

The earliest NMR images were obtained independently by Lauterbur[17] (who named the technique *zeugmatography*) and Damadian[49] (who used the name FONAR for his technique). The basic principle of all NMR imaging techniques* is to apply a magnetic field gradient that inhomogeneously broadens the resonance line. As a result, different parts of the line correspond to nuclei in slightly different fields resonating at slightly different frequencies. Since there is now a relation between atomic spatial position and NMR frequency, different parts of the NMR line arise from nuclei in different places. By carefully decomposing the resulting signal into its individual frequency components, it is possible to determine the actual number of nuclei in each picture region (pixel) in the sample. One can then easily form on a television monitor a picture of the nuclear distribution in which the visual intensity (brightness) is proportional to the number of resonating nuclei within a pixel.

Many of the problems associated with Fourier transform NMR (see Section 5.2.2) also occur in NMR imaging. It is important to perform an imaging sequence in a highly homogeneous magnet in order to avoid having to apply large field gradients and having then to minimize distortions due to spurious variations in the magnetic field. In order to obtain a planar image, it is necessary to apply a second large field gradient perpendicular to the first along with frequency-selective (narrow bandwidth) excitation rf pulses, which excite the nuclei within their bandwidth (corresponding to a thin planar region). Improved planar resolution can be obtained by applying a perpendicular *alternating* field gradient.[51] There is then one plane in which the field gradient is fixed; nuclear signals from outside this plane vary in time and tend to cancel one another.

---

[49] R. Damadian, L. Minkoff, M. Goldsmith, M. Stanford, and J. Koutcher, *Science* **194**, 1430 (1976).
[50] P. Mansfield and P. G. Morris, *Adv. Magn. Reson., Suppl.* **2** (1982).
[51] W. S. Hinshaw, *Phys. Lett. A.* **48A**, 87 (1974).

---

* A comprehensive reivew of NMR imaging applications to medicine is found in Mansfield and Morris.[50]

5.4. Electron Paramagnetic Resonance

5.4.1. Electron Resonance Spectra

In the case of NMR the magnetic probe is the nucleus, a compact object that has widely spaced internal energy levels and interacts weakly with its surroundings. As a result, the NMR Hamiltonian is dominated by terms which are known and constant (such as the magnetic moment and the spin value). Thus the identification of the particular nucleus involved in a defect is easy in NMR. In contrast, there is no such useful free information involved in EPR. All electrons are alike in their invariant intrinsic properties, the spin magnetic moment and spin angular momentum. In an unfavorable situation, then, the existence of an electron spin resonance may tell only that there are unpaired electron spins in the crystal and little else.

EPR can also provide a value for the number of unpaired spins. With care, the number determined will even be close to the correct number. The measurement is almost always made by comparing the areas under the unsaturated absorption curves for the sample and for a standard. The curves should be taken under conditions of equal cavity Q and microwave field distribution, which usually means that the data are taken simultaneously. Samples with a large dielectric constant are particularly difficult to study accurately because they cause distortion in the field distribution. The spectra must not interfere with each other, but if they are very different it may be necessary to correct for differences in transition probabilities. Signals with Lorentzian lines require care in that a sizable fraction of the area lies in the long, low-intensity tails of the absorption line. Ideally, then, the standard should (1) have a relatively narrow line to avoid interference and to avoid base-line-shift problems over the width of the line, (2) have one spin per chemical formula unit so that the number of spins in the sample can be determined by weighing, (3) have a large enough size so it can be handled without losing an appreciable fraction of its volume in the process, and (4) have a low enough number of spins that it is a small perturbation on the losses in the cavity. It is truly unfortunate that these conditions are mutually exclusive; a compromise must be made on at least one of the first three since the fourth is essential. Wertz and Bolton[52] discuss a number of possible standards. It is particularly difficult to find a good standard which does not saturate too easily at low temperatures. For low-temperature application compromises are again necessary.

The primary advantage of EPR over NMR in the study of defects is that of sensitivity. Whereas NMR experiments on defects at the 1–0.1% level may be difficult, EPR is most commonly done at the 1–10 parts per million

---

[52] J. E. Wertz and J. R. Bolton, "Electron Spin Resonance: Elementary Theory and Practical Applications." McGraw-Hill, New York, 1972.

range. Indeed, in favorable cases sensitivities at the 10^{-11}-concentration range are possible. This higher sensitivity is mainly a result of the larger quantum of energy absorbed when an electron rather than a nucleus is flipped in a given magnetic field. Very often with EPR, sample purity is the limiting factor. A forest of low-level background signals may swamp weak details. For example, consider a paramagnetic defect containing carbon. Hyperfine satellites due to ^{13}C would prove the participation of carbon in defect, but these satellites are 200 times weaker than the lines from defects containing ^{12}C. As a result of interference from other less abundant defects, these hyperfine lines might not be seen.

5.4.2. Electron–Crystal Interactions

As usual in magnetic resonance, most of the information sought lies in the interaction of the spin with its surroundings. We now list some of these interactions. They are mentioned in the order of their importance in the analysis of spectra, rather than the more usual listing in the order of their strength. For more details there exists an extensive literature.[1,3,52,53]

The most useful interaction is the hyperfine interaction between the electron and nuclei. We have already discussed the effect of this interaction on NMR spectra in Section 5.3.3. We now discuss its effect on EPR spectra. It is primarily the importance of this interaction that justifies the inclusion of EPR in a volume devoted to nuclear methods. In general, each electron can interact with more than one nucleus so that the term in the Hamiltonian is

$$\sum_{i=1}^{N} \mathcal{H}_{en_i}, \qquad (5.4.1)$$

where the sum is over the N nuclei. The hyperfine interactions with the nuclei surrounding a paramagnetic ion or atom are often referred to as superhyperfine interactions. Each term in the sum may be expressed

$$\mathcal{H}_{en} = \frac{8\pi}{3} \gamma_e \gamma_n \hbar^2 \mathbf{I} \cdot \mathbf{S}\, \delta(\mathbf{r}) + \frac{\gamma_e \gamma_n \hbar^2}{r^3} \left[\frac{3(\mathbf{I} \cdot \mathbf{r})(\mathbf{S} \cdot \mathbf{r})}{r^2} - \mathbf{I} \cdot \mathbf{S} \right]. \qquad (5.4.2)$$

The angular momenta $\mathbf{I}$ and $\mathbf{S}$ are measured in units of $\hbar$, and $\mathbf{r}$ is the vector from the nucleus to the electron. If the electron is in an atomic orbital around the nucleus, the first term, called the contact term, is nonzero only for the s states and the second is nonzero only for non-s states, as we have seen earlier. For multinuclear centers care must be taken in distinguishing between these terms. For example, an electron in a p state about one nucleus may have a contact-term interaction with another nucleus. This interaction

[53] G. E. Pake and T. L. Estle, "The Physical Principles of Electron Paramagnetic Resonance." Benjamin, New York, 1973.

occurs because the p function is not a free-atom p function, but must be orthogonalized to the s functions on the second nucleus.

The number and intensity of the EPR lines contain such information as the number and spin of the nuclei which interact with the defect electron. A single nucleus of spin I results in $2I + 1$ equally intense and, neglecting quadrupole and second-order effects, equally spaced EPR lines. An electron interacting with several equivalent nuclei results in a series of EPR lines with the central lines more intense than those in the wings. In the case of spin $\frac{1}{2}$ nuclei the line intensities go as the coefficients of the terms in the expansion of the function $(1 + x)^n$, where n is the number of nuclei. The angular dependence of the splittings can be used to deduce the location of the nuclei relative to the electron wave function.

The stronger terms that affect the spin system are the spin–orbit interaction and electrostatic interactions between the unpaired electrons and the other nuclear and electron charges. There is no simple universal theoretical treatment of the effect of these terms. For different cases different terms are dominant. For example, heavy ions with incomplete f shells will require a treatment in which the spin–orbit interaction is much stronger than the crystal field due to neighboring ions. On the other hand, a light ion with magnetism due to d electrons requires a treatment in which the crystal field is taken as dominant and the spin–orbit coupling is treated as a subsequent perturbation. The existence of these terms is manifested in the spectra in two ways. The magnetic resonance value of the magnetic field becomes dependent on the orientation of the magnetic field with respect to the crystal axes. Ions with spins greater than $\frac{1}{2}$ can exhibit more than one resonance field value. These effects can be included in the Hamiltonian by making the spectroscopic g factor a tensor and by adding a spin–spin interaction tensor,

$$\mathcal{H} = \beta \mathbf{H} \cdot \mathbf{g} \cdot \mathbf{S} + \mathbf{S} \cdot \mathbf{D} \cdot \mathbf{S}. \tag{5.4.3}$$

From these effects the easy information to extract, and the most useful, is the number of equivalent, but not identical, sites per unit cell, the symmetry of these sites, and the spin of the defect. A detailed analysis of the spectral data yields values for the strength of the first few terms in the expansion of the crystal field.

5.4.3. Spectrometer Design Considerations

The paramount considerations in designing an EPR spectrometer are sensitivity, flexibility, and ease of operation. The criterion of sensitivity is the minimum number of spins detectable by an EPR spectrometer. Since Feher's[54] classic article, there have been numerous calculations of this

---

[54] G. Feher, *Bell Syst. Tech. J.* **36**, 449 (1957).

minimum number, N_{min}. These calculations compare the noise voltage at the detector to the change in signal voltage that results when H_0 satisfies the resonance condition. At resonance the real and imaginary parts of the sample susceptibility change, resulting in changes in the amplitude and phase of the power transmitted through or reflected from the cavity. The results of these calculations[55] are all quite similar,

$$N_{min} = \frac{CV_c k_B T_s \Delta H}{g^2 \beta^2 S(S+1) H_0 Q_0} \left(\frac{F k_B T \Delta f}{P_0} \right)^{1/2}, \qquad (5.4.4)$$

where C is a constant of order unity which depends on the particular system, V_c the volume of the cavity, k_B Boltzmann's constant, T_s the sample temperature, ΔH the width of the absorption line in gauss, g the spectroscopic g factor, β the Bohr magneton, S the spin value in units of $\hbar$, H_0 the resonant value of the magnetic field, Q_0 the unloaded cavity Q, T the detector–amplifier temperature, Δf = system bandwidth, P_0 the microwave power incident on the cavity, and F a noise figure to account for various noise sources other than Johnson noise in the detector.

This result for N_{min} assumes T_s is not so low that deviations from the Curie law occur. Further, the derivation holds only if the system is not saturated; that is, P_0 must be low enough that the spin system is in equilibrium with the lattice at temperature T_s. This condition may be a stringent one. For many defect systems, particularly at low temperatures, P_0 may be limited to perhaps 10^{-6} W, although the klystron generator may be able to provide 1 W.

The noise figure F contains contributions from the microwave source, the detector, and the amplifier following the detector. Each of the terms in F will depend on the frequency of the information that modulates the microwave signal. Buckmaster and Dering[56-59] have investigated the form of F. For a bridge spectrometer carefully tuned for absorption they find

$$F = \frac{Q_0^2 \sigma P_0}{24 \, k_B T} + \frac{F_{MW} + F_{Amp} - 1}{G}, \qquad (5.4.5)$$

with σ the rms FM noise deviation of the microwave source, F_{MW} the noise figure of the microwave detector, F_{Amp} the noise figure of the amplifier, and

---

[55] C. P. Poole, "Electron Spin Resonance," 2nd ed. Wiley (Interscience), New York, 1982.
[56] H. A. Buckmaster and J. C. Dering, *IEEE Trans. Instrum. Meas.* **IM-16**, 13 (1967).
[57] H. A. Buckmaster and J. C. Dering, *J. Sci. Instrum.* **43**, 554 (1966).
[58] H. A. Buckmaster and J. C. Dering, *J. Sci. Instrum.* **44**, 430 (1967).
[59] H. A. Buckmaster and J. C. Dering, *Proc. 14th Colloq. AMPERE, Ljubljana,* p. 1017 (1966).

5.4. ELECTRON PARAMAGNETIC RESONANCE

G the conversion gain of the detector. All other symbols are as defined in Eq. (5.4.4). The first term involves the transformation of FM noise from the source into AM noise at the detector by the frequency-dependent cavity characteristics. Since the FM-noise characteristics of the microwave source are not a function of the detector–amplifier temperature, this term in F must be proportional to $1/T$ to cancel the T in Eq. (5.4.4). At sufficiently high power levels this term will dominate the others. In this case $N_{\min}$ becomes independent of P_0, Q_0, and T. For lower power levels F_{MW}/G and F_{Amp}/G become important. From Eqs. (5.4.4) and (5.4.5) we see that the lowest value of $N_{\min}$ corresponds to the largest P_0 for which the resonance is not saturated.

Were it not for the second important consideration in EPR spectrometer design, flexibility, the commercially built units would undoubtedly be more popular with physicists. These units are typically quite sensitive and easy to operate, but they are not designed with ease of modification in mind. Even if the desired modification is topologically possible, most people are somewhat loath to cut into a multithousand-dollar commercial instrument. The most usual spectrometer changes involve the sample and thus the cavity region of the instrument. Among these changes are the application of additional electric or magnetic fields ranging in frequency from dc to the γ-ray range. Charged-particle irradiation is sometimes required. Mechanical manipulations may involve applied stress, sample orientation, and sample changing. Since these spectrometer changes involve mainly the cavity arm of the spectrometer, it is sometimes possible to use a commercial unit with a home-built cavity, Dewar system, and some additional electronics. Home-built spectrometers have the advantages that the waveguide components and electronics can be more easily modified and individual components (e.g., a lock-in amplifier) may be more easily borrowed for another experiment.

The third important EPR spectrometer design consideration, ease of operation, has already been mentioned as a positive feature of commercial spectrometers. For each adjustment of the spectrometer there must be some method of ascertaining that the adjustment goal has been achieved. For example, there must be a tuning procedure that ensures that the klystron is tuned to the sample–cavity resonance frequency. Similarly, a microwave phase must be adjusted to select either absorption or dispersion mode of operation (detection of either the imaginary or the real part of the susceptibility) and a tuning procedure must be available to determine that the correct phase has been obtained.

In the following sections we discuss various aspects of spectrometer design, keeping in mind the three considerations just discussed. Only a brief discussion of spectrometer design is given here. More comprehensive treatments may be found in the books of Algers[60] and Poole.[55]

The vast majority of EPR spectrometers utilize slow passages through the resonance lines rather than pulses. We shall limit this discussion to such slow-passage spectrometers. Readers interested in electron spin echoes are referred to the article of that title by Mims.[61]

5.4.3.1. Block Diagrams. The essential components of an EPR spectrometer are a microwave power source, a region where the microwave power interacts with the sample, a detector of changes in the microwave power, and a magnet to apply H_0 to the sample. There are numerous versions of each element as well as many other important, if not essential, components. In almost all cases the microwave power is supplied by a microwave klystron, the interaction region is a resonant cavity, and the detector is a microwave crystal diode. H_0 is supplied by an electromagnet and is augmented by a modulation field, H_m at frequency f_m.

There are a large number of possible spectrometer configurations. Alger, in his useful book,[60] lists ten basic spectrometer forms. The three examples that follow are all frequently used and span most of the range of complexity. In the figures a double line indicates a waveguide connection and a single line a cable connection. Arrows denote the direction of signal flow.

The spectrometer represented in Fig. 18 is simple but reasonably effective. The wavemeter is used to determine the microwave frequency. The first attenuator sets the power level to the cavity arm of the tee and to the arm containing the phase shifter, short, and second attenuator. The matching device along with the cavity forms a matched termination for the waveguide. Magnetic resonance causes small changes in the impedance and causes a small reflection. This reflection is added to the reflection from the opposite arm, which is adjustable in amplitude and phase. The latter reflection biases the detector into its most sensitive region and, through phase adjustment, selects the absorption or dispersion signal.

The klystron is coupled to the rest of the spectrometer through a ferrite isolator. This microwave one-way street prevents reactive elements in the system from affecting the klystron.

The matching device is not essential to the operation of the spectrometer, but is desirable to achieve the maximum S/N ratio. The cavity arm is often long, relatively flexible, and subject to vibrations. If the cavity arm has a standing wave, these vibrations are more effective in modulating the power delivered to the detector. Furthermore, the EPR signal is maximized, for the

[60] R. S. Alger, "Electron Paramagnetic Resonance: Techniques and Applications." Wiley (Interscience), New York, 1968.

[61] W. B. Mims, in "Electron Paramagnetic Resonance" (S. Geschwind, ed.), p. 263. Plenum, New York, 1972.

5.4. ELECTRON PARAMAGNETIC RESONANCE

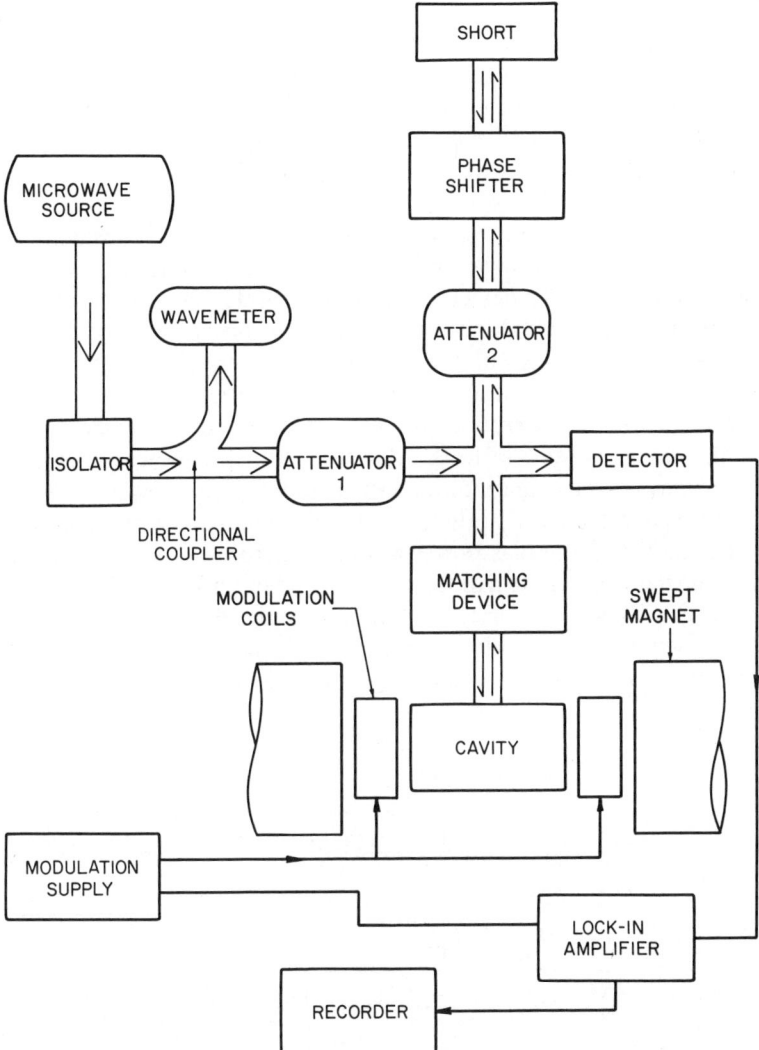

FIG. 18. Block diagram of a simple tee-bridge EPR spectrometer.

types of spectrometers discussed here,[54] if the cavity impedance is matched to the waveguide impedance.

The principal deficiencies of this spectrometer lie in its high- and low-power operation. At maximum power one half of the klystron's potential is largely wasted in attenuator 2 of the bias arm. At lowest power the power available to the bias arm will be insufficient for proper crystal bias. Further, one half the signal reflected from the cavity is wasted because of the division at the tee. Another weakness of the spectrometer is its sensitivity, at the detector, to AM noise on the bias power. This problem is not particularly serious since the AM-noise characteristics of the usual sources are much better than their FM-noise specifications. Finally, for a spectrometer of this sort, the noise figure of the detector and the modulation frequency are not independent. This point is discussed in Section 5.4.3.5.

Figure 19 shows a homodyne spectrometer which is considerably improved over that of Fig. 18. The first directional coupler extracts a small fraction of the power for detector biasing and for the wavemeter. Most of the power is available for high-power operation of the cavity since a circulator has replaced the tee. A circulator is a device that delivers power presented at one input to the next input in sequence. Thus, in the homodyne spectrometer, power from the klystron is presented to the cavity, and power reflected from the cavity is sent on to the detectors. For low-power operation the detector bias power is maintained independent of the power to the cavity. The phase characteristics of the tee at the detectors are such that the signal power is added to the bias power at one detector and is subtracted at the other. The detectors drive a differential amplifier, and thus AM noise on the bias power will be canceled to the extent that the system is truly balanced. The first attenuator should be one for which the phase shift is essentially independent of attenuation level (e.g., the 382 series of Hewlett-Packard,

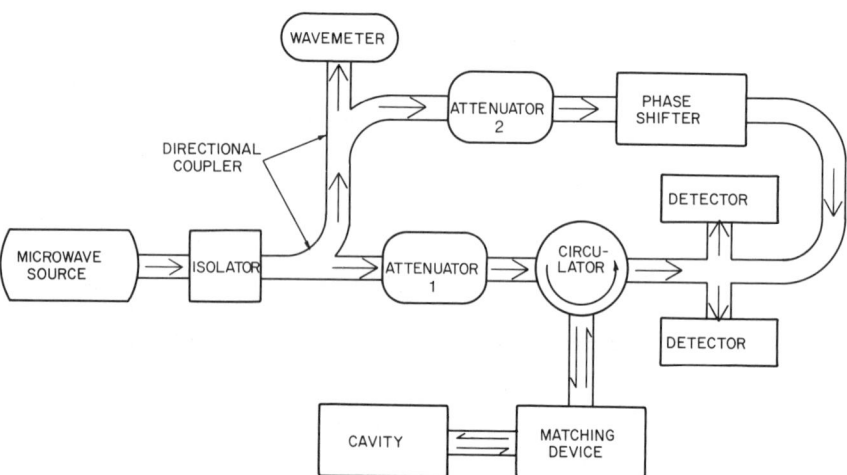

FIG. 19. Block diagram of a homodyne spectrometer. The upper arm provides microwave bias power to the detectors. A circulator has replaced the tee of Fig. 18 in this version of the homodyne spectrometer. The magnet, modulation equiment, lock-in, and recorder of Fig. 18 have not been included in this drawing.

5.4. ELECTRON PARAMAGNETIC RESONANCE

Palo Alto, CA 94304). Such a choice allows the spectrometer to be tuned for absorption or dispersion at high power levels even though the actual spectra may be recorded under low-power conditions. The homodyne spectrometer still suffers from the dependence of detector noise on modulation frequency.

The superheterodyne spectrometer outlined in Fig. 20 is considerably more complex than the previous examples. The gain realized from this complexity is that the detector noise is now independent of the modulation frequency. The i.f. oscillator frequency, which is the difference frequency between the microwave source and the local oscillator frequencies, is now the frequency at which the detector noise characteristics are important. This frequency is often chosen to be 30 or 60 MHz, where the detectors are quietest. Further discussion is in Section 5.4.3.5.

At low power levels the superheterodyne spectrometer is excellent. At high power levels the complexity is wasted since microphonics induce signals at the modulation frequency that dominate the detector noise. The cross-over point is roughly at the milliwatt level.

5.4.3.2. Choice of Operating Frequency. Strictly on the basis of Eq. (5.4.4), it would appear that the operating frequency should be chosen as high as possible since V_c decreases and H_0 increases with frequency. In

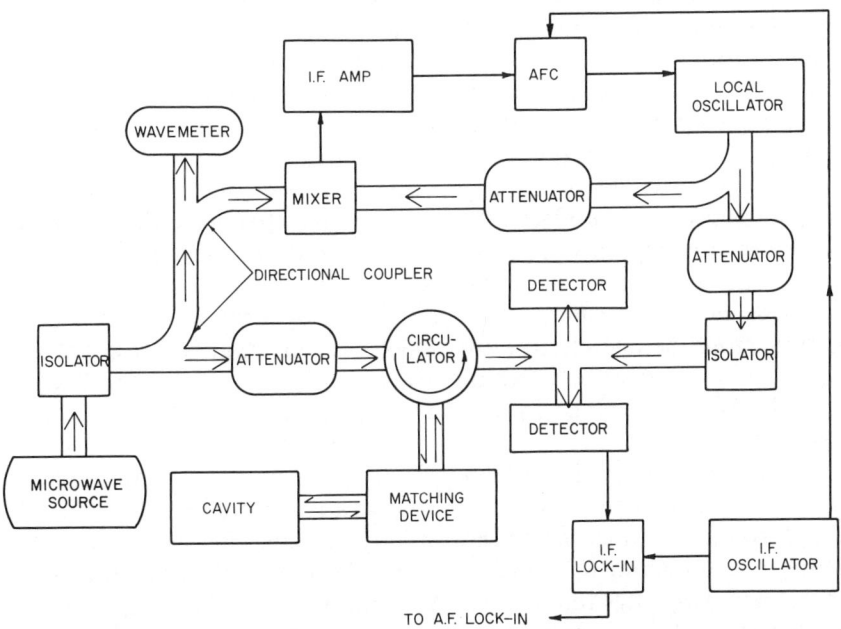

FIG. 20. Block diagram of a superheterodyne spectrometer. This version of the spectrometer utilizes two separate microwave sources which are phase-locked to have a frequency difference equal to the intermediate frequency.

practice most work is done in the range 10–33 GHz. The higher frequency is appropriate if small samples must be studied or if better resolution is desirable. The most common frequency for EPR is 10 GHz. Components, both new and used, are cheapest at this frequency. The samples used at this lower frequency can be larger and thus easier to study with other techniques, e.g., optical absorption. A maximum H_0 of 0.5 T suffices for most experiments at 10 GHz, but this field must be available over a fairly large volume. A gap between the magnet pole faces of 7.5 cm or even 10 cm is most useful in providing room for cavities, modulation coils, Dewars, and other items. There is, of course, a trade-off between magnet gap and field homogeneity. Typically, the liquid samples of the chemist require higher homogeneity than do solid samples. Since commercial EPR spectrometers are strongly influenced by the needs of chemists, the commercial systems often have a smaller than optimal magnet gap. In some cases the choice of operating frequency will depend on the properties of an existing magnet. For example, a magnet with 30.5-cm pole diameter, 6.5-cm gap, and 1-T peak field would indicate 20 GHz as an appropriate compromise among sensitivity, field strength, and versatility.

5.4.3.3. Microwave Power Sources. The reflex klystron has usually been the microwave power source used in EPR spectrometers. Newer solid state sources are simpler in that they require only one dc power supply rather than the three needed by a klystron. Both klystrons and solid state sources can be obtained in fixed and variable frequency versions. In general, variable frequency sources are used because microwave sample probes are not easily tuned. The FM-noise performance of the source is very important, particularly in a spectrometer tuned to the dispersion mode since a shift in source frequency has the same effect as a shift in the cavity frequency due to the real part of the susceptibility. The solution of the FM-noise problem is considered in the next section. In the case of a superheterodyne spectrometer, the local oscillator power may either be supplied by a second klystron or be derived from the one klystron by a sideband generator. Klystron voltages should be supplied by good commercial regulated-power supplies. Cooling is probably best obtained from a water-cooled flange, oil bath, or a low-velocity fan.

5.4.3.4. Frequency Stabilization. Both short- and long-term frequency stability of the microwave power source are important. The σ in the expression for the noise figure [Eq. (5.4.5)] quantifies the short-term instability of the microwave source. The long-term instability, or drift, determines how long after tuning the spectrometer the number of spins calculated from Eq. (5.4.4) can be detected. The derivation of Eq. (5.4.4) assumes the microwave-source frequency equals the cavity frequency. When the source frequency deviates from the cavity frequency, the system becomes

5.4. ELECTRON PARAMAGNETIC RESONANCE

more sensitive to the short-term frequency fluctuations than predicted by Eq. (5.4.4).

Two basic methods of frequency stabilization are used: stabilization directly to the sample cavity frequency and stabilization to an external standard. The former is simplest, but precludes the observation of the dispersion mode since changes in sample cavity frequency due to the spin resonance are treated as signal-source frequency shifts that are eliminated by the circuitry. The dispersion signal is not as susceptible to saturation as is the absorption, and saturation is often a problem with defect systems. Thus inability to study dispersion signals is a significant shortcoming of systems that lock the source to the cavity frequency.

Better stabilization is possible when the source is locked to an external standard because higher-Q circuits may be utilized. Excellent results may be obtained with commercial instruments that allow a source to be phase-locked to a harmonic of a crystal oscillator. Such a source can be tuned over a narrow range by varying a capacitor in the crystal oscillator; usually a voltage-variable capacitor is used. A bank of crystals with voltage tuning of the crystal oscillator allows continuous frequency coverage. A frequency-counter reading of the crystal–oscillator frequency plus knowledge of the harmonic used yields the microwave frequency.

A very elegant system for absorption studies utilizes phase-locked crystal control for spectral purity and voltage tunes the crystal oscillator with a second system to follow slow drifts in the cavity frequency.

A comparison of Johnson noise and the noise due to source instability shows that, with representative conditions, an rms FM instability of 10–100 Hz makes the two noise sources of comparable magnitude. To achieve such a low level of FM instability, the phase-locking type of frequency stabilizer is required.

5.4.3.5. Detectors. At first glance it might appear that the detector, as the first stage in the detection electronics, would be the dominant component in determining the system noise. If this view were correct, low-noise first stages such as parametric amplifiers and masers would be well worth the trouble and money involved in their use. Actually, since in EPR a small change in a large signal is sought, generator noise and system microphonics limit the gains that are possible with exotic first stages, in contrast to NMR. Thus most spectrometers use simple crystal diodes or bolometer detectors. Low-noise GaAs FET preamplifiers, though still rather expensive, are worth serious consideration. Hoentzsch et al.[62] have emphasized the advantage of using such a preamplifier in conjunction with a homodyne spectrometer when low modulation frequencies are necessary.

[62] Ch. Hoentzsch, J. R. Niklas, and J. M. Spaeth, *Rev. Sci. Instrum.* **49**, 1100 (1978).

The bolometer is limited, by its slow thermal-response time, to use with modulation frequencies below about 1 kHz. The only advantage of bolometers lies in that fact that they have a low or vanishing component of the so-called $1/f$ noise (i.e., noise proportional to $f^{-\alpha}$, where $\alpha \sim 1$). Thus bolometers are useful if very low modulation frequencies must be used and the spectrometer is not of the superheterodyne type or does not utilize a low-noise FET preamplifier. Bolometers can be obtained which are identical in shape to crystals diodes, allowing interchangable use.

Crystal diodes (usually simply called crystals) are the workhorse detectors in EPR spectrometers. The biggest variation among crystal diodes is in their $1/f$ noise properties. Conversion gain and high-frequency noise show less variation. The development of significant markets for police radar, automatic braking systems, and motion-detection intrusion alarm systems has been beneficial to the EPR practitioner. These systems are all essentially homodyne systems, and detector manufacturers have responded with diodes designed with low $1/f$ noise.

Figure 21 represents rms noise versus lock-in frequency for several sets of detector diodes. The data were taken with a homodyne spectrometer with a normal bias power of 0.1 mW and zero power to the circulator. The data on the new MA 40075 and DDB 6224 diodes are for diodes fresh out of the manufacturer's package. The MA 40075 diodes date from 1974 and the DDB 6224 diodes were purchased in 1977 (as of 1982 the DDB 6224 units were still the detectors recommended for homodyne spectrometers by Alpha Industries, Woburn, MA 01801). A clear improvement for the newer diodes can be seen in the range 1–0.01 kHz, but the advantage of high-frequency modulation, a low-noise preamplifier, or superheterodyne detecton is still obvious. The narrowing of the spread at 0.01 kHz may be due to noise in the audio preamplifier becoming dominant.

The third set of data on the graph, from the old MA 40075 crystals, illustrates a point emphasized by Alger.[60] The noise performance of a diode can be slowly degraded through various microdisasters. Checks against "standard" diodes occasionally should be made. Some good diodes should probably be reserved for use when only the best sensitivity will suffice.

In principle, each diode has a bias level and resulting source impedance that will result in the best S/N ratio when matched to the audio preamplifier. In practice, an adjustment of bias level and amplifier impedance for each type of diode is probably adequate.

5.4.3.6. *Modulation.* There is no clear-cut choice for the best modulation frequency. This fact suggests that a system be designed that is as versatile as possible with regard to modulation capabilities. For any given experiment the choice of frequency may be dominated by one of several considerations.

Clearly, from the content of earlier sections, it can be seen that a high modulation frequency is desirable from the standpoint of low noise. Indeed,

5.4. ELECTRON PARAMAGNETIC RESONANCE

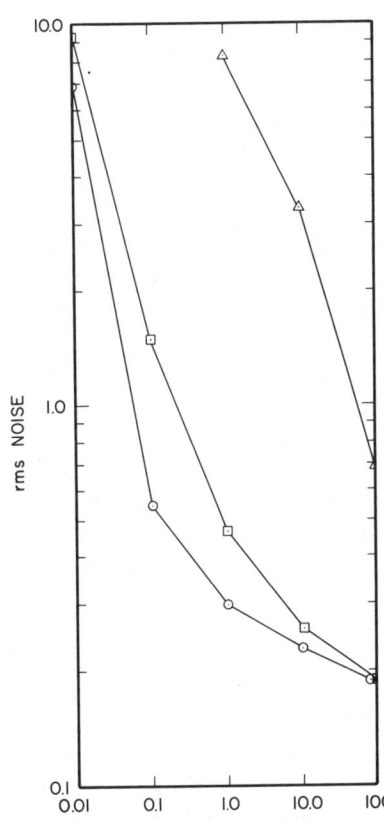

FIG. 21. Root-mean-square noise voltage in arbitrary units measured across various crystal detectors biased with 0.1 mW of microwave power. The noise is measured in a 1-Hz bandwidth about the various frequencies shown. The circles indicate the data for fresh DDB 6224 crystals, the squares for fresh MS 40075 crystals, and the triangles for an old used MA 40075 crystals.

avoidance of noise at very low frequencies is the whole reason for using modulation. In practice, detector noise, klystron noise, and system microphonics are well suppressed at a frequency of 100 kHz, and this modulation frequency is often used.

Other considerations point to the advantages of low-frequency modulation. At low frequency, cavity and Dewar walls may be located between the modulation coils and the sample with less concern for skin depth problems. The large modulation amplitudes needed for broad lines are more easily achieved at low frequency. Low-frequency-modulation coils usually rotate with the magnet for angular-dependent studies. The collinearity of H_0 and H_m is thus preserved.

The study of narrow EPR lines with the highest resolution requires low-frequency modulation. Due to the nonlinear response of the spin system to the modulaton, sideband signals are produced. These signals mix with the microwave reference signal in the detector to produce beat signals. As the magnetic field H_0 is swept, various of these beats fall within the bandpass of the spectrometer. Thus, in addition to the desired central line, spurious sideband lines are produced. At 100 kHz these sideband lines are spaced at 36 mG.

It is neither practical nor desirable to keep the modulation frequency always below the reciprocals of the spin-system relaxation times. The impracticality follows because the spin–lattice relaxation time T_1 may be seconds or even minutes long. Weger,[63] in particular, has shown that for inhomogeneously broadened lines the strongest signals often result from higher-frequency modulaton. On the other hand, some experiments designed to probe the spin-system dynamics require modulation frequencies of order $1/T_1$. Modulations other than field modulation, such as microwave power modulation or light modulation in particular, fall into this category. In short, a variable-frequency lock-in is highly desirable for a versatile EPR spectrometer.

5.4.3.7. Cryogenics. Cryogenic facilities are essential for EPR defect studies. To minimize the effects of lifetime broadening of the EPR line, it may be necessary to lower the temperature in order to increase T_1 appropriately. Further, some defects can be produced or maintained only at low temperature. There will also be an increase in the S/N ratio at lower temperatures if T_1 is not too long.

Though refrigerators or heat-exchanger storage-Dewar systems are used, most laboratories use standard glass or metal double-walled cryostats and cryogenic fluids. The choice also exists as to whether to put the cryostat in the sample cavity or the sample cavity in the cryostat. The latter is more often the choice, though the former has its advantages. If the tip of the glass cryostat is inserted into the cavity, at most only the sample need be manipulated remotely from outside. Such an arrangement is, however, limited to small samples and suffers from microphonics, both due to bubbles and from the motion of the cryostat in the cavity. Such a system is also not well suited to variable-temperature operation.

The most versatile system employs a cavity which is sealed in a separate can at the end of a stainless-steel waveguide. The can is immersed in the cryogenic fluid. With exchange gas in the can the temperature of the cavity and sample are close to the bath temperature, but there are no microphonics due to bubbles in the cavity as would be the case if the unsealed cavity were simply submerged in the bath. Further, with the exchange gas removed, the cavity and sample can be heated above the bath temperature if the thermal link between the cavity and the enclosure is designed to be sufficiently weak.

External modulation coils can be used at a higher frequency with the glass system than with the metal system. This fact would be important if the spectrometer were not of the superheterodyne type and high modulation

[63] M. Weger, *Bell Syst. Tech. J.* **39**, 1013 (1960).

5.4. ELECTRON PARAMAGNETIC RESONANCE

amplitudes were needed. For lower modulation amplitudes, and thus lower power dissipation, it is often feasible to place the high-frequency modulation coils inside the Dewar. Rose-Innes[64] gives a concise discussion of the advantages and disadvantages of glass and metal Dewar systems in Sections 3.3 and 3.6 of his book on cryogenic techniques.

5.4.3.8. Cavities. The region where the microwave interacts with the sample is subject to more constraints than any other part of the spectrometer. The following is a partial list of relevant considerations:

(1) the volume available in a Dewar system,
(2) the effects of temperature variation,
(3) coupling to the waveguide system,.
(4) changing samples,
(5) shielding of modulation fields,
(6) orientation of $\mathbf{H}_m$ and $\mathbf{H}_1$ relative to $\mathbf{H}_0$,
(7) possible locations for cuts and/or wires,
(8) introduction of various sorts of radiation and/or other perturbations, and
(9) degenerate modes.

These considerations, plus the large variety of possible solutions, have resulted in a rich literature of designs including not only cavities but also slow-wave structures. If one has a "special problem," the literature must be consulted. The book by Alger[60] and the instrument journals are particularly recommended. Here we discuss only the versatile cavities most commonly used.

Figures 22a and 22b show the field configurations for the rectangular TE_{102} and the cylindrical TE_{011} cavities. (The TE means that the electric field is purely transverse to the axis of the waveguide from which the cavity is formed by the addition of end plates; similarly, TM indicates a mode with purely transverse magnetic field. The subscripts indicate the number of half-wavelengths in the field pattern in various directions. The directions, in order, for the rectangular cavity are the wide transverse direction, the narrow transverse direction, and the longitudinal direction. For the cylindrical cavity the directions are the angular direction, the radial direction, and the longitudinal direction.) These two cavities are probably used as much as all other choices combined. The currents in the cavity walls may be easily visualized from the magnetic field pattern since the field dies off as one goes into the metal and, since

$$\mathbf{j} \propto \nabla \times \mathbf{B}, \tag{5.4.6}$$

$\mathbf{j}$ is proportional to and perpendicular to $\mathbf{B}$ at the surface. Any cuts or holes

[64] A. C. Rose-Innes, "Low Temperature Laboratory Techniques," 2nd ed. English Univ. Press, London, 1973.

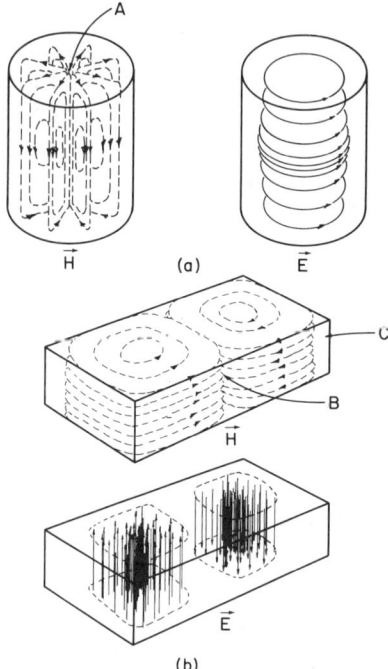

FIG. 22. Field configurations in the (a) cylindrical TE_{011} cavity and (b) rectangular TE_{102} cavity. Solid lines depict the electric field lines and dashed lines indicate the magnetic field lines. A, B, and C indicate places where holes may be made in the cavities with minimum loss of cavity Q. (Adapted with permission from R. S. Alger.[60])

in the cavity must be made with these currents in mind. Narrow openings, such as light-irradiation slots or demountable joints, require only that the cut be parallel to the current lines. Larger holes, such as used for sample insertion, should be centered at current null points. For the rectangular TE_{102} cavity such points are located at the short edges and in the middle of the long, narrow face. The center of the end plates of the cylindrical TE_{011} cavity can accommodate a particularly large hole without appreciable loss of Q. The fact that this cavity has neither radial or axial currents nor radial or axial electric fields is unique and allows considerable flexibility in placing cuts, wires, and other auxiliary items in the cavity. Berlinger and Müller[65] and Brower[66] have made a particularly convincing case for the versatility of the cylindrical TE_{011} cavity. The advantages of this cavity are so great that it is worthwhile to find ways around its two faults, degeneracy and size. In the ideal closed cylinder, the TE_{011} and TM_{111} modes are degenerate. The TM_{111} mode must be shifted in frequency so that the sample does not cause uncontrolled mode mixing. One way to lift the degeneracy is with a gap between the sidewalls and the bottom plate as in Fig. 23. This gap is not

[65] W. Berlinger and K. A. Müller, *Rev. Sci. Instrum.* **48**, 1161 (1977).
[66] K. L. Brower, *Rev. Sci. Instrum.* **48**, 135 (1977).

5.4. ELECTRON PARAMAGNETIC RESONANCE

FIG. 23. Cross section of a cylindrical TE_{011} cavity with degeneracy-lifting structure at the bottom. The added volume at the bottom is not effective in determining the frequency of the TE_{011} mode since that region is not excited by the currents that flow in the upper volume. In practice, very little extra volume is needed to adequately lift the TE_{011}–TM_{111} degeneracy. The hole in the top plate represents the waveguide-to-cavity coupling iris. The hole is offset from the center so that the H field in the waveguide can excite the H field in the cavity.

"seen" by the TE_{011} mode which has no radial currents. Wavemeters use this technique and also heavily damp the modes which couple to the extra volume. Heavy damping is not desirable for the EPR cavity since any asymmetry in the sample causes some currents to flow in the damped region and thus results in loss of cavity Q for the desirable mode.

For some ratios of cavity diameter to length, other modes are also degenerate with the TE_{011}. Avoiding these modes unfortunately aggravates the size problem for this cavity. A ratio of diameter to length of about 1.6 is desirable from a mode standpoint, but leads to a cavity that is quite demanding in terms of diameter, and ultimately, in magnet pole gap. In principle, dielectric loading could be used to reduce the cavity size. Teflon* and quartz have low enough dielectric loss to be useful in this respect. Dielectric purity, microphonics, and the added constraints on design are negative aspects of this dielectric-loading approach.

If, as usual, H_0 can be rotated in the horizontal plane, H_1 should be vertical, so as always to be perpendicular to H_0. Vertical sample insertion is usually dictated by cryogenic considerations. These last two contraints essentially equate the Dewar inside-diameter requirements of the two cavity types we are considering since the rectangular cavity must be oriented with its long axis perpendicular to the Dewar axis.

Copper or plated-brass cavities are the easiest to construct and attach to other parts of the system, but are less than ideal in terms of modulation flexibility. With such cavities high-frequency modulation must be supplied

---

* Registered trademark, E. I. DuPont de Nemours and Co.

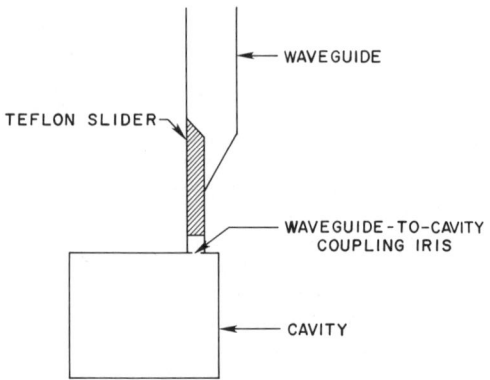

Fig. 24. Cross section of a device to match a resonant cavity to the waveguide characteristic impedance. The movable piece of Teflon is used to vary the amount of waveguide that is beyond cutoff at the operating frequency.

by internal conductors. The currents in these conductors interact with the static H_0 field, and the resulting forces tend to introduce spurious signals at the modulation frequency. Even at 100 kHz these signals can limit the spectrometer sensitivity. Although low-frequency modulation fields can penetrate thick metal cavity walls, the induced currents and H_0 again yield offset signals. This effect is, of course, particularly troublesome when high modulation amplitudes are used to study wide lines. Cavities with dielectric walls coated with a thin layer of silver or gold are very effective in reducing these modulation-pickup signals. Although such structures cannot be soldered easily to the rest of the system, they can be vacuum sealed with a variety of products. Rubber cement, in particular, is surprisingly effective. This material even makes a strong low-temperature bond to Teflon,[67] a fact which is also very useful in mounting samples. Sections of the cavity made of aluminum or beryllium (i.e., low-Z elements) are sometimes used to introduce high-energy radiation for defect production.

5.4.3.9. Matching Devices. The desirability of matching the cavity to the waveguide was mentioned in Section 5.4.3.1. The device should be capable of matching a wide range of cavity Q and should have both short- and long-term stability. In practice, the coupler first introduced by Gordon[68] has much to recommend it. In Gordon's scheme the cavity coupling hole, or iris as it is often called, is large enough to overcouple the cavity in most circumstances; that is, the impedance of the iris–cavity is lower than the characteristic impedance of the waveguide. As shown in Fig. 24, the wave-

[67] J. F. Reichert, *Rev. Sci. Instrum.* **43**, 1727 (1972).
[68] J. P. Gordon, *Rev. Sci. Instrum.* **32**, 658 (1961).

guide connected to the cavity is tapered down to a section that is beyond cutoff at the operating frequency. (A waveguide is beyond cutoff when the waveguide dimensions are so small that at the given frequency there is no traveling-wave solution for the field equations. The solution in this case exhibits an exponential decrease in field strength with distance.) A dielectric insert, usually Teflon, is tapered at one end to reduce reflections and is partially inserted in the small waveguide to bring this guide below cutoff. By decreasing the amount of waveguide left beyond cutoff, the coupling to the cavity is increased. Vibrations in the system, as well as changes in position with changes of cryogenic fluid depth, must be minimized. Microwave leakage through the position-control mechanism should be eliminated. Berlinger and Müller[65] and Isaacson[69] have given useful discussions of how to achieve these goals.

5.4.4. Representative Studies

A beautiful example of useful EPR data is provided by the U_2 center in KCl, which was first studied by Delbecq et al.[70] Figure 25 shows a plot of the derivative of the EPR absorption due to the centers versus H_0 for H_0 along the $\langle 100 \rangle$ direction. As the direction of H_0 is varied, the number and spacing of the lines in the two groups vary. The splitting between the two groups is large and independent of the direction of H_0. The number and relative strength of the lines within each group indicate that the electron interacts with four neighboring spin $\frac{3}{2}$ nuclei. The splitting between groups implies there is one proton at the center of the defect. Using only very general arguments, it is possible to conclude that the defect is an atomic hydrogen interstitial surrounded by a tetrahedron of chloride ions. Details of the magnitude of the splittings yield information about the electron wave function.

An example which demonstrates the limitations of magnetic resonance as well as its strength is illustrated in Fig. 26. In this figure the EPR resonant fields are plotted as a function of magnetic field direction. The sample is rutile (TiO_2) intentionally doped with aluminum and hydrogen.[71] The magnetic field is in the plane perpendicular to the "c" axis. In this crystal there are two titanium sites per unit cell that are magnetically equivalent when the magnetic field is along a $\langle 100 \rangle$ or $\langle 010 \rangle$ direction. The figure shows four lines that differ only in an angular displacement. This information alone is sufficient to ensure that the defect has spin $\frac{1}{2}$ and is not simply substitutional for either Ti^{4+} or O^{2-}. The fact that the lines are only 0.5 G

[69] R. A. Isaacson, Rev. Sci. Instrum. **47**, 973 (1976).
[70] C. J. Delbecq, B. Smaller, and P. H. Yuster, Phys. Rev. **104**, 599 (1956).
[71] W. D. Ohlsen, unpublished work.

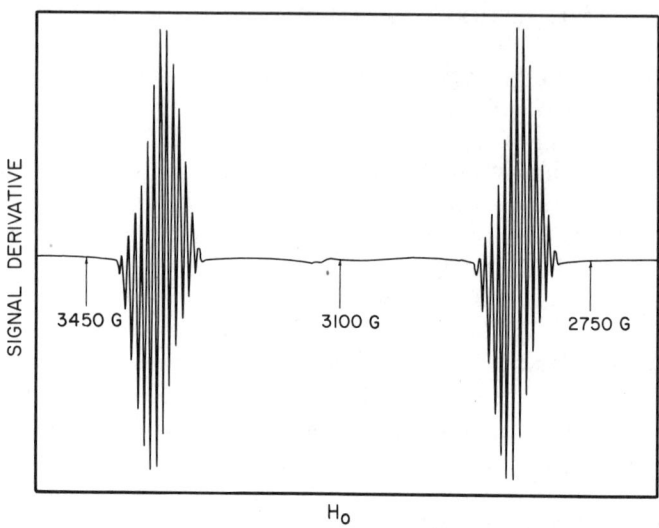

FIG. 25. The EPR derivative signal of interstitial atomic hydrogen in KCl. The splitting between the two groups of lines is due to the hyperfine interaction of the electron with the hydrogen nucleus. The splittings within the two groups of lines are due to the superhyperfine interactions with surrounding chlorine nuclei.

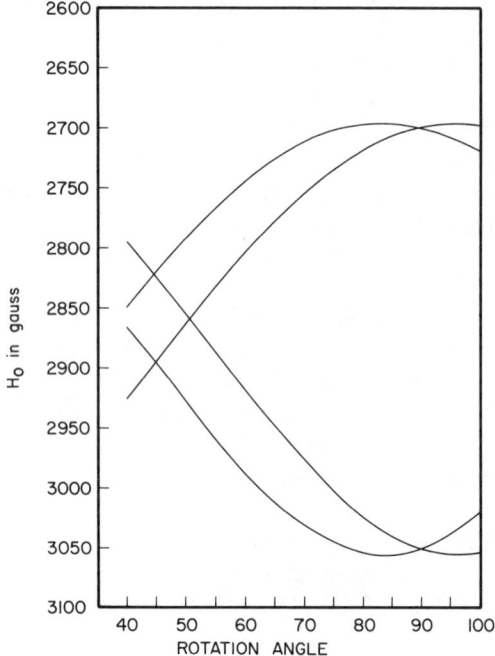

FIG. 26. Plot of the resonance fields versus orientation of the magnetic field in the basal plane for a defect in rutile. The magnetic field is along $\langle 100 \rangle$ at the rotation angle of 45° and along $\langle 110 \rangle$ for 90°. The pattern shows that the defect has electron spin $\frac{1}{2}$, nuclear spin zero, and, probably, interstitial position.

5.4. ELECTRON PARAMAGNETIC RESONANCE

broad and show no splitting essentially eliminates either hydrogen or aluminum, both of which have nuclear moments, from participation in the center. The fourfold multiplicity of lines can be shown to be consistent with an interstitial site or several two center defects. Without a clear hyperfine signature, it is impossible to identify this center which probably involves some impurity in the crystal.

The first example given had optimal information contained in the number of lines, in their positions, and in their relative intensities. The second example contained insufficient data for defect identification but enough data to be useful in choosing between models suggested by other experiments.

Figure 27 shows an example of the multiplicity of resonance lines that results when an electron system with spin greater than $\frac{1}{2}$ interacts with the crystal field. Again the crystal is TiO_2.[72] In this case, hyperfine data, which are not plotted, identify the defect as a Mo impurity. From the coalescence of pairs of lines when H_0 is along the "a" and "c" axes, it is clear the Mo is substitutional for a Ti. The fact that there are two values of the field that result in resonances even with the field along the "a" and "c" axes indicates a spin greater than $\frac{1}{2}$. In this particular case the spin is $\frac{3}{2}$, which identifies the Mo as Mo^{3+}. The crystal field splits the four levels associated with spin $\frac{3}{2}$ into two pairs. The pairs are further split by the magnetic field, and, for the proper field value, resonance transitions occur. The variation in relative line strengths as a function of temperature leads, through the Boltzmann factor, to the same value of the crystal field splitting as was obtained by matching the angular-dependence data.

A fourth type of example could easily be displayed. Any practitioner of EPR could provide the spectrum that would consist of a jumble of overlapping lines so rich in "information" as to defy analysis. Such spectra serve only as a warning about overly simplistic defect models.

We conclude this section with a few studies in which EPR data are central to the understanding of materials of obvious practical importance. These studies emphasize the use of EPR as a materials science tool.

The first work is the study by Brower[73] of substitutional nitrogen donors in silicon. Earlier investigators had found that silicon nitride formation resulted from attempts to dope silicon with nitrogen during crystal growth, but that donor nitrogen could be introduced by ion implantation followed by thermal anneals above 500°C. These nitrogen dopings yielded complicated samples with less than 5%, if any, of the N atoms on substitutional sites. A small fraction of the nitrogen acted as donors with ionization energy 0.017 ± 0.002 eV.

[72] W. D. Ohlsen, *Phys. Rev. B* **7**, 4058 (1973).
[73] K. L. Brower, *Phys. Rev. Lett.* **44**, 1627 (1980).

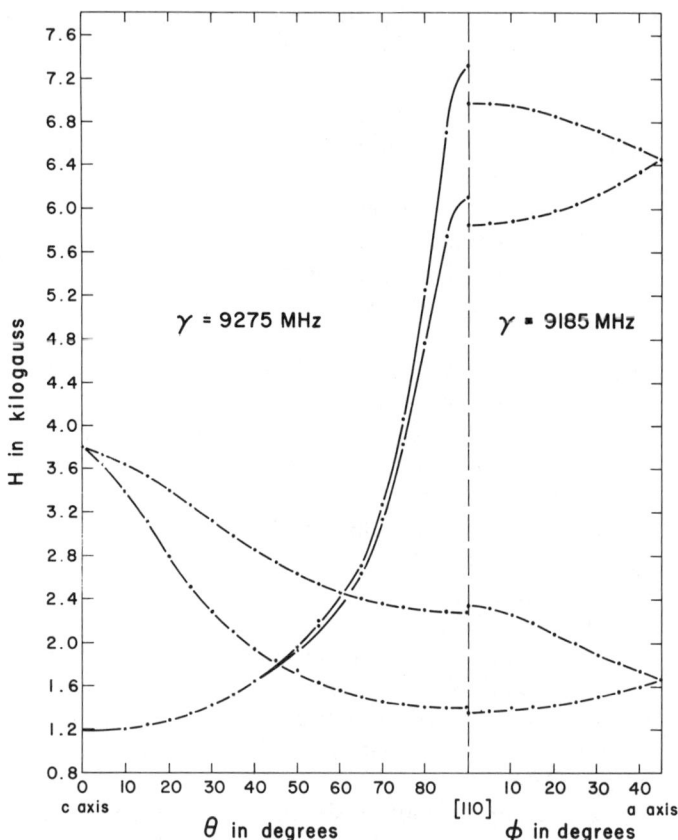

FIG. 27. Plot of the resonance fields versus magnetic-field orientation relative to the crystal axes for Mo^{3+} in rutile. θ measures the angle between H_0 and the c axis in the c axis-$\langle 110 \rangle$ axis plane, ϕ measures the angle between H_0 and the $\langle 110 \rangle$ axis in the basal plane. The data were taken at two microwave frequencies γ because the different orientations of the sample in the cavity resulted in two different cavity resonant frequencies. The Mo^{3+} has electronic spin $\frac{3}{2}$ and substitutes for a Ti^{4+}. Charge compensation is nonlocal.

Brower[73] found that implanting silicon with 4×10^{15} ions cm^{-2} (0.995 parts ^{28}Si to 0.005 parts N_2) at 160 keV and then pulsed ruby-laser annealing the sample resulted in isolated substitutional nitrogen-associated donors. The dependence of the donor EPR-line field values on the orientation of the field relative to the crystal axes indicates that the donor wave function is not s-like, but has axial symmetry about a $\langle 111 \rangle$ axis, that is, along a N–Si bond axis. From the ^{14}N and ^{29}Si hyperfine splittings Brower could calculate that 73% of the wave function is made up of orbitals located on the single Si and 9% is on the N. The N part of the orbital is 28% s-like and 72% $p_{\parallel}$-like, and

5.4. ELECTRON PARAMAGNETIC RESONANCE

the corresponding numbers for the Si part are 12% and 88%. Stress alignment and reorientation could be followed with the EPR spectra. The fact that alignment was perpendicular to the stress indicates an antibonding orbital for the electron. An activation energy of 0.084 ± 0.005 eV and preexponential factor of $\simeq 4 \times 10^{-9}$ s were found for thermal reorientation. Light with $h\nu \geq 0.58$ eV caused randomization of stress-induced polarizations indicating an optical ionization energy of 0.58 eV for this donor.

The extraction of all this defect information is made possible by the detail in the EPR spectrum. However, the most important contribution of EPR to this study is indicated by Brower's comment on the implantation-beam composition: "Preliminary studies indicated that these N impurities are introduced during ion implantation as a result of contamination of the ^{28}Si beam with $\approx 0.5\%$ $^{14}N_2$." In other words, the original purpose of the study had nothing to do with nitrogen, and without the clear ^{14}N hyperfine signature on the EPR lines, the participation of nitrogen in determining the properties of the crystal might not have been suspected.

The second of our studies is one of a series of papers by Davies, Nicholls, and co-workers[74] dealing with luminescence centers. The host crystal is ZnSe, and the center is associated with copper doping. Earlier optical polarization studies on an analogous center in ZnS were best described by a center of C_{3v} symmetry, though C_{2v} could not be ruled out. The luminescence transition was thought to be due to an internal transition of the center since the wavelength was observed to be independent of excitation intensity and of time after excitation.

Optically detected magnetic resonance (ODMR) was employed in this study and in the others of the series. Changes in the spin distributions caused by magnetic resonance transitions lead, through spin-selection rules, to changes in emission intensity. The microwaves were power modulated, and phase-sensitive detection was applied to the luminescence signal. This technique is discussed further in Section 5.6.4.

Although the study did not lead to a unique model for the defect, several significant results were obtained. First, the four-line hyperfine structure that was observed strongly suggested a copper center. The inability to fit other terms in the spin Hamiltonian to models involving Li, Na, K, or Br, which are the other possible spin $\frac{3}{2}$ nuclei, left only copper as the paramagnetic entity in the center. Next, the data showed that the center has C_{2v} symmetry, which established that the copper was associated with another ion. The EPR spectra required that the Cu^{2+} ion resides either in a zinc substitutional site or at an interstitial site surrounded by four selenium ions. Besides the Cu^{2+} resonance, a shallow donor resonance was observed by ODMR utilizing the

---

[74] J. L. Patel, J. J. Davies, and J. E. Nicholls, *J. Phys. C* **14**, 5545 (1981).

same luminescence signal. Thus, despite the data presented earlier, it appears that the luminescence process involves a donor-acceptor recombination rather than a transition involving only the Cu^{2+}. It is clear that without the intimate association of the emission spectrum with the paramagnetic resonance, which is central to ODMR, some of these findings would not have been possible.

Our final study, by Schallenberger and Hausmann,[75] concerns the intrinsic defects in zinc oxide crystals. The study utilized 3-MeV electron irradiation to produce vacancies and interstitials. Four types of the defects so produced are paramagnetic. Three of the centers are spin $\frac{1}{2}$, and one is spin 1. All of these centers exhibit tensor g values, and one of them has nonisotropic hyperfine interactions with several nearby zinc nuclei. The increase and decrease of the strengths of the various resonances as a result of irradiating the sample with light of various wavelengths was observed.

In most intrinsic defect studies the basic electronic structure of the perfect crystal is assumed to be reasonably well understood, and the emphasis is on characterizing the defects. The study of Schallenberger and Hausmann[75] had a more ambitious goal. The defects were viewed as probes for measuring the details of the chemical bonding of zinc oxide. These authors were able to find models for their defects which explained the observations. The models required that d-d bonding between next-nearest neighbors, both O-O and Zn-Zn, plays a significant role in the perfect crystal. Such a model of the bonding explains several features of the perfect crystal; the success in fitting the EPR data gives added credence to the model.

5.5. Additional Perturbations

5.5.1. Light

Light as a perturbation is introduced into a magnetic resonance defect experiment for a number of reasons. The light may be essential in forming defects, as in the case of interstitial hydrogen formed by the photodissociation of OH^- ions, or it may produce an excited state of the defect of interest, e.g., triplet excited states. Photogeneration of electron-hole pairs is used as a technique for reducing an inconveniently long T_1.

Light can be introduced into a cryostat through light pipes from the top or through windows in the side or bottom. For experiments involving polarized light, the optical activity of stressed windows must be considered. Appropriately placed holes or arrays of parallel slots are usually used to get light into or out of a cavity. Helices or transparent dielectric cavities allow good optical access to the sample.

[75] B. Schallenberger and A. Hausmann, *Z. Phys. B* **44**, 143 (1981).

5.5. ADDITIONAL PERTURBATIONS

5.5.2. Stress

Application of stress can be essential for detection of the magnetic resonance of a defect. The R center in alkali halides is the classic example of a defect requiring application of stress for EPR detection. An R center consists of an equilateral triangle of anion vacancies in the $\langle 111 \rangle$ plane with three trapped electrons. Krupka and Silsbee[76] were the first to see and explain the EPR of this defect, whose ground state in an otherwise perfect crystal has a twofold vibronic degeneracy. Random strains lift this degeneracy and, since the g value depends on the level splitting, broaden the EPR line to unobservability. At high stress the g value approaches a limit; thus a high applied stress narrows the line to the point where it can be detected.

In other studies, the application of stress may be of great, if not crucial, importance since the number of lines seen may be either increased or decreased. An increase can occur if the stress lowers the symmetry. For example, a nucleus of spin $I > \frac{1}{2}$ sitting at a site of cubic symmetry has $2I$ transition with a $\Delta I = \pm 1$ selection rule as discussed in Section 5.3.2. These transitions all occur at the same magnetic field value. A stress that lowers the site symmetry causes electric field gradients at the nucleus that interact with a nuclear quadrupole moment to split the single nuclear resonance line into $2I$ lines.

Stress can also cause a decrease in the number of magnetic resonance lines in systems that have defects with several electronically equivalent, but magnetically inequivalent sites. The O_2^- center in the alkali halides is such a system.[77] The six possible $\langle 110 \rangle$-type orientations are equally likely in an unstrained crystal, and each orientation of the molecular axis relative to the magnetic field yields a different set of lines. Stress may be applied, which lifts the electronic equivalence sufficiently that at low temperatures only the orientation with lowest energy is appreciably populated. Such an effect, of course, requires reasonable reorientation probabilities at the low temperature. The results of stress experiments may aid in formulating a defect model and can furnish electron–lattice coupling constants.

Techniques for applying hydrostatic stress are discussed by Benedek.[78] Uniaxial stress may be applied either with rigid rods into the cryostat or with wire-and-lever arrangements. As has been previously noted, the TE_{011} cylindrical cavity is particularly accessible for such manipulations. It is more complicated to put a sample in tension, but the same effects can be obtained by gluing a thin sample to a substrate that has a lower expansion coefficient and then cooling the resulting sandwich.

---

[76] D. C. Krupka and R. H. Silsbee, *Phys. Rev.* **152**, 816 (1966).

[77] W. Känzig, *J. Phys. Chem. Solids* **23**, 479 (1962).

[78] G. B. Benedek, "Magnetic Resonance at High Pressure." Wiley (Interscience), New York, 1963.

5.5.3. Electric Fields

The application of electric fields as a perturbation to magnetic resonance systems is not a common procedure. The largest of externally applied electric fields is small compared to internal fields, and the splittings that result are, correspondingly, small. A shift in resonance field that varies linearly in electric field is obtained only for defects that lack inversion symmetry. However, since normal magnetic resonance experiments are not sensitive to this lack of symmetry, valuable additional information may be obtained in some cases. The authoritative book by Mims[79] should be consulted for details of electric field effects related to EPR.

5.6. Double-Resonance Techniques

Since the defects one wishes to study are often present in low concentrations, the sensitivity limitations of conventional spectrometers can be an important problem. The problem, of course, occurs more often for NMR than for EPR. We now introduce a variety of elegant techniques in which higher sensitivity is obtained by exciting two resonance transitions in the sample at the same time. These techniques are not applicable in every case. Each technique has requirements that must be met by the system under study. Furthermore, these techniques all involve a considerable increase in experimental complexity.

5.6.1. Nuclear–Nuclear Double Resonance

The sample to be studied may have two nuclear resonances, one too weak to see and one strong enough for study. The spins associated with the weak resonance are usually called the S spins, and those related to the strong resonance are called the I spins. Often it is the weak resonance that is of interest. Nuclear–nuclear double-resonance (NNDR) methods allow one to detect the satisfying of the resonance condition for the S-spin system by the appearance of an effect on the I-spin resonance signal.

Here we give only a brief indication of the theory of one type of NNDR experiment. For details, the lucid book by Slichter,[1] a principal architect of the field, is highly recommended.

The basic problem that must be overcome in an NNDR experiment is that of coupling the two spin systems so that they can exchange energy. Mutual spin flips, which change the magnetizations in a static field, do not usually conserve energy since the Zeeman splittings of the two nuclei are in

---

[79] W. B. Mims, "The Linear Electric Field Effect in Paramagnetic Resonance." Oxford Univ. Press (Clarendon), London and New York, 1976.

5.6. DOUBLE-RESONANCE TECHNIQUES

general different. If two resonant rf magnetic fields are applied to the system, a different, and certainly not obvious, view may be taken. Each spin (assumed to be $\frac{1}{2}$) is viewed in its *own* rotating reference frame as defined in Section 5.1.1. The I spins' rf field has a negligible effect on the S spins since it is not near the S spins' resonant frequency. Similarly, the S spins' rf field can be neglected when considering the I spins. Consider the case where each spin system's net magnetization is aligned along its associated H_1. Such an alignment can be achieved by applying a 90° pulse to prepare the magnetization perpendicular to H_0 and then phase-shifting the radio frequency field by 90° to align H_1 along the magnetization. Reversing the S spin in the field of H_{1S} requires an energy exchange of $\hbar\gamma_S H_{1S}$. Similarly, reversing an I spin involves an energy change $\hbar\gamma_I H_{1I}$. These energies are equal if the rf magnetic field strengths are adjusted so that $\gamma_I H_{1I} = \gamma_S H_{1S}$. This condition is known as the Hahn condition in honor of its discoverer.[23] The matching condition is not too stringent for H_1s of the same magnitude as the resonance linewidths. With the Hahn condition satisfied, the dipole–dipole interaction can cause energy-conserving mutual spin flips, which change the two magnetizations.

What sort of equilibrium is achieved by the two coupled spin systems? The answer has its roots in the crucial paper by Redfield[5] which leads to the conclusion that there is a temperature T_{spin} of a spin system in the rotating frame that is related to the magnetization and effective field by Curie's law,

$$\mathbf{M} = C\mathbf{H}_{\text{eff}}/T_{\text{spin}}. \tag{5.6.1}$$

In a double-resonance experiment the I- and S-spin systems usually come to equilibrium at a common spin temperature. If the S spins are continually being heated, this heating will be coupled to the I spins and will show up eventually as a decrease in the I magnetization. ("Eventually" must be in a time short compared to the I system spin–lattice relaxation time.) The S spins can be heated by rapidly reversing H_{1S}, that is, by phase-shifting the S-spin radio-frequency field by 180°. After such a reversal, the magnetization is opposite to H_{1S}, and the spin system has a negative spin temperature. Since a negative temperature is "hotter" than any positive temperature, the S spins heat up the I spins, and the I spins cool the S spins. After the S spins have cooled sufficiently, the field reversal is carried out again. After many cycles the low-heat-capacity S spins have heated the high-heat-capacity I spins, and the fact that H_{1S} was tuned to the S spins' resonance can be detected by seeing a reduction in the magnitude of the I spins FID relative to that obtained when the H_{1S} field is not in resonance with the S spins.

Figure 28 shows NNDR data.[80] The sample is NaF doped with 0.25% LiF. The I spins are the ^{19}F spins, and the S spins are those ^{23}Na spins that are

---

[80] K. F. Nelson and W. D. Ohlsen, *Phys. Rev.* **180**, 36 (1969).

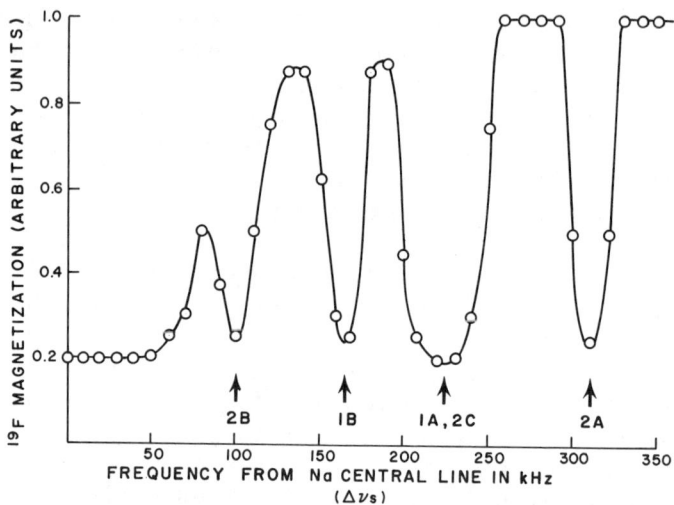

FIG. 28. Normalized ^{19}F FID magnitude as a function ^{23}Na-resonance irradiation frequency. The ^{23}Na transitions marked 1 stem from nearest Na$^+$ to a Li$^+$, and the transitions marked 2 stem from next-nearest Na$^+$ neighbors. The data are for a direction of the magnetic field relative to the crystal axes that results in resonably good resolution of the various resonances. The depression of the ^{19}F FID at the left of the plot results from coupling the ^{19}F spins to unshifted and only slightly shifted ^{23}Na spins. (From Nelson and Ohlsen.[80])

near Li$^+$ ions. The substitution of an Li$^+$ ion for a Na$^+$ ion removes the cubic symmetry for the Na$^+$ ions near the Li$^+$, and these Na spins have their resonances shifted by the quadrupole interaction. The NNDR experiment yields quadrupole interaction values that are compared to theoretical values in a test of computation techniques applicable to alkali halides.

5.6.2. Electron–Nuclear Double Resonance

In an electron–nuclear double-resonance (ENDOR) experiment, H_0 is first adjusted so that the microwave generator is in resonance with an EPR line. Then an rf generator is frequency-swept through the resonant frequency of the nuclei. The nuclear-spin transitions, and thus their resonance frequencies, are detected as changes in the electron resonance signal.

ENDOR can be viewed either as a method for increasing the resolution of EPR spectra or as a technique for improving NMR sensitivity for those relatively few nuclei that are located near paramagnetic defects. These views exclude the so-called distance ENDOR,[81] which involves nuclei far from the paramagnetic defect and which yields little, if any, defect information. In any sort of ENDOR, nuclear-spin transitions cause an effect on the EPR

[81] J. Lambe, N. Laurence, E. C. McIrvine, and R. W. Terhune, *Phys. Rev.* **122**, 1161 (1961).

spectrometer detection circuit. The EPR spectrometer acts as a sensitive detector for the NMR. The NMR lines detected are shifted from their usual pure-crystal position by the nuclear–defect interactions.

In general, the Hamiltonian of one electron interacting with N nuclei has the form

$$\mathcal{H} = \mathcal{H}_e + \sum_{i=1}^{N} \mathcal{H}_{en_i} + \sum_{i=1}^{N} \mathcal{H}_{n_i}. \quad (5.6.2)$$

$\mathcal{H}_e$ is the electron Hamiltonian without hyperfine terms. $\mathcal{H}_{en_i}$ is the electron–ith-nucleus interaction and $\mathcal{H}_{n_i}$ is the rest of the Hamiltonian for the ith nucleus. The increased resolution of ENDOR relative to conventional EPR stems from the fact that the energy involved in an electron-spin flip is dependent on all N of the interaction terms, whereas a particular nuclear flip involves only that particular interaction term. Because of the various possible nuclear-spin states the EPR line is split into many, perhaps overlapping, lines. The two possible electron states result in a splitting of the NMR line into only two widely spaced lines.

If one of the nuclei that interacts with the electron is due to an impurity atom that is part of the defect, there is a good chance that the interaction with this nucleus is strong enough to cause a resolved splitting of the EPR line. In this fortunate case the value of the nuclear spin can be obtained from normal EPR. The ENDOR spectrum yields, in addition, the nuclear moment, allowing a positive identification of the element involved.

The ENDOR effect requires a partial saturation of the EPR transition so that the populations of the levels connected by the microwaves are not related by the lattice temperature. For clarity, consider the simple four-level system depicted in Fig. 29. The diagram is appropriate for a single electron of spin $S = \frac{1}{2}$ interacting with a spin $I = \frac{1}{2}$ nucleus through a term $a(\mathbf{S} \cdot \mathbf{I})$ with a positive. The states are labeled by the high-field eigenvalues of S_z and I_z. The microwave radiation is represented by $h\nu$, and W_1 and W_2 represent the strongest relaxation terms. W_2 arises from lattice modulation of that part of the interaction term which has the form $(a/2)(I^+S^- + I^-S^+)$, where I^+, S^+, I^-, and S^- are spin-raising and -lowering operators.

The steady-state relative populations in the limit of strong saturation are shown next to each level in Fig. 29, where ϵ is the relative population difference between electron-spin states in thermal equilibrium. That is, if the microwave radiation were absent, the upper levels would have populations $1 - \frac{1}{2}\epsilon$ and the lower levels would have populations $1 + \frac{1}{2}\epsilon$. The populations for strong saturation result from four constraints. The microwave saturation results in the levels on the left-hand side of the diagram having equal populations. A W_1 process forces a population difference ϵ between the two right-hand levels that it couples. Similarly, the W_2 process will result in a

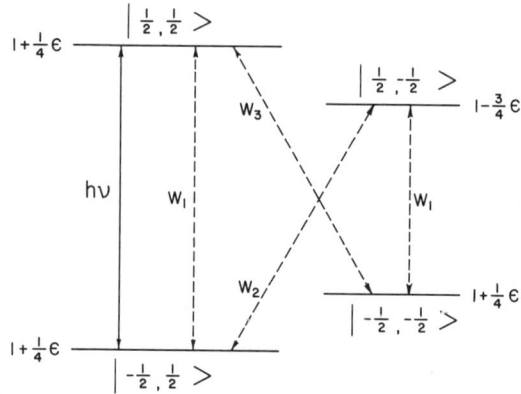

FIG. 29. The four levels for a system of an electron interacting with a spin $\frac{1}{2}$ nucleus. The levels are labeled by the z components of their spin angular momenta. ϵ is the thermal-equilibrium relative population difference between the electron-spin states. The steady-state relative populations shown by each level assume that the transition probability due to $h\nu$ is much, greater than the relaxation probabilities W_1 and W_2 and that W_3 is much weaker than W_1 or W_2.

population difference ϵ between the two levels that it links. Finally, the total number of spins is conserved, so the sum of the four levels must add up to 4 as in the thermal-equilibrium case. The EPR signal amplitude depends on the population difference between the $|\frac{1}{2}, \frac{1}{2}\rangle$ and $|-\frac{1}{2}, \frac{1}{2}\rangle$ states. This population difference can be altered by nuclear resonance transitions between the levels $|\frac{1}{2}, \frac{1}{2}\rangle$ and $|\frac{1}{2}, -\frac{1}{2}\rangle$. The nuclear transitions can be effective in two ways. An adiabatic fast passage through the resonance or a 180° pulse on resonance would exchange the populations of the $|\frac{1}{2}, \frac{1}{2}\rangle$ and $|\frac{1}{2}, -\frac{1}{2}\rangle$ states, which would yield a transient increase in the EPR signal. Alternatively, saturation of the nuclear transition would lead to the involvement of W_2 processes in determining the level populations and resultant EPR signal amplitude. In the latter case it would be natural to use amplitude modulation of the NMR power and use lock-in detection of changes in the EPR signal. The modulation period should not be short compared to the thermal-relaxation times of the system. Since these times may be reasonably long, it is desirable to use a superheterodyne spectrometer or a spectrometer with a low-noise GaAs FET preamplifier, both of which maintain a good sensitivity at low modulation frequencies.

There are many choices and compromises involved in the choice of an ENDOR coil system. The coil may be located inside the cavity with due care being taken to minimize the component of the microwave electric field which is along the coil wire. Such a choice is simple, but decreases the cavity Q and increases microphonic noise. Alternatively, the coil can be wound outside a cavity whose walls are much thinner than the skin depth ("thin") at the nuclear frequency and much greater than the skin depth ("thick") for the microwaves. This choice restricts the cavity construction and yields lower rf magnetic fields. A third choice is to wind the coil outside a cavity that is "thick" at the nuclear frequency and is cut in such a manner that rf currents are induced to flow in the inside walls of the cavity.[82,83] Somewhat more complexity and loss of vacuum integrity are the prices paid for this choice.

The problem of driving the ENDOR coil also can be attacked in several ways. The ENDOR coil can be used as the L of the LC circuit in a power oscillator. This approach is simple, but not well suited to constant H_1 and linear frequency scans. Constant H_1 is very desirable since peaks in the rf power dissipated in the microwave cavity region can easily cause spurious peaks in the output of the EPR spectrometer. Most systems use a swept oscillator driving a broad-band power amplifier. This system can yield a constant H_1 by either of two approaches. The first method has the advantage of simplicity but the disadvantage of low H_1 for a given amount of current. The coil is made of a single loop of wire. This loop has a small impedance compared to the 50-Ω impedance of the coaxial line driving the coil. Instead of using the loop to terminate the cable, the loop is inserted in series with the cable center wire and the cable continues on to a 50-Ω matched load. Thus the line appears to be almost a frequency-independent 50-Ω load for the amplifier. The second method of obtaining a frequency-independent H_1 can yield much higher H_1 values but is considerably more complex. In this method the current flowing in the coil is sensed, and feedback control is used to adjust an impedance-matching system. Gruber *et al.*[84] have described such a system.

It is impossible to present any general calculation of the intensity of ENDOR signals, but a reasonable rule of thumb is that the ENDOR signals will be of the order of 1% of the strength of the normal EPR signal. For more details of ENDOR spectroscopy the reader is referred to the book on the subject by Kevan and Kispert.[85]

5.6.3. Dynamic Nuclear Polarization

Dynamic nuclear polarization is the name given to a number of techniques in which saturating electron-spin transitions of a sample result in a nuclear polarization that is much greater than the equilibrium Boltzmann value. Such increased polarization may be necessary to raise a resonance of dilute defect nuclei to detectability.

[82] W. C. Holton and H. Blum, *Phys. Rev.* **125**, 89 (1962).
[83] M. Giordano, F. Momo, and A. Stogiu, *J. Phys. E.* **12**, 815 (1979).
[84] H. Gruber, J. Forrer, A. Schweijer, and H. H. Gunthard, *J. Phys. E* **7**, 569 (1974).
[85] L. Kevan and L. Kispert, "Electron Spin Double Resonance Spectroscopy." Wiley, New York, 1976.

Reference to Fig. 29 shows that saturation of the EPR transition results in a nuclear alignment in steady state that is more appropriate for an electronic than a nuclear magnetic moment. That is, the $I_z = -\frac{1}{2}$ states have a relative population $2 - \frac{1}{2}\epsilon$, and the $I_z = +\frac{1}{2}$ states have a relative population $2 + \frac{1}{2}\epsilon$. As indicated in Section 5.6.2, the electron-spin thermal-equilibrium Boltzmann populations would be $2 - \epsilon$ and $2 + \epsilon$. Thus we have achieved a significantly enhanced population difference for the two nuclear states ($I_z = +\frac{1}{2}$ and $I_z = -\frac{1}{2}$). This difference is of the same magnitude as the population difference for pure electron-spin states in thermal equilibrium. There are a large number of other schemes which also achieve large nuclear polarizations by exciting microwave transitions of the system. A more detailed discussion of these dynamic nuclear polarization techniques has been given by Jeffries.[86]

Dynamic nuclear polarization is usually not the best magnetic resonance method for studying defects in solids. Rather than do NMR on the polarized nuclei, it is more effective to study the back effect of the NMR on the polarizing system as is done in ENDOR. An exception to this rule occurs if the electron spins that are being saturated are mobile as in the case of conduction electrons in a semiconductor. Only if the position vector of the nucleus relative to the electron is fixed can the details of the interaction of the nucleus with some other defect be read out of the details of the electron resonance.

Under some conditions it is possible to introduce paramagnetic impurities to enhance by dynamic nuclear polarization the resonance of nuclei associated with a defect. Brun and co-workers[87] have demonstrated this possibility using Cr^{3+} impurities to polarize ^{17}O dilute nuclear spins in TiO_2 and Al_2O_3. In these studies the Cr^{3+} was the only impurity, but the low isotopic abundance of ^{17}O, 0.037%, demonstrates the applicability of this technique to an impurity–defect system.

One can envision yet more exotic schemes in which techniques of nuclear–nuclear double resonance are combined with dynamic nuclear polarization. It is doubtful that such methods would have wide applicability.

5.6.4. Other Double-Resonance Techniques

The most important double-resonance technique not discussed so far is that of optically detected magnetic resonance (ODMR). In this case one of the resonances is not a spin resonance, but an electronic resonance, i.e., an

---

[86] C. D. Jeffries, in "Electron Paramagnetic Resonance" (S. Geschwind, ed.), p. 217. Plenum, New York, 1972.

[87] P. Bösiger, E. Brun, and D. Meier, *Proc. 19th Congr. AMPERE, Heidelberg*, p. 545 (1976). Ch. Gabathuler, E. Hundt, and E. Brun, *Proc. 17th Congr. AMPERE, Turku*, p. 499 (1972).

5.6. DOUBLE-RESONANCE TECHNIQUES

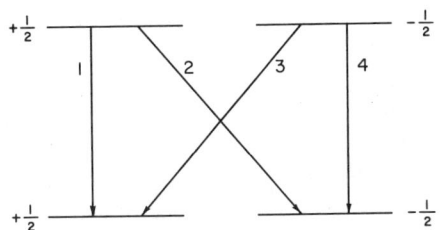

FIG. 30. A plot of the emissive optical transitions possible between a magnetic doublet excited state and a magnetic doublet ground state. All four transitions are linearly polarized when viewed in a direction perpendicular to the magnetic field. Transitions 2 and 3 are circularly polarized when viewed along the field.

optical transition. Use is made of the fact that the absorption or emission of light is in some manner dependent on the populations of the magnetic sublevels of the system. Magnetic resonance, by changing the populations of the magnetic sublevels, changes the optical properties. Since optical photons are detected, rather than those in the microwave or radio-frequency range, the sensitivity is greatly increased. The minimum number of defects that may be studied by ODMR depends on the parameters of the system, but 10^5 might be a typical number. Such sensitivity renders excited-state EPR possible even for systems that do not have the long lifetimes of metastable triplet states.

The selectivity of optically detected EPR is as important as is its sensitivity; in fact, the sensitivity would be useless without optical selectivity. Not even electronic-grade semiconductors are pure enough for 10^5 centers to dominate the unwanted background signals that would be present were it not for the fact that the light signal selected is specific to the defect of interest.

As a specific example of how optically detected EPR might be done, consider a system with a magnetic doublet ground state and a magnetic doublet excited state as depicted in Fig. 30. Lines 1–4 indicate optical emission processes. Viewed along the magnetic field, transitions 2 and 3 are seen and are circularly polarized with 2 left-handed and 3 right-handed. Normally there will be a population difference between the two excited states either because of selective excitation or because of thermalization. Thus magnetic resonance of the excited state will change the populations and result in a change in amplitude of the emission signals that can be separated with a circular-polarization analyzer. Other processes and related subjects, all dealing with EPR, are treated in much greater detail in an excellent article by Geschwind.[88]

Though ODMR is more often done with electron-spin resonances, the

---

[88] S. Geschwind, *in* "Electron Paramagnetic Resonance" (S. Geschwind, ed.), p. 353. Plenum, New York, 1972.

same principles are involved in nuclear studies, sometimes labeled ODNMR. Already in 1950 Kastler[89] proposed optical detection of the NMR of ^{199}Hg in vapor. The nuclear resonance of nuclei near a defect can be studied when these nuclei have a hyperfine interaction with the optical center.[90,91]

Electron–electron double-resonance (ELDOR) techniques also exist. Referring again to Fig. 29, there are two allowed microwave transitions for this system; one is between the two levels on the left-hand side of the figure, and the other is between the two levels on the right-hand side. An allowed–allowed ELDOR experiment involves detecting the change in the EPR signal of one of these transitions caused by exciting the other transition. Forbidden–allowed and allowed–forbidden ELDOR involves one or the other of the microwave transitions being a "diagonal" transition in which, according to our high-field labels, both the electron spin and the nuclear spin change state.

The ELDOR technique yields information about the hyperfine interaction constants and the relaxation processes, but is difficult to instrument. The microwave sample structure must accommodate two different microwave fields, one of which is frequency swept in the course of the experiment. The subject is treated in the book by Kevan and Kisper.[85]

Acknowledgments

The authors wish to thank Professor H. T. Stokes, Dr. C. E. Hayes, and T. A. Case for useful discussions regarding NMR instrumentation. The authors were supported by the National Science Foundation under Grants DMR 76-18966 and DMR 74-17545 during the writing of this article.

---

[89] A. Kastler, *J. Phys. Radium* **11,** 255 (1950).
[90] D. P. Burum, R. M. Shelby, and R. M. Macfarlane, *Phys. Rev. B: Condens. Matter* **25,** 3009 (1982).
[91] D. Paget, *Phys. Rev. B: Condens. Matter* **25,** 4444 (1982).

6. NUCLEAR MAGNETIC RESONANCE RELAXATION TIME METHODS FOR STUDYING ATOMIC AND MOLECULAR MOTIONS IN SOLIDS

By David C. Ailion

Department of Physics
The University of Utah
Salt Lake City, Utah

6.1. Introduction

One of the major applications of nuclear magnetic resonance (NMR) has been the study of atomic and molecular motions in solids. Because of its nuclear specificity, sensitivity to the microscopic environment of a diffusing atom, and ability to study a variety of atomic and molecular motions over a wide temperature range, NMR has been strikingly successful in studying atomic diffusion and molecular reorientations. Part 5 of this volume describes how nuclear and electron paramagnetic resonance (EPR) can be used to determine structural features of defects in solids. In this part, NMR relaxation time techniques* for elucidating the dynamic properties of mobile atoms and defects are discussed.

In contrast to non-NMR techniques, NMR has the important advantage that the resonance signal is characteristic of the particular nucleus being studied. For this reason, NMR can normally distinguish the diffusion of a specific nuclear species from that of other nuclei. (An exception may occur in heteronuclear systems in which case the $T_{1D\rho}$ technique,[3] which is discussed later, is very useful.) In contrast, techniques like dielectric loss[4] and

---

[1] D. C. Ailion, in "New Techniques and Applications of Magnetic Resonance" (R. Blinc and M. Vilfan, eds.), pp. 97–112. E. Kardelj University Press, Ljubljana, Yugoslavia, 1982.
[2] D. C. Ailion, in "Nuclear and Electron Resonance Spectroscopies Applied to Materials Science" (E. L. Kaufmann and G. K. Shenoy, eds.), pp. 55–68. Elsevier/North-Holland, New York, 1981.
[3] H. T. Stokes and D. C. Ailion, *Phys. Rev. B* **18**, 141 (1978).
[4] A. S. Nowick, in "Point Defects in Solids" (J. H. Crawford, Jr. and L. M. Slifkin, eds.), Vol. 1, pp. 151–200. Plenum, New York, 1972.

* These techniques and their applications to atomic motions are reviewed briefly in Ailion.[1,2]

ionic conductivity[5] measure properties that do not belong exclusively to the nuclei whose motions are of interest.

A second very important feature of NMR relaxation methods is that they detect motions that give rise to fluctuating local magnetic fields and can measure both local motions, such as molecular reorientations, and nonlocal motions, like translational diffusion. On the other hand, many of the non-NMR techniques are restricted to one type of motion or the other, but can not be used for both. For example, techniques like radioactive tracers[6] and ionic conductivity measure mass flow from one part of the sample to another and can be used to study nonlocal diffusion but not bound (local) diffusion or molecular reorientations, whereas techniques like dielectric relaxation and ionic thermocurrent (ITC)[7] are normally applied to bound motions, but are not used for nonlocal translational diffusion. Furthermore, many of these techniques are restricted to a narrow range of motional frequencies occurring in a restricted temperature region. In contrast, NMR can study a wide variety of local and nonlocal atomic motions over a wide temperature range *without* changing techniques. One can then directly observe changes in activation energy arising from the onset of a second motional mechanism without having the data affected differentially by artifacts due to different techniques.

A further important advantage of NMR is its sensitivity to those microscopic features that affect atomic motions. This advantage arises from the short-range nature of many of the interactions that affect the spin system. (For example, the dipole–dipole energy, which may determine the linewidth and the spin–lattice relaxation time, is proportional to the inverse sixth power of the internuclear distance.[8]) Thus NMR can be used to study the immediate environment of the nuclei and thereby elucidate the microscopic details governing atomic motions.

NMR has some other advantages. It is not restricted to substances having a radioactive isotope of convenient half-life, as is the radioactive tracer method. Also, the results of ionic conductivity and tracer measurements may be affected by the presence of grain boundaries that can provide a "short circuit" path for a mobile atom. NMR, on the other hand, tends to be much less affected by grain-boundary diffusion since it is a bulk technique and the majority of diffusing atoms are not typically near grain boundaries.

Probably, the principal disadvantage of NMR is that the signal-to-noise (S/N) ratio is proportional to the number of nuclei. Thus it is very difficult to use NMR to study *directly* the properties of impurities and other defects.

[5] A. B. Lidiard, *Handb. Phys.* **20**, 246 (1957).
[6] D. Lazarus, *Solid State Phys.* **10**, 71 (1960).
[7] C. Bucci, R. Fieschi, and G. Guidi, *Phys. Rev.* **148**, 816 (1966).
[8] C. P. Slichter, "Principles of Magnetic Resonance," 2nd ed. Springer-Verlag, Berlin and New York, 1978.

Fortunately, a number of elegant indirect techniques exist for studying defects by observing their effects on the NMR signal and on the relaxation time. However, the substance must contain a suitable abundant nuclear species with a magnetic moment sufficiently large to be detectable by NMR. Even the techniques for studying the motion of weakly magnetic[3] (i.e., low gyromagnetic ratio γ) or dilute atoms detect effects on the resonance of a strongly magnetic species. This second species must be sufficiently strong to be observed directly; it need not be mobile, but it must be present. These techniques are discussed in this part and in Part 5.

6.2. Conventional Relaxation Time Methods

Most relaxation time measurements in solids require a standard pulse spectrometer, preferably having a short recovery time and capable of operation over a range of frequencies. For $T_{1\rho}$ and $T_{1D\rho}$ operation the spectrometer should be capable of producing long (milliseconds to seconds) rf pulses. Since hardware considerations in constructing or purchasing such a system are discussed in Part 5 of this volume, they will not be repeated here. Rather, in this part, I focus on the techniques, rather than the hardware, for studying atomic motions in solids. In this section, methods are presented for measuring the conventional relaxation times, T_2 and T_1, defined in Part 5.

6.2.1. Motional Narrowing of the Linewidth and the Spin–Spin Relaxation Time (T_2)

6.2.1.1. *Motional Narrowing.* Probably the most straightforward method for studying ionic diffusion in solids is to observe motional narrowing of the linewidth. In order to understand this phenomenon, we must briefly review the origin of line broadening in solids. The basic phenomenon of magnetic resonance and the interactions that affect line shapes and line positions are discussed in the previous part.

Consider a solid consisting of identical atoms, each of which has spin $\frac{1}{2}$ and thus no quadrupole moment. If we place this system in a large static magnetic field $H_0\mathbf{k}$ and in a perpendicular alternating field $\mathbf{H}_1 \cos \omega t$, we will observe a resonance absorption line as ω is swept through the resonance frequency $\omega_0(=\gamma H_0)$. The linewidth arises from each nucleus seeing a magnetic field that is the resultant of $\mathbf{H}_0$ and the average local dipolar field arising from the neighboring spins. Since the instantaneous value and direction of this dipolar field depends on the actual orientations of the neighbors, there will be a spread in the value of the dipolar fields through the crystal. This spread gives rise to variations in the resonance frequencies of the individual nuclei of the order of 1–10 G typically. This linewidth can be observed conveniently in a pulse experiment in which $\mathbf{H}_1$ is turned on just

6. NMR RELAXATION TIME METHODS FOR ATOMIC MOTIONS

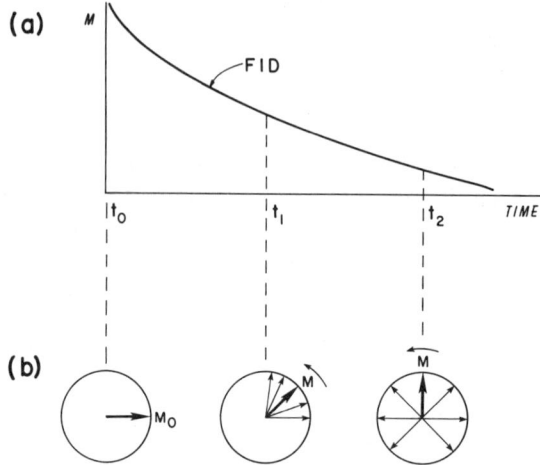

FIG. 1. (a) Free-induction decay (FID) and (b) corresponding spin isochromats during the decay. The isochromats correspond to individual magnetic moments in the xy plane. At t_0 they are all in phase and the resultant xy magnetization is M_0, at t_1 they are somewhat out of phase and the magnetization M has decreased, at t_2 they are further out of phase and the magnetization has decreased further.

long enough to tilt the magnetization into the xy plane (a 90° or $\pi/2$ pulse). After the rf pulse is turned off, the magnetization in the xy plane will precess about $\mathbf{H}_0$, which is in the z direction. Since the individual nuclei are precessing at slightly different frequencies, their magnetization contributions will get out of phase (Fig. 1) and the resultant xy magnetization will decay to zero with a time constant T_2, called the spin–spin relaxation time. This magnetization response to a 90° pulse is called the free-induction decay (FID). Since the FID is the Fourier transform of the line-shape function,[9] T_2 is proportional to the inverse of the linewidth. In a solid, typical T_2 values are in the range 1–100 μs.

Now, in order to understand motional narrowing, consider what happens if the nuclei are diffusing at a rate (τ_c^{-1}) more rapid than the rigid-lattice spin–spin relaxation rate (T_2^{-1}) between lattice sites having different values of the instantaneous dipolar field. For a symmetric resonance line, the resultant field experienced by any one nucleus will be greater than H_0 as often as it will be less than H_0, thereby resulting in a decreased average spread in the dipolar field. Thus we get a decrease in linewidth or, alternatively, an increase in T_2. This line narrowing occurs only when the mean atomic jump time τ_c is less than T_2, in which case it can be shown[8] that the linewidth, and thus T_2^{-1}, are proportional to τ_c. Using a simple argument,

---

[9] A. Abragam, "The Principles of Nuclear Magnetism." Oxford Univ. Press (Clarendon), London and New York, 1961.

6.2. CONVENTIONAL RELAXATION TIME METHODS

Slichter, in his book[8], shows for motional narrowing that τ_c is related to T_2 by

$$1/T_2 = (\delta\omega_d)^2 \tau_c, \qquad (6.2.1)$$

where $\delta\omega_d$ is the spread in frequency corresponding to the spread in dipolar local fields. We see from this formula that T_2 is proportional to τ_c^{-1} in the motionally narrowed region. Since τ_c is related to the activation energy E_A by

$$\tau_c = \tau_0 \exp(E_A/k_B T), \qquad (6.2.2)$$

where k_B is Boltzmann's constant, the activation energy can be determined from the slope of a plot of ln T_2 versus reciprocal temperature. By measuring T_2 and the temperature of the onset of motional narrowing (at which $\tau_c \simeq T_2$), τ_0 can be determined, at least within an order of magnitude. Figure 2 shows a typical plot of ln T_2 and ln τ_c versus $1/T$ (as well as other relaxation times to be defined later); the onset of motional narrowing can be easily seen by the abrupt change in slope occurring when $\tau_c \sim T_2$. It is clear that motional information can be obtained with this technique only at temperatures for which $\tau_c < T_2$ (~ 1–100 μs in solids).

It is possible in some cases to use motional narrowing to obtain dynamic information about the actual motional process, specifically when the atom of interest is part of a reorienting molecular group. Since the second

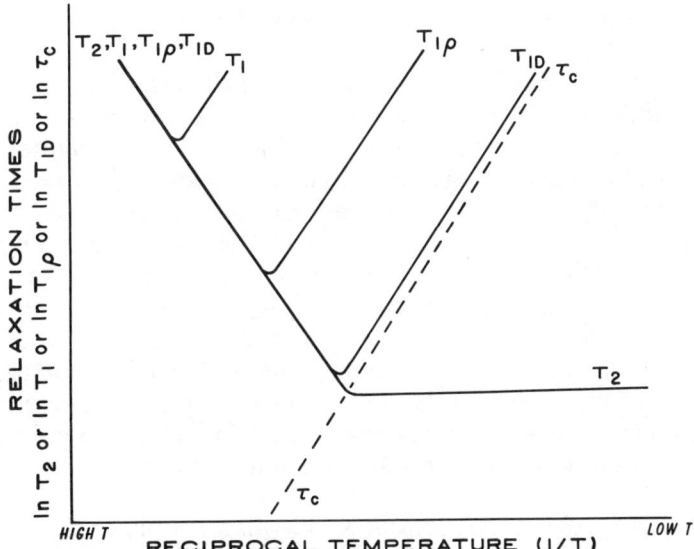

FIG. 2. Temperature plot of relaxation times (T_2, T_1, $T_{1\rho}$, and T_{1D}) and atomic jump time τ_c.

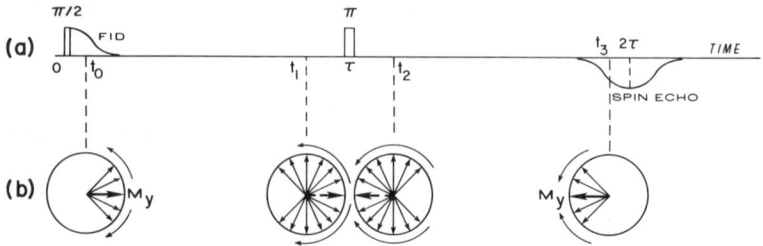

FIG. 3. (a) Pulse sequence and NMR signals for a spin echo sequence. The sequence consists of a $\pi/2$ pulse followed after a time τ by a π pulse, resulting in the formation of an echo a time 2τ after the $\pi/2$ pulse. (b) Isochromats relative to the magnetization M_y at various points in the sequence. The first picture, at t_0 during the FID, shows the magnetic moment components beginning to get out of phase. The second picture, at $t_1 = \tau^-$ just before the π pulse, shows the individual components, now largely out of phase, continuing to precess *away* from the direction of M_y (indicated by a dashed vector). The third picture, at $t_2 = \tau^+$ just after the π pulse, shows that the effect of the π pulse is to rotate the isochromat plane and to cause the magnetization components to precess *towards* the new direction of M_y. The fourth picture, at t_3, shows the refocusing effect of the π pulse and the formation of a spin echo.

moment of the resonance line can be measured and calculated[8] in both the rigid lattice and at temperatures for which a particular reorientational motion results in an averaging of some of the dipolar linewidth, a comparison of experimental and theoretical second moments can determine the actual motional process[10] as well as the structure in the motionally narrowed region.[11]

6.2.1.2. *Methods for Measuring T_2.* We shall now discuss techniques for measuring T_2. The obvious method is to measure the slope of the FID. However, in the presence of substantial motional narrowing, T_2 (and thus the linewidth) may be limited by inhomogeneity of the external magnetic field rather than internal dipolar fields. (In this case, the time constant of the FID is called T_2^* and is not characteristic of the motionally narrowed dipolar interaction.) What we want to measure is the true dipolar T_2 (i.e., the value in a perfectly homogeneous magnetic field), which can be accomplished without attempting to improve the magnet's homogeneity by the use of a simple but ingenious technique called the *spin-echo* technique.[12] With this technique a pulse sequence consisting of a $\pi/2$ pulse followed by a π pulse (Fig. 3) is used to reverse the dephasing due to the static inhomogeneous field but not the dephasing due to the time-varying dipolar interactions. Consider first the case of noninteracting spins in an inhomogeneous field. The $\pi/2$ pulse will rotate the magnetization from the z direction to the xy

---

[10] E. R. Andrew and R. G. Eades, *Proc. R. Soc. London, Ser. A* **218**, 537 (1953).
[11] H. T. Stokes, T. A. Case, D. C. Ailion, and C. H. Wang, *J. Chem. Phys.* **70**, 3572 (1979).
[12] E. L. Hahn, *Phy. Rev.* **80**, 580 (1950).

6.2. CONVENTIONAL RELAXATION TIME METHODS

plane, where it will then dephase. If we define the direction of H_1 to be the x direction in a rotating reference frame in which H_1 is fixed, then the magnetization immediately after the pulse is in the y direction in the rotating frame. In this rotating frame some of the magnetization components will be precessing faster and some slower than M_y. As can be seen in Fig. 3b, the application of a π pulse at a later time τ will reverse the sense of this dephasing in the rotating frame, thereby resulting in refocusing the magnetization to form an echo after an additional time interval τ. If the original dephasing is due to inhomogeneous fields which have not changed during the total time interval 2τ, then the total dephasing will be entirely reversed by the π pulse, and the echo will have the same amplitude as the original FID. If, on the other hand, some of the field variation responsible for the original dephasing has changed during the time 2τ (e.g., if the dephasing is due to time-varying local fields), then not all of the magnetization will be recovered, and the spin echo will have a reduced amplitude. By repeating the pulse sequence for different values of τ and then plotting the heights of the spin echoes as a function of τ, we can determine the true dipolar T_2, since the spin-echo amplitude[8] is proportional to $\exp(-2\tau/T_2)$. However, it is more efficient to measure spin echoes in a single-shot experiment using the Carr–Purcell sequence,[13] described below.

If there is rapid diffusion in an inhomogeneous magnetic field, the spin-echo amplitude is further reduced[8] by a factor dependent on the diffusion constant D and the static magnetic field gradient, $\partial H_0/\partial z$, so that

$$M(2\tau) = -M(0) \exp\left(-\frac{2\tau}{T_2}\right) \exp\left(-\frac{2\tau^3 D}{3}\left(\gamma \frac{\partial H_0}{\partial z}\right)^2\right). \quad (6.2.3)$$

For the case of, say, a liquid in an inhomogeneous field in which $\partial H_0/\partial z$ is not known, the decay of the spin-echo amplitude due to this second factor may prevent the measurement of T_2.

Fortunately, a technique called the Carr–Purcell sequence[13] exists for measurement of T_2 even in the presence of rapid diffusion in an inhomogeneous field. This sequence, shown in Fig. 4, consists of a normal $\pi/2 - \pi$ pulse sequence followed by an extra π pulse after each time interval 2τ. The result is a sequence of echoes, as shown. Even in the absence of the diffusion inhomogeneity effect, the Carr–Purcell sequence is highly efficient and allows the measurement of T_2 in a single shot, rather than after a number of repetitions. To produce n echoes using the repetition technique, n pulse sequences would be required; these would provide the same precision as could be obtained from one n-pulse Carr–Purcell sequence. If signal averaging is then used with n of these Carr–Purcell sequences, an improvement in

---

[13] H. Y. Carr and E. M. Purcell, *Phys. Rev.* **94**, 630 (1954).

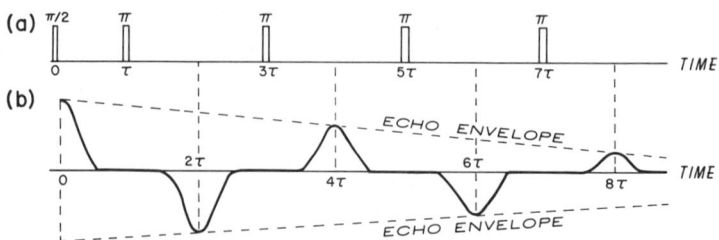

FIG. 4. (a) Carr–Purcell pulse sequence. (b) Spin echoes arising from the Carr–Purcell sequence.

signal-to-noise (S/N) of $\sqrt{n}$ would be obtained in the same time required for one T_2 determination using the repetition method.

To see how the diffusion-field-gradient factor can be made negligible, consider the expression for $M(2\tau)$ in Eq. (6.2.3) to be the initial magnetization for a second 2τ interval. One can then calculate $M(4\tau)$, which again can be used as the initial magnetization for the next interval. In this way, the magnetization for the nth echo can be obtained[8] and is given by

$$M(2n\tau) = (-1)^n M(0) \exp\left(\frac{-2n\tau}{T_2}\right) \exp\left(-\frac{(2n\tau)\tau^2 D}{3}\left(\gamma\frac{\partial H_0}{\partial z}\right)^2\right). \quad (6.2.4)$$

We are able to remove effects of diffusion in the inhomogeneous field by varying τ but holding $2n\tau$ constant so that only the second factor changes in the magnetization measurements. The accumulation of errors in the Carr–Purcell sequence arising from slight departures from perfect π pulses can be prevented[8] by the insertion of a 90° phase shift between the $\pi/2$ pulse and successive π pulses, as suggested by Meiboom and Gill.[14]

Expressions (6.2.3) and (6.2.4) can be used to determine the diffusion constant D directly, rather than to infer D from τ_c deduced from measured T_2 values. By applying a known external gradient in H_0 (larger than the field inhomogeneity), the second factor in the equation can dominate the decrease in echo amplitude in which case D is easily determined. The physical reason for the dependence of the echo on D is that, if atomic diffusion occurs along the direction of the field gradient, the atoms will dephase rapidly and T_2 will be shortened. The decrease in dephasing time is directly dependent on D and can be used to measure D directly. This method has been improved[15] to allow the measurement of smaller values of the diffusion coefficient by applying a pulsed (rather than static) field gradient that is turned off during the rf pulse. Thus the linewidth will be small at the time of the rf pulse, thereby allowing the use of an H_1 that is not particularly large

[14] S. Meiboom and D. Gill, *Rev. Sci. Instrum.* **29**, 6881 (1958).
[15] E. O. Stejskal and J. E. Tanner, *J. Chem. Phys.* **42**, 288 (1965).

6.2. CONVENTIONAL RELAXATION TIME METHODS

and detection systems that have moderate bandwidth. This technique has been applied to F⁻ self-diffusion[16] in the superionic conductor PbF_2 and to the study of divacancy diffusion[17] in metallic sodium.

The value of directly determining D can be appreciated by realizing that the correlation factor f, which is the ratio of the actual diffusion coefficient to the value it would have if jumps were random and uncorrelated,[18] can be used to determine the diffusion mechanism. Therefore, a determination of f by using a comparison of NMR with, for example, radioactive tracer results, would require the direct determination of D with some precision by NMR.

6.2.2. Spin–Lattice Relaxation (T_1)

6.2.2.1. Relaxation due to Fluctuating Dipolar Interactions.

A second well-established method for studying diffusion is to look for a minimum in a temperature plot of the spin–lattice relaxation time* T_1. Spin–lattice relaxation characterizes the process by which a spin system reaches thermal equilibrium with its surroundings and can be due to any process which can transfer energy between the spins and the lattice. In a solid, T_1 is usually much longer than T_2 and typically may range in value from a few milliseconds in metals like Cu to thousands of seconds in insulators of high purity. Furthermore, it is normally dependent on temperature and on frequency ω_0. It is easily measured since it characterizes the rate of buildup of the z component of the magnetization after the xy dephasing following the FID. One may think of spin–spin and spin–lattice relaxation in a solid as a two-step process—the first (T_2) describes the process of establishing a spin temperature (i.e., a Boltzmann distribution among the Zeeman levels), whereas the second (T_1) describes the process by which this temperature changes as it approaches the lattice temperature.

It is possible to understand in various ways the process by which atomic motions affect T_1. One way is to use the language of perturbation theory.[19] If the Zeeman Hamiltonian is the unperturbed Hamiltonian $\mathcal{H}_0$ and the dipolar is the perturbing Hamiltonian $\mathcal{H}'$, then atomic diffusion results in fluctuations in the dipolar Hamiltonian since it causes the distances between pairs of spins to change. We may then use the formalism of time-dependent

---

[16] R. E. Gordon and J. H. Strange, *J. Phys. C* **11**, 3213 (1978).
[17] G. Brünger, O. Kanert, and D. Wolf, *Solid State Commun.* **33**, 569 (1980).
[18] N. Peterson, *in* "Diffusion in Solids: Recent Developments" (A. S. Nowick and J. J. Burton, eds.), pp. 115–170. Academic Press, New York, 1975.
[19] A. Messiah, "Quantum Mechanics," Vol. II, Chap. XVI and XVII. Wiley, New York, 1966.

---

* T_1 is defined and discussed in detail in Part 5, in Slichter,[8] in Abragam,[9] and in all standard books on magnetic resonance.

perturbation theory to calculate the transition probability for flipping a spin (T_1^{-1}) due to the time-dependent dipolar Hamiltonian. Since frequency components in the perturbation which equal the energy level spacing divided by $\hbar$ contribute the most to the transition probability T_1^{-1}, the condition for the T_1 minimum is that $1/\tau_c$ equals ω_0 or $\omega_0 \tau_0 \approx 1$. Since only τ_c depends upon temperature, this condition describes the temperature of the minimum for fixed ω_0. (One should note that a plot of T_1 versus ω_0 for fixed temperature will likewise show a minimum when $\omega_0 \tau_c \approx 1$.)

Bloembergen, Purcell, and Pound (BPP) showed[20] that a fluctuating dipolar field, resulting from atomic diffusion or molecular reorientations, would induce transitions among Zeeman levels and would, accordingly, relax the spin system. Assuming an exponential correlation function, they showed that T_1 due to motions is given by an expression* of the form

$$\frac{1}{T_1} \sim \frac{\omega_d^2 \tau_c}{1 + \omega_0^2 \tau_c^2}, \qquad (6.2.5)$$

where ω_d is a frequency corresponding to a typical dipolar splitting. We note that this formula indicates that, in a temperature plot of T_1, there will be a minimum value for T_1 at the temperature for which $\tau_c = \omega_0^{-1}$, in agreement with our earlier reasoning. Furthermore, at temperatures above the minimum (where $\omega_0 \tau_c \ll 1$) T_1 is proportional to τ_c^{-1}, whereas, below the minimum ($\omega_0 \tau_c \gg 1$), T_1 is proportional to τ_c. Since $\ln \tau_c$ is proportional to the activation energy E_A, according to Eq. (6.2.2), the slope of $\ln T_1$ versus reciprocal temperature will be proportional to $-E_A$ and to E_A at temperatures above and below the minimum, respectively. Furthermore, a fit to this curve (or simply a measurement of the temperature of the minimum) will determine the preexponential factor τ_0. Finally, this equation predicts that on the high-temperature side of the minimum ($\omega_0 \tau_c \ll 1$) T_1 is independent of frequency, whereas on the low-temperature side ($\omega_0 \tau_c \gg 1$) T_1 is proportional to ω_0^2. Thus a measurement of the frequency dependence of T_1 can determine the applicability of Eq. (6.2.5) or the validity of assuming an exponential correlation function. (In Section 6.5 we present an example[21] in which the BPP theory does not apply.)

We note that the minimum occurs when $\tau_c \sim 1/\omega_0 \sim 10^{-8}$ s and motional information can be obtained on either side of the minimum, provided

---

[20] N. Bloembergen, E. M. Purcell, and R. V. Pound, *Phys. Rev.* **73**, 679 (1948).
[21] M. Vilfan, R. Blinc, and J. W. Doane, *Solid State Commun.* **11**, 1073 (1972).

---

* Actually, there are additional terms of similar form. For interactions involving identical spins there is an extra term proportional to $1/(1 + 4\omega_0^2 \tau_0^2)$; for heteronuclear systems, there are also terms having $[1 + (\omega_I - \omega_S)^2 \tau_c^2]$ and $[1 + (\omega_I + \omega_S)^2 \tau_c^2]$ in the denominator. (See Chapter VIII of Abragam.[9])

6.2. CONVENTIONAL RELAXATION TIME METHODS

the transition probability (T_1^{-1}) for flipping a spin is greater than that due to other mechanisms (i.e., T_1 due to diffusion is less than T_1 due to other mechanisms). For typical NMR fields the temperature range over which T_1 can detect motions is comparable to that of motional narrowing. A typical T_1 minimum is shown in Fig. 2.

6.2.2.2. Relaxation due to Fluctuating Nondipolar Interactions.

6.2.2.2.1. QUADRUPOLAR RELAXATION. Many nuclei (of spin greater than $\frac{1}{2}$) possess an electric quadrupole moment as well as a magnetic moment. As a result, if these nuclei experience an electric field gradient, their Zeeman energy levels are shifted and are no longer equally spaced. For this reason extra resonance lines (satellites) appear at frequencies slightly different than the Larmor frequencies. Since the position of these lines depends upon electric field gradients, which in turn are very sensitive to the crystal structure in the vicinity of the quadrupolar nuclei,[22] measurements of line positions in quadrupolar systems have been widely used in structure determinations. In a powder sample an effect of these extra lines is to broaden substantially the apparent NMR linewidth, which will decrease the S/N ratio as well as the FID decay time in a pulse experiment.

Atomic motions can give rise to fluctuating field gradients, which, in turn, contribute to spin–lattice relaxation. For the case where the fluctuating quadrupolar interactions can be treated as perturbations inducing transitions among Zeeman levels, the quadrupolar spin–relaxation time is given by expressions like the BPP relation, Eq. (6.2.5), except that dipolar frequency ω_d is replaced in the numerator by a quadrupole splitting frequency ω_q. As an example, Fig. 5 shows[23] T_1 of ^{23}Na in NaCN. This relaxation arises from the fluctuating field gradient at a Na site due to head-to-toe reorientations of neighboring CN$^-$ ions. The value of τ_c can be determined at the minimum (by setting $\tau_c \approx \omega_0^{-1}$) and compared to values obtained with other techniques (e.g., dielectric relaxation[24]).

Another feature of quadrupolar relaxation is that the motion of *charged* atoms that have very small magnetic moments can be studied because of their contribution to the quadrupolar relaxation of nearby nuclei. An interesting example[25] is an observation of the localized motion of off-center Ag$^+$ defects in RbCl. This experiment, which was performed by measuring quadrupolar relaxation of ^{85}Rb, ^{87}Rb, and ^{35}Cl in samples containing only 350 ppm Ag$^+$, would have been impossible to perform by direct observation of the very weakly magnetic nucleus Ag$^+$.

---

[22] T. P. Das and E. L. Hahn, *Solid State Phys.*, Suppl. **1** (1958).
[23] H. T. Stokes, T. A. Case, and D. C. Ailion, to be published.
[24] M. Julian and F. Lüty, *Ferroelectrics* **16**, 201 (1977).
[25] O. Kanert, R. Küchler, and M. Mali, *J. Phys.* (*Orsay, Fr.*) **41**, C6-404 (1980).

450   6. NMR RELAXATION TIME METHODS FOR ATOMIC MOTIONS

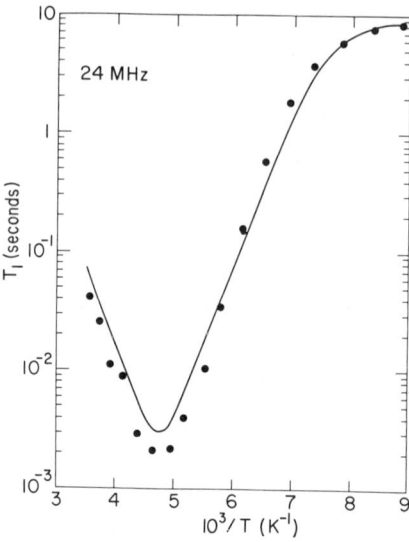

Fig. 5. T_1 versus $10^3/T$ for ^{23}Na in NaCN. These data will appear in Stokes et al.[23]

6.2.2.2.2. CHEMICAL-SHIFT ANISOTROPY (CSA) RELAXATION. Another interaction, which is of considerable importance in chemistry, is the so-called chemical shift. This interaction is derived from the fine-structure interaction between the nuclear magnetic moment and the magnetic field due to the atomic electron's orbital angular momentum. In a molecule or a solid in which electric charges are nearby, the electron orbit precesses so that the average angular momentum is "quenched," and the average interaction is zero. If, however, a magnetic field H_0 is applied, as in an NMR experiment, then the electron's angular momentum is unquenched to some extent, and there will be a small additional field at the site of the nucleus. This chemical-shift field results in a small shift in the Larmor frequency, which depends on the orbital wave function, which in turn is strongly dependent on the electron's immediate environment. We thus can obtain for a molecule a spectrum of lines; the study of such spectra in liquids has been widely used to determine the structure of molecules. In solids, one does not easily observe these spectra since the chemical shifts (typically 1–100 mG) are usually much smaller than the dipolar linewidths. However, by applying a series of rf pulses in different directions, it is possible to simulate motional narrowing and to reduce dipolar linewidths in solids artificially so that very small chemical shifts can be measured.[26]

Since the chemical shift is typically anisotropic, it can contribute to T_1 for a molecule reorienting between orientations having different chemical shifts

[26] J. S. Waugh, L. M. Huber, and U. Haeberlen, *Phys. Rev.* **20**, 180 (1968).

6.2. CONVENTIONAL RELAXATION TIME METHODS

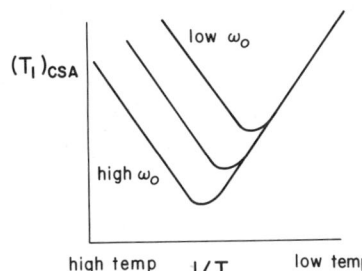

FIG. 6. T_1 versus $10^3/T$ for CSA at different frequencies.

or for an atom diffusing between chemically shifted sites. Since the chemical shift is so small, its relaxation contribution is normally not observed in solids and is usually masked by the larger dipolar relaxation. Furthermore, the chemical shift is often invariant for a 180° rotation;[23] hence such a jump will not then contribute to CSA relaxation even if the chemical shift is larger than the dipolar interaction. For these reasons, observations of CSA relaxation, though difficult to observe, can provide detailed information about the relative symmetries of the initial and final orientations or sites.

For CSA relaxation, the ω_d^2 in the numerator of Eq. (6.2.5) will be replaced by a chemical-shift coupling constant. Since the chemical shift is proportional to ω_0, T_1 due to CSA will have a frequency dependence that is very different from that shown in Fig. 2. For CSA, T_1 is then described by terms* like

$$\frac{1}{T_1} \sim \frac{\omega_0^2 \tau_c}{1 + \omega_0^2 \tau_c^2}. \qquad (6.2.6)$$

The frequency dependence of T_1 due to CSA is given in Fig. 6. We note that the relaxation rate is enhanced by going to higher fields, in contrast to the case of dipolar relaxation. The experimental verification of these features in KCN is described in Section 6.5.

It should be emphasized that very weak dipolar and quadrupolar couplings and a strong chemical shift would be required for CSA relaxation to dominate T_1.

6.2.2.3. Methods for Measuring T_1. As discussed in the book by Fukushima and Roeder,[27] all T_1 methods use basically two pulses or groups of pulses. The first is to disturb the magnetization from equilibrium, and the second is to monitor its return to equilibrium.

---

[27] E. Fukushima and S. B. W. Roeder, "Experimental Pulse NMR: A Nuts and Bolts Approach." Addison-Wesley, Reading, Massachusetts, 1981.

* See p. 316 of Abragam.[9]

The most obvious method for measuring T_1 is simply to apply two $\pi/2$ pulses and to monitor the magnetization following the second as a function of the time τ between them. The first pulse tips the magnetization out of the original equilibrium direction (z direction), and the second monitors the recovery of the z component of the magnetization, which is described by

$$M_z(\tau) = M_0[(1 - \exp(\tau/T_1)] \tag{6.2.7}$$

and shown in Fig. 7a. This method is sometimes called the saturation–recovery technique.

A second method is to apply a $\pi-\pi/2$ pulse sequence (rather than the $\pi/2-\pi/2$ sequence described above). This method is called the inversion–recovery technique since the preparation pulse (the π pulse) actually reverses the magnetization. Since the magnetization measured by the second pulse (Fig. 7b) obeys the equation

$$M_z(\tau) = M_0 - 2M_0 \exp(-\tau/T_1), \tag{6.2.8}$$

this technique doubles the effective S/N ratio (since M_z changes by $2M_0$ as contrasted to a change of only M_0 by the saturation–recovery method).

However, the apparent S/N advantage of the $\pi-\pi/2$ sequence is offset in solids, where $T_1 > T_2$, by the fact that one must wait a time longer than T_1 for the magnetization to be fully recovered before repeating the $\pi-\pi/2$ sequence. No such requirement exists for the $\pi/2-\pi/2$ sequence since the

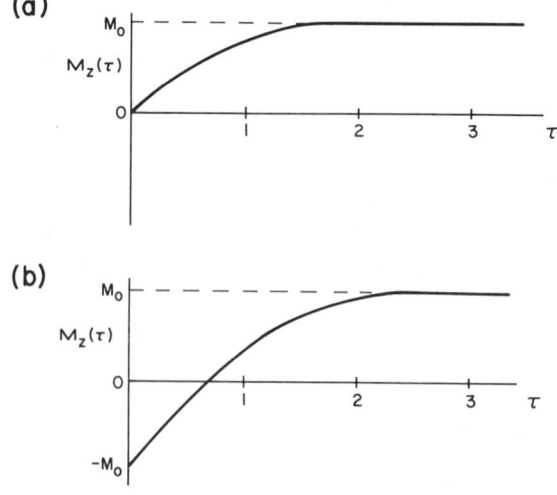

FIG. 7. (a) Magnetization following a saturation recovery ($\pi/2 - \pi/2$) sequence, according to Eq. (6.2.7). (b) Magnetization following an inversion recovery ($\pi - \pi/2$) sequence, according to Eq. (6.2.8).

preparation pulse fully destroys the z magnetization, even if it has not fully recovered. Thus signal averaging can be performed far more efficiently with the $\pi/2 - \pi/2$ method provided the power amplifier can be operated at the resulting higher duty cycle. A second advantage of the saturation–recovery technique is that the bandwidth of the excitation pulse (a $\pi/2$ pulse) is twice as large as that of the π pulse in the inversion–recovery scheme since the π pulse is twice as long; accordingly, the saturation–recovery scheme may be more effective for broad-line investigations. Even greater bandwidth can be obtained by using excitation pulses whose widths are less than 90°.

A variation of the saturation–recovery scheme is to replace the preparation $\pi/2$ pulse by a closely spaced group (or "saturating comb") of $\pi/2$ pulses, which is followed a time τ later by a $\pi/2$ observation pulse. (For experimental convenience this second pulse could be the first pulse in the next saturating comb.) The saturating comb guarantees the zeroing of the magnetization even if the pulses are not exactly $\pi/2$. This technique is particularly useful if the H_1 field is not homogeneous. Also, the saturating comb is quite useful in initializing the populations of the central energy levels ($m_S = \pm\frac{1}{2}$) of a quadrupolar system.[28]

6.3. Ultraslow Motion Techniques

In all motional studies it is desirable to extend the frequency range of motions that can be studied. The ability to study motions over a wide frequency range, corresponding to a wide temperature range, allows the possibility of studying separately different types of motions that may not even be observable if only a narrow range is available. Also, it is often important to study motions associated with a physical phenomenon, such as a phase transition, that occurs in a particular temperature region. Many techniques exist for studying rapid diffusion that results in appreciable mass flow (e.g., radioactive tracers and ionic diffusion); however, relatively few techniques are sensitive to motions that are infrequent (or "ultraslow" as they are sometimes misnamed). Conventional NMR, as we have seen, is limited either to motions sufficiently frequent to narrow the line (typically, $\tau_c < 10^{-5}$ s) or to motions that can result in an observable minimum in T_1 (occurring at $\tau_c \sim 10^{-8}$ s typically). Motions having frequencies several orders of magnitude higher or lower than that for the minimum can usually be observed.

The temperature range over which diffusion can be studied by NMR can be extended to appreciably lower temperatures by the ultraslow motion

---

[28] A. Avogadro and A. Rigamonti, in "Magnetic Resonance and Related Phenomena" (V. Hovi, ed.), pp. 255–259. North-Holland Publ., Amsterdam, 1973.

techniques* that allow the observation of less frequent atomic motions than can be observed by T_1 measurements. The basic idea can be understood by considering what happens to a T_1 minimum if the field H_0 and thus ω_0 are greatly reduced. We see from Eq. (6.2.5) that, if we reduce ω_0, the minimum will correspond to a longer value of τ_c and thus to lower frequency motions. For a typical diffusion mechanism, atomic jumps become less frequent as the temperature is lowered. Thus the T_1 minimum in a low field will occur at a lower temperature than would the T_1 minimum at a higher field.

There are experimental problems associated with attempting to study relaxation in very weak fields where the NMR signal might be very small. Also there are theoretical difficulties, since the theory underlying Eq. (6.2.5) is a perturbation theory that treats the fluctuating dipolar Hamiltonian due to diffusion as a perturbation on the Zeeman Hamiltonian; such an assumption may not be valid in very weak applied fields. We shall now discuss solutions to both these problems.

6.3.1. Spin–Lattice Relaxation in the Rotating Frame ($T_{1\rho}$)

The experimental problem can be handled easily. An attractive approach is to measure $T_{1\rho}$, the spin–lattice relaxation time in the rotating frame of the rf field H_1. According to Redfield,[30] a spin system in an rf field strong enough to saturate the resonance line (i.e., $\gamma^2 H_1^2 T_1 T_2 \gg 1$) will form a canonical distribution among energy levels separated by γH_{eff} rather than γH_0, where the effective field $\mathbf{H}_{\text{eff}}$ in the rotating frame is given by

$$\mathbf{H}_{\text{eff}} = H_1 \mathbf{i} + (H_0 - \omega/\gamma)\mathbf{k} \qquad (6.3.1)$$

In this case, the spins are characterized by a "spin temperature in the rotating frame," which means that Curie's law will hold, but with $\mathbf{M}_{\text{eq}} = C\mathbf{H}_{\text{eff}}/T$. The significance is that, at exact resonance ($H_0 = \omega/\gamma$), the equilibrium magnetization $\mathbf{M}_{\text{eq}}$ will be parallel to $\mathbf{H}_1$ rather than to $\mathbf{H}_0$. The $T_{1\rho}$ minimum will then occur when $\gamma H_1 \tau_c \cong 1$, for $\mathbf{H}_0$ at exact resonance. Since H_1 is typically of order 1–50 G, this minimum arises from lower frequency motions than does the normal T_1 minimum. If H_1 is much larger than the dipolar local field, the dipolar fluctuations can still be treated by perturbation theory (this approach is called the "weak-collision" theory[31]). If, on the other hand, H_1 is comparable to, or less than, the dipolar local field,

---

[29] D. C. Ailion, *Adv. Magn. Reson.* 177 (1971).
[30] A. G. Redfield, *Phys. Rev.* **98**, 1787 (1955).
[31] D. C. Look and I. J. Lowe, *J. Chem. Phys.* **44**, 2995 (1966).

* A general review of ultraslow motion techniques can be found in Ailion.[29]

6.3. ULTRASLOW MOTION TECHNIQUES

perturbation theory fails, and a different theory, called the "strong-collision" theory [32], must be used.

6.3.1.1. Weak-Collision Theory. All of the weak-collision calculations are applications of the BPP theory in which the fluctuating dipolar Hamiltonian is treated as a perturbation on the Zeeman system. The relaxation time is inversely related to W_{nm}, the probability per unit time of a transition between spin states $|m\rangle$ and $|n\rangle$. Since each W_{nm} is proportional to j_{nm}, the spectral density of the correlation function arising from atomic motions, these calculations reduce to assuming a form for the correlation function and then calculating the corresponding spectral density.* In calculating the rotating-frame spin–lattice relaxation time $T_{1\rho}$, the basic Zeeman Hamiltonian for a single nuclear species I in an effective field $\mathbf{H}_{\text{eff}}$ (given by Eq. (6.3.1)) is given by

$$\mathcal{H}_Z^\rho = -\gamma_I H_{\text{eff}} I_z^\rho, \qquad (6.3.2)$$

where I_z^ρ is the component along $\mathbf{H}_{\text{eff}}$ of the total spin and is given by a sum over all the atoms

$$I_z^\rho = \sum_j I_{zj}^\rho. \qquad (6.3.3)$$

For exact resonance $\mathbf{H}_{\text{eff}}$ is in the x direction in the rotating frame, in which case Eq. (6.3.2) becomes

$$\mathcal{H}_Z^\rho = -\gamma_I \hbar H_1 I_x. \qquad (6.3.4)$$

Look and Lowe[31] assumed, as did BPP, that the correlation function is an exponentially decreasing function of the jump time τ_c, which is equivalent to assuming that the atomic jumping corresponds to the dipolar energy fluctuating randomly between two values. They then showed that $T_{1\rho}$ would exhibit a low frequency minimum, which, for exact resonance, obeys a formula of the form†

$$\frac{1}{T_{1\rho}} = K \frac{\tau_c}{1 + 4\omega_1^2 \tau_c^2}, \qquad (6.3.5)$$

where $\omega_1(=\gamma H_1)$ is proportional to the energy-level splitting in the rotating frame. Thus the $T_{1\rho}$ minimum occurs when $\tau_c \sim (2\omega_1)^{-1}$, which may be

---

[32] C. P. Slichter and D. C. Ailion, *Phys. Rev. A* **135**, 1099 (1964).

* A detailed discussion of the use of correlation functions and spectral densities in calculating relaxation times in given in Chapter 5 of Slichter.[8]

† There is an important typographical error in Look and Lowe[31] and in Ailion[29] in that the 4 was omitted from the denominator of the expressions for $T_{1\rho}$ in both references.

several orders of magnitude longer than the value of τ_c at a typical T_1 minimum. Accordingly, a $T_{1\rho}$ minimum is deeper and corresponds to slower atomic motions occurring at lower temperatures than is a T_1 minimum, as shown in Fig. 2.

6.3.1.2. Strong-Collision Theory. In order to get even more effective relaxation so as to observe even lower frequency motions, one can reduce H_1 (e.g., by adiabatic demagnetization in the rotating frame[33] (ADRF)) to a value comparable to the dipolar local field, or even to zero. In this case, perturbation theories like BPP and the weak-collision theory will not be valid since the perturbation (the dipolar Hamiltonian) is comparable to or larger than the unperturbed Hamiltonian (the Zeeman Hamiltonian). Nevertheless, we can still understand the nature of the relaxation behavior in zero field. The underlying physical requirement for all relaxation time minima is that the most effective relaxation (and thus the minimum) will occur when the principal frequency components of the fluctuating Hamiltonian are comparable to the energy-level splitting (in units of $\hbar$). We have seen, for Zeeman relaxation in the laboratory or rotating frames, that this requirement is equivalent to requiring that $1/\tau_c \sim \omega_0$ or $1/\tau_c \sim \omega_1$, respectively. For the case of relaxation due to a fluctuating dipolar Hamiltonian in zero field, the energy-level splittings $\hbar\omega_d$ are determined by dipolar interactions. Thus the minimum in zero field occurs when $1/\tau_c \sim \omega_d$ or $\tau_c \sim T_2$, which is at the onset of motional narrowing. The zero-field relaxation time, commonly called the *dipolar* relaxation time[34] T_{1D}, is also shown in Fig. 2.

The strong-collision theory, originally developed by Slichter and Ailion[32] for low-field relaxation, is *not* a perturbation theory, but is based on the assumption that sufficient time elapses between diffusion jumps to enable the dipolar and rotating frame Zeeman interactions to cross-relax to a common-spin temperature between each jump. This temperature is then disturbed by the sudden change in dipolar energy resulting from a jump, but reequilibrates to a new value prior to the next jump. This spin-temperature assumption is equivalent to requiring that $\tau_c \gg T_2$, which means that the strong-collision theory will be valid in the "rigid-lattice" below the $T_{1\rho}$ minimum and the temperature for motional narrowing. The ADRF process transfers long-range Zeeman order due to spins aligned preferentially along H_{eff} to short-range dipolar order arising from spins aligned along their individual local fields. A single diffusion jump per spin will then have a major effect on the relaxation so that, in zero H_1, we would expect T_{1D} to be of order τ_c. In the weak-collision regime, much less of the order is dipolar

---

[33] C. P. Slichter and W. C. Holton, *Phys. Rev.* **122**, 1701 (1961).
[34] J. Jeener and P. Broekaert, *Phys. Rev.* **157**, 232 (1967).

6.3. ULTRASLOW MOTION TECHNIQUES

and much more is Zeeman; accordingly, it will take many jumps to relax the magnetization and $T_{1\rho}$ will be much larger than τ_c. Both these effects are shown in Fig. 2.

The actual strong-collision result for $T_{1\rho}$ at exact resonance due to dipolar fluctuations of a homonuclear (i.e., having only one nuclear species) system in a field $\mathbf{H}_1$ is[32]

$$\frac{1}{T_{1\rho}} = \frac{2(1-p)}{\tau_c} \frac{H_D^2}{H_1^2 + H_D^2} = \frac{1}{T_{1D}} \frac{H_D^2}{H_1^2 + H_D^2}, \qquad (6.3.6)$$

where H_D is the dipolar local field and $1-p$ is a calculatable geometrical factor of order unity that characterizes the fractional change in energy resulting from a diffusion jump. Because of this factor, T_{1D} shows a 10–20% dependence on the orientation of the field $\mathbf{H}_0$ relative to the crystal axes, and, furthermore, this anisotropy depends on diffusion mechanism.[35] We note that, when H_1 equals zero, $T_{1\rho}$ is of order τ_c, as we have seen intuitively. Furthermore, for $H_1 = 0$, $T_{1\rho}$ is identical to the dipolar relaxation time T_{1D} characterizing energy exchanges between the dipolar system and the lattice. Diffusion can be studied by this technique provided the relaxation rate due to diffusion is greater than the relaxation rate due to other T_1 processes. The condition for observability of diffusion effects is then that $\tau_c < T_1$. Since T_1 values are typically 10^{-3}–10^3 s, this condition is much less stringent than the condition for motional narrowing ($\tau_c < T_2$).

We should note that the strong-collision formula, Eq. (6.3.6), can be used to obtain τ_c from the measured relaxation time, assuming $\tau_c \gg T_2$. BPP-type theories can be used to determine τ_c in the motionally narrowed region where $\tau_c \ll T_2$. In the vicinity of the T_{1D} minimum neither theory is valid. Nevertheless, Wolf[36] has generated a common formalism for treating both the strong- and weak-collision regimes.

6.3.1.3. Experimental Techniques for Measuring $T_{1\rho}$. Essentially all methods for measuring $T_{1\rho}$ start by tipping the magnetization from its original direction (the z direction) to a direction parallel to $\mathbf{H}_{\text{eff}}$ in the rotating frame. For exact resonance the magnetization would then be aligned parallel to $\mathbf{H}_1$ (in contrast to the situation immediately after a $\pi/2$ pulse when the magnetization would be perpendicular to $\mathbf{H}_1$). $T_{1\rho}$ can then be determined by measuring the magnetization as a function of time τ in the aligned state since $T_{1\rho}$ is given by

$$M_x = M_0 \exp(-\tau/T_{1\rho}), \qquad (6.3.7)$$

---

[35] D. C. Ailion and P. Ho, *Phys. Rev.* **168**, 662 (1968).
[36] D. Wolf, "Spin Temperature and Nuclear Spin Relaxation in Matter." Oxford Univ. Press (Clarendon), London and New York, 1979.

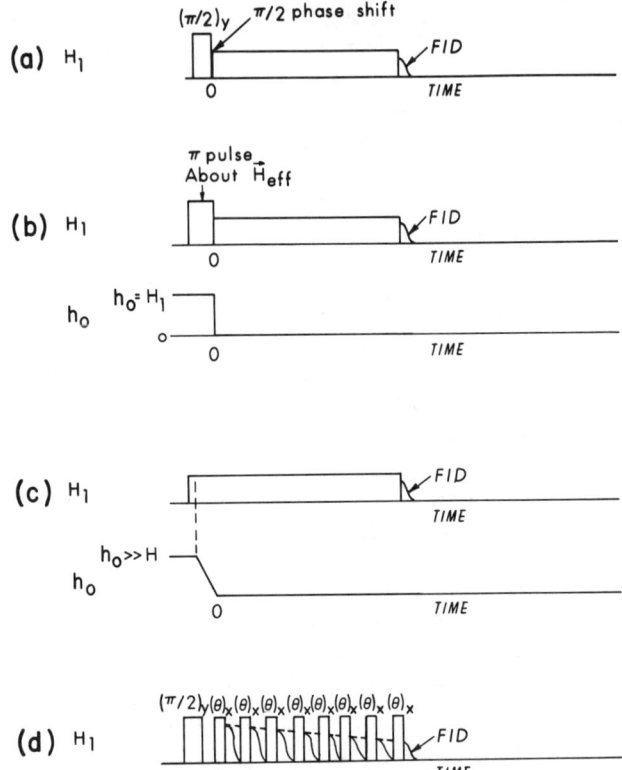

FIG. 8. Pulse sequences for measuring $T_{1\rho}$; (a), (b), and (c) all have a spin-locking sequence followed by a long rf pulse. The spin locking in (a) consists of a $\pi/2$ pulse followed by a $\pi/2$ phase shift; in (b), spin locking is achieved by a π pulse about H_{eff} (at an angle of 45° with respect to the z direction); in (c), H_1 is turned on when the z field (or frequency) is far off resonance, and spin locking is achieved as a result of an adiabatic return to resonance. In (d), spin locking is achieved, as in (a), by a $\pi/2$ pulse followed by a $\pi/2$ phase shift; $T_{1\rho}$ is measured by monitoring the FID following short pulses of duration θ.

where M_x is the height of the FID following the turn-off of the variable length H_1 pulse.

Probably, the most straightforward method for achieving the spin-locked (i.e., magnetization parallel to H_1) configuration is to apply a $\pi/2$ pulse immediately followed by a $\pi/2$ phase shift, which is then followed by an H_1 pulse of arbitrary height and variable duration.[37] (See Fig. 8a.)

Alternatively, spin locking can be achieved by applying an rf field H_1 and simultaneously pulsing H_0 off resonance[38] by a field h_0 exactly equal to H_1,

[37] S. R. Hartmann and E. L. Hahn, *Phys. Rev.* **128**, 2042 (1962).
[38] G. P. Jones, D. C. Douglass, and D. W. McCall, *Rev. Sci. Instrum.* **36**, 1460 (1965).

6.3. ULTRASLOW MOTION TECHNIQUES

so that H_{eff} makes an angle 45° with respect to H_1. After the magnetization has precessed through precisely 180° about this H_{eff}, it will be aligned along H_1; accordingly, sharply switching off H_0 at this time will result in spin locking. This sequence is shown in Fig. 8b.

Both of the above techniques require precise jitter-free timing of the pulse widths, amplitudes, and phase shifts. A technique which is much less dependent on precision in the rf pulse timing is the ADRF technique first developed by Slichter and Holton[33] and used by Ailion and Slichter[39] in their original ultraslow motion experiments. In this approach, shown in Fig. 8c, the field (or frequency) is turned far off resonance by an amount h_0 and then an rf field H_1 is turned on. The magnetization then precesses about H_{eff} and the transverse component disappears in a time of order T_2. By requiring that $h_0 \gg H_1$, H_{eff} is essentially along the z direction so that very little magnetization is lost as a result of the turnon of H_1. However, the magnetization is now locked along the direction of H_{eff}. The static field is then allowed to return to resonance adiabatically, and H_{eff} changes from essentially the z direction to the x direction in the rotating frame. To be adiabatic the return to resonance must be sufficiently slow that the magnetization remains parallel to the changing H_{eff} throughout the sweep. Near resonance, this condition is equivalent to requiring that the inequality

$$\frac{dH_0}{dt} \ll \gamma H_1^2 \tag{6.3.8}$$

be satisfied (for $H_1 \gg H_D$). If the field sweep is not adiabatic and order is lost, then the resultant magnetization and signal-to-noise ratio will be reduced.

A difficulty arises in trying to measure $T_{1\rho}$ in small rf fields by the previous techniques since it is very difficult to satisfy the adiabatic condition, Eq. (6.3.8), for small H_1. In this case, a variant[29] of the above techniques can be used. Instead of the H_1 being held constant during the experiment, the spin-lock sequence is followed by an adiabatic demagnetization of H_1 from a value large compared to the local field to the desired value for the $T_{1\rho}$ measurement. After waiting a variable length of time, the loss in dipolar order resulting form the relaxation can be monitored by adiabatically remagnetizing H_1 back to its original value and then sharply turning off H_1. $T_{1\rho}$ is measured by monitoring the magnitude of the FID as a function of time in the demagnetized state. A disadvantage of this technique is that the extra time required for the additional demagnetization and remagnetization may make difficult the measurement of very short $T_{1\rho}$. However, the use of this technique successfully circumvents the problems of loss of signal occurring when measuring $T_{1\rho}$ in small H_1 fields.

---

[39] D. C. Ailion and C. P. Slichter, *Phys. Rev. A* **137**, 235 (1965).

All the above techniques for measuring $T_{1\rho}$ require the production of long, powerful H_1 pulses, which, in turn, require power amplifiers capable of long duty cycle operation. To prevent deterioration of the pulse shape, it is important that the power amplifier's power supply be capable of delivering a steady high current without appreciable decrease in voltage. Another difficulty arising from the use of long, high-power rf pulses is that the power dissipated may cause appreciable heating of the sample. It may be possible to reduce such unwanted heating by decreasing the duty cycle (and thus the efficiency of data collection) by lengthening the dead time between pulse sequence repetitions. However, there may be sufficient heating during a single pulse that the sample's instantaneous temperature at the time of the FID is considerably different from the average temperature as measured by a thermocouple. In this case it may be necessary to work with a lower H_1.

The last three techniques require that the spin-locking sequence be followed by a single, long rf pulse, after which the FID amplitude is measured. $T_{1\rho}$ is then determined by repeating this sequence for different rf pulse lengths τ, according to Eq. (6.3.7). Since a time much larger than T_1 should elapse between a repetition of the pulse sequence, the process of measuring $T_{1\rho}$ can be very tedious for samples having a long T_1, even without signal averaging. To avoid this inconvenience, Rhim et al.[40] have developed a method for measuring $T_{1\rho}$ in a single shot (Fig. 8d). The essential feature of their technique is the recognition that if the long pulse is replaced by a string of short rf pulses (of angle $\theta < \pi/2$) and the *average rf field* is large compared to the local field, then the resulting time constant for the decrease of the magnetization following the individual θ pulses will approach $T_{1\rho}$ in the limit of small θ. Since the decreasing FIDs can be sampled following the individual θ pulses, this technique can measure $T_{1\rho}$ in a single shot. One disadvantage is that the decay time is a complicated function[41] of θ and equals $T_{1\rho}$ only for small θ.

6.3.2. Dipolar Relaxation (T_{1D})

As we have just seen, low-frequency atomic diffusion or molecular reorientations can be studied by measuring $T_{1\rho}$, the spin–lattice relaxation time in the rotating frame of the rf field. Even lower frequency motions can be studied by measuring the dipolar relaxation time T_{1D}, which corresponds to the value of $T_{1\rho}$ in the limit $H_1 = 0$. According to the strong-collision formula [Eq. (6.3.6)], T_{1D} is related to the atomic jump time τ_c by

$$\frac{1}{T_{1D}} = \frac{2(1-p)}{\tau_c}. \tag{6.3.9}$$

[40] W. K. Rhim, D. P. Burum, and D. D. Elleman, *Phys. Rev. Lett.* **37**, 1764 (1976).
[41] W. K. Rhim, D. P. Burum, and D. D. Elleman, *J. Chem. Phys.* **68**, 692 (1978).

6.3. ULTRASLOW MOTION TECHNIQUES

As we have stated earlier, the T_{1D} minimum will occur at the onset of motional narrowing. At the minimum the value of T_{1D} will be of order T_2 (see Fig. 2).

6.3.2.1. Strong-Collision Theory for Heteronuclear Systems.

A problem arises in the interpretation of T_{1D} measurements in heteronuclear systems in which one nuclear species is diffusing and the others are stationary. For simplicity, consider a system containing two species of comparable abundance and gyromagnetic ratio, whose spin quantum numbers are labeled I and S. Assume further that the I and S spins are in strong contact. (A good example would be the alkali halide LiF, in which every lithium atom is surrounded by fluorines and vice versa.) The problem is that, if we irradiate the I spins and measure T_{1D}, we will obtain the same result as we would if we had measured T_{1D} by irradiating the S spins, independent of which spin is actually diffusing.[42] Hence, from a T_{1D} measurement alone, we will be unable to determine which species is diffusing. This feature arises because there is normally strong coupling between parts of the dipolar interaction involving different species (i.e., the $I-I$ and $S-S$ dipolar terms are strongly coupled through the $I-S$ interaction). So, if an S spin undergoes a jump, there will be rapid cross-relaxation between $I-I$ and $S-S$ terms, which will result in the local heating being transferred rapidly to the entire spin system. Thus in the strong-collision limit, the entire dipolar reservoir is describable by a single temperature prior to a diffusion jump, but either spin species alone is not. (We should realize that there is no similar difficulty in identifying the diffusing species in a T_1 measurement since the I and S Zeeman reservoirs will normally be uncoupled or weakly coupled because their resonance frequencies are probably quite different.)

However, there is a way to determine in a T_{1D} experiment which spin species is diffusing for the case where the S spins are weakly magnetic (small γ_S) compared to the more strongly magnetic I spins (large γ_I). If the relaxation is due to the motion of S spins, T_{1D} will have an enormous anisotropy, whereas for diffusion of I spins the anisotropy will be very small [arising only from the $1 - p$ factor of Eq. (6.3.9)]. To see this, consider the generalization of the strong-collision result [Eq. (6.3.9)] for a heteronuclear system.

In the strong-collision theory, the relaxation time is calculated by considering the fractional change in dipolar energy per jump for N jumping atoms with a time τ between jumps. T_{1D} is given by

$$\frac{1}{T_{1D}} = \frac{N}{\tau} \frac{E_{Df} - E_{Di}}{E_{Di}} = \frac{N}{\tau} \frac{\langle \mathcal{H}_{Df}^{(0)} \rangle - \langle \mathcal{H}_{Di}^{(0)} \rangle}{\langle \mathcal{H}_{Di}^{(0)} \rangle}, \qquad (6.3.10)$$

---

[42] H. T. Stokes and D. C. Ailion, *Phys. Rev. B* **16**, 4746 (1977).

where E_{Di} and E_{Df} represent the dipolar energy before and after a jump. $\mathcal{H}_{Di}^{(0)}$ and $\mathcal{H}_{Df}^{(0)}$ similarly represent the secular parts of the dipolar Hamiltonian and the brackets indicate expectation values. (The secular part is the part that commutes with the Zeeman Hamiltonian.) If the system can be characterized by a spin temperature before the jump, the expectation values can be expressed as traces of the quantum mechanical operators[29,32] and we obtain the following result:

$$\frac{1}{T_{1D}} = \frac{N}{\tau} \frac{\text{Tr}(\mathcal{H}_{Di}^{(0)})^2 - \text{Tr}(\mathcal{H}_{Di}^{(0)} \mathcal{H}_{Df}^{(0)})}{\text{Tr}(\mathcal{H}_{Di}^{(0)})^2}, \quad (6.3.11)$$

where $\mathcal{H}_{Di}^{(0)}$ and $\mathcal{H}_{Df}^{(0)}$ are the secular dipolar Hamiltonians before and after a jump, respectively. For a system consisting of two spin species, $\mathcal{H}_D^{(0)}$ can be written

$$\mathcal{H}_D^{(0)} = \mathcal{H}_{DII}^{(0)} + \mathcal{H}_{DIS}^{(0)} + \mathcal{H}_{DSS}^{(0)}, \quad (6.3.12)$$

where the terms on the right-hand side represent I-I, I-S, and S-S dipolar interactions, respectively. The assumption of strong I and weak S spins then results in the following inequality:

$$\mathcal{H}_{DII}^{(0)} \gg \mathcal{H}_{DIS}^{(0)} \gg \mathcal{H}_{DSS}^{(0)}. \quad (6.3.13)$$

If T_{1D} is due to the diffusion of I atoms, then only changes in $\mathcal{H}_{DII}^{(0)}$ need be considered in the numerator of Eq. (6.3.11), and we have the result

$$\frac{1}{T_{1D}} = \frac{2(1 - p_{II})}{\tau_I}, \quad (6.3.14)$$

where τ_I represents the time between jumps of the I spins and $1 - p_{II}$ is proportional to changes in the I-I dipolar energy resulting from I jumps. This result is identical to the strong-collision result [Eq. (6.3.9)] for homonuclear systems.

However, if T_{1D} is due to the diffusion of S atoms, then the numerator of Eq. (6.3.11) involves changes in $\mathcal{H}(_{DIS}^{(0)})$, with the result that

$$\frac{1}{T_{1D}} = \frac{1 - p_{SI}}{\tau_S} \frac{H_{DIS}^2}{H_{DII}^2}, \quad (6.3.15)$$

where H_{DIS} and H_{DII} are the I-S and I-I contributions to the dipolar local fields and τ_S is the time between jumps of the s spins. The precise definitions of these local fields and the p-parameters are given in Stokes and Ailion.[3]

Equation (6.3.15) predicts a strong anisotropy, arising from the factor H_{DIS}^2/H_{DII}^2, which is absent for motion by I atoms. The observation of this anisotropy would confirm that the diffusion arises from the motion of weakly magnetic spins. Figure 9, taken from Stokes and Ailion,[3] shows the

6.3. ULTRASLOW MOTION TECHNIQUES

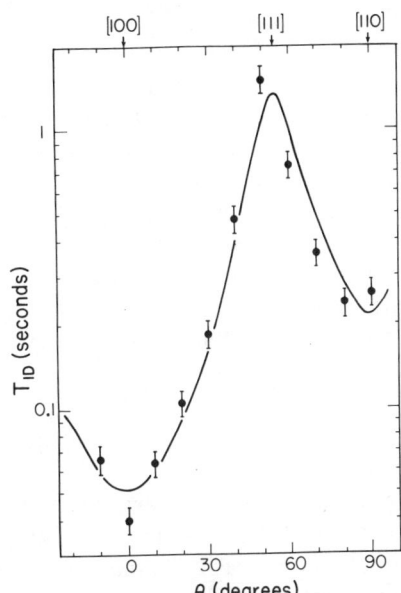

FIG. 9. T_{1D} anisotropy in KF:0.1% CaF$_2$ at 227°C. The crystal is rotated about its [110] axis. The solid curve is the calculated anisotropy of H^2_{DII}/H^2_{DIS}. (From Stokes and Ailion.[3])

verification of this idea for a single crystal of KF doped with 0.1% CaF$_2$. In this case, mobile potassium vacancies are created to compensate for the extra charge of the Ca^{++}. We thus have a case in which the diffusion of weak spins (potassium) dominates T_{1D}. The solid curve is the calculated anisotropy of the local field factor of Eq. (6.3.15).

6.3.2.2. Experimental Techniques for Measuring T_{1D}. There are basically two commonly used methods for measuring T_{1D}. Each involves transferring order to the dipolar system and then monitoring the decrease in dipolar order arising from dipolar relaxation. The first technique consists of following a spin-locking sequence by an adiabatic demagnetization–remagnetization cycle of the rf field. The second consists of creating a state of dipolar order by a pair of phase-shifted pulses and then monitoring the decrease in dipolar order by a third pulse.

6.3.2.2.1. ADIABATIC DEMAGNETIZATION IN THE ROTATING FRAME (ADRF). This method is very similar to the method for measuring $T_{1\rho}$, described in Section 6.3.1.3, in which $\mathbf{H}_1$ is adiabatically demagnetized from a large to a small $\mathbf{H}_1$ value following a spin-locking sequence. The principal difference is that $\mathbf{H}_1$ is now reduced adiabatically to zero so that there is no contribution from Zeeman relaxation. The adiabatic reduction in $\mathbf{H}_1$ transfers order from Zeeman order in the rotating frame to dipolar order (in alignment of the spins along the local fields). The subsequent adiabatic remagnetization converts the remaining dipolar order (not destroyed by dipolar relaxation) back to Zeeman order where it is observed in the form of an FID following the turnoff of the rf pulse. Dipolar relaxation in the demagnetized state results in a reduced FID following the H_1 turnoff. This sequence is shown in Fig. 10a.

Since the rf pulse in effect has been shut off during the variable time τ, many of the difficulties in using long pulses in $T_{1\rho}$ experiments will not occur in a T_{1D} experiment. There will be no requirement that the power amplifier be able to reproduce the shape of long pulses without deterioration, nor will there be appreciable rf heating. On the other hand, the need for two additional adiabatic sweeps will make difficult the measurement of a very short T_{1D}. For small H_1 (less than the local field H_D), the adiabatic condition becomes

$$\frac{dH_1}{dt} \ll \gamma H_D^2. \qquad (6.3.16)$$

For typical experimental conditions, it is then difficult to measure a T_{1D} less than 0.1–1 ms. Another advantage of the ADRF technique is that the apparatus can easily be made to measure $T_{1\rho}$ versus H_1 simply by partially demagnetizing H_1 to small (but nonzero) values.

The ADRF technique is implemented by simply inserting a trapezoidal waveform into one of the transmitter amplifier stages. The waveform can be adjusted to reduce the gain of the amplifier by biasing it towards cutoff.

6.3.2.2.2. PHASE-SHIFTED PULSE PAIR. Another method for measuring T_{1D}, which avoids the difficulty in measuring a short T_{1D} arising from requiring two adiabatic processes to transfer order between the Zeeman and dipolar systems, is the phase-shifted pulse-pair technique, originally devel-

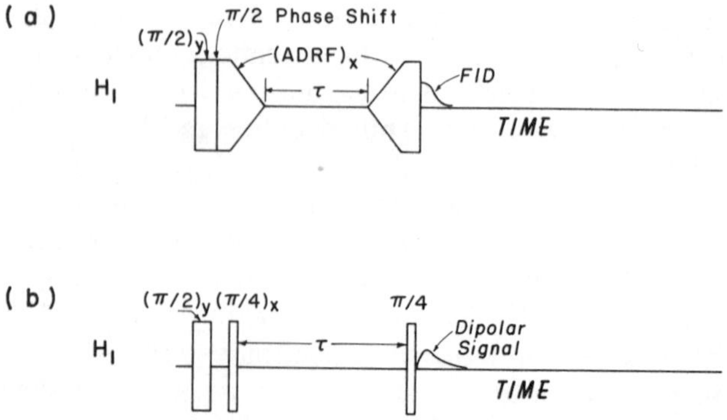

FIG. 10. Pulse sequences for measuring T_{1D}. (a) The ADRF sequence showing demagnetization of H_1 following spin locking. (b) The phase-shifted pulse-pair sequence. The first two pulses are separated by a time that is approximately T_2 for maximum signal.

oped by Jeener and Broekaert.[34] In this technique, a state of dipolar order is created by the application of two closely spaced pulses which are 90° out of phase. This dipolar order is then observed by applying a third pulse at an arbitrary phase in the xy plane. The dipolar signal that appears after the third pulse is proportional to the time derivative of the FID. The first pulse is a $\pi/2$ pulse, whereas the second and third pulses are $\pi/4$ pulses for maximum efficiency of Zeeman-to-dipolar order transfer and thus maximum size of the dipolar signal. To see the dipolar signal, the receiver reference signal is 90° out of phase with respect to the third pulse. This pulse sequence is shown in Fig. 10b.

A major disadvantage of this technique is that the maximum efficiency for transfer of order is only 56%, in contrast to the 100% efficiency theoretically attainable with ADRF. On the other hand, this technique can measure very short relaxation times and does not require power amplifiers capable of delivering long pulses. In summary, the ADRF technique is superior when measuring longer relaxation times or when $T_{1\rho}$ measurement is also desired, whereas the phase-shifted pulse-pair method is better for very short T_{1D} measurement.

6.3.3. Dipolar Relaxation in the Rotating Frame ($T_{1D\rho}$)

As we have seen in Section 6.3.2.1, a measurement of the anisotropy of T_{1D} in a single crystal can be used to verify that a weakly magnetic spin species is diffusing. There are two weaknesses in this approach: single crystals are not always available, and, possibly more serious, even though S spins are diffusing, they may not be dominating T_{1D} if they are sufficiently weakly magnetic or in low abundance. This second feature can be seen in Eq. (6.3.15), since for $H_{DIS} \ll H_{DII}$, we have that $T_{1D} \gg \tau_S$ for weak (S) spin motion. In contrast, for strong (I) spin motion, $T_{1D} \sim \tau_I$. These results arise from the fact that the heat capacities of the terms involving S spins are much smaller than the I–I heat capacities. Accordingly, it may be very difficult, or even impossible, in many cases to study the diffusion of sufficiently weak S spins in a T_{1D} experiment.

In order to determine the diffusing species in a slow-motion experiment, it is necessary to identify a parameter that is sensitive to which species is diffusing. If this parameter can be varied by the experimenter, then the diffusing species can be identified and studied. Furthermore, for the case of diffusion by spins that are so weakly magnetic as to preclude their study through direct T_{1D} measurement, it is desirable to vary the parameter so as to enhance the temperature change of the entire system resulting from the weak spins' motions. A way to perform such experiments is to observe the dipolar relaxation in the rotating frame of one of the species, say, the I spins.

In this frame, only part of the original dipolar Hamiltonian will be secular. We use the symbol $T_{1D\rho}$ for the relaxation time of this secular part $\mathcal{H}_D^{(00)}$, in analogy to T_{1D}, which is the relaxation time for $\mathcal{H}_D^{(0)}$. We may regard $T_{1D\rho}$ to be the dipolar relaxation time *in the rotating frame*, in contrast to T_{1D}, which is the dipolar relaxation time in the laboratory frame. (In the original paper,[3] the symbol $T_{1D'}$ was used instead of $T_{1D\rho}$.)

To see how $T_{1D\rho}$ can be used to resolve diffusion effects of different spin species as well as to enhance the effects of diffusion by the weak S spins, let us consider the terms in $\mathcal{H}_D^{(00)}$. According to Goldman,* $\mathcal{H}_D^{(00)}$ is related to the terms in the original Hamiltonian $\mathcal{H}_D^{(0)}$ by the expression

$$\mathcal{H}_D^{(00)} = -\frac{1}{2}(1 - 3\cos^2\theta_I)\mathcal{H}_{DII}^{(0)} + \cos\theta_I \mathcal{H}_{DIS}^{(0)} + \mathcal{H}_{DSS}^{(0)}, \qquad (6.3.17)$$

where θ_I is the angle between $\mathbf{H}_{\text{eff}}$ of Eq. (6.3.1) and $\mathbf{H}_0$. We see that varying θ_I (which is easily done by varying the resonance frequency) varies the relative strengths of the dipolar terms in Eq. (6.3.17) involving $I-I$, $I-S$, and $S-S$ interactions. We then find that the relaxation time $T_{1D\rho}$ of the rotating frame dipolar Hamiltonian $\mathcal{H}_D^{(00)}$ depends on θ_I; this dependence can be used to determine which species is diffusing.

Stokes and Ailion[3] derived strong-collision theory expressions for $T_{1D\rho}$ for both types of motion: I diffusion and S diffusion. In analogy to Eq. (6.3.11), $\mathcal{H}_D^{(0)}$ was simply changed to $\mathcal{H}_D^{(00)}$ to obtain

$$\frac{1}{T_{1D\rho}} = \frac{N}{\tau}\frac{\text{Tr}(\mathcal{H}_{Di}^{(00)})^2 - \text{Tr}(\mathcal{H}_{Di}^{(00)}\mathcal{H}_{Df}^{(00)})}{\text{Tr}(\mathcal{H}_{Di}^{(00)})^2}. \qquad (6.3.18)$$

Since $\mathcal{H}_D^{(00)}$ depends on θ_I, so does $T_{1D\rho}$. However, for the case of motion by the strong I spins, much of the dependence on θ_I disappears and one gets that $T_{1D\rho}$ equals T_{1D} and is described by the strong collision result [Eq. (6.3.14)] for all θ_I except near the magic angle θ_m (defined by $\cos^2\theta_m = \frac{1}{3}$), where there is a small change. (See Stokes and Ailion[3] for the detailed behavior of $T_{1D\rho}$ for diffusion by I spins.) On the other hand, for diffusion by weak S spins, the result is

$$\frac{1}{T_{1D\rho}(\theta_I)} = \frac{1}{\tau_S}\frac{\cos^2\theta_I H_{DIS}^2(1-p_{SI}) + 2H_{DSS}^2(1-p_{SS})}{[\frac{1}{2}(3\cos^2\theta_I - 1)]^2 H_{DII}^2 + \cos^2\theta_I H_{DIS}^2 + H_{DSS}^2}, \qquad (6.3.19)$$

where p_{SI} and p_{SS} are geometrical factors of order unity (analogous to p of the conventional strong collision theory). For weak S spins the second term in

---

* These transformations are discussed on p. 37 of Goldman.[43]

[43] M. Goldman, "Spin Temperature and Nuclear Magnetic Resonance in Solids." Oxford Univ. Press (Clarendon), London and New York, 1970.

6.3. ULTRASLOW MOTION TECHNIQUES

the numerator is negligible except for $\theta_I = 90°$ (exact resonance). Accordingly, we shall neglect it except at $\theta_I = 90°$.

Now consider the strong dependence of $T_{1D\rho}$ on θ_I. Suppose we are far off resonance ($\theta_I = 0$). Using Eq. (6.3.13) to neglect small terms, we obtain

$$\frac{1}{T_{1D\rho}(0)} = \frac{1 - p_{SI}}{\tau_S} \frac{H^2_{DIS}}{H^2_{DII}} = \frac{1}{T_{1D}}, \qquad (6.3.20)$$

which is identical to Eq. (6.3.15) for T_{1D} for diffusion by S spins.

However, at the magic angle θ_m, the first term in the denominator vanishes and Eq. (6.3.19) becomes simply (neglecting again the H^2_{DSS} terms)

$$\frac{1}{T_{1D\rho}(\theta_m)} = \frac{1 - p_{SI}}{\tau_S}. \qquad (6.3.21)$$

Thus, $T_{1D\rho}(\theta_m) \sim \tau_S$ and is much smaller than T_{1D}. By adjusting the field to the magic angle, we have effectively enhanced the relaxation rate of Eq. (6.3.20) by the heat capacity ratio (H^2_{DII}/H^2_{DIS}) of the $I-I$ to $I-S$ interactions. Viewed another way, the heat capacity of the entire system is sharply reduced at θ_m since the large $I-I$ interaction is now gone. Because there is a much smaller heat capacity at this angle, there will now be a much larger temperature change of the spin system resulting from the diffusion jump.

At exact resonance ($\theta_I = 90°$), the second term in the numerator of Eq. (6.3.19) will dominate. Then Eq. (6.3.19) reduces to

$$\frac{1}{T_{1D\rho}(90°)} = \frac{2(1 - p_{SS})}{\tau_S} \frac{4H^2_{DSS}}{H^2_{DII}}. \qquad (6.3.22)$$

In this case, $T_{1D\rho}(90°) \gg \tau_S$ and will in fact be much larger than T_{1D}. Physically, this feature results from the fact that, at exact resonance, the $I-S$ coupling term will be zero and $T_{1D\rho}$ will be insensitive to the motion of the S spins.

Before proceeding with the experimental verification of these ideas, let us consider the actual pulse sequence (shown in Fig. 11) used to measure $T_{1D\rho}$. A spin-locking sequence followed by an ADRF is applied on resonance to the I spins. This step transfers Zeeman order, which was originally along H_0, to dipolar order of the $\mathcal{H}_D^{(0)}$ reservoir. A large off-resonance pulse is now applied. The dipolar system is then properly described in the rotating frame (by $\mathcal{H}_D^{(00)}$ rather than the lab frame $\mathcal{H}_D^{(0)}$). However, right after the pulse is turned on, $\mathcal{H}_D^{(00)} \approx \mathcal{H}_D^{(0)}$, as we can see from Eq. (6.3.17) for $\theta_I = 0$. Accordingly, there is no change in the dipolar order of the system resulting from the turnon of the pulse; however, the dipolar reservoir must now be described in the rotating reference frame. After establishing a spin temperature in the rotating frame, the frequency of the rf pulse is then reduced

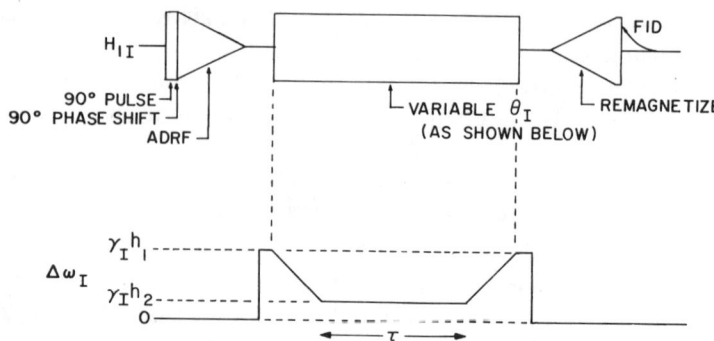

FIG. 11. Pulse sequence for measuring $T_{1D\rho}$. The lower part of the figure shows the variation of $\Delta\omega_I = \gamma_I H_0 - \omega_I$, the off-resonance frequency of H_{1I}. (From Stokes and Ailion.[3])

adiabatically toward resonance, thereby reducing $\mathbf{H}_{\text{eff}}$ and changing θ_I (and thus $\mathcal{H}_D^{(00)}$). By reversing this sequence we transfer the remaining order back to lab frame Zeeman order where it appears as an FID following the turnoff of the rf pulse. If one increases the time τ in the state at θ_I, the magnitude of the resulting FID will be reduced due to $T_{1D\rho}$ relaxation. A plot of magnetization versus τ on a semilog plot will yield a straight line of slope $-1/T_{1D\rho}(\theta_I)$. To vary θ_I, one merely needs to vary the frequency of the pulse (and thus h_2) in the variable time interval τ.

This pulse sequence bears a superficial resemblance to that used in measuring $T_{1\rho}$, since in both cases, the relaxation is measured in the presence of a strong rf field. The principal difference is that in a $T_{1\rho}$ experiment the magnetization is aligned along $\mathbf{H}_{\text{eff}}$ (or $\mathbf{H}_1$ for exact resonance), whereas in a $T_{1D\rho}$ experiment the individual magnetic moments are aligned preferentially along the local field directions. The sharp turnon of the initially off-resonance pulse merely changes the coordinate frame used to describe the state of the system, but does not initially change the order since the individual local fields (and thus the individual magnetic moments) are randomly oriented with respect to $\mathbf{H}_{\text{eff}}$. (If the turnon of this off-resonance pulse were slow enough to be adiabatic, order would then be transferred from dipolar to Zeeman in the rotating frame, with the result that $T_{1\rho}$ would indeed be measured instead of $T_{1D\rho}$.) The subsequent adiabatic changing of θ_I amounts to an adiabatic reduction in the individual local fields.

In performing a $T_{1D\rho}$ experiment it is quite important to be particularly careful about certain experimental features. Since the relaxation time depends so sensitively on $\mathbf{H}_{\text{eff}}$ in the rotating frame, it is crucial to have good frequency stability during the variable time interval τ. For this purpose a good sweepable-frequency synthesizer is essential. Furthermore, since the

6.4. METHODS FOR CALIBRATION

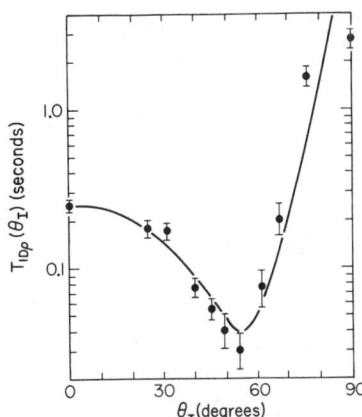

FIG. 12. $T_{1D\rho}$ as a function of θ_I in KF:0.1% CaF_2 at 200°C. Note that the data point at $\theta_I = 0$ is T_{1D}. (From Stokes and Ailion.[3])

relaxation time has a particularly strong H_1 dependence near θ_m, it is much more important in a $T_{1D\rho}$ than in a T_{1D} experiment to have good H_1 homogeneity. With poor H_1 homogeneity, only a fraction of the nuclei will experience θ_m; the result will be a smearing out of the relaxation components corresponding to different θ_I near θ_m and a reduction in the size of the apparent T_{1D} decrease at θ_m.

Figure 12 (taken from Stokes and Ailion[3]) shows experimental verification of these ideas for KF:0.1% CaF_2 in which the dominant diffusing species is the weakly magnetic K^+ ions. The solid curve is calculated from Eq. (6.3.19). All the strong dependence of $T_{1D\rho}$ on θ_I (described earlier) appears, thereby indicating that the weak S spins (in this case ^{39}K) are diffusing rather than the strong I spins (^{19}F here).

An interesting potential application of this $T_{1D\rho}$ technique is the study of impurity diffusion.[1] In this case the S spins are weak because their concentration N_S is small (but γ_S may be large). As in the case of small γ_S, diffusion by the weak S spins should again be characterized by a sharp dip[1] in $T_{1D\rho}$ at the magic angle. Fior this application to be fruitful, an abundant easily observable I-spin species must exist and be in good thermal contact with the S spins.

6.4. Methods for Calibration and Other Auxiliary Experiments

6.4.1. Determination of Exact Resonance

In T_{1D} and $T_{1D\rho}$ experiments it is often critical to have H_0 adjusted to exact resonance (i.e., the off-resonance field $h_0 = 0$). There are two reasons for the importance of exact resonance in these experiments. First, if the static field is off resonance, the apparent dipolar relaxation rate, as measured in zero H_1, will be smaller than the true rate T_{1D}^{-1} by a factor[29] $H_D^2/(h_0^2 + H_D^2)$, similar to the factor $H_D^2/(H_1^2 + H_D^2)$ in Eq. (6.3.6). Second, and more important, is that the order is totally destroyed after a long time in the demagnetized state only if the applied fields are zero. Even for a time in the demagnetized state that is long compared to T_{1D} (or $T_{1\rho}$), the measured signal after the remagnetization in ADRF will be zero[29,39] only for $h_0 = 0$. Even for small distances off resonance the magnetization in the rotating frame will decay to an equilibrium value which may be substantially different from zero.[29,39] If the experimenter is unaware of being off resonance, he may have a large systematic error in the measured relaxation time resulting from attempts to fit the decaying magnetization to a curve that goes to zero after a long time.

Probably, the most accurate method for obtaining the center of the resonance line in solids is to create a state of dipolar order, either by ADRF or by the phase-shifted pulse pair, and then allow a time τ that is long compared to T_{1D} to elapse before observing the residual order. The external field is then adjusted until the magnetization goes through a null corresponding to exact resonance. If T_{1D} is of order T_1, then a difficulty with this technique is that, during the time τ, a buildup of magnetization may occur along the z direction. This magnetization is then rotated by the remagnetization pulse in ADRF so as to have a component in the xy plane and thus a spurious signal at the turnoff of the pulse. This difficulty can be eliminated easily by following the remagnetization time with an additional time interval during which the field $\mathbf{H}_1$ is held constant. If the length of this interval is adjusted to be long compared to T_2, but short compared to T_{1D}, all components of the magnetization perpendicular to $\mathbf{H}_1$ will have decayed to zero, and there will not be an additional signal for exact resonance.

6.4.2. Calibrated $\mathbf{H}_1$ Monitor

It is very useful to have a calibrated monitor of $\mathbf{H}_1$ that can provide an accurate display of the shape of the rf pulse without detuning the power stage. Furthermore, it is convenient if this display is calibrated in gauss.

An easy approach is to put a capacitative voltage divider across the sample coil (Fig. 13). If a small capacitance C_1 is connected to the hot side, there will be very little detuning due to the monitor. Furthermore, if a large capacitance C_2 is connected to the ground side, there will be little further change in voltage by the addition of a monitor cable or a scope probe across C_2. If C_1 is, say, 5 pF and C_2 is 0.005 μF, the rf voltage across the monitor will be reduced from several thousand volts to a few volts. It is often convenient to not connect any external condenser to the hot side, in which case C_1 is merely the stray capacitance.

6.4. METHODS FOR CALIBRATION

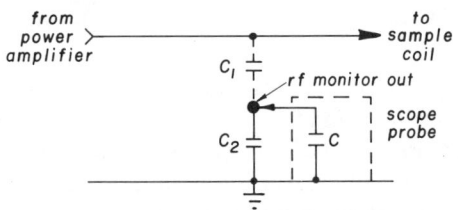

FIG. 13. H_1 monitor circuit. C_1 may be simply stray capacitance, $C_2 \gg C_1$, and C represents the scope-probe capacitance. By making $C_2 \gg C$, there will be negligible detuning due to C.

A simple way to calibrate H_1 is to use a liquid sample and observe the oscillations in the peak magnetization of an FID as the pulse width is increased to correspond successively to a $\pi/2, 3\pi/2, 5\pi/2, \ldots$ pulse. If t_w is the measured time interval between two successive nulls corresponding to a π and a 2π pulse, H_1 is determined in gauss from the relation

$$\gamma H_1 t_w = \pi. \tag{6.4.1}$$

By comparing this value to the measured peak-to-peak height in volts of the rf envelope of the portion of the H_1 pulse between π and 2π, as viewed on the H_1 monitor, calibration of the monitor voltage is obtained in gauss of H_1.

It is often more convenient to place the voltage-divider circuit of Fig. 13 across the primary of the impedance-matching transformer between the power amplifier and the sample coil, rather than to have this circuit in the magnet gap. In this case, a disadvantage is that recalibration may be necessary if the frequency or the sample Q is changed, as can result from looking at a different nucleus or using a different sample or changing the temperature. An alternative scheme is to not use the capacitive divider of Fig. 13, but rather to use a small pickup coil that must be mounted so as to pick up some of the magnetic flux in the sample coil. Since the current in this coil is directly proportional to H_1 (rather than to the voltage across an impedance-matching circuit), the resulting calibration will not change with sample changes. It must be emphasized that all components in the monitor circuit, whether a capacitive divider or pickup coil, must be mounted securely and permanently.

6.4.3. Determination of the Local Field in the Rotating Frame

In order to determine the atomic jump time τ_c from an experimentally measured $T_{1\rho}$ using a strong-collision formula like Eq. (6.3.6), it is important to know the value of the local field H_D. Similarly, in a T_{1D} experiment in which the motion of the S spins is dominant or in a $T_{1D\rho}$ experiment (for $\theta_I \neq \theta_m$), it may be important to know H_{DII} and H_{DSS}. If these local field terms arise from dipolar interactions and if the distance between atoms is known, they can be calculated. Nevertheless, it is useful, particularly in $T_{1\rho}$ experiments in which quadrupole interactions are present, to be able to measure directly the local field.

If quadrupolar and other interactions are sufficiently small that the strong-collision theory applies, the local field can be measured easily. There are two methods[29] for doing this. One method is to measure the magnetization following a partial ADRF in which the resultant magnetization is plotted as a function of the value of H_1 at which the demagnetization sweep is stopped. The local field is directly determined by fitting the resultant magnetization M for different values of H_1 to the formula[33]

$$M = M_0 H_1/(H_1^2 + H_L^2)^{1/2}, \qquad (6.4.2)$$

where H_L is the local field (equaling H_D for purely dipolar interactions). A second method is to plot $T_{1\rho}$ as a function of H_1 for fixed τ_c (and thus temperature) using the strong-collision formula, Eq. (6.3.6). A plot in the ultraslow-motion strong-collision region of the diffusion contribution to $T_{1\rho}$ versus H_1^2 yields a straight line,[29,39] which can be extrapolated to negative H_1 to yield a $T_{1\rho} = 0$ intercept at $H_1^2 = -H_L^2$.

6.5. Applications of Nuclear Magnetic Resonance Techniques to Motional Studies in Solids

In this section a number of examples are presented of applications of the previously described techniques for studying translational diffusion and molecular reorientations. Since the literature contains many such examples, I present just a few here that illustrate different interactions and different relaxation mechanisms. The purpose is to show the kinds of physical information obtainable from these NMR techniques.

6.5.1. Diffusion

As described earlier, a typical NMR study of diffusion will involve the observation of a relaxation time minimum. Under the usual circumstances of dipolar coupling, the theory[20] of BPP will apply, and a symmetric minimum will be observed in a temperature plot of T_1. Below the minimum, T_1 will be proportional to ω_0^2. Under these circumstances, this theory can be used to obtain the atomic jump time τ_c at each temperature. Inapplicability of the BPP theory would be indicated by the observation of either an asymmetric minimum or a frequency dependence different from ω_0^2 on the low-temperature side. Similarly, for dipolar interactions, the strong collision theory would apply in the limit of weak applied fields, and a

6.5. NMR APPLICATIONS TO MOTIONAL STUDIES

symmetric minimum in $T_{1\rho}$ or T_{1D} would then be observed. It would occur at the temperature for which $\tau_c \sim T_2$ for T_{1D} and at somewhat higher temperatures for $T_{1\rho}$.

6.5.1.1. Translational Diffusion.

Figure 14 shows a semilog plot of the temperature dependence of $T_{1\rho}$ is powdered metallic lithium[39] at $H_1 = 1.3$ G. Since the dipolar local field $H_D = 1.2$ G, the strong-collision theory applies. The principal feature is that the $T_{1\rho}$ minimum is symmetric and occurs at the onset of motional narrowing. Furthermore, the minimum value of $T_{1\rho}$ is within an order of magnitude of T_2, as expected from the strong-collision theory. Figure 15 shows a plot of the atomic jump time τ_c obtained from the experimental points. As can be seen, the data exhibit Arrhenius behavior over eight orders of magnitude and correspond to extremely slow diffusion (of the order of one jump per second for each atom) at the lowest temperatures.

6.5.1.2. Localized (Bound) Diffusion.

One of the great advantages of NMR is that it can observe localized motions of atoms, such as molecular reorientations or the motion of atoms bound to an impurity, as easily as it can observe nonlocal translational diffusion. In contrast, non-NMR techniques typically measure either local *or* nonlocal motions, but very few have the capability of measuring local *and* nonlocal motions over a wide temperature range using the same techniques. Since NMR is not so limited, it has the capability of studying many different diffusion mechanisms that dominate the diffusion in different temperature regions. Thus one can in principle use NMR to measure the binding energy of a mobile nucleus to a charged impurity by comparing activation energies in different temperature regions for local and for nonlocal motions (provided there are not major structural differences between these two regions.)

As an example, consider a substance like CaF_2 or SrF_2 that is doped with a substitutional trivalent impurity. In order to maintain charge neutrality, an excess number of fluorine interstitial ions are formed.[44] (Thus the concentration of these fluorine interstitials equals the concentration of the impurities.) At low temperatures the fluorines are bound to the impurity atoms and their motion is between neighboring sites. At higher temperatures the fluorines can break the bonds and diffuse over large distances. In an NMR experiment, the transition from the temperature region where the motion of bound (local) fluorines dominates the relaxation to the higher-temperature region where the motion of unbound (nonlocal) fluorine dominates is characterized by a sudden increase in activation energy. This increase is equal to the binding energy, neglecting the difference in barrier heights between local potential wells and barrier heights between nonlocal wells.

[44] S. H. N. Wei and D. C. Ailion, *Phys. Rev. B* **19**, 4470 (1979).

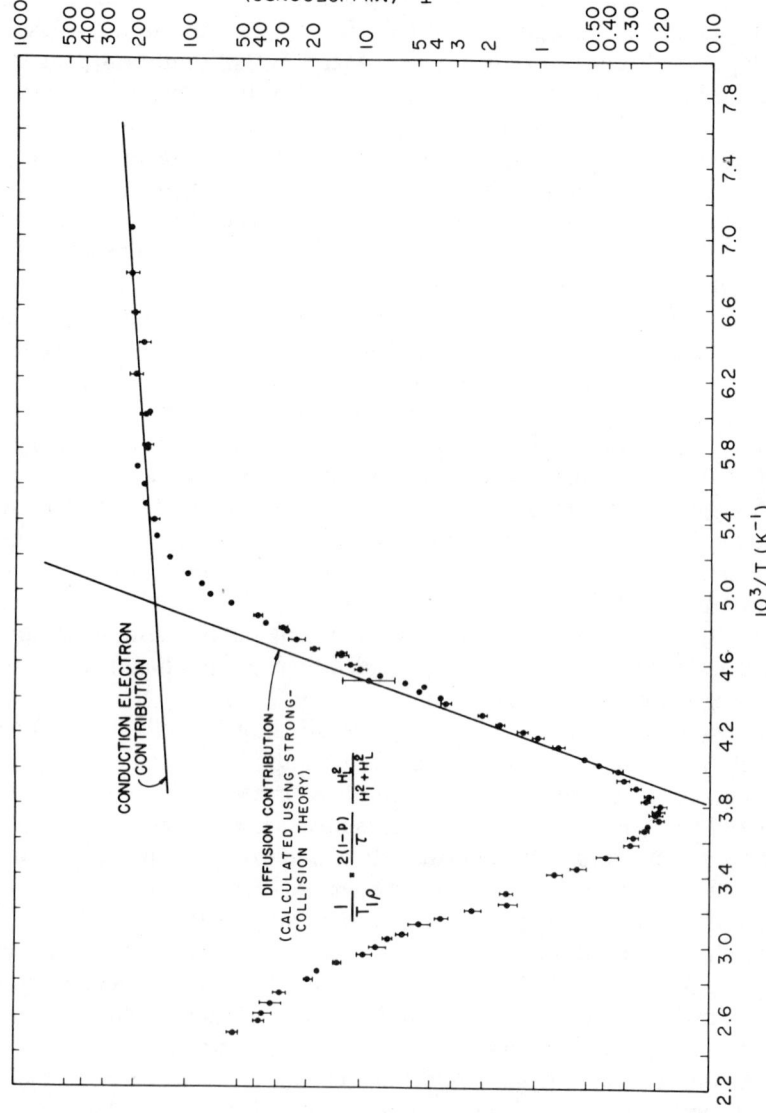

FIG. 14. $T_{1\rho}$ versus reciprocal temperature in metallic lithium with $H_1 = 1.3$ G. (From Ailion and Slichter.[39])

6.5. NMR APPLICATIONS TO MOTIONAL STUDIES

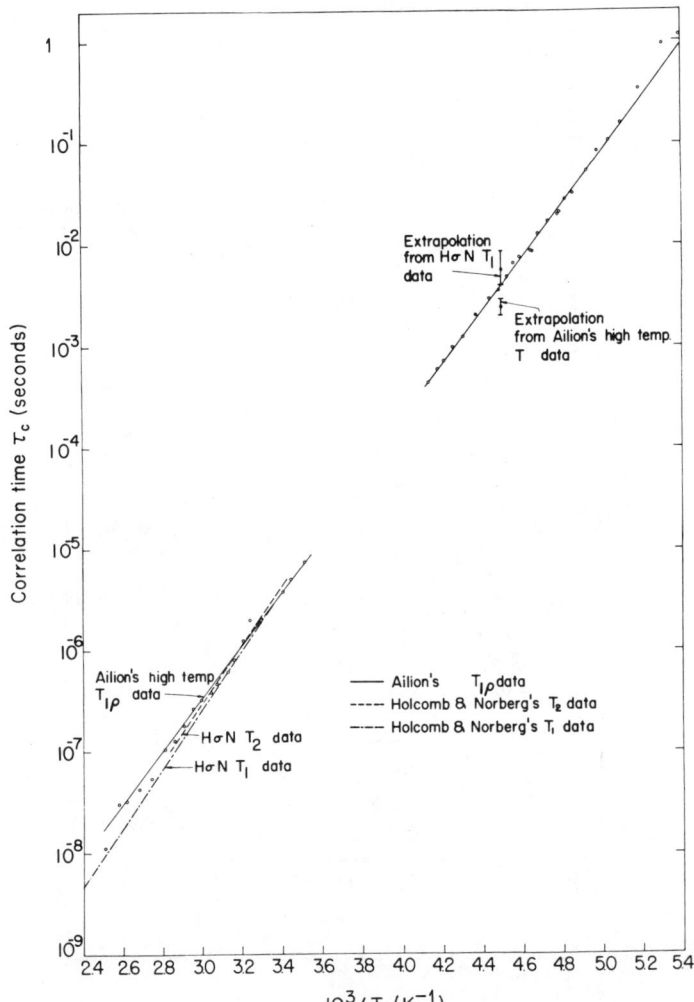

FIG. 15. τ_c versus reciprocal temperature in metallic lithium. The experimental points are obtained from the data of Fig. 14 using the strong-collision theory in the low-temperature region and the weak-collision theory in the high-temperature region. The "gap" in the middle corresponds to the region around the $T_{1\rho}$ minimum where neither theory is valid. (From Ailion and Slichter.[39])

Figure 16, taken from Wei and Ailion,[44] shows T_{1D} measurements on SrF_2 doped with various concentration of Y^{3+}. Up to dopant concentrations greater than 0.1% where clustering of impurities may be important, the relaxation rate (T_{1D}^{-1}) is proportional to dopant concentration, thereby

476   6. NMR RELAXATION TIME METHODS FOR ATOMIC MOTIONS

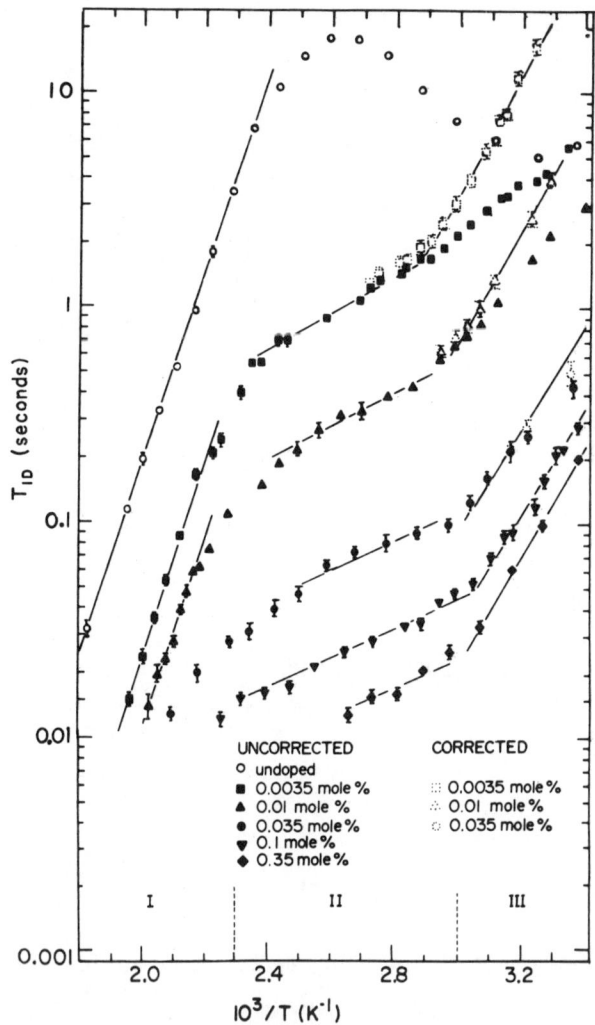

FIG. 16. T_{1D} versus reciprocal temperature for SrF_2 doped with various concentrations of Y^{3+}. The low-temperature data (Region III) were corrected by subtracting the background relaxation measured for the undoped sample. (From Wei and Ailion.[44])

supporting the idea that the relaxation is due to fluorine interstitials created to compensate for the charge of the impurity. The sharp increase in slope above 435 K ($10^3/T = 2.3$ K^{-1}) is attributed to the nonlocal motion of unbound fluorine interstitials (Region I), whereas below this temperature the relaxation is due to the motion of F^- interstitials bound to the Y^{3+}. A somewhat curious feature of these data is the sudden decrease in slope at

333 K ($10^3/T = 3.0$ K^{-1}) as the temperture is raised from Region III to Region II. The activation energy in Region III agrees with other measurements[45] for F$^-$ in a nearest-neighbor site. The transition to Region II may result from motions by next-nearest-neighbor fluorines[44] or, possibly, from the formation of clusters. It should be emphasized that the relaxation in Region I is in the strong-collision regime since it involves interstitialcy jumps of all the F$^-$ spins. In Regions II and III where the diffusion involves the local motions of only a small fraction of the spins, the relaxation of the bulk spins is much weaker (T_{1D} is longer) and occurs by the transfer of dipolar energy from the neighboring atoms to those more distant by mutual spin flips (spin diffusion). (See Chapter V of Abragam[9] for a discussion of spin diffusion.)

6.5.1.3.. Diffusion Studies for Systems Characterized by Nonexponential Correlation Functions.

As described in Section 6.2, the theory of BPP assumes that fluctuations in dipolar energy due to translational diffusion or molecular reorientations can be described by an exponential correlation function. This assumption results in a symmetric T_1 minimum in a plot of T_1 versus reciprocal temperature as well as a T_1 proportional to ω_0^2 on the low-temperature side of the minimum, as we have seen. However, there are numerous examples in the literature where this thoery does not apply. Examples are found in superionic conductors,[46] liquid crystals,[21] polymers,[47] and plastic crystals.[48]

In this section, an example is given of T_1 observations in the nematic liquid crystals p-azoyanisole (PAA) and 4-n-methoxybenzylidene-4'-n-butylanilene (MBBA). These molecules are highly anisotropic and can be aligned easily by an external magnetic field such that the molecular principal axis will be approximately parallel to the magnetic field. In this ordered state there are two important mechanisms for spin–lattice relaxation. Order fluctuations have been shown[49] to modulate the dipolar coupling so as to give rise to a relaxation rate which has a frequency dependence at fixed temperature given by

$$1/T_1 = A/\sqrt{\omega_0} + B. \quad (6.5.1)$$

On the other hand, molecular diffusion[50,51] can also modulate intermolecu-

---

[45] E. L. Kitts, Jr., M. Ikeya, and J. H. Crawford, Jr., *Phys. Rev. B* **8**, 5840 (1973).
[46] R. E. Walstedt, R. Dupree, J. P. Remeika, and A. Rodriquez, *Phys. Rev. B* **15**, 3442 (1977).
[47] F. Devreux, K. Holczer, M. Nechtschein, T. C. Clarke, and R. L. Green, in "Physics in One Dimension" (J. Bernasconi and, T. Schneider, eds.), pp. 194–200. Springer-Verlag, Berlin and New York, 1981.
[48] H. T. Stokes, T. A. Case, D. C. Ailion, and C. H. Wang, *J. Chem. Phys.* **70**, 3563 (1979).
[49] J. W. Doane and D. L. Johnson, *Chem. Phys. Lett.* **6**, 291 (1970).
[50] J. F. Harmon and B. N. Muller, *Phys. Rev.* **182**, 400 (1969).
[51] H. C. Torrey, *Phys. Rev.* **92**, 962 (1953).

478   6. NMR RELAXATION TIME METHODS FOR ATOMIC MOTIONS

lar couplings so as to produce a frequency-dependent T_1 which, in the low-frequency limit ($\omega_0 \tau_c \ll 1$), varies as

$$1/T_1 = C - F\sqrt{\omega_0}. \tag{6.5.2}$$

Figure 17 shows experimental and theoretical results in the liquid crystal MBBA. Figure 17a is a plot of ($1/T_1$ versus $1/\sqrt{\omega_0}$ and shows strong departure from the results predicted[49] for order fluctuations [Eq. (6.5.1)]. On the other hand, there is very good agreement in Fig. 17b between experimental values of $1/T_1$ plotted against $\sqrt{\omega_0}$ and the results of calculations[50,51] assuming the molecular diffusion mechanism. These results indicate that, in MBBA, molecular diffusion rather than order fluctuations dominates the NMR relaxation. In contrast, in the liquid crystal PAA, the opposite is true—the data[21] can be fit to Eq. (6.5.1) but not to Eq. (6.5.2), indicating that in PAA order fluctuations, not diffusion, dominate the relaxation. This

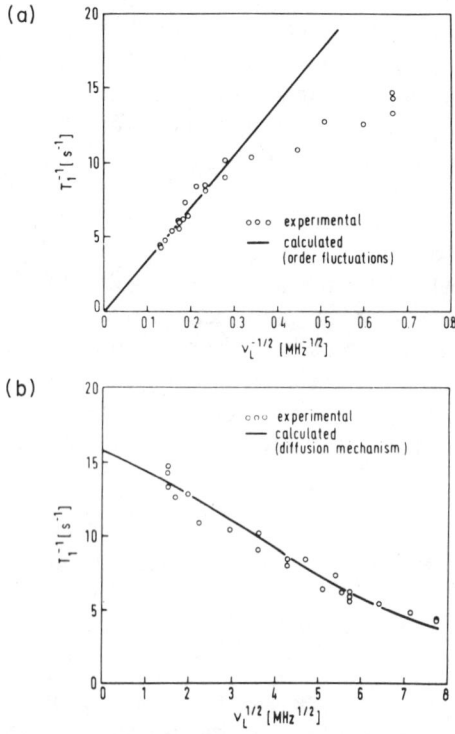

FIG. 17. Frequency dependence of the proton spin–lattice relaxation time T_1 in MBBA at $T = 291$ K. (a) Plot of $1/T_1$ versus $v_L^{-1/2}$; $[=(\omega_0/2\pi)^{-1/2}]$; the solid line is calculated for the order-fluctuation mechanism using Eq. (6.5.1). (b) Plot of $1/T_1$ versus $v_L^{1/2}$; the solid line is a theoretical[51] plot for molecular diffusion that shows approximately the linear dependence predicted by Eq. (6.5.2). (From Vilfan et al.[21] Copyright 1972, Pergamon Press, Ltd.)

section illustrates the importance of measuring the frequency dependence of the relaxation time in determining the relaxation mechanism.

6.5.2. Molecular Reorientations

One of the big advantages of NMR over other techniques is that it can be used to study the motional dynamics of a reorienting molecule for which there is no mass transport as easily as it can be applied to translational diffusion. The chemical literature contains numerous examples* of applications of NMR T_1 measurements to the study of reorienting molecules, typically in liquids. Often the BPP theory applies, and a symmetric minimum is observed, with $T_1 \propto \omega_0^2$ on the low temperature side. Since this situation is so similar to examples already discussed, further elaboration is not given here. Rather, two examples are presented of reorienting molecules in solids that illustrate particularly interesting relaxation mechanisms.

6.5.2.1. Relaxation due to Motion between Potential Wells of Unequal Depth.
Diffusion observed with NMR normally arises from an atomic jump from a potential energy well over a barrier to another well of the same depth. However, if the two wells are of unequal depth, it is possible to see anomalous NMR effects on T_1 and on T_{1D}.

Consider what happens if an atom jumps from a shallow well over an energy barrier of height H to another well whose minimum is Δ below that of the shallow well (Fig. 18). We can see qualitatively what should happen by recognizing that the relative concentration of atoms in the shallow well should be reduced by the Boltzmann factor $\exp(-\Delta/k_B T)$. Accordingly, the spin–lattice relaxation rate in Eq. (6.2.5) should be replaced by

$$\frac{1}{T_1} \sim \exp\left(\frac{-\Delta}{k_B T}\right) \frac{(\omega_d)^2 \tau_c}{1 + \omega_0^2 \tau_c^2}. \tag{6.5.3}$$

This result has important consequences in an NMR experiment. The most obvious effect is to reduce the relaxation rate and correspondingly raise T_1 by the Boltzmann factor. Thus this process will result in a very long T_1, even at the minimum. A second striking effect is to make the T_1 minimum asymmetric. Since $\tau_c \propto \exp(H/k_B T)$, we see from Eq. (6.5.3) that at temperatures above the minimum the slope of $\ln T_1$ versus $1/T$ is proportional to the energy difference $H - \Delta$, whereas, below the minimum, the slope is proportional to $H + \Delta$. (For processes that have a temperature-dependent factor involving an energy of formation E_F, H is replaced by the activation energy $E_A = E_F + H$. Also there may be a binding energy factor for some defects. However, the above results will still hold, except that H is replaced

---

* These are found typically in journals like the *Journal of Chemical Physics*, the *Journal of Magnetic Resonance*, and the *Journal of the American Chemical Society*.

480   6. NMR RELAXATION TIME METHODS FOR ATOMIC MOTIONS

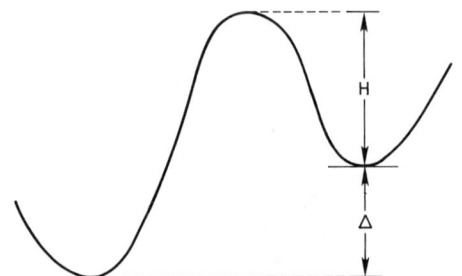

FIG. 18. Two unequal potential wells. (From Ailion.[2])

by E_A.) Additional effects are that the temperature of the minimum is shifted slightly and the frequency dependence of the T_1 minimum is also reduced.[52]

These effects have been observed[52] for molecular reorientations in an organic compound [trans, trans-muconodinitrile (TMD)] and are reproduced in Fig. 19. Note the very long relaxation times (~ 20 s) at the minima as well as the asymmetry in the T_1 curves above versus below the minima. These data were analyzed using the above concepts and values of H and Δ were determined for the unequal potential wells.

Such observations make possible the independent measurement of the parameters H and Δ, which characterize the detailed shape of the potential-energy calculated using different force models.[53] However, such observations are not common in NMR because the shallow T_1 minimum due to motion between unequal wells can be masked easily by other relaxation mechanisms. Furthermore, these effects are probably not observable in metals where the T_1 is dominated by the strong conduction electron relaxation. Moreover, to be seen in insulating materials like the above would require samples of very high purity.

6.5.2.2. Chemical-Shift Anisotropy (CSA) Relaxation due to a Reorienting Molecular Group. As we have discussed in Section 6.3, molecular reorientations can cause fluctuations in the chemical shift which can also contribute to the spin lattice relaxation. Equation (6.2.6) predicts a very different frequency dependence for CSA relaxation than for dipolar relaxation (see Fig. 6). However, CSA relaxation is rarely observed in solids since it is often masked by stronger relaxation mechanisms. If dipolar relaxation and CSA relaxation are both present, CSA relaxation will be observable only

---

[52] M. Polak and D. C. Ailion, *J. Chem. Phys.* **67**, 3029 (1977).
[53] M. Polak, *J. Chem. Phys.* **67**, 5572 (1977).

6.5. NMR APPLICATIONS TO MOTIONAL STUDIES

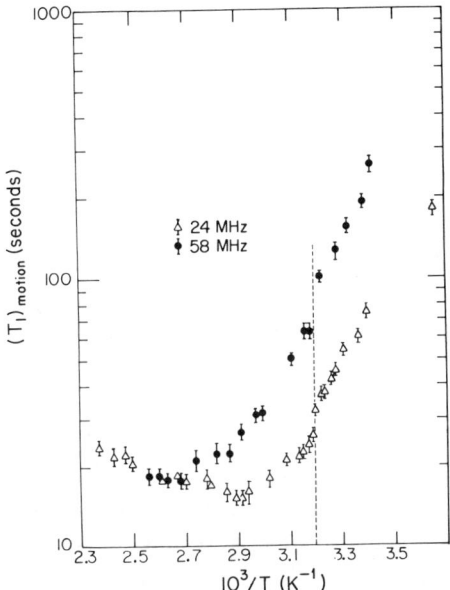

FIG. 19. Proton T_1 versus reciprocal temperature in TMD at two different frequencies due to motions between greatly unequal wells. Note the unequal slopes above and below the minima and the large values of T_1. (From Polak and Ailion.[52])

at very high fields since dipolar relaxation will dominate at low fields. However, in the rare cases where CSA relaxation can be observed in solids, it can provide considerable insight into the microscopic features of the motional mechanism.

Consider a T_1 minimum for which the dipolar and CSA contributions are equal. If the frequency is raised to the region where CSA relaxation is dominant *or* lowered to the region where dipolar relaxation is dominant, the value of T_1 at the minimum will be lowered.[54] Thus a careful measurement of the frequency dependence of the T_1 minimum can identify and allow the study of CSA relaxation.

Figure 20 shows measurements[54] at three different frequencies of ^{13}C relaxation in the elastically ordered phase of KCN enriched with 90% ^{13}C. (^{13}C has spin $\frac{1}{2}$ and a large magnetic moment, in contrast to the abundant isotope ^{12}C which has zero magnetic moment.) The actual curves in Fig. 20 are due to a combination of dipolar relaxation and CSA relaxation, both

---

[54] H. T. Stokes, T. A. Case, and D. C. Ailion, *Phys. Rev. Lett.* **47**, 268 (1981).

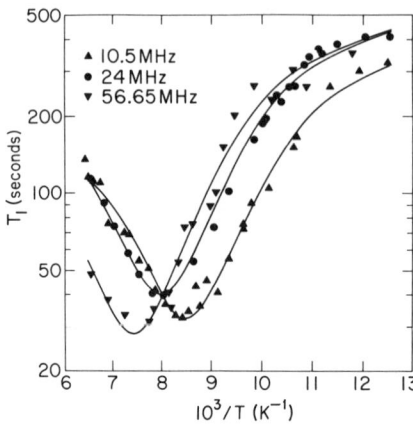

FIG. 20. T_1 versus reciprocal temperature for ^{13}C in KCN. The solid lines are fit to the data by assuming that the observed relaxation arises from dipolar and CSA mechanisms. (From Stokes et al.[54])

resulting from reorienting CN⁻ molecules. The low-frequency (10 MHz) data are due primarily to dipolar relaxation, whereas CSA relaxation is favored at the high frequency (56.65 MHz). At the intermediate frequency of 24 MHz the two mechanisms have comparable contributions. Normally, such CSA relaxation would be masked by much stronger dipolar relaxation. However, in this system the neighboring nucleus is potassium, which has a very small magnetic moment. Thus the dipolar relaxation is quite weak and arises primarily from the intramolecular interaction of ^{13}C with the nearest ^{14}N nucleus. Since both the intramolecular dipolar and CSA interactions are invariant for 180° jumps, these data tell us that the NMR relaxation must arise from CN⁻ jumps through angles other than 180°. A careful analysis of the spectral density determined by the values of the T_1 minima indicates that the *observed* relaxation arises from small-angle CN⁻ reorientations (∼ 2°) resulting from small shifts in the position of a CN⁻ ion's potential energy well which occur when neighboring CN⁻ ions undergo reorientations. These data result in the remarkable conclusion that CN⁻ ions are slightly misoriented with regard to the orthorhombic *b* axis in KCN. We should emphasize that CSA relaxation is observable in this case mainly because the neighboring nuclei have small magnetic moments and correspondingly small dipolar interactions.

Acknowledgments

The author benefitted from discussions with T. A. Case and Professor R. Blinc. Much gratitude is expressed to Professor W. Ohlsen for critically reading this manuscript. During the writing of this paper, the author was supported by the U.S. National Science Foundation under Grant DMR 76-18966.

AUTHOR INDEX

Numbers in parentheses are reference numbers and indicate that an author's work is referred to although the name is not cited in the text.

A

Aalders, J., 170
Abdulla, U., 341
Abragam, A., 361, 362, 366(2), 369(2), 390(2), 392, 395, 401(2), 406(3), 442, 447, 448, 451, 477
Adams, M. D., 46
Ailion, D. C., 382, 387, 388, 394, 397, 398(41), 439, 441(3), 444, 449, 450(23), 451(23), 454, 455, 456, 457, 459, 461, 462, 463, 466, 468, 469, 470(29, 39), 472(29, 39), 473, 474, 475, 476, 477, 480, 481, 482(54)
Ait-Salem, M., 220
Alam, A., 95
Alefeld, B., 206, 215, 220
Alexander, R. B., 329, 339, 341
Alger, R. S., 410, 416, 419, 420
Als-Nielsen, J., 191
Amelinckx, S., 276
Amsel, G., 250
Andersen, H. H., 225, 227(6), 231, 245(22)
Andersen, J. U., 280, 282, 293, 295(68, 69), 299(31), 338, 347(31), 352(31)
Anderson, C. A., 267
Anderson, W. A., 378, 390
Andreani, R., 275
Andreasen, O., 338
Andrew, E. R., 373, 389(10), 444
Andrews, G. A., 30
Anger, H. O., 112
Appleton, B. R., 286, 290, 310(56), 312, 313(93), 333, 356
Arnold, G. W., 233
Arponen, J., 83
Aucouturier, M., 48
Auld, B. A., 34, 35(3), 48, 54, 59(3), 62, 65, 496, 499, 501(1), 504
Auslow, J. A., 526

Avogadro, A., 453
Axmann, A., 170

B

Bacon, G. E., 211, 212, 275
Baeri, P., 311, 312, 317, 318(98), 320(98)
Baker, P. S., 32
Ball, D. J., 238, 239
Bansil, A., 80
Barat, M., 252, 256(75)
Barber, D. J., 257, 263
Barfoot, K. M., 225
Barr, L. W., 71, 73
Barrett, C. S., 276
Barrett, J. H., 286, 289, 290, 292, 348, 349(58), 350
Bates, D. D., 213, 214(22)
Bauer, E., 236
Bauer, G. S., 156, 169, 170
Bazhin, A. I., 271
Becker, E. D., 385
Becker, K., 35
Becker, P., 159
Bedwell, M. O., 124
Behrisch, R., 248
Beirens, L. C. M., 236, 237(41), 240(41), 241(41), 242(41)
Bell, F., 255
Benedek, G. B., 429
Benninghoven, A., 258, 264, 268(92), 269
Bergersen, B., 83, 85
Berko, S., 80, 88, 89, 90, 94, 95(10), 106(3), 111, 116(10), 117(10), 118(10), 120(10), 137
Berlinger, W., 420, 423
Berry, B. S., 276
Bessiere, M., 170
Betz, H. D., 252, 255
Bhalla, C. P., 252

AUTHOR INDEX

Bierman, D. J., 271, 273(129)
Bird, J. R., 247
Birkholz, W., 47
Birr, M., 215
Bisson, P. E., 106, 113(78), 114(78), 115(78)
Blaise, G., 264, 265, 266(113), 267
Blandin, A., 267
Blattner, R. J., 233
Blau, M., 44
Bleaney, B., 361, 362, 406(3)
Blewitt, T. H., 19
Blinc, R., 448, 477(21), 478(21), 482
Bloch, F., 378, 379
Bley, F., 170
Bloembergen, N., 372, 389(8), 448
Blum, H., 435
Blume, R. J., 387
Bøgh, E., 282, 295(39), 308(39), 345, 349, 356
Bohr, N., 286
Bolton, J. R., 405, 406(52)
Bonderup, E., 293, 295(68, 69)
Borders, J. A., 233, 312
Børgesen, P., 231
Borie, B., 154
Bösiger, P., 436
Bøttiger, J., 231, 245(22), 299
Bragg, W. H., 230
Brandt, W., 85
Braun, M., 273
Brennan, J. G., 249
Bretenberger, E., 51
Brice, D. K., 229, 312, 319, 320, 321(102)
Briggs, J. S., 256
Brockhouse, B. N., 187, 188
Brodsky, A., 32
Broekaert, P., 382, 456, 465
Brongersma, H. H., 236, 237, 238, 240(41), 241, 242
Brooks, R. D., 75
Brower, K. L., 420, 425, 426
Brown, F., 235
Brown, J. M., 28
Brown, M. D., 249
Brown, W. L., 331, 332(114)
Browne, E., 81
Brues, A. M., 28
Brun, E., 436
Brünger, G., 447
Bucci, C., 440

Buchenan, U., 190
Buck, T. M., 236, 238, 239(49), 258(45)
Buckmaster, H. A., 408
Bugeat, J. P., 342
Burhop, E. H. S., 240
Burkel, E., 171
Burum, D. P., 438, 460
Bussman, W., 72, 73
Butler, J. W., 231, 245(21)
Buyrn, A. B., 81

C

Cameron, J. R., 38
Campbell, J. L., 102, 103
Campisano, S. U., 233, 305, 311, 312(89), 317, 318(98), 320(98)
Canter, K. F., 89
Carbotte, J. P., 82
Carr, H. Y., 445, 446
Carstanjen, H.-D., 282, 339(44), 342
Carswell, D. J., 3, 4, 74
Carter, G., 263
Case, T. A., 382, 444, 449, 450(23), 451(23), 477, 481, 482(54)
Caspers, L. M., 89
Chadwick, A. V., 73
Chajechi, T., 47
Chakraborty, B., 80, 90, 95(10), 116(10), 117(10), 118(10), 120(10), 137
Chami, A. C., 342
Chapman, G. E., 261, 262
Chason, M. K., 86, 90, 105, 134, 135(74, 89), 137(51), 138(74)
Chemin, J. F., 282
Chen, W. K., 71
Chen, Y. S., 236, 258(45)
Chernow, F., 312
Chiao, T., 253
Christensen, A. N., 170
Chu, W. K., 224, 225, 230, 231, 235, 278, 297(21), 300(21), 303, 309
Chudley, C. T., 198
Ciavola, G., 311, 312(89)
Clark, H. M., 5, 7(4), 8, 74
Clark, W. G., 391
Clarke, T. C., 477
Cobas, A., 239
Coburn, J. W., 263, 264
Cohen, J. B., 170

AUTHOR INDEX

Cohen, R. L., 40
Colenbrander, B. G., 350, 351(136), 356(136)
Colligon, J. S., 257
Comas, F., 30
Conlon, F. B., 38
Conners, D. C., 85
Cooley, J. W., 390
Copley, J. R. D., 217
Corbett, J. W., 275
Cotterill, R. M. J., 86(39), 87, 93, 94(56)
Cottrell, A. H., 280
Council, K. A., 137
Crawford, Jr., J. H., 477
Creber, D. K., 280, 282(32)
Crouthamel, C. E., 75
Cue, N., 252, 253(72)

D

Dairiki, J., 81
Damadian, R., 404
Damask, A. C., 280
Dannefaer, S., 101
Das, T. P., 366, 395, 396(4), 449
Dauwe, C., 101
Davies, J. A., 275, 276(10), 277(10), 278(10), 280, 282, 286, 288, 299, 301, 314(94), 315, 316(96), 323(10), 330, 338, 347, 351(77), 352, 353, 356
Davies, J. J., 427
Dawber, P. G., 186
Dearnaley, G., 75
DeBonte, W. J., 233, 234(32)
Dederichs, P. H., 276
de Fontaine, D., 153
de Graaf, L. A., 202(11), 203, 204(11), 206
Delbecq, C. J., 423
Denley, D., 276
de Novion, C. H., 170
Dering, J. C., 408
Descouts, P., 106, 113(78), 114(78), 115(78)
Dettman, K., 257
DeVoe, J. R., 74
Devreux, F., 477
de Vrijen, J., 148, 170
Dickmann, J. E., 121
Diehl, J., 275
Dienes, G. J., 280
Dmitriev, P. P., 26
Doane, H. W., 477, 478(49)

Doane, J. W., 448, 477(21), 478(21)
Doebler, R. E., 81
Dolling, G., 179, 213
Dorikens, M., 97, 101, 131(64)
Dorikens-Vanpraet, L., 97, 101, 131(64)
Dorn, J. E., 93
Douglass, D. C., 458
Doyama, M., 92
Doyle, B. L., 225, 227(7)
Dupanloup, A., 106, 113(78), 114(78), 115(78)
Dupree, R., 477
Dutkiewicz, V., 253

E

Eades, R. G., 444
Eady, J. A., 102
Eckold, G., 210
Eckseler, H., 73
Eckstein, W., 237, 248
Edwards, C. L., 30
Eichholz, G. G., 95
Eisen, F. H., 309, 310
Eldrup, M., 136, 137
Elich, J. J., 262
Ellegaard, C., 293, 295(66)
Elleman, D. D., 402, 460
Ellett, Jr., J. D., 386
Elliott, R. J., 182, 184, 186, 198, 219
Emmoth, B., 273
Erickson, R. L., 239, 240
Erdtmann, G., 74
Erginsoy, C., 283, 286, 310(56)
Eriksson, L., 282, 330
Ernst, R. R., 390
Erskine, J. C., 90
Esbensen, H., 293, 295(68, 69)
Eschbach, H. L., 225
Estle, T. L., 406
Estrup, P. J., 275
Evans, Jr., C. A., 233, 268
Evans, R. D., 1, 36, 74

F

Farmery, B. W., 261, 262(95)
Farrar, T. C., 385
Fayard, M., 170
Feher, G., 407, 411(54)

Feldman, L. C., 256, 282, 290, 346, 348, 349, 354, 355, 356, 358, 359
Fellgett, P., 374
Feng, J. S. Y., 231, 230(16), 231
Fenzl, H. J., 190
Fergues, E., 75
Feuerstein, A., 225
Fieschi, R., 440
Filius, H., 89
Firsov, O. B., 231
Fischer, H. A., 44, 45
Fleischer, R. L., 48
Flotow, H. E., 206
Fluss, M. J., 80, 83, 86, 88, 90, 95(10), 105, 116(10), 117, 118, 120, 121, 134, 135, 137, 138(74, 75), 140, 143(100)
Flynn, C. P., 198
Foley, E. B., 271, 273(130), 274(130)
Folkmann, F., 254, 256(79)
Follstaedt, D. M., 317, 318(98), 320(98), 331
Ford, M. R., 33
Forrer, J., 435
Foti, G., 233, 305, 317, 318(98), 320(98)
Fox, R., 73
Frank, F. C., 263
Frank, W., 84
Freyer, K., 47
Friedlander, G., 74
Fujiwara, K., 275
Fukushima, E., 373, 387(13), 388(13), 389, 390(13), 451
Funke, K., 210
Furgeson, G. A., 202(11), 203, 204(11)
Furukawa, S., 305, 306, 307(83), 308(84)

G

Gaarde, C., 254
Gemmell, D.S, 275, 276(9), 278(9), 281(9), 282(9), 283(9), 295, 319(74), 323(9), 330(9)
Geschwind, S., 437
Gevers, R., 276
Ghoshal, S. N., 21, 22
Gibby, M. G., 386, 403
Gibson, J. B., 280
Gibson, W. M., 286, 310(56), 331, 332(114)
Gilbody, H. B., 240
Gill, D., 446
Gingerich, R. R., 102
Giordano, M., 435

Gissler, W., 198, 206, 211
Goland, A. N., 84, 134, 136(90), 280
Golanski, A., 315, 316(96)
Goldburg, W. I., 373, 389(11)
Goldman, M., 466
Goldschmidt, H. J., 322, 339(106)
Goldsmith, M., 404
Gordon, J. P., 422
Gordon, R. E., 447
Gorham-Bergerson, E., 198
Goto, T., 295
Gottschalk, A., 112
Grahmann, H., 225
Grasso, F., 233, 305
Green, R. L., 477
Gruber, H., 435
Guidi, G., 440
Guinier, A., 281
Gunthard, H. H., 435
Gupta, R. P., 83, 86, 88(38)
Gustafson, D. R., 121
Gutowsky, H. S., 392
Guttman, L., 322
Gyulai, J., 250, 319

H

Haeberlen, U., 373, 386, 389(9, 12), 401(12), 402(9, 12), 403, 450
Haghgooie, M., 80, 106(3), 111(3)
Hagstrum, H. D., 239, 257, 272
Hahn, E. L., 366, 382, 395, 396(4), 431(23), 444, 449, 458
Hahn, R. L., 246
Haiman, O., 75
Hall, T. M., 84, 134, 136
Halpern, A. M., 252
Hamilton, J. F., 44
Hansen, H. E., 84, 89
Hansen, M., 328
Hansen, W. W., 379
Hardy, W. H., 124
Harmon, J. F., 477, 478(50)
Hartmann, S. R., 382, 431(23), 458
Harvey, B. G., 74
Hasiguti, R. R., 275
Haugen, H. K., 315, 316(95, 96)
Hausmann, A., 428
Hautojärvi, P., 78, 83, 84
Hayes, C. E., 388, 394
Hayes, W., 219

AUTHOR INDEX

Hede, B., 83
Hehenkamp, T., 92
Heidemann, A., 215, 220
Heiland, W., 236, 238, 240(40), 241(50), 243, 244(40)
Heintze, V., 247
Helwig, J. T., 137
Hennequin, J. F., 267
Herlach, D., 104, 138(72)
Herrmanne, N., 264
Herschbach, K., 335, 336(120a)
Herzig, C., 72(36), 73
Hikosaka, K., 305, 306, 307(83), 308(83)
Hinshaw, W. S., 404
Hinthorne, J. R., 267
Hirth, J. P., 280
Hirvonen, J. K., 225, 228
Ho, P., 457
Hodges, C. H., 83
Hoentzsch, C., 415
Hofer, W. O., 263
Hoff, H. A., 72(36), 73
Hoffman, K., 137
Hoffmann, K., 80, 95(10), 116(10), 117(10), 118(10), 120(10)
Holcomb, D. F., 392
Holczer, K., 477
Holden, N. E., 14
Hollander, J. M., 2, 7(2), 8(2), 74
Holmberg, B., 339
Holton, W. C., 435, 456, 459, 472
Homewood, C. A., 75
Hone, D. W., 267
Hopkins, F., 253
Howe, L. M., 280, 281, 282(25, 27), 286, 287(28), 288, 292(28), 293, 294(28), 295, 296, 315, 316, 322(28), 327, 328(28), 333, 334(116), 335, 336(28), 337, 344(28), 345
Huang, Y. M., 136
Huber, L. M., 373, 386, 389(9), 402(9), 403(9), 450
Hubler, G. K., 225
Hufschmidt, M., 247
Hume-Rothery, W., 322
Huus, T., 254

I

Ibel, K., 152
Ikeya, M., 477

Isaacson, R. A., 423
Ishiwara, H., 305, 306, 307, 308
Itoh, N., 293, 295, 317

J

Jackman, T. E., 101, 280, 282(32), 353
Jackson, D. P., 280, 282(31), 299(31), 347(31), 350, 352(31), 353, 356
Jackson, J. J., 335, 336(120a)
Jacques, H., 335
Jain, K. C., 84
Jamieson, H. C., 84
Jardine, L. J., 81
Jay, B., 206
Jeavons, A., 112, 113
Jeener, J., 382, 456, 465
Jeffrey, R. N., 121
Jeffries, C. D., 436
Jenkins, C. H., 356
Jenson, F. E., 253
Johansson, N. G. E., 330
Johnson, D. L., 477, 478(49)
Jones, G. P., 458
Jorgensen, T., 51
Julian, M., 449
Jurela, Z., 267
Just, W., 169, 170

K

Kaim, R. E., 342
Kalbitzer, S., 225
Kalus, J., 210
Kaminsky, M., 264
Kamke, D., 247
Kanert, O., 447, 449
Känzig, W., 429
Kaplan, M., 193
Käppeler, F., 231, 245(20)
Kastler, A., 438
Katz, W., 268
Kauffman, R. L., 348
Kaufmann, E. N., 342
Kay, E., 263, 264
Keaton, P. W., 225, 227(7)
Kehl, G., 45
Kehr, K. W., 206
Keil, E., 309

Keller, J., 250
Kelly, R., 271
Kemp, K., 254
Kennedy, J. W., 74
Kerkdijk, C. B., 271
Kerkow, H., 290
Kerr, D. P., 101
Kersten, H. H., 350, 351(136), 356(136)
Kesternich, W., 84
Kevan, L., 435, 438
Khoo, T. L., 90
Kirkegaard, P., 137
Kirkendall, E. O., 235
Kispert, L., 435, 438
Kittel, C., 372
Kitts, Jr., E. L., 477
Klank, A. C., 19
Kleber, M., 255
Kleeman, R., 230
Kleppmann, W. G., 219
Knapp, J. A., 331
Knight, W. D., 398
Knights, J. C., 393
Knoll, G. F., 75, 95
Knudsen, H., 231, 245(22)
Koester, L., 150, 211, 212
Kögel, G., 131
Koju, T., 319
Konstantinov, I. O., 26
Korringa, J., 399
Kösel, G., 104
Kostorz, G., 156, 170
Kotai, E., 250
Koutcher, J., 404
Krakauer, H., 80
Krall, H. R., 51
Krasnov, N. N., 26
Kraus, J., 271
Kraütle, H., 235
Krivoglaz, M. A., 153, 155
Kruger, P., 75
Krumhansl, J. A., 182, 186
Krupka, D. C., 429
Kubica, P., 82, 83
Küchler, R., 449
Kushner, R. A., 264

L

Laakkonen, J., 84
Laegsgard, E., 282

Lakatos, K., 186
Lamb, W. E., 239
Lambe, J., 432
Land, D. J., 249
Lanford, W. A., 250, 251
Laskar, A. L., 71
Lassen, N. O., 293, 295(66)
Lau, S. S., 235, 319
Laurence, N., 432
Lauterbur, P. C., 377, 404
Law, J., 252
Lazarus, D., 440
Leath, P. L., 182
Lechner, R. E., 210
LeClaire, A. D., 71, 73
L'Ecuyer, J., 301
Lederer, C. M., 2, 7(2), 8(2), 74
Lee, M., 373, 389(11)
Lefebvre, S., 170
Legnini, D. G., 86, 90, 105, 134, 135, 137, 138(74), 140, 143(100)
Lehman, C., 262
Lehmann, C., 283
Leibfried, G., 283
Lerner, J. L., 105, 135(74), 138(74)
Leskovar, B., 126
Lever, R. F., 228
Leymonie, C., 44
Liau, Z. L., 235
Lichten, W., 252, 256
Lichtenberger, P. C., 101, 106
Lidiard, A. B., 440
Liebl, H., 265
Ligeon, E., 342
Lindhard, J., 231, 259, 276, 281(20), 283, 284, 285, 286, 290, 291, 306(20)
Linker, G., 231, 245(20)
Lippel, P., 80, 95(10), 116(10), 117(10), 118(10), 120(10), 137
Litmark, U., 42
Little, A., 253
Lloyd, L. T., 42
Lo, C. C., 126
Lodhi, A. S., 230(17), 231
Lomer, W. M., 92
Look, D. C., 454, 455
Lori, J., 299
Lothe, J., 280
Lovesey, S. W., 173, 175(1), 181(1), 182, 190(1), 196, 201(1)

AUTHOR INDEX

Lowe, I. J., 380, 389, 454, 455
Luguijo, E., 270, 273(125), 303, 304(80), 305(80)
Lushbaugh, C., 30
Lüty, F., 449
Lutze-Birk, A., 47
Lynn, K. G., 89, 124

M

McCaldin, J. O., 233
McCall, D. W., 458
McCaughan, D. V., 264, 273
McCracken, G. M., 257, 261(90)
MacDonald, A. B., 90
Macek, J. H., 256
McFall, W. D., 43, 71
Macfarlane, R. M., 438
McHugh, J. A., 264, 267(112)
McIrvine, E. C., 432
McKee, B. T. A., 84, 90, 136
McKenzie, D. R., 154
MacKenzie, I. K., 86(39), 87, 90, 93(39), 102, 106, 121
MacKintosh, W. D., 235
McMullen, T., 83
McNair, D., 238, 239(49)
McRae, E. G., 275
Mader, J., 80, 106(3), 111(3)
Mader, J. J., 80, 111
Madison, D. H., 252
Maggiore, C. J., 225, 227(7)
Maier, K., 104, 138
Mali, M., 449
Manninen, M., 83
Mansel, W., 335, 336(120)
Mansfield, P., 389, 402, 404
Mantl, S., 84, 103
Manuel, A. A., 106, 113(78), 114(78), 115
Maradudin, A. A., 191
Marin, G., 243
Marion, J. B., 226, 246(11)
Marrello, V., 233
Marshall, W., 173, 175(1), 181(1), 182(1), 190(1), 196, 201(1)
Martinson, I., 273
Mashkova, E. S., 237
Massalski, T. B., 276
Massey, H. S. W., 240
Matschke, F. E. P., 237
Matsunami, N., 280, 282(31), 286(28), 287, 292, 293, 294, 295, 299(31), 301, 317, 328(28), 335(28), 336(28), 337, 344(28), 347(31), 352(31), 356
Maul, J., 264
Mayer, J. W., 224, 231, 233, 235, 278, 282, 297(21), 300(21, 41), 301(41), 303, 304(80), 305(80), 309(21), 319, 323(41), 325, 330, 359
Mayers, J., 106, 112(77)
Mehring, M., 386, 401, 402(45), 403
Meiboom, S., 446
Meier, D., 436
Meijer, F., 236, 238(43), 241(43)
Merkle, K. L., 295, 317, 319(74)
Merzbacher, E., 252
Messiah, A., 447
Metz, H., 104
Metzger, H., 171
Meyer, H., 335, 336(120)
Meyers, O., 231, 245(20)
Mezey, G., 250
Mikkelson, R. C., 295, 319(74)
Milgram, M., 280
Miller, G. L., 236, 258(45)
Miller, J. M., 74
Miller, J. W., 356
Mills, A. P., 89
Mims, W. B., 410, 430
Minarik, J., 19
Minier, C., 275
Minier, M., 275
Minjarends, P. E., 94
Minkoff, L., 404
Mitchell, I. V., 225, 282
Mitchell, J. B., 93, 321(103), 322
Moisy-Maurice, V., 170
Molchanov, V. A., 237
Molière, G., 284, 285, 286
Möller, W., 247
Momo, F., 435
Monahan, J. E., 62
Mook, H. A., 213, 214
Moore, J. A., 345
Morgan, K Z., 33
Morgan, D. V., 283, 286, 291, 323(63), 328(63)
Morita, K., 293, 295(67)
Morris, P. G., 404
Mortlock, A. J., 48
Morton, G. A., 51
Mory, J., 317

Moss, M., 263
Mostoller, M., 189, 193
Moszynski, M., 126
Mott, W. E., 75
Mullen, J. G., 43, 70, 73
Muller, B. N., 477, 478(50)
Müller, H., 235
Müller, K. A., 420, 423
Mundy, J. N., 42, 43, 71, 72(36), 73
Murphy, V. T., 264
Myllylä, R., 138

N

Nagatomo, M., 305, 306(83), 307(83), 308(83)
Nagy, T., 250
Nakamura, K., 233
Namba, S., 312
Narayan, J., 312, 313(93)
Neame, K. D., 75
Nechtschein, M., 477
Neiminen, R. M., 84
Nelson, K. F., 431, 432
Nelson, R. S., 233
Nicholls, J. E., 427
Nicklow, R. M., 191, 192
Nicolet, M.-A., 224, 231, 233, 235, 278, 297(21), 300(21), 303, 309(21)
Niehus, H., 236
Nielsen, B., 84, 89
Nielsen, V., 259
Nieminen, R., 83
Niklas, J. R., 415
Noggle, T. S., 356
Norberg, R. E., 392
Norris, D. I. R., 280
Northrup, D. C., 75
Norton, P. R., 280, 282(31, 32), 299, 347(31), 351(77), 352(31), 353, 356
Nourtier, A., 267
Nowick, A. S., 276, 439
Nowicki, L. J., 45

O

Oechsner, H., 264
Oetzmann, H., 225
Ohlsen, W. D., 423, 425, 431, 432, 482
Onderdenlinden, D., 262
Ottaviani, G., 233
Overman, R. T., 74

P

Packard, M. G., 379
Paget, D., 438
Pagh, B., 84
Pake, G. E., 392, 406
Palmer, D. W., 247, 342
Panke, H., 255
Parker, W., 74
Patel, J. L., 427
Paulus, T. J., 124
Peercy, P. S., 225, 227(7)
Peisl, J., 171
Peltner, H. E., 73
Perfetti, P., 276
Perkins, A., 82
Perlman, I., 2, 7(2), 8(2), 74
Perreard, E., 106, 113(78), 114(78), 115(78)
Peter, M., 106, 113(78), 114(78), 115(78)
Petersen, K., 84, 89, 93, 94
Peterson, N. L., 21, 42, 43, 70, 71, 73, 447
Pettigrew, G. L., 38
Petty, R. J., 339, 341
Pickett, W. E., 80
Picraux, S. T., 275, 276(11), 278(11), 282, 295(38), 296(38), 317, 318, 319, 320, 323(11), 330, 331, 332, 338, 339, 340, 342, 359
Pierce, J. R., 51
Pietsch, H., 290
Pines, A., 386, 403
Poate, J. M., 233, 234, 329, 331
Polak, M., 480, 481
Poole, C. P., 408, 410
Poston, J. W., 95
Pound, R. V., 372, 389(8), 448
Powers, D., 230(17), 231
Powles, J. G., 389
Prescot, J. R., 51
Pretorius, R., 235
Price, P. B., 48
Pronko, P. P., 295, 312, 313(93), 317, 319(74)
Punkkinen, M., 389
Purcell, E. M., 372, 389(8), 445, 446, 448

Q

Quenneville, A. F., 280, 282(25, 27), 295, 296(75), 322(28), 327, 333, 334(116), 335, 337, 345

AUTHOR INDEX

Quéré, Y., 282, 286, 310(59), 317, 319
Quittner, P., 63

R

Radelaar, S., 148
Rainville, M. H., 281, 315, 316(95)
Rausch, E. O., 271
Raynor, G. V., 322
Rebout, D., 48
Redfield, A. G., 367, 368(5), 388(5), 431, 454
Rehn, L. E., 335
Reichert, J. F., 422
Reidinger, F., 195, 209(4)
Reimer, J. A., 393
Remaut, G., 276
Remeika, J. P., 477
Remsberg, L. P., 59, 69
Renouf, T. J., 48
Rhim, W. K., 402, 460
Rhines, F. N., 21
Rhodes, H. E., 400
Ricci, E., 246
Richter, D., 220
Rigamonti, A., 453
Rimini, E., 233, 282, 303, 304, 305, 309, 311, 312(89), 317, 318(98), 320(98), 323(41), 325
Robinson, D. A. H., 236, 258(45)
Robinson, J. T., 73
Robinson, L. C., 71
Robrock, K.-H., 335
Rodgers, J. W., 342
Rodriquez, A., 477
Roeder, S. B. W., 373, 387(13), 388(13), 389, 390(13), 451
Rogers, A. W., 44, 45, 47(1)
Roosend, H. E., 262
Rose-Innes, A. C., 419
Roth, J., 248, 319
Roth, M., 170
Rothman, S. J., 14, 42, 43, 45, 71, 73
Rowe, J. M., 195, 202, 203, 204, 206
Rowland, T. J., 372
Rubin, R., 206
Ruderman, M. A., 372
Rusch, T. W., 239
Rush, J. J., 202(11), 203, 204(11), 206
Rushworth, A. J., 219
Rutherford, E., 223
Ryan, J. F., 219

S

Sachot, R., 106, 113(78), 114(78), 115(78)
Samuel, D., 250
Samuelson, G., 387
Sander, L., 92
Saris, F. W., 271, 282, 333, 334(116), 339, 350, 351(136), 356, 357(142)
Schaefer, H. E., 104
Schaefer, J., 388, 390
Schallenberger, B., 428
Scharff, M., 231, 259
Schartner, K. H., 271
Schelten, J., 152
Scherzer, B. M. U., 231
Schilling, W., 275, 280, 322(22), 335(22)
Schiøtt, H. E., 231, 293, 295(68, 69)
Schmatz, W., 152, 153, 167, 169, 179
Schneider, J., 165
Schober, T., 321(103), 322
Schofer, H. R., 190
Schow, O. E., 356
Schreiber, L. B., 402
Schulte, C. W., 101
Schulz, F., 264
Schumacher, D., 275
Schumacher, H., 169
Schwahn, D., 167, 170
Schweijer, A., 435
Scott, T. L., 19
Seeger, A., 84, 275, 280, 281
Seegers, D., 97, 131(64)
Seidman, D., 276
Seitz, E., 169
Seymour, R. S., 154
Shapiro, S. M., 195, 208, 209
Sharma, S. C., 88
Shaw, D., 374, 377, 389, 390(14)
Shelby, R. M., 438
Sherwood, J. N., 73
Shewmon, P. G., 21
Shihab-Eldin, A. A., 81
Shockley, W., 51
Siegbahn, K., 75, 95
Siegel, R. W., 80, 83, 84, 85, 86, 87, 88, 95(10), 116(10), 117(10), 118(10), 120(10), 134, 135, 137
Siesel, R. W., 90
Sigmund, P., 224, 257, 259, 260, 261, 262, 263, 266
Sigurd, D., 233, 235

Silsbee, R. H., 429
Silverman, P. J., 256, 348, 354, 355(138)
Simms, D. L., 268, 271, 273, 274(130)
Simons, D. G., 249
Sinfelt, J. H., 400
Sinha, S. K., 85
Sinnott, M. J., 48
Sitter, C. W., 280, 282(32)
Sköld, K., 211
Slatis, H., 74
Slichter, C. P., 361, 369(1), 376, 385(16), 395(1), 396(1), 398, 399(1), 400, 401, 403, 406(1), 430, 440, 442(8), 443, 444(8), 445(8), 446(8), 447, 455, 456, 457(32), 459, 462(32), 470(39), 472(33, 39), 473(39), 474, 475
Slodzian, G., 267
Smaller, B., 423
Smedskjaer, L., 86(39), 87, 93(39), 104
Smedskjaer, L. C., 83, 85, 86, 88(38), 90, 105, 121, 134, 135, 137, 138(74, 75), 140, 143
Smeenk, R. G., 356, 357(142)
Smith, D. P., 237, 239, 240
Smith, H. G., 189, 191, 192(13)
Smith, H. M., 51
Snead, Jr., C. L., 134, 136(90)
Snodgrass, F. W., 213, 214(22)
Snyder, W. S., 33
Solt, G., 171
Soszka, W., 350, 351(136), 356(136)
Spaeth, J. M., 415
Sparks, Jr., C. J., 154
Spindler, E., 255
Spokas, J. J., 376, 385(16)
Springer, T., 152, 194, 195, 197, 206, 220
Stanford, M., 404
Stanley, J., 42
Steeds, J. W., 263
Stehling, W., 255
Stejskal, E. O., 388, 390, 446
Stengärd, I., 354, 355(138), 356
Steunenberg, R. K., 46
Stewart, A. D. G., 263
Stewart, A. T., 82, 84
Steyerl, A., 161, 211, 212(20)
Stogiu, A., 435
Stöhr, J., 276
Stohr, J. F., 48
Stokes, H. T., 382, 400, 439, 441(3), 444, 449, 450, 451(23), 461, 462, 463, 466, 468, 469, 477, 481, 482
Stoneham, A. M., 198
Stott, M. J., 83, 85
Strange, J. H., 447
Stumpe, E., 264
Stump, N., 198, 211
Sturak, B., 47
Sutton, R. B., 75
Svensson, E. C., 187, 188
Swanson, M. L., 280, 282, 286(28), 287(28), 292(28), 293(28), 294(28), 295, 296(75), 322(27), 327, 328(28), 333, 334(116), 335, 336(28, 45), 337, 344(28), 345

T

Taglauer, E., 236, 238, 240(40), 241(50), 243, 244(40)
Tam, S. W., 85, 88
Tanner, J. E., 446
Tarr, C. E., 380
Täubner, F., 290
Taylor, D. S., 182
Taylor, D. W., 83, 184
Taylor, L. S., 29, 74
Terhune, R. W., 432
Thackery, P. A., 233
Thomas, E. W., 268, 271
Thompson, D. A., 314, 315, 316(95, 96)
Thompson, M. W., 261, 262(95, 96), 263
Thrane, N., 93, 94(56)
Tolk, N. H., 238, 268, 271, 273, 274(130)
Tomlin, D. H., 48
Torrey, H. C., 477, 478(51)
Träff, J. H. O. L., 86(39), 87, 93(39)
Trautvetter, H. P., 250, 251(69)
Treado, P. A., 231, 245(21)
Trentler, H. C., 47
Triftshäuser, W., 84, 90, 103, 104
Tromp, R. M., 356, 357(142)
Trumpy, G., 84, 86(39), 87, 93(39)
Tse, C. W., 43, 71
Tseng, W. F., 319
Tsong, I. S. T., 257, 263
Tu, K. N., 235
Tukey, J. W., 390
Tuli, J. K., 81
Tully, J. C., 238

AUTHOR INDEX

Turkenburg, W. C., 350, 351, 356(136)
Turner, P. A., 233, 234(32)
Tzalmona, A., 397, 398(41)

U

Uggerhøj, E., 338
Ullrich, B. M., 224
Urbach, F., 44

V

Van Bueren, H. G., 275, 280(8)
van der Ligt, G. C. J., 236, 237(41), 240(41), 241(41), 242(41)
van der Veen, J. F., 282, 356, 357
van der Weg, W. F., 235, 236, 268, 270, 271, 273(125, 129)
van Dijk, C., 148, 170
van Gurp, G. J., 235
van Hove, L., 196
Van Landuyt, J., 276
van Veen, A., 89
Van Vliet, D., 283, 285, 286, 291, 292, 323(63, 64), 328(63, 64)
Vaughan, R. W., 393, 402
Vehanen, A., 84
Verbeek, H., 237
Vianden, R., 342
Vijayaraghavan, P. R., 191, 193(13)
Vilfan, M., 448, 477(21), 478
Vineyard, G. H., 70, 210, 280
Vinhas, L. A., 206
Vogl, G., 275, 335, 336(120)
Vollmers, K. W., 389
von Guerard, B., 171

W

Wagner, W., 170
Wakabayashi, N., 149, 189, 206, 220
Wakoh, S., 80
Walker, F. W., 14
Walker, R. M., 48
Walker, R. S., 314(94), 315
Wallis, L., 47
Walstedt, R. E., 477
Walter, C. M., 71
Walters, P. A., 106, 112(77)
Wang, C. H., 444, 477
Wang, P.-K., 400

Warburton, W. K., 88, 137
Watkins, G. D., 333, 339
Watson, R. L., 253
Waugh, J. S., 373, 386, 389(9), 402(9), 403, 450
Weger, M., 418
Wei, S. H. N., 473, 475, 476, 477(44)
Weijsenfeld, C. H., 263
Weil, R., 42
Wenzl, H., 321(103), 322
Werner, G., 44, 45
Werner, H. W., 236, 238(43), 241(43)
Werner, K., 163, 169, 171
Wert, C., 42
Wertheim, G. K., 40
Wertz, J. E., 405, 406(52)
West, R. N., 78, 79(2), 85, 95, 106, 112(77)
Wheatley, G. H., 236, 238, 239(49), 258(45)
White, C. W., 238, 268, 271, 273, 274(130), 312, 313, 333
Whitton, J. L., 299, 303, 321, 322
Wiggers, L. W., 339
Wilkinson, M. K., 191, 193(13)
Williams, J. S., 227, 232, 233
Williams, P., 268
Wilson, I. H., 261, 262(95), 263
Wilson, S. R., 312, 313(93), 333
Wittmaack, K., 264
Wolf, D., 447, 457
Wolicki, E. A., 231, 245, 247(58), 249
Wrobel, J. R., 295
Wu, C.-S., 74, 75

Y

Yahalom, J., 233, 234(32)
Young, F. C., 226, 246(11)
Young, Jr., F. W., 333
Yuan, L. C. L., 74, 75
Yukawa, S., 48
Yuster, P. H., 423

Z

Zehner, D. M., 356
Zeitler, E., 309
Zeyen, C. M. E., 171
Ziegler, J. F., 42, 224, 225, 227(6), 228, 230(16), 231, 250, 251(69)
Zinken, A., 190
Zinn, W., 309
Zuhr, R. A., 348

SUBJECT INDEX

A

Absolute counting, *see* Counting
Absorption
 absorbers, high Z, 7
 coefficients, 5, 6, 8
 longslit attenuation, 109
 self-absorption, 6, 8, 66
Accelerator systems, channeling experiments, 297
Activation energy determination, 443, 448, 477
Adiabatic condition, 459, 464
Adiabatic demagnetization in rotating field, *see* NMR, ultraslow motion techniques
Adsorbed atoms, *see* Channeling applications
Advantages and disadvantages of NMR diffusion techniques, 439–441
Alloys, applications of positrons to, 94
Alpha particles
 calibrated sources, 65
 decay, emission, energy, 2
 detection, photographic film, 44–48
 detection, proportional counters, 50, 60
 detection, scintillators, 50
 detection, silicon surface barrier, 53
 range, 3
 sample preparation, 60
Analog to digital converters, 55, 56, 99–101, 110, 123–124
Angular correlation (of annihilation radiation), 79, 95, 96, 106–122
 angular correlation, 2-D, 79, 109–115
 data analysis, 2-D, 115–121
 deconvolution, 2-D, 119
 electronics
  1-D, 107, 111
  2-D, 110
 longslit attenuation, 109
 longslit, 1-D, 79, 107
 peakcounting, 86, 119
 position sensitive detectors, 98, 109–115
 positron sources, 81, 121–122
 resolution, 94–97, 107, 108, 115
 scatter of annihilation radiation, 109
 scintillators, 97, 98, 107, 109–111
 stabilization, 115
 stopping profile of positron, 104, 109
 umklapp, 2-D, 116, 117
Angular distribution of sputtered atoms, 262
Annealing studies, positrons, 93
Annihilation of positrons, 78
Attenuation factors, 166
Auger electron, 4, 61, 251, 256, 272
Autoradiography, 44–49
 diffusion studies, 48
 nuclear track materials, 48
 photographic emulsion, 44–48

B

Background, 177, 218
 epithermal and fast neutrons, 163
 from incoherent elastic scattering, 160, 163
 in PIXE, 253–254
 from sample and apparatus, 163
Backscattering, *see* Rutherford backscattering
Backscattering spectrometer, 215
Band modes, 186, 189
Barns, *see* Nuclear cross section
Becquerel, *see* Radiation units
Beta particles
 calibrated sources, 65
 detection, proportional counters, 60
 detection, pulse counters, 50, 60
 detectors, liquid scintillators, 52, 60
 detectors, solid state and plastic, 50
 range, 5,6
Bloch state of positron, 79, 80, 83
Blocking, 290, 350, 357
Bloembergen, Purcell, and Pound theory, 448–449, 456
Born approximation, 147, 149, 150
Bragg scattering, 147, 148, 158, 159, 176
Bragg's rule, 230

SUBJECT INDEX

Bremsstrahlung, 4, 5, 61
  in particle bombardment, 254
Brillouin zone peak, 154

C

Carrier free isotope, 20, 21, 23, 25, 42
Carr–Purcell sequence, see Measuring spin–spin relaxation time T_2
Centroid of counting time interval, 68
Channeled fraction of beam, 295, 308, 327
Channeling, apparatus
  accelerator system, 297
  basic, 278
  beam lines, 298
  current integration, 297, 299
  detectors, see Detectors, for channeling
  experimental techniques, 297–303
  goniometer, 298
  target chamber, 298–299
  target preparation, 302, 303
  vacuum system, 298, 345
Channeling, applications
  adsorbed atoms, 345, 350, 354–358
  amorphous layers, 303, 304, 308
  dislocations, bubbles and voids, 315–319
  disorder, 303, 311–313
  displacement of host crystal atoms, 308–322
  displacement of solute atoms, 323, 333–335
  effect of films, 304, 305
  interstitial atoms, 277, 278, 333–339
  lattice defects, 275, 283, 297, 303, 322, 345
  location of solute atoms, 322
  ordered alloys, 329
  phase changes and precipitation, 281, 321, 331, 333
  polycrystalline layers, 305–308
  radiation damage, 311–316, 339
  random sites, 345
  sensitivity, 281, 290
  substitutional impurities, 277, 328–339
  surface studies, see Surface studies, channeling
  vacancy trapping, 343, 344
Channeling, characteristic angles, 283
  critical angle, 287, 289, 303, 304
  Lindhardt angle, 283, 290
Channeling, definition, 283
Channeling, surface peak, 301, 347

Channeling, theory
  Bohr potential, 286
  computer simulation, 289, 291, 292, 349, 352, 353
  continuum model, 283
  continuum potentials, 276, 283–285
  flux distribution of channeled ions, 291
  Lindhardt potential, 284–286
  Molière potential, 284–286
  particle trajectories, 286–289
  screening radius, 276, 283, 284, 290
  string potential, 291
  thermal vibrations, 286
  Thomas–Fermi potential, 276
Channeling, yield
  aligned, 289
  angular dependence, 289, 293, 301, 306, 307, 348
  from solutes, 323, 324
  calculation of, 289, 290
  from displaced atoms, 293
  in double alignment, 290
  minimum, 289, 290
  amorphous layer, 304
  normalized, 276, 289–295, 327
Charged particle reactions, see Nuclear reactions
Chemical processing in isotope production, 22–25
  of irradiated material, 23
  of target material, 25
  of unirradiated material, 24
Chemical shift anisotropy relaxation, 450–451, 480–482
Chemical shift, NMR, 398, 400–402
  shielding tensor, 402
Close encounter processes, 300
Coherent scattering, 148, 150, 152, 196, 203, 206–211
  vibrational modes, 172
Coincidence counting, see Angular correlation (of annihilation radiation); Lifetime (of positron), 52, 58, 59
Cold neutrons, 161, 164, 216
Compton effect, see also Gamma rays
  scattering in PIXE backgrounds, 254–255
Computer control in neutron scattering, 165
Computers in counting systems, 55, 56, 139–143
Concentration profiles, Gaussian, 40

Contamination, neutron spectrum, 163, 164, 179
Convolution approximation, 210
Correlation factor in diffusion, 70, 447
Correlation function, 175
　neutron scattering, 195–206
　pair-correlation, σ_{coh}, 197, 210
　self-correlation, σ_{inc}, 197, 200
Correlation time τ_c, see Jump frequency determination
Counting
　absolute, 61, 66
　amplification stability, 51
　background, 66, 67
　countrate and resolution for Dopper broadening, 97
　differential, 64
　instabilities, see Instabilities
　integral, 52, 64
　minimum efficient counting time, 67
　pileup, 100
　pulse, decay time, line-width, pileup, 51
　relative, 64
　resolution or dead time, 65, 68–70
　statistics, 64, 66–67
　sum peaks, 62
Critical angles, 287, 289, 303, 304
Curie, see Radiation units

D

Data analysis, positrons
　angular correlation, see Angular correlation (of annihilation radiation)
　Doppler broadening, see Doppler broadening (of annihilation radiation)
　lifetime, see Lifetime (of positron)
　vacancy formation enthalpy, 89
　vacancy–impurity binding, 92
　voids, 93, 94
Data handling, 387–391
　Fourier transforms, 388–391
　integration, 387–391
　sample and holders, 387
Dead time, see Counting
Debye–Waller factor, 149, 150, 152, 206, 207, 208, 218
Dechanneling, 281, 289, 293–297
　dechanneled fraction, 281, 295
　by defects, 295, 315–319
　diffusion calculation, 293, 294
　lattice vibrations and electronic collisions, 294
　rate, 281
Decontamination, 37
Deconvolution of positron spectra
　angular correlation, 2-D, see Angular correlation (of annihilation radiation)
　Doppler broadening, see Doppler broadening (of annihilation radiation)
　lifetime spectrum, see Lifetime (of positron)
Defect distribution profile, 309, 315
Defect dynamics, 182
　heavy impurities, band modes, 186
　light impurities, localized modes, 190
　mass defect parameter, 182, 184
　theory, 182–183
Defects, neutron scattering cross sections, 152
Delocalized positron, see Bloch state of positron
Depth profile, 231–233, 244–250
　channeling, damage and displaced atoms, 309, 320
　PIXE, 256
　RRA, 250
　SCANIIR, 268
　SIMS, 266
　sputtering, 263
Detection of neutral particles, 237
Detectors
　for channeling, 300–301
　for EPR, see EPR, spectrometer design
　neutrons, 160
　scintillation detectors, 50–60
　semiconductor detectors, 50, 53–54, 60–61, 62
Detectors, annihilation radiation
　angular correlation, see Angular correlation (of annihilation radiation)
　Doppler broadening, see Doppler broadening (of annihilation radiation)
　lifetime, see Lifetime (of positron)
Diffuse scattering of neutrons, 147, 154, 158, 161–167
　elastic, 148
　lattice distortion studies, 169
　limit of sensitivity, vacancies, 161
　spectrometer, 161–163
　thermal, 148
　time-of-flight, 161

SUBJECT INDEX

Diffusion, *see also* NMR, measurement of diffusion coefficient
  autoradiography measurement, 48–49
  BaF_2 or CaF_2, 208
  coefficients, 40, 70
  grain boundary, 48
  hydrogen, 194, 203–206, 211
  jump model, 198–206
  Kirkendall markers, 235
  radiation enhanced, 231
  residence time, 199, 205
  thin films, 233–236
Dipolar relaxation in rotating frame ($T_{1D\rho}$), 465–469
  spin locking, ADRF sequence for, 467
  strong collision theory for, 466–467
Dipolar relaxation in lab frame (T_{1D}), 456, 460–465
  strong collision theory for heteronuclear systems, 461–465
Dislocations, positrons, 83, 84
Disordered systems, 94
Doppler broadening (of annihilation radiation), 79, 80, 95, 96, 98, 99–106
  data analysis, 101–104
  deconvolution, 101
  detectors, 97–98
  electronics, 99–101
  energy resolution, 99, 100
  lineshape, 86, 89, 90, 101–103
  lineshape parameter, generalized, 103
  lineshape, defect specific, 103
  pileup, 100
  sample-detector configuration, 105, 107
  source contribution, 104
  stabilization, 139–143
Double alignment, channeling, 281, 290, 291, 331
Double scattering, 164

E

Elastic scattering, 149–151
ELDOR, *see* NMR, double resonance techniques
Electron capture, 2–4
  in PIXE backgrounds, 254–255
Electron–crystal interactions, EPR, 406–407
  electrostatic interaction, 407

hyperfine interaction, 406
  spin–orbit interactions, 407
  superhyperfine interaction, 406
Electron–nuclear interactions, 398–403
  chemical shifts, 398, 400–402
  Knight shift, 398–400
Electron paramagnetic resonance, *see* EPR
ENDOR, *see* NMR, double resonance techniques
Energy dependence of dechanneling, 281, 315–319
  of χ_{min}, 306
Energy spectrum of sputtered atoms, 258–261
Energy to depth conversion
  in nuclear reaction analyses, 247
  in Rutherford backscattering, 227–230
EPR (electron paramagnetic resonance), 405–428
  defect models, 423
  as materials science tool, examples of, 425–428
  minimum number of detectable spins, 407, 408
  representative studies, 423–428
  spectra, 405, 406
EPR, spectrometer design, 407–423
  block diagrams, 410–413
  bolometer detector, 415–416
  cavities, 418–423
  cryogenics and variable temperatures, 418–423
  crystal diode detector, 415–416
  detectors, 415–416
  frequency stabilization, 414, 415
  homodyne spectrometer, 412–413
  matching devices, 422–423
  microwave power sources, 414
  modulation, 416–418
  noise figure, 408, 409
  operating frequency, 413
  simple spectrometer, 410–412
  superheterodyne spectrometer, 413
ESA, *see* Spherical energy analyzer
Excitation of x-rays by particles, 251–253

F

Faraday cup, 298–299
Fast neutron reactions, *see* Nuclear reactions

Fermi surface, 94
umklapp, 116, 117
Film
  autoradiographic, 44–49
  dosimeter, 35
Fission, see Nuclear reactions
Flux, 17, 18, 25
  fast neutron, 19, 20
  thermal neutron, 19
Flux distribution
  in channeling, 291–293, 309
  in surface studies, 346
FONAR, see Imaging, NMR
Fourier transform, see Data handling, 389–391
Free induction decay (FID), 365
Frequency modes, neutron scattering, see Band modes; Localized modes; Normal modes; Resonant modes
Fume hoods, 32, 33

G

Gamma rays, see also Angular correlation (of annihilation radiation); Doppler broadening (of annihilation radiation); Lifetime (of positron)
  absorption, 8
  longslit, 109
  calibrated sources, 65
  Compton effect, 7, 8, 61, 62, 66
  detection
    photographic film, 44, 60
    pulse counters, 50, 61
  emission, 2, 6
  energy spectrum, 6, 61
    analysis, 62
  heating, 20
  pair production, 6, 7, 8, 61, 66
  photoelectric effect, 7, 61, 62, 66
  sample preparation, 61
  specific ionization, 8
Geiger–Müller counters, 34
Glove boxes, 33
Green's function matrix, 175, 183
  of defects, 183

H

Half life, 9
  isotope separation, 71
Hamiltonian, for EPR systems, 406, 407, 433

Heavy impurities, see Band modes
Hole burning experiment, NMR, 392, 393
Huang scattering, 166
Hyperfine interaction, EPR, 406
Hyperfine interaction, NMR, see Knight shift

I

Imaging, NMR, 403, 404
  FONAR, 404
  spin imaging, 403–404
  zeugmatography, 404
Impact parameter, 289, 301
Incoherent scattering, 148
  processes, 196, 198–206, 210, 211
Instabilities
  microscopic method, use of, 142–143
  postacquisition stabilization, 141, 142
  reference spectra, use of, 140, 141
  stabilization of, 139–143
Instrument resolution function, neutrons, 203
Integral counting, see Counting
Intensity calculations for neutron scattering, 166
Internal conversion, 6
Ion chambers, dosimeters, 34, 35
Ion implantation, 303, 308–313, 317–320, 331
Ion neutralization spectrometry (INS), 257
Ion scattering spectrometry (ISS), 224
Irradiation facilities, 19, 20; see also Sample preparation for irradiation
Isomeric transition, 6
Isotope conversion efficiency, 20
Isotope effects in diffusion, 70
  half life separation technique, 71
  spectrometry technique, 71–73

J

Jump frequency determination, 443–444, 456–457, 461–465

K

Kanzaki forces, 156, 170
Knight shift, 398–400
  pseudodipolar coupling, 400
  pseudoexchange coupling, 400

L

Lattice distortions, neutron scattering, 154, 169

SUBJECT INDEX 499

Lattice dynamics of perfect crystals, 173–174
Lattice site location, 277
   angular scans, ion channeling, 278, 323, 331, 333, 352
   displaced octahedral site, 338
   flux peaking, 323, 327
   interstitial sites, octahedral, 277, 324, 339
   interstitial sites, tetrahedral, 324, 338, 343
   triangulation technique, 277
LEED measurements, complementary to channeling studies, 354
Lifetime (of positron), 79, 95, 97, 122–139
   amplitude walk, 124, 126
   coincidence fast–fast, slow–fast, 124
   constant fraction discriminator, 124–125
   data analysis, 132–138
   deconvolution, 132, 133
   detectors, 98, 123–131
   electronics, 123–124
   global fitting, 136, 137
   mean lifetime by peakshift, 133
   mistriggering of constant fraction discriminator, 131
   phototube, quantum efficiency of, 126
   phototubes, 124–130
   positron sources, 138–139
   pulseshaping, fast timing, 124, 127
   scattering effects, 131
   scintillator, PILOT-U, 97, 125, 126, 130
   side channel selection, 130
   source contribution, 134, 135
   stabilization, 139–143
   time-to-amplitude converter, 123–125
   timing resolution, 125–130, 131, 133–134, 136
Light impurities, *see* Localized modes
Lindhardt potential, *see* Channeling, theory
Line broadening, neutrons, 195, 197
Line broadening, NMR, 364–366
   electrostatic broadening, 364
   homogeneous broadening, 365
   inhomogeneous broadening, 364, 365
Liquid scintillation detectors, 51, 52, 60, *see also* Pulse counting systems
Local field, 456, 462, 466, 471–472
Localized modes, 190–193
Localized positron, *see* Trapping of positrons
Lorentzian, width, 197, 201, 205, 218, 219
Low energy ion scattering (LEIS), 224, 236–244, 263
   applications, 241–244

   calibration, 240
   equipment, 236
   spectra, 236–239

M

Magic angle, 467
Mass defects, 183–186
Materials science studies, EPR, 425–428
Measuring spin relaxation (T_1)
   inversion recovery technique, 452–453
   saturating-comb technique, 453
   saturation recovery technique, 452–453
Measuring spin–spin relaxation time T_2, 444–447
   Carr–Purcell sequence, 444–445
   Meiboom–Gill modification, 446
   shape of FID, 444–445
   spin echo technique, 444–445
Measuring T_{1D}, 464–465
   ADRF method applied to, 463–464
   phase shifted pulse pair method applied to, 464–465
Measuring $T_{1D\rho}$, 468–470
Measuring $T_{1\rho}$, 457–460
   ADRF method applied to, 459
   spin locking method applied to, 459–460
   Meiboom–Gill modification, *see* Measuring spin–spin relaxation time T_2
Minimum number of spins detectable, EPR, 407–409, dependence on FM noise of microwave source, 408–409; *see also* NMR, double resonance techniques
Molière potential, *see* Channeling, theory
Monochromator, 213
Monochromator crystals (neutron scattering), 178–179
Motional narrowing, 441–444, 473–477
Multichannel analyzers, 56, 57; *see* Angular correlation (of annihilation radiation), electronics; Doppler broadening (of annihilation radiation), electronics; Lifetime (of positron), electronics
Multiple scattering, *see also* Scattering
   ions, 241
   neutrons, 212, 217

N

Neutrals, sputtering of, 237–244, 257, 267
Neutron detection, 176, 177
Neutron flux, 19

Neutron scattering, experimental
  analyzer and monochromator crystals, 178, 179
  Bragg, 158
  diffuse, 161
  Huang, 166
  small angle neutron scattering (SANS), 159
  small Q values, 152, 156, 165–167
NMR (nuclear magnetic resonance)
  absorption, 363
  basic phenomena, 361–364
  electric field perturbation, 430
  gyromagnetic ratio, 362
  Larmor frequency, 362
  light perturbation, 428
  limitations of magnetic resonance, 423–424
  lineshift, 372
  linewidth, 372
  motional narrowing, 372
  position of resonance line, 371
  resonance condition, 362
  rotating reference frame, 362–363, 368
  stress perturbation, 429
NMR, applications
  bound diffusion, 473
  CaF_2, 473–477
  chemical shifts, 400–402
  domain structure determination, 397–398
  F^- self-diffusion, 447
  hydrogen NMR in amorphous-Si:H film, 392, 393
  impurity diffusion, 469
  $KCN(^{13}C)$, 482
  Knight shifts, 398–400
  K^+ self-diffusion, 469
  lineshape studies, 391–394
  liquid crystals, nematic (PAA) (MBBA), 477–478
  localized diffusion, 473
  molecular diffusion, 477–480
  molecular reorientations, 479–482
  NMR imaging, 403, 404
  platinum NMR in small particles, 400
  between potential wells of unequal depth, 479–482
  quadrupole resonance, 394–398
  sodium in Na beta alumina, 393, 394
  SrF_2, 473–477
  translational diffusion, 473
NMR, artificial line narrowing, 402–403
  double resonance, 403
  magic angle, 373
  multipulse sequence, 403
NMR, calibration techniques
  exact resonance, 469–470
  of local field (H_D), 471–472
  RF pulse monitor (H_1), 470–471
NMR, double resonance techniques, 403, 430–438
  dynamic nuclear polarization, 435–436
  electron–electron double resonance (ELDOR), 438
  electron–nuclear double resonance (ENDOR), 432–435
  ENDOR coil system, 434–435
  Hahn condition, 431
  nuclear–nuclear double resonance (NNDR), 430–432
  optically detected magnetic resonance (ODMR), 427, 428, 436–438
  optically detected nuclear magnetic resonance (ODNMR), 436–438
NMR, measurement of diffusion coefficient, 446–447
  ADRF applied to the measurement of D, 456–457
  divacancy, 447
  fluorine, self-diffusion in PbF_2, 447
NMR, spectrometer design, 373–391
  block diagram, 375
  computers and interfaces, 387
  detector, 385–387
  diode detector, 385
  frequency source, 381–382
  gate, 382
  magnet, 376–378
  magnetic field stabilization, 378
  phase sensitive detection, 385–387
  power amplifier, 383–384
  preamp, 384–385
  pulse generator, 382–383
  pulsed NMR design, 375–391
  pulsed versus CW spectrometers, 374–375
  receiver, 384–387
  recovery time, 383–384, 389
  sample probe, 378–381
  signal averaging devices, 387
  signal-to-noise considerations, 379, 381, 386, 387–391
NMR, ultraslow motion techniques
  adiabatic demagnetization in a rotating frame (ADRF), 456, 459, 463–464

Ag$^+$ diffusion in RbCl, 449
  frequency dependence, 448, 451, 477–478
  lab frame (T_1), 447–449
  metallic Li, 473
  NaCN, 449
  off-resonance pulsing, 459
  rotating frame ($T_{1\rho}$), 454, 456
  spin lattice relaxation in a rotating frame ($T_{1\rho}$), 455
  strong collision theory, 456–457, 461–467
  strong collision theory for homonuclear systems, 456–457
  weak collision theory, 455–456
  weakly magnetic spins' motions, 461–467
NNDR, see NMR, Double resonance techniques
Noise, $1/f$ and low frequency modulation, 416–417, 434
Nonlocal diffusion, see NMR, Applications of, bound diffusion
Normal modes, 174, 183
Nuclear cross section, 15, 16
  barns, definition, 15
  charged particles, 21
  excitation function, 16
  fast neutrons, 21
  thermal neutrons, 19, 20
Nuclear reaction analysis (NRA), 244–256
  channeling, 301, 333, 342
  charged particle activation analysis (CPAA), 245–246
  cross-section, 247
  neutron activation analysis (NAA), 245–246
  particle induced x-ray emission (PIXE), 250–251, 256
  prompt reaction analysis (PRA), 246–250
  resonant reaction analysis (RRA), 248–250
Nuclear reaction technique, channeling, 301, 333, 342
Nuclear reactions
  (α,α), 248
  (d,α), 248
  (d,n), 247
  (d,p), 247
  (n,α), (n,p), 19, 20
  (n,γ), 14, 19, 20
  (p,α), 248
  (p,$\alpha\gamma$), 248, 250
  (p, γ), 247, 249
  (^{15}n,$\alpha\gamma$), 250
  charged particle, 15, 21
  fission, 21
  spallation, 21
  yield, 17
Nuclear reactor, 19–21
Nuclear tracks, see Autoradiography
Nuclei
  chart of, 15
  positron emitters, 81
Number of spins, counting experiments, 405

O

ODMR, see Optically detected magnetic resonance (ODMR)
ODNMR, see NMR, double resonance techniques
Optically detected magnetic resonance (ODMR), 427–428, 436–438
Optical spectra in SCANIIR, 268
Orbital angular momentum, 398, 401, see also Chemical shift, NMR

P

Pair production effect, see Gamma rays
Perturbations, externally applied to defects, 428–430
  electric fields, 430
  light, 428
  stress, 429
Phase factor, 149, 155, 174
Photoelectric effect, see Gamma rays
Photographic emulsions, see Autoradiography
Photomultiplier,
  application to angular correlation, 107, 109–111
  application to positron lifetime, 123–130
  photocathode, 51
  pulse counting, 50, 51
  quantum efficiency, 126
  RC time constant, 51
Pickup coil
  crossed, 378–380
  single, 378–381
  voltage from, 369
Planar channeling, 286
Polarized neutrons, 148
Positronium, 78, 80
  para- and ortho-, 78
  pickoff, 81

Positrons, *see also* Angular correlation; Bloch state of positron; Doppler broadening; Lifetime; Sources, positron; Trapping of positrons
 annihilation, 6, 58
 electron momenta determined with, 79, 94
 electron enhancement around the, 80
 emission, 3
 range, 6
Prevacancy effects, positrons, 84
Primary extinction, 147, 159
Proportional flow counters, 34, 50
 alpha, beta particles, 60
 dead time, 68
Pulse counting systems, 55–59, *see also* Angular correlation, electronics; Doppler broadening, electronics; Lifetime, electronics
 amplifiers, preamplifiers, 55
 analog to digital converters, 55, 56
 discriminators, scalers, timers, 55
 drift, 66
 high voltage supply, 55
 liquid scintillators, 52
 multichannel analyzers, 56–57, 62
 pulse height analyzers, 52, 55
 single-channel analyzers, 55, 57
Pulse pile up, 225

Q

Quadrupole moment, 395
Quadrupolar relaxation, 449
Quadrupole interaction, 429, 432
Quasielastic scattering, 201, 210, 219
 time-of-flight measurements, 203, 213–214
Quasimolecular transitions in PIXE background, 256

R

Radiation hazards, 27–33
 dose rate effect, 31
 dose, equivalent, 28
 dose, response, threshold, 28
 exposure, 28, 31
 lethal dose, 29, 30
 maximum body burden, 32
Radiation protection, 33–39
 fume hoods and glove boxes, 33
 personnel monitoring, 35
 radioactive waste, decontamination, 36
 shielding, 36
 spills, 38
 standards, licensing requirements, 29, 40
 survey instrumentation, 33
Radiation units
 activity (becquerel, curie), 9
 dose (rad, gray), 29
 dose equivalent (rem, sievert), 28
 exposure (roentgen), 28
Radioactive decay, 1–10
 chains, 9
 constants, 9, 10, 68, 71
Radioactive sources, *see also* Sources, positron
 diffusion, thin layer, 40
 packaging, 38
 preparation, 40, 41–43
 weightless, 3, 60
Radioisotopes
 carrier free, 23
 commercially available, 11, 12–14, 26
 impurities, 11
 production, 11–26
Random component of beam, 295, 309, 327
Random spectrum in channeling, 301, 309
Relaxation, magnetic resonance, 364–367
 Bloch equations, 367–371
 spin–lattice relaxation, 366, 367
 spin–spin relaxation, 364, 366
Resolution, 94–97, 217–219
 angular correlation, *see* Angular correlation (of annihilation radiation)
 doppler broadening, *see* Doppler broadening (of annihilation radiation)
 in neutron scattering, 151
 lifetime, *see* Lifetime (of positron)
 resolution time, *see* Counting
Resolution, angular in neutron scattering, 178–179
Resolution, depth
 in prompt reaction analysis, 247, 248
 in RBS, 228–229, 232–233
 in resonant reaction analysis, 250
 in sputtering, 264
Resolution, energy
 Doppler broadening, 99, 100
 of Ge(Li) system, 53, 62
 of NaI(Tl) system, 51

SUBJECT INDEX

Resonant modes, 186–190
Rotational reference frame, 454–456, 465–466
Rutherford backscattering, 300–301
　application, 231–235
　in channeling, 300–301
　collisions, 276
　cross-section, 223–225, 228–231
　for defect distribution profiles, 311
　detection sensitivity, 229
　edge, spectrum, 226, 229, 233
　equipment, 224
　law, 303
　limitations, 230
　spectrum, 228, 230, 232, 300, 301

S

Sample preparation for irradiation, 19, 20, 24, 25
Scattering
　multiple events, 281, 304, 309
　single events, 281, 309
Scattering cross section, 147
　coherent, 148–152, 195–198, 203, 207–211
　constant Q scans, 180–181, 186
　for defects, 152
　for disordered systems, 152–154
　effect of lattice dilatation, 154, 156, 169
　elastic, 198
　elastic Born approximation, 149–151
　elastic dipole force tensor P, 156, 158
　frequency dependence, 172
　incoherent, 148–149, 172, 181, 195–206, 210
　inelastic, 199
　inelastic coherent, 174–176
　theoretical limitations, defects, 182–183
Scattering length, coherent, 148–150, 152
Scattering vector Q, 149, 151, 152
　experimental values, 160, 163, 164
Screening radius, see Channeling, theory
Secondary electron emission (SEE), 257
Secondary ion mass spectroscopy (SIMS), 264–268
　instrumentation for SIMS, 265
Secular equilibrium, 10
Shielding (neutron scattering), 178, see also Radiation protection
Single alignment, channeling, 290
Single binary collision peaks (SBC), 237

Small angle scattering of neutrons (SANS), 159
　apparatus, 159–160
　results, Al–Zn, 167
Soller, slit collimator, 176
Source contribution, positrons, 104, 134, 135
Sources, positron
　angular correlation, see Angular correlation (of annihilation radiation)
　Doppler broadening, see Doppler broadening (of annihilation radiation)
　isotopes, 81
　lifetime, see Lifetime (of positron)
　slow beam, 82
　source contribution, 104, 134, 135
Specific activity, 9, 17
Spectrum unfolding in resonant nuclear reactions, 249
Spherical energy analyzer (ESA), 236
Spin diffusion, 392
Spin echo technique, see Measuring spin–spin relaxation time T_2
Spin temperature, 456–457
Spin–lattice relaxation, 366. 367, 447–449
Spin–spin relaxation, 364, 366, 441–443
Sputtering, 257–264
　in LEIS, 244
　Sigmund theory, 257–259
　surface topography, 263
Stopping cross-section, 229–230, 263
Stopping power, energy loss factor, 227, 229, 230, 231, 246
Strong collision theory, 456–457, 461–466
Structure factor, 210
Superhyperfine interaction, EPR, 406
Surface composition analysis by neutral and ion impact radiation (SCANIIR), 268–274
Surface impurities, 227, 230, 240, 263
Surface studies with positrons, 89
Surface studies, channeling, 282, 345–357
　adsorbates, coverage determination, 351
　adsorbed atoms, effect of, 280, 350, 357
　adsorbed hydrogen, 353
　basic principles, 346–351
　blocking dip, 350, 357
　distance of closest approach, 346
　enhanced vibrational amplitude, 347, 353, 354
　impurity atom location, 356–358

lateral displacement, 280, 347, 355
LEED studies, 354
relaxation, platinum, 347, 351–354
shadow cone, 346, 347
surface Debye temperature, 352
surface peak yield, 301, 347–351, 354
surface relaxation, 281, 347, 352, 357

T

Thermal neutron, see Nuclear cross section; Nuclear reactions
Thermal vibrations, effect on channeling, 287, 289, 348
Thermalization of positrons, 82
Thomas–Fermi potential, 276
Time-of-flight
in LEIS, 236
neutron spectrometer, 213–216
spectra from proton diffusion, 203
technique in neutron scattering, 161–163
Transmission electron microscopy and channeling studies of defects, 314, 315, 320
Transmutation reaction, 20, 22, 23
Trapping of positrons, 81, 83
annealing studies, 93
defect specificity, 93
detrapping, 83
in dislocations, 83, 84
prevacancy effects, 84
specific trapping rate, 90, 91
specific trapping rate, temperature dependence, 88, 91
transition or motion limited, 83
trapping model, 85, 88
at vacancies, 83, 89
at vacancy–impurity sites, 92
in voids, 89, 93, 94
Triangulation procedure, 277

Triple axis neutron spectrometer, 176–182, 213
T_1 minima, 448–449, 481–482

V

Vacancy formation enthalpy, positrons, 89
Vacancy–impurity binding, positrons, 92
Vibration modes, 172
Voids, positrons, 93, 94

W

Wave vector $\mathbf{Q}$, 151, 156, 167

X

X-rays
bremsstrahlung, 4
characteristic, 4
detectors, 50, 53
specific ionization, 48
X-rays in PIXE, 250

Y

Yield, see also Specific activity
flourescence in PIXE, 252
LEIS, 238
NRA, 245
RBS, 226, 229
secondary ions, 267–268
sputtering, 257, 259, 266, 272

Z

Zeeman Hamiltonian, 448
Zeeman levels, 448
Zeugmatography, see Imaging, NMR